Additive Manufacturing Technologies

Ian Gibson • David Rosen • Brent Stucker

Additive Manufacturing Technologies

3D Printing, Rapid Prototyping, and Direct Digital Manufacturing

Second Edition

Ian Gibson
School of Engineering
Deakin University
Victoria, Australia

David Rosen
George W. Woodruff School of Mechanical Engineering
Georgia Institute of Technology
Atlanta, GA USA

Brent Stucker
Department of Industrial Engineering, J B Speed
University of Louisville
Louisville, KY USA

ISBN 978-1-4939-2112-6 ISBN 978-1-4939-2113-3 (eBook)
DOI 10.1007/978-1-4939-2113-3
Springer New York Heidelberg Dordrecht London

Library of Congress Control Number: 2014953293

Printed on acid-free paper

Springer is part of Springer Science+Business Media (www.springer.com)

Preface

Thank you for taking the time to read this book on Additive Manufacturing (AM). We hope you benefit from the time and effort it has taken putting it together and that you think it was a worthwhile undertaking. It all started as a discussion at a conference in Portugal when we realized that we were putting together books with similar aims and objectives. Since we are friends as well as colleagues, it seemed sensible that we join forces rather than compete; sharing the load and playing to each other's strengths undoubtedly means a better all-round effort and result.

We wrote this book because we have all been working in the field of AM for many years. Although none of us like to be called "old," we do seem to have 60 years of experience, collectively, and have each established reputations as educators and researchers in this field. We have each seen the technologies described in this book take shape and develop into serious commercial tools, with tens of thousands of users and many millions of parts being made by AM machines each year. AM is now being incorporated into curricula in many schools, polytechnics, and universities around the world. More and more students are becoming aware of these technologies and yet, as we saw it, there was no single text adequate for such curricula. We believe that the first edition of this book provided such a text, and based upon the updated information in this 2nd edition, we hope we've improved upon that start.

Additive Manufacturing is defined by a range of technologies that are capable of translating virtual solid model data into physical models in a quick and easy process. The data are broken down into a series of 2D cross-sections of a finite thickness. These cross-sections are fed into AM machines so that they can be combined, adding them together in a layer-by-layer sequence to form the physical part. The geometry of the part is therefore clearly reproduced in the AM machine without having to adjust for manufacturing processes, like attention to tooling, undercuts, draft angles, or other features. We can say therefore that the AM machine is a What You See Is What You Build (WYSIWYB) process that is particularly valuable the more complex the geometry is. This basic principle drives nearly all AM machines, with variations in each technology in terms of the techniques used for creating layers and in bonding them together. Further variations include speed, layer thickness, range of materials, accuracy, and of course cost. With so many variables, it is clear to see why this book must be so long and detailed. Having said that, we still feel there is much more we could have written about.

The first three chapters of this book provide a basic overview of AM processes. Without fully describing each technology, we provide an appreciation for why AM is so important to many branches of industry. We outline the rapid development of this technology from humble beginnings that showed promise but still requiring much development, to one that is now maturing and showing real benefit to product development organizations. In reading these chapters, we hope you can learn the basics of how AM works.

The next nine chapters (Chaps. 4–12) take each group of technologies in turn and describe them in detail. The fundamentals of each technology are dealt with in terms of the basic process, whether it involves photopolymer curing, sintering, melting, etc., so that the reader can appreciate what is needed in order to understand, develop, and optimize each technology. Most technologies discussed in this book have been commercialized by at least one company; and these machines are described along with discussion on how to get the best out of them. The last chapter in this group focused on inexpensive processes and machines, which overlaps some of the material in earlier chapters, but we felt that the exponentially increasing interest in these low-cost machines justified the special treatment.

The final chapters deal with how to apply AM technology in different settings. Firstly, we look at selection methods for sorting through the many options concerning the type of machine you should buy in relation to your application and provide guidelines on how to select the right technology for your purpose. Since all AM machines depend on input from 3D CAD software, we go on to discuss how this process takes place. We follow this with a discussion of post-processing methods and technologies so that if your selected machine and material cannot produce exactly what you want, you have the means for improving the part's properties and appearance. A chapter on software issues in AM completes this group of chapters.

AM technologies have improved to the extent that many manufacturers are using AM machine output for end-product use. Called Direct Digital Manufacturing, this opens the door to many exciting and novel applications considered impossible, infeasible, or uneconomic in the past. We can now consider the possibility of mass customization, where a product can be produced according to the tastes of an individual consumer but at a cost-effective price. Then, we look at how the use of this technology has affected the design process considering how we might improve our designs because of the WYSIWYB approach. This moves us on nicely to the subjects of applications of AM, including tooling and products in the medical, aerospace, and automotive industries. We complete the book with a chapter on the business, or enterprise-level, aspects of AM, investigating how these systems enable creative businesses and entrepreneurs to invent new products, and where AM will likely develop in the future.

This book is primarily aimed at students and educators studying Additive Manufacturing, either as a self-contained course or as a module within a larger course on manufacturing technology. There is sufficient depth for an undergraduate or graduate-level course, with many references to point the student further along the path. Each chapter also has a number of exercise questions designed to test the reader's knowledge and to expand their thinking. A companion instructor's guide is being developed as part of the 2nd edition to include additional exercises and their solutions, to aid educators. Researchers into AM may also find this text useful in helping them understand the state of the art and the opportunities for further research.

We have made a wide range of changes in moving from the first edition, completed in 2009, to this new edition. As well as bringing everything as up to date as is possible in this rapidly changing field, we have added in a number of new sections and chapters. The chapter on medical applications has been extended to include discussion on automotive and aerospace. There is a new chapter on rapid tooling as well as one that discusses the recent movements in the low-cost AM sector. We have inserted a range of recent technological innovations, including discussion on the new Additive Manufacturing File Format as well as other inclusions surrounding the standardization of AM with ASTM and ISO. We have also updated the terminology in the text to conform to terminology developed by the ASTM F42 committee, which has also been adopted as an ISO international standard. In this 2nd edition we have edited the text to, as much as possible, remove references to company-specific technologies and instead focus more on technological principles and general understanding. We split the original chapter on printing processes into two chapters on material jetting and on binder jetting to reflect the standard terminology and the evolution of these processes in different directions. As a result of these many additions and changes, we feel that this edition is now significantly more comprehensive than the first one.

Although we have worked hard to make this book as comprehensive as possible, we recognize that a book about such rapidly changing technology will not be up-to-date for very long. With this in mind, and to help educators and students better utilize this book, we will update our course website at http://www.springer.com/978-1-4419-1119-3, with additional homework exercises and other aids for educators. If you have comments, questions, or suggestions for improvement, they are welcome. We anticipate updating this book in the future, and we look forward to hearing how you have used these materials and how we might improve this book.

As mentioned earlier, each author is an established expert in Additive Manufacturing with many years of research experience. In addition, in many ways, this book is only possible due to the many students and colleagues with whom we have collaborated over the years. To introduce you to the authors and some of the others who have made this book possible, we will end this preface with brief author biographies and acknowledgements.

Singapore, Singapore Ian Gibson
Atlanta, GA, USA David Rosen
Louisville, KY, USA Brent Stucker

Acknowledgements

Dr. Brent Stucker thanks Utah State and VTT Technical Research Center of Finland, which provided time to work on the first edition of this book while on sabbatical in Helsinki; and more recently the University of Louisville for providing the academic freedom and environment needed to complete the 2nd edition. Additionally, much of this book would not have been possible without the many graduate students and postdoctoral researchers who have worked with Dr. Stucker over the years. In particular, he would like to thank Dr. G.D. Janaki Ram of the Indian Institute of Technology Madras, whose coauthoring of the "Layer-Based Additive Manufacturing Technologies" chapter in the *CRC Materials Processing Handbook* helped lead to the organization of this book. Additionally, the following students' work led to one or more things mentioned in this book and in the accompanying solution manual: Muni Malhotra, Xiuzhi Qu, Carson Esplin, Adam Smith, Joshua George, Christopher Robinson, Yanzhe Yang, Matthew Swank, John Obielodan, Kai Zeng, Haijun Gong, Xiaodong Xing, Hengfeng Gu, Md. Anam, Nachiket Patil, and Deepankar Pal. Special thanks are due to Dr. Stucker's wife Gail, and their children: Tristie, Andrew, Megan, and Emma, who patiently supported many days and evenings on this book.

Prof. David W. Rosen acknowledges support from Georgia Tech and the many graduate students and postdocs who contributed technically to the content in this book. In particular, he thanks Drs. Fei Ding, Amit Jariwala, Scott Johnston, Ameya Limaye, J. Mark Meacham, Benay Sager, L. Angela Tse, Hongqing Wang, Chris Williams, Yong Yang, and Wenchao Zhou, as well as Lauren Margolin and Xiayun Zhao. A special thanks goes out to his wife Joan and children Erik and Krista for their patience while he worked on this book.

Prof. Ian Gibson would like to acknowledge the support of Deakin University in providing sufficient time for him to work on this book. L.K. Anand also helped in preparing many of the drawings and images for his chapters. Finally, he wishes to thank his lovely wife, Lina, for her patience, love, and understanding during the long hours preparing the material and writing the chapters. He also dedicates this book to his late father, Robert Ervin Gibson, and hopes he would be proud of this wonderful achievement.

Contents

Introduction and Basic Principles

1

Abstract

The technology described in this book was originally referred to as rapid prototyping. The term rapid prototyping (RP) is used in a variety of industries to describe a process for rapidly creating a system or part representation before final release or commercialization. In other words, the emphasis is on creating something quickly and that the output is a prototype or basis model from which further models and eventually the final product will be derived. Management consultants and software engineers both use the term rapid prototyping to describe a process of developing business and software solutions in a piecewise fashion that allows clients and other stakeholders to test ideas and provide feedback during the development process. In a product development context, the term rapid prototyping was used widely to describe technologies which created physical prototypes directly from digital data. This text is about these latter technologies, first developed for prototyping, but now used for many more purposes.

1.1 What Is Additive Manufacturing?

Additive manufacturing is the formalized term for what used to be called rapid prototyping and what is popularly called 3D Printing. The term rapid prototyping (RP) is used in a variety of industries to describe a process for rapidly creating a system or part representation before final release or commercialization. In other words, the emphasis is on creating something quickly and that the output is a prototype or basis model from which further models and eventually the final product will be derived. Management consultants and software engineers both also use the term rapid prototyping to describe a process of developing business and software solutions in a piecewise fashion that allows clients and other stakeholders to test ideas and provide feedback during the development process. In a product development context, the term rapid prototyping was used widely to

I. Gibson et al., *Additive Manufacturing Technologies*,
DOI 10.1007/978-1-4939-2113-3_1

describe technologies which created physical prototypes directly from digital model data. This text is about these latter technologies, first developed for prototyping, but now used for many more purposes.

Users of RP technology have come to realize that this term is inadequate and in particular does not effectively describe more recent applications of the technology. Improvements in the quality of the output from these machines have meant that there is often a much closer link to the final product. Many parts are in fact now directly manufactured in these machines, so it is not possible for us to label them as "prototypes." The term rapid prototyping also overlooks the basic principle of these technologies in that they all fabricate parts using an additive approach. A recently formed Technical Committee within ASTM International agreed that new terminology should be adopted. While this is still under debate, recently adopted ASTM consensus standards now use the term additive manufacturing [1].

Referred to in short as AM, the basic principle of this technology is that a model, initially generated using a three-dimensional Computer-Aided Design (3D CAD) system, can be fabricated directly without the need for process planning. Although this is not in reality as simple as it first sounds, AM technology certainly significantly simplifies the process of producing complex 3D objects directly from CAD data. Other manufacturing processes require a careful and detailed analysis of the part geometry to determine things like the order in which different features can be fabricated, what tools and processes must be used, and what additional fixtures may be required to complete the part. In contrast, AM needs only some basic dimensional details and a small amount of understanding as to how the AM machine works and the materials that are used to build the part.

The key to how AM works is that parts are made by adding material in layers; each layer is a thin cross-section of the part derived from the original CAD data. Obviously in the physical world, each layer must have a finite thickness to it and so the resulting part will be an approximation of the original data, as illustrated by Fig. 1.1. The thinner each layer is, the closer the final part will be to the original. All commercialized AM machines to date use a layer-based approach, and the major ways that they differ are in the materials that can be used, how the layers are created, and how the layers are bonded to each other. Such differences will determine factors like the accuracy of the final part plus its material properties and mechanical properties. They will also determine factors like how quickly the part can be made, how much post-processing is required, the size of the AM machine used, and the overall cost of the machine and process.

This chapter will introduce the basic concepts of additive manufacturing and describe a generic AM process from design to application. It will go on to discuss the implications of AM on design and manufacturing and attempt to help in understanding how it has changed the entire product development process. Since AM is an increasingly important tool for product development, the chapter ends with a discussion of some related tools in the product development process.

Fig. 1.1 CAD image of a teacup with further images showing the effects of building using different layer thicknesses

1.2 What Are AM Parts Used for?

Throughout this book you will find a wide variety of applications for AM. You will also realize that the number of applications is increasing as the processes develop and improve. Initially, AM was used specifically to create visualization models for products as they were being developed. It is widely known that models can be much more helpful than drawings or renderings in fully understanding the intent of the designer when presenting the conceptual design. While drawings are quicker and easier to create, models are nearly always required in the end to fully validate the design.

Following this initial purpose of simple model making, AM technology has developed over time as materials, accuracy, and the overall quality of the output improved. Models were quickly employed to supply information about what is known as the "3 Fs" of Form, Fit, and Function. The initial models were used to help fully appreciate the shape and general purpose of a design (Form). Improved accuracy in the process meant that components were capable of being built to the tolerances required for assembly purposes (Fit). Improved material properties meant that parts could be properly handled so that they could be assessed according to how they would eventually work (Function).

To say that AM technology is only useful for making models, though, would be inaccurate and undervaluing the technology. AM, when used in conjunction with other technologies to form process chains, can be used to significantly shorten product development times and costs. More recently, some of these technologies have been developed to the extent that the output is suitable for end use. This explains why the terminology has essentially evolved from rapid prototyping to additive manufacturing. Furthermore, use of high-power laser technology has meant that parts can now also be directly made in a variety of metals, thus extending the application range even further.

1.3 The Generic AM Process

AM involves a number of steps that move from the virtual CAD description to the physical resultant part. Different products will involve AM in different ways and to different degrees. Small, relatively simple products may only make use of AM for visualization models, while larger, more complex products with greater engineering content may involve AM during numerous stages and iterations throughout the development process. Furthermore, early stages of the product development process may only require rough parts, with AM being used because of the speed at which they can be fabricated. At later stages of the process, parts may require careful cleaning and post-processing (including sanding, surface preparation, and painting) before they are used, with AM being useful here because of the complexity of form that can be created without having to consider tooling. Later on, we will investigate thoroughly the different stages of the AM process, but to summarize, most AM processes involve, to some degree at least, the following eight steps (as illustrated in Fig. 1.2).

1.3.1 Step 1: CAD

All AM parts must start from a software model that fully describes the external geometry. This can involve the use of almost any professional CAD solid modeling software, but the output must be a 3D solid or surface representation. Reverse engineering equipment (e.g., laser and optical scanning) can also be used to create this representation.

1.3.2 Step 2: Conversion to STL

Nearly every AM machine accepts the STL file format, which has become a de facto standard, and nowadays nearly every CAD system can output such a file format. This file describes the external closed surfaces of the original CAD model and forms the basis for calculation of the slices.

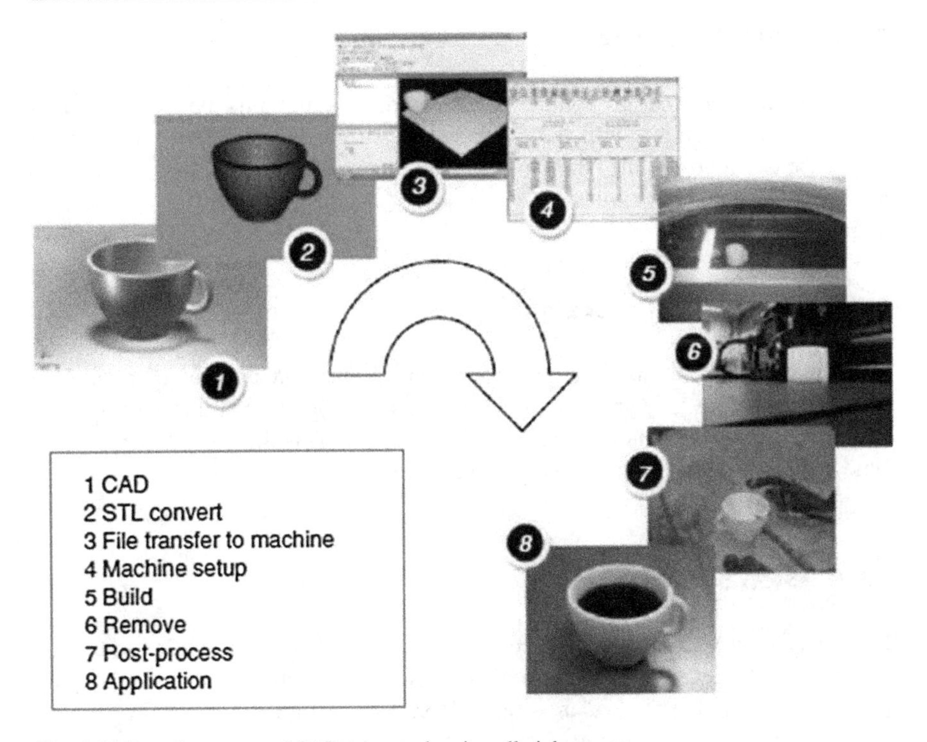

Fig. 1.2 Generic process of CAD to part, showing all eight stages

1.3.3 Step 3: Transfer to AM Machine and STL File Manipulation

The STL file describing the part must be transferred to the AM machine. Here, there may be some general manipulation of the file so that it is the correct size, position, and orientation for building.

1.3.4 Step 4: Machine Setup

The AM machine must be properly set up prior to the build process. Such settings would relate to the build parameters like the material constraints, energy source, layer thickness, timings, etc.

1.3.5 Step 5: Build

Building the part is mainly an automated process and the machine can largely carry on without supervision. Only superficial monitoring of the machine needs to take place at this time to ensure no errors have taken place like running out of material, power or software glitches, etc.

1.3.6 Step 6: Removal

Once the AM machine has completed the build, the parts must be removed. This may require interaction with the machine, which may have safety interlocks to ensure for example that the operating temperatures are sufficiently low or that there are no actively moving parts.

1.3.7 Step 7: Post-processing

Once removed from the machine, parts may require an amount of additional cleaning up before they are ready for use. Parts may be weak at this stage or they may have supporting features that must be removed. This therefore often requires time and careful, experienced manual manipulation.

1.3.8 Step 8: Application

Parts may now be ready to be used. However, they may also require additional treatment before they are acceptable for use. For example, they may require priming and painting to give an acceptable surface texture and finish. Treatments may be laborious and lengthy if the finishing requirements are very demanding. They may also be required to be assembled together with other mechanical or electronic components to form a final model or product.

While the numerous stages in the AM process have now been discussed, it is important to realize that many AM machines require careful maintenance. Many AM machines use fragile laser or printer technology that must be carefully monitored and that should preferably not be used in a dirty or noisy environment. While machines are generally designed to operate unattended, it is important to include regular checks in the maintenance schedule, and that different technologies require different levels of maintenance. It is also important to note that AM processes fall outside of most materials and process standards; explaining the recent interest in the ASTM F42 Technical Committee on Additive Manufacturing Technologies, which is working to address and overcome this problem [1]. However, many machine vendors recommend and provide test patterns that can be used periodically to confirm that the machines are operating within acceptable limits.

In addition to the machinery, materials may also require careful handling. The raw materials used in some AM processes have limited shelf-life and may also be required to be kept in conditions that prevent them from unwanted chemical reactions. Exposure to moisture, excess light, and other contaminants should also be avoided. Most processes use materials that can be reused for more than one build. However, it may be that reuse could degrade the properties if performed many times over, and therefore a procedure for maintaining consistent material quality through recycling should also be observed.

1.4 Why Use the Term Additive Manufacturing?

By now, you should realize that the technology we are referring to is primarily the use of additive processes, combining materials layer by layer. The term additive manufacturing, or AM, seems to describe this quite well, but there are many other terms which are in use. This section discusses other terms that have been used to describe this technology as a way of explaining the overall purpose and benefits of the technology for product development.

1.4.1 Automated Fabrication (Autofab)

This term was popularized by Marshall Burns in his book of the same name, which was one of the first texts to cover this technology in the early 1990s [2]. The emphasis here is on the use of automation to manufacture products, thus implying the simplification or removal of manual tasks from the process. Computers and microcontrollers are used to control the actuators and to monitor the system variables. This term can also be used to describe other forms of Computer Numerical Controlled (CNC) machining centers since there is no direct reference as to how parts are built or the number of stages it would take to build them, although Burns does primarily focus on the technologies also covered by this book. Some key technologies are however omitted since they arose after the book was written.

1.4.2 Freeform Fabrication or Solid Freeform Fabrication

The emphasis here is in the capability of the processes to fabricate complex geometric shapes. Sometimes the advantage of these technologies is described in terms of providing "complexity for free," implying that it doesn't particularly matter what the shape of the input object actually is. A simple cube or cylinder would take almost as much time and effort to fabricate within the machine as a complex anatomical structure with the same enclosing volume. The reference to "Freeform" relates to the independence of form from the manufacturing process. This is very different from most conventional manufacturing processes that become much more involved as the geometric complexity increases.

1.4.3 Additive Manufacturing or Layer-Based Manufacturing

These descriptions relate to the way the processes fabricate parts by adding material in layers. This is in contrast to machining technology that removes, or subtracts material from a block of raw material. It should be noted that some of the processes are not purely additive, in that they may add material at one point but also use subtractive processes at some stage as well. Currently, every commercial process works in a layer-wise fashion. However, there is nothing to suggest that this is an essential approach to use and that future systems may add material in other ways and yet still come under a broad classification that is appropriate to this text. A slight variation on this, Additive Fabrication, is a term that was popularized by Terry Wohlers, a well-known industry consultant in this field and who compiles a widely regarded annual industry report on the state of this industry [3]. However, many professionals prefer the term "manufacturing" to "fabrication" since "fabrication" has some negative connotations that infer the part may still be a "prototype" rather than a finished article. Additionally, in some regions of the world the term fabrication is associated with sheet metal bending and related processes, and thus professionals from these regions often object to the use of the word fabrication for this industry. Additive manufacturing is, therefore, starting to become widely used, and has also been adopted by Wohlers in his most recent publications and presentations.

1.4.4 Stereolithography or 3D Printing

These two terms were initially used to describe specific machines. Stereolithography (SL) was termed by the US company 3D Systems [4, 5] and 3D Printing (3DP) was widely used by researchers at MIT [6] who invented an ink-jet printing-based technology. Both terms allude to the use of 2D processes (lithography and printing) and extending them into the third dimension. Since most people are very familiar with printing technology, the idea of printing a physical three-dimensional object should make sense. Many consider that eventually the term 3D Printing will become the most commonly used wording to describe AM technologies. Recent media interest in the technology has proven this to be true and the general public is much more likely to know the term 3D Printing than any other term mentioned in this book.

1.4.5 Rapid Prototyping

Rapid prototyping was termed because of the process this technology was designed to enhance or replace. Manufacturers and product developers used to find prototyping a complex, tedious, and expensive process that often impeded the developmental and creative phases during the introduction of a new product. RP was found to significantly speed up this process and thus the term was adopted. However, users and developers of this technology now realize that AM technology can be used for much more than just prototyping.

Significant improvements in accuracy and material properties have seen this technology catapulted into testing, tooling, manufacturing, and other realms that are outside the "prototyping" definition. However, it can also be seen that most of the other terms described above are also flawed in some way. One possibility is that many will continue to use the term RP without specifically restricting it to the manufacture of prototypes, much in the way that IBM makes things other than business machines and that 3M manufactures products outside of the mining industry. It will be interesting to watch how terminology develops in the future.

Where possible, we have used additive manufacturing or its abbreviation AM throughout this book as the generic term for the suite of technologies covered by this book. It should be noted that, in the literature, most of the terms introduced above are interchangeable; but different terminology may emphasize the approach used in a particular instance. Thus, both in this book and while searching for or reading other literature, the reader must consider the context to best understand what each of these terms means.

1.5 The Benefits of AM

Many people have described this technology as revolutionizing product development and manufacturing. Some have even gone on to say that manufacturing, as we know it today, may not exist if we follow AM to its ultimate conclusion and that we are experiencing a new industrial revolution. AM is now frequently referred to as one of a series of disruptive technologies that are changing the way we design products and set up new businesses. We might, therefore, like to ask "why is this the case?" What is it about AM that enthuses and inspires some to make these kinds of statements?

First, let's consider the "rapid" character of this technology. The speed advantage is not just in terms of the time it takes to build parts. The speeding up of the whole product development process relies much on the fact that we are using computers throughout. Since 3D CAD is being used as the starting point and the transfer to AM is relatively seamless, there is much less concern over data conversion or interpretation of the design intent. Just as 3D CAD is becoming What You See Is What You Get (WYSIWYG), so it is the same with AM and we might just as easily say that What You See Is What You Build (WYSIWYB).

The seamlessness can also be seen in terms of the reduction in process steps. Regardless of the complexity of parts to be built, building within an AM machine is generally performed in a single step. Most other manufacturing processes would require multiple and iterative stages to be carried out. As you include more features in a design, the number of these stages may increase dramatically. Even a relatively simple change in the design may result in a significant increase in the time required to build using conventional methods. AM can, therefore, be seen as a way to more effectively predict the amount of time to fabricate models, regardless of what changes may be implemented during this formative stage of the product development.

Similarly, the number of processes and resources required can be significantly reduced when using AM. If a skilled craftsman was requested to build a prototype according to a set of CAD drawings, he may find that he must manufacture the part in a number of stages. This may be because he must employ a variety of construction methods, ranging from hand carving, through molding and forming techniques, to CNC machining. Hand carving and similar operations are tedious, difficult, and prone to error. Molding technology can be messy and obviously requires the building of one or more molds. CNC machining requires careful planning and a sequential approach that may also require construction of fixtures before the part itself can be made. All this of course presupposes that these technologies are within the repertoire of the craftsman and readily available.

AM can be used to remove or at least simplify many of these multistage processes. With the addition of some supporting technologies like silicone-rubber molding, drills, polishers, grinders, etc. it can be possible to manufacture a vast range of different parts with different characteristics. Workshops which adopt AM technology can be much cleaner, more streamlined, and more versatile than before.

1.6 Distinction Between AM and CNC Machining

As mentioned in the discussion on Automated Fabrication, AM shares some of its DNA with CNC machining technology. CNC is also a computer-based technology that is used to manufacture products. CNC differs mainly in that it is primarily a subtractive rather than additive process, requiring a block of material that must be at least as big as the part that is to be made. This section discusses a range of topics where comparisons between CNC machining and AM can be made. The purpose is not really to influence choice of one technology over another rather than to establish how they may be implemented for different stages in the product development process, or for different types of product.

1.6.1 Material

AM technology was originally developed around polymeric materials, waxes, and paper laminates. Subsequently, there has been introduction of composites, metals, and ceramics. CNC machining can be used for soft materials, like medium-density fiberboard (MDF), machinable foams, machinable waxes, and even some polymers. However, use of CNC to shape softer materials is focused on preparing these parts for use in a multistage process like casting. When using CNC machining to make final products, it works particularly well for hard, relatively brittle materials like steels and other metal alloys to produce high accuracy parts with well-defined properties. Some AM parts, in contrast, may have voids or anisotropy that are a function of part orientation, process parameters or how the design was input to the machine, whereas CNC parts will normally be more homogeneous and predictable in quality.

1.6.2 Speed

High-speed CNC machining can generally remove material much faster than AM machines can add a similar volume of material. However, this is only part of the picture, as AM technology can be used to produce a part in a single stage. CNC machines require considerable setup and process planning, particularly as parts become more complex in their geometry. Speed must therefore be considered in terms of the whole process rather than just the physical interaction of the part material. CNC is likely to be a multistage manufacturing process, requiring repositioning or relocation of parts within one machine or use of more than one machine. To make a part in an AM machine, it may only take a few hours; and in fact multiple parts are often batched together inside a single AM build. Finishing may take a few days if the requirement is for high quality. Using CNC machining, even 5-axis high-speed machining, this same process may take weeks with considerably more uncertainty over the completion time.

1.6.3 Complexity

As mentioned above, the higher the geometric complexity, the greater the advantage AM has over CNC. If CNC is being used to create a part directly in a single piece, then there may be some geometric features that cannot be fabricated. Since a machining tool must be carried in a spindle, there may be certain accessibility constraints or clashes preventing the tool from being located on the machining surface of a part. AM processes are not constrained in the same way and undercuts and internal features can be easily built without specific process planning. Certain parts cannot be fabricated by CNC unless they are broken up into components and reassembled at a later stage. Consider, for example, the possibility of machining a ship inside a bottle. How would you machine the ship while it is still inside the bottle? Most likely you would machine both elements separately and work out a way to combine them together as an assembly and/or joining process. With AM you can build the ship and the bottle all at once. An expert in machining must therefore analyze each part prior to it being built to ensure that it indeed can be built and to determine what methods need to be used. While it is still possible that some parts cannot be built with AM, the likelihood is much lower and there are generally ways in which this may be overcome without too much difficulty.

1.6.4 Accuracy

AM machines generally operate with a resolution of a few tens of microns. It is common for AM machines to also have different resolution along different orthogonal axes. Typically, the vertical build axis corresponds to layer thickness and this would be of a lower resolution compared with the two axes in the build plane. Accuracy in the build plane is determined by the positioning of the build mechanism, which will normally involve gearboxes and motors of some kind. This mechanism may also determine the minimum feature size as well. For example, SL uses a laser as part of the build mechanism that will normally be positioned using galvanometric mirror drives. The resolution of the galvanometers would determine the overall dimensions of parts built, while the diameter of the laser beam would determine the minimum wall thickness. The accuracy of CNC machines on the other hand is mainly determined by a similar positioning resolution along all three orthogonal axes and by the diameter of the rotary cutting tools. There are factors that are defined by the tool geometry, like the radius of internal corners, but wall thickness can be thinner than the tool diameter since it is a subtractive process. In both cases very fine detail will also be a function of the desired geometry and properties of the build material.

1.6.5 Geometry

AM machines essentially break up a complex, 3D problem into a series of simple 2D cross-sections with a nominal thickness. In this way, the connection of surfaces in 3D is removed and continuity is determined by how close the proximity of one cross-section is with an adjacent one. Since this cannot be easily done in CNC, machining of surfaces must normally be generated in 3D space. With simple geometries, like cylinders, cuboids, cones, etc., this is a relatively easy process defined by joining points along a path; these points being quite far apart and the tool orientation being fixed. In cases of freeform surfaces, these points can become very close together with many changes in orientation. Such geometry can become extremely difficult to produce with CNC, even with 5-axis interpolated control or greater. Undercuts, enclosures, sharp internal corners, and other features can all fail if these features are beyond a certain limit. Consider, for example, the features represented in the part in Fig. 1.3. Many of them would be very difficult to machine without manipulation of the part at various stages.

1.6.6 Programming

Determining the program sequence for a CNC machine can be very involved, including tool selection, machine speed settings, approach position and angle, etc. Many AM machines also have options that must be selected, but the range, complexity, and implications surrounding their choice are minimal in comparison. The worst that is likely to happen in most AM machines is that the part will not be built very well if the programming is not done properly. Incorrect programming of a CNC machine could result in severe damage to the machine and may even be a human safety risk.

1.7 Example AM Parts

Figure 1.4 shows a montage of parts fabricated using some of the common AM processes. Part a. was fabricated using a stereolithography machine and depicts a simplified fuselage for an unmanned aerial vehicle where the skin is reinforced with

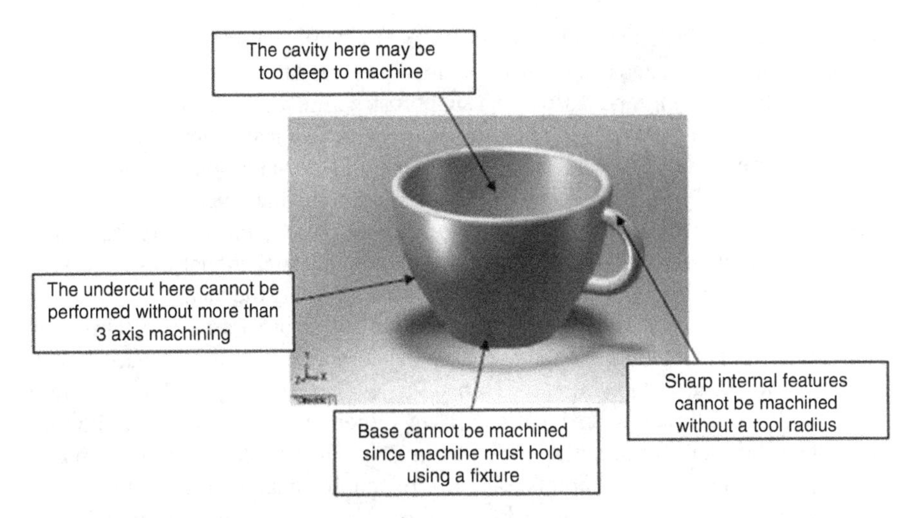

Fig. 1.3 Features that represent problems using CNC machining

Fig. 1.4 Montage of AM parts

a conformal lattice structure (see Chap. 4 for more information about the process). A more complete description of this part is included in the Design for Additive Manufacturing chapter. Parts b. and c. were fabricated using material jetting (Chap. 7). Part b. demonstrates the capability of depositing multiple materials simultaneously, where one set of nozzles deposited the clear material, while another set deposited the black material for the lines and the Objet name. Part c. is a section of chain. Both parts b. and c. have working revolute joints that were fabricated using clearances for the joints and dissolvable support structure. Part d. is a metal part that was fabricated in a metal powder bed fusion machine using an electron beam as its energy source (Chap. 5). The part is a model of a facial implant. Part e. was fabricated in an Mcor Technologies sheet lamination machine that has ink-jet printing capability for the multiple colors (Chap. 9). Parts f. and g. were fabricated using material extrusion (Chap. 6). Part f. is a ratchet mechanism that was fabricated in a single build in an industrial machine. Again, the working mechanism is achieved through proper joint designs and dissolvable support structure. Part g. was fabricated in a low-cost, personal machine (that one of the authors has at home). Parts h. and i. were fabricated using polymer powder bed fusion. Part h. is the well-known "brain gear" model of a three-dimensional gear train. When one gear is rotated, all other gears rotate as well. Since parts fabricated in polymer PBF do not need supports, working revolute and gear joints can be created by managing clearances and removing the loose powder from the joint regions. Part i. is another conformal lattice structure showing the shape complexity capability of AM technologies.

1.8 Other Related Technologies

The most common input method for AM technology is to accept a file converted into the STL file format originally built within a conventional 3D CAD system. There are, however, other ways in which the STL files can be generated and other technologies that can be used in conjunction with AM technology. This section will describe a few of these.

1.8.1 Reverse Engineering Technology

More and more models are being built from data generated using reverse engineering (RE) 3D imaging equipment and software. In this context, RE is the process of capturing geometric data from another object. These data are usually initially available in what is termed "point cloud" form, meaning an unconnected set of points representing the object surfaces. These points need to be connected together using RE software like Geomagic [7], which may also be used to combine point clouds from different scans and to perform other functions like hole-filling and smoothing. In many cases, the data will not be entirely complete. Samples may, for example, need to be placed in a holding fixture and thus the surfaces adjacent to this fixture may not be scanned. In addition, some surfaces may obscure others, like with deep crevices and internal features; so that the representation may not turn out exactly how the object is in reality. Recently there have been huge improvements in scanning technology. An adapted handphone using its inbuilt camera can now produce a high-quality 3D scan for just a few hundred dollars that even just a few years ago would have required an expensive laser-scanning or stereoscopic camera system costing $100,000 or more.

Engineered objects would normally be scanned using laser-scanning or touch-probe technology. Objects that have complex internal features or anatomical models may make use of Computerized Tomography (CT), which was initially developed for medical imaging but is also available for scanning industrially produced objects. This technique essentially works in a similar way to AM, by scanning layer by layer and using software to join these layers and identify the surface boundaries. Boundaries from adjacent layers are then connected together to form surfaces. The advantage of CT technology is that internal features can also be generated. High-energy X-rays are used in industrial technology to create high-resolution images of around 1 μm. Another approach that can help digitize objects is the Capture Geometry Inside [8] technology that also works very much like a reverse of AM technology, where 2D imaging is used to capture cross-sections of a part as it is machined away layer by layer. Obviously this is a destructive approach to geometry capture so it cannot be used for every type of product.

AM can be used to reproduce the articles that were scanned, which essentially would form a kind of 3D facsimile (3D Fax) process. More likely, however, the data will be modified and/or combined with other data to form complex, freeform artifacts that are taking advantage of the "complexity for free" feature of the technology. An example may be where individual patient data are combined with an engineering design to form a customized medical implant. This is something that will be discussed in much more detail later on in this book.

1.8.2 Computer-Aided Engineering

3D CAD is an extremely valuable resource for product design and development. One major benefit to using software-based design is the ability to implement change easily and cheaply. If we are able to keep the design primarily in a software format for a larger proportion of the product development cycle, we can ensure that any design changes are performed virtually on the software description rather than physically on the product itself. The more we know about how the product is going to perform *before* it is built, the more effective that product is going to be. This is also the most cost-effective way to deal with product development. If problems are only noticed after parts are physically manufactured, this can be very costly. 3D CAD can make use of AM to help visualize and perform basic tests on candidate designs prior to full-scale commitment to manufacturing. However, the more complex and performance-related the design, the less likely we are to gain sufficient insight using these methods. However, 3D CAD is also commonly linked to other software packages, often using techniques like finite element method (FEM) to calculate the mechanical properties of a design, collectively known as Computer-Aided Engineering (CAE) software. Forces, dynamics, stresses, flow, and other properties can be calculated to determine how well a design will perform under certain conditions. While such software cannot easily predict the exact behavior of a part, for analysis of critical parts a combination of CAE, backed up with AM-based experimental analysis, may be a useful solution. Further, with the advent of Direct Digital Manufacture, where AM can be used to directly produce final products, there is an increasing need for CAE tools to evaluate how these parts would perform prior to AM so that we can build these products right first time as a form of Design for Additive Manufacturing (D for AM).

1.8.3 Haptic-Based CAD

3D CAD systems are generally built on the principle that models are constructed from basic geometric shapes that are then combined in different ways to make more complex forms. This works very well for the engineered products we are familiar with, but may not be so effective for more unusual designs. Many consumer products are developed from ideas generated by artists and designers rather than engineers. We also note that AM has provided a mechanism for greater freedom of expression. AM is in fact now becoming a popular tool for artists and sculptors, like, for example, Bathsheba Grossman [9] who takes advantage of the geometric freedom to create visually exciting sculptures. One problem we face today is that some computer-based design tools constrain or restrict the creative processes and that there is scope for a CAD system that provides greater freedom. Haptic-based CAD modeling systems like the experimental system shown in Fig. 1.5 [10], work in a similar way to the commercially available Freeform [11] modeling system to provide a design environment that is more intuitive than other standard CAD systems. They often use a robotic haptic feedback device called the Phantom to

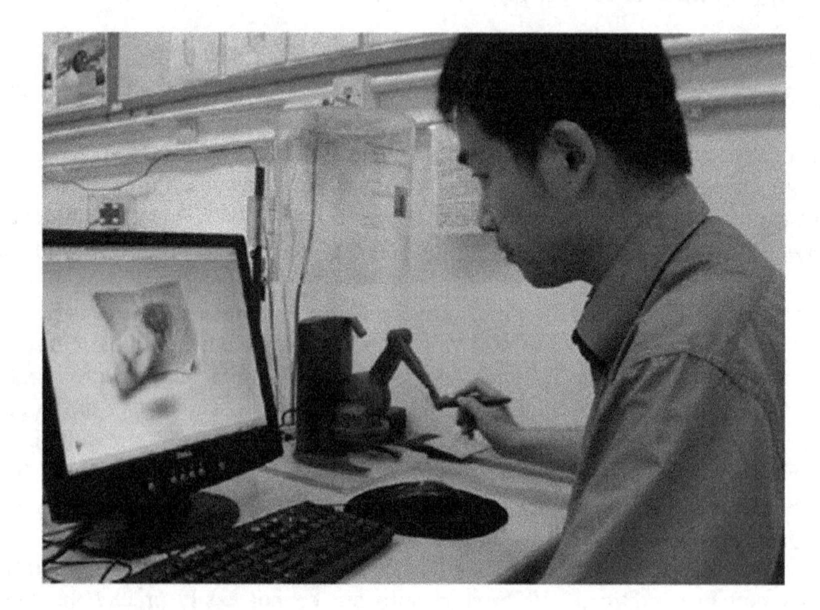

Fig. 1.5 Freeform modeling system

provide force feedback relating to the virtual modeling environment. An object can be seen on-screen, but also felt in 3D space using the Phantom. The modeling environment includes what is known as Virtual Clay that deforms under force applied using the haptic cursor. This provides a mechanism for direct interaction with the modeling material, much like how a sculptor interacts with actual clay. The results using this system are generally much more organic and freeform surfaces that can be incorporated into product designs by using additional engineering CAD tools. As consumers become more demanding and discerning we can see that CAD tools for non-engineers like designers, sculptors, and even members of the general public are likely to become much more commonplace.

1.9 About this Book

There have been a number of texts describing additive manufacturing processes, either as dedicated books or as sections in other books. So far, however, there have been no texts dedicated to teaching this technology in a comprehensive way within a university setting. Recently, universities have been incorporating additive manufacturing into various curricula. This has varied from segments of single modules to complete postgraduate courses. This text is aimed at supporting these curricula with a comprehensive coverage of as many aspects of this technology as possible. The authors of this text have all been involved in setting up programs in their home universities and have written this book because they feel that there are no books to date that cover the required material in sufficient breadth and depth. Furthermore, with the increasing interest in 3D Printing, we believe that this text can also provide a comprehensive understanding of the technologies involved. Despite the increased popularity, it is clear that there is a significant lack of basic understanding by many of the breadth that AM has to offer.

Early chapters in this book discuss general aspects of AM, followed by chapters which focus on specific AM technologies. The final chapters focus more on generic processes and applications. It is anticipated that the reader will be familiar with 3D solid modeling CAD technology and have at least a small amount of knowledge about product design, development, and manufacturing. The majority of readers would be expected to have an engineering or design background, more specifically product design, or mechanical, materials or manufacturing engineering. Since AM technology also involves significant electronic and information technology components, readers with a background in computer applications and mechatronics may also find this text beneficial.

1.10 Exercises

1. Find three other definitions for rapid prototyping other than that of additive manufacturing as covered by this book.

2. From the web, find different examples of applications of AM that illustrate their use for "Form," "Fit," and "Function."
3. What functions can be carried out on point cloud data using Reverse Engineering software? How do these tools differ from conventional 3D CAD software?
4. What is your favorite term (AM, Freeform Fabrication, RP, etc.) for describing this technology and why?
5. Create a web link list of videos showing operation of different AM technologies and representative process chains.
6. Make a list of different characteristics of AM technologies as a means to compare with CNC machining. Under what circumstances do AM have the advantage and under what would CNC?
7. How does the Phantom desktop haptic device work and why might it be more useful for creating freeform models than conventional 3D CAD?

References

1. ASTM Committee F42 on Additive Manufacturing Technologies. http://www.astm.org/COMMITTEE/F42.htm
2. Burns M (1993) Automated fabrication: improving productivity in manufacturing. Prentice Hall, Englewood Cliffs
3. Wohlers TT (2009) Wohlers report 2009: rapid prototyping & tooling state of the industry. Annual worldwide progress report. Wohlers Associates, Detroit
4. Jacobs PF (1995) Stereolithography and other RP and M technologies: from rapid prototyping to rapid tooling. SME, New York
5. 3D Systems. http://www.3dsystems.com
6. Sachs EM, Cima MJ, Williams P, Brancazio D, Cornie J (1992) Three dimensional printing: rapid tooling and prototypes directly from a CAD model. J Eng Ind 114(4):481–488
7. Geomagic Reverse Engineering software. http://www.geomagic.com
8. CGI. Capture geometry inside. http://www.cgiinspection.com
9. Grossman B. http://www.bathsheba.com
10. Gao Z, Gibson I (2007) A 6 DOF haptic interface and its applications in CAD. Int J Comput Appl Technol 30(3):163–171
11. Sensable. http://www.sensable.com

Development of Additive Manufacturing Technology

2

Abstract

It is important to understand that AM was not developed in isolation from other technologies. For example it would not be possible for AM to exist were it not for innovations in areas like 3D graphics and Computer-Aided Design software. This chapter highlights some of the key moments that catalogue the development of Additive Manufacturing technology. It will describe how the different technologies converged to a state where they could be integrated into AM machines. It will also discuss milestone AM technologies and how they have contributed to increase the range of AM applications. Furthermore, we will discuss how the application of Additive Manufacturing has evolved to include greater functionality and embrace a wider range of applications beyond the initial intention of just prototyping.

2.1 Introduction

Additive Manufacturing (AM) technology came about as a result of developments in a variety of different technology sectors. Like with many manufacturing technologies, improvements in computing power and reduction in mass storage costs paved the way for processing the large amounts of data typical of modern 3D Computer-Aided Design (CAD) models within reasonable time frames. Nowadays, we have become quite accustomed to having powerful computers and other complex automated machines around us and sometimes it may be difficult for us to imagine how the pioneers struggled to develop the first AM machines.

This chapter highlights some of the key moments that catalogue the development of Additive Manufacturing technology. It will describe how the different technologies converged to a state where they could be integrated into AM machines. It will also discuss milestone AM technologies. Furthermore, we will discuss how the application of Additive Manufacturing has evolved to include

I. Gibson et al., *Additive Manufacturing Technologies*,
DOI 10.1007/978-1-4939-2113-3_2

greater functionality and embrace a wider range of applications beyond the initial intention of just prototyping.

2.2 Computers

Like many other technologies, AM came about as a result of the invention of the computer. However, there was little indication that the first computers built in the 1940s (like the Zuse Z3 [1], ENIAC [2] and EDSAC [3] computers) would change lives in the way that they so obviously have. Inventions like the thermionic valve, transistor, and microchip made it possible for computers to become faster, smaller, and cheaper with greater functionality. This development has been so quick that even Bill Gates of Microsoft was caught off-guard when he thought in 1981 that 640 kb of memory would be sufficient for any Windows-based computer. In 1989, he admitted his error when addressing the University of Waterloo Computer Science Club [4]. Similarly in 1977, Ken Olsen of Digital Electronics Corp. (DEC) stated that there would never be any reason for people to have computers in their homes when he addressed the World Future Society in Boston [5]. That remarkable misjudgment may have caused Olsen to lose his job not long afterwards.

One key to the development of computers as serviceable tools lies in their ability to perform tasks in real-time. In the early days, serious computational tasks took many hours or even days to prepare, run, and complete. This served as a hindrance to everyday computer use and it is only since it was shown that tasks can be completed in real-time that computers have been accepted as everyday items rather than just for academics or big business. This has included the ability to display results not just numerically but graphically as well. For this we owe a debt of thanks at least in part to the gaming industry, which has pioneered many developments in graphics technology with the aim to display more detailed and more "realistic" images to enhance the gaming experience.

AM takes full advantage of many of the important features of computer technology, both directly (in the AM machines themselves) and indirectly (within the supporting technology), including:

- *Processing power:* Part data files can be very large and require a reasonable amount of processing power to manipulate while setting up the machine and when slicing the data before building. Earlier machines would have had difficulty handling large CAD data files.
- *Graphics capability:* AM machine operation does not require a big graphics engine except to see the file while positioning within the virtual machine space. However, all machines benefit from a good graphical user interface (GUI) that can make the machine easier to set up, operate, and maintain.
- *Machine control:* AM technology requires precise positioning of equipment in a similar way to a Computer Numerical Controlled (CNC) machining center, or even a high-end photocopy machine or laser printer. Such equipment requires

controllers that take information from sensors for determining status and actuators for positioning and other output functions. Computation is generally required in order to determine the control requirements. Conducting these control tasks even in real-time does not normally require significant amounts of processing power by today's standards. Dedicated functions like positioning of motors, lenses, etc. would normally require individual controller modules. A computer would be used to oversee the communication to and from these controllers and pass data related to the part build function.
- *Networking:* Nearly every computer these days has a method for communicating with other computers around the world. Files for building would normally be designed on another computer to that running the AM machine. Earlier systems would have required the files to be loaded from disk or tape. Nowadays almost all files will be sent using an Ethernet connection, often via the Internet.
- *Integration:* As is indicated by the variety of functions, the computer forms a central component that ties different processes together. The purpose of the computer would be to communicate with other parts of the system, to process data, and to send that data from one part of the system to the other. Figure 2.1 shows how the above mentioned technologies are integrated to form an AM machine.

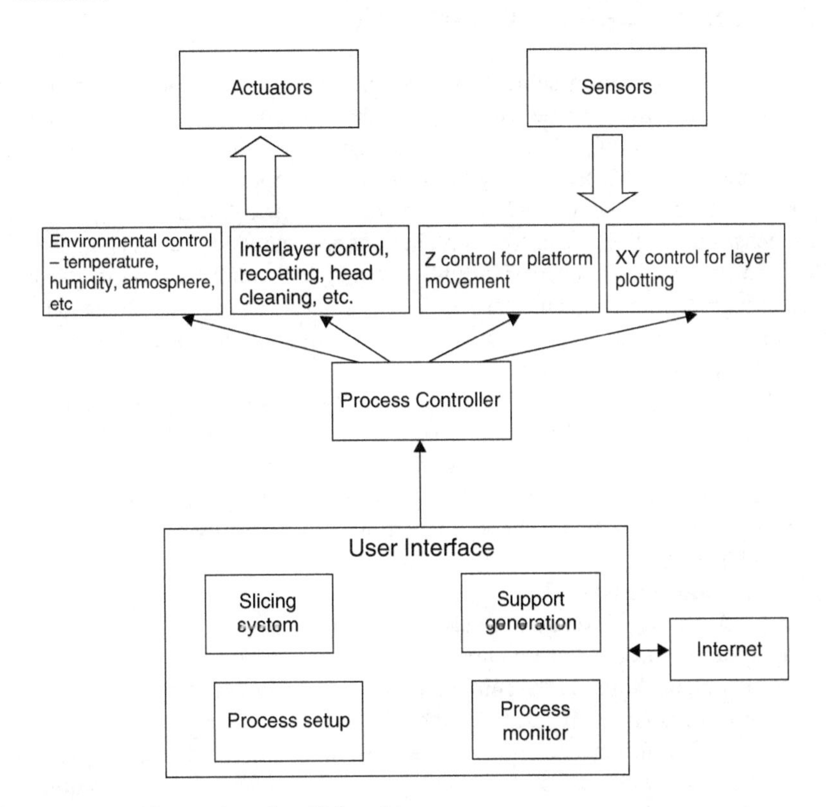

Fig. 2.1 General integration of an AM machine

Earlier computer-based design environments required physically large mainframe and mini computers. Workstations that generally ran the graphics and input/output functions were connected to these computers. The computer then ran the complex calculations for manipulating the models. This was a costly solution based around the fact that the processor and memory components were very expensive elements. With the reduction in the cost of these components, Personal Computers (PCs) became viable solutions. Earlier PCs were not powerful enough to replace the complex functions that workstation-based computers could perform, but the speedy development of PCs soon overcame all but the most computationally expensive requirements.

Without computers there would be no capability to display 3D graphic images. Without 3D graphics, there would be no CAD. Without this ability to represent objects digitally in 3D, we would have a limited desire to use machines to fabricate anything but the simplest shapes. It is safe to say, therefore, that without the computers we have today, we would not have seen Additive Manufacturing develop.

2.3 Computer-Aided Design Technology

Today, every engineering student must learn how to use computers for many of their tasks, including the development of new designs. CAD technologies are available for assisting in the design of large buildings and of nano-scale microprocessors. CAD technology holds within it the knowledge associated with a particular type of product, including geometric, electrical, thermal, dynamic, and static behavior. CAD systems may contain rules associated with such behaviors that allow the user to focus on design and functionality without worrying too much whether a product can or cannot work. CAD also allows the user to focus on small features of a large product, maintaining data integrity and ordering it to understand how subsystems integrate with the remainder.

Additive Manufacturing technology primarily makes use of the output from mechanical engineering, 3D Solid Modeling CAD software. It is important to understand that this is only a branch of a much larger set of CAD systems and, therefore, not all CAD systems will produce output suitable for layer-based AM technology. Currently, AM technology focuses on reproducing geometric form and so the better CAD systems to use are those that produce such forms in the most precise and effective way.

Early CAD systems were extremely limited by the display technology. The first display systems had little or no capacity to produce anything other than alphanumeric text output. Some early computers had specialized graphic output devices that displayed graphics separate from the text commands used to drive them. Even so, the geometric forms were shown primarily in a vector form, displaying wireframe output. As well as the heavy demand on the computing power required to display the graphics for such systems, this was because most displays were monochrome, making it very difficult to show 3D geometric forms on screen without lighting and shading effects.

CAD would not have developed so quickly if it were not for the demands set by Computer-Aided Manufacture (CAM). CAM represents a channel for converting the virtual models developed in CAD into the physical products that we use in our everyday lives. It is doubtful that without the demands associated with this conversion from virtual to real that CAD would have developed so far or so quickly. This, in turn, was fuelled and driven by the developments in associated technologies, like processor, memory, and display technologies. CAM systems produce the code for numerically controlled (NC) machinery, essentially combining coordinate data with commands to select and actuate the cutting tools. Early NC technologies would take CAM data relating to the location of machined features, like holes, slots, pockets, etc. These features would then be fabricated by machining from a stock material. As NC machines proved their value in their precise, automated functionality, the sophistication of the features increased. This has now extended to the ability to machine highly complex, freeform surfaces. However, there are two key limitations to all NC machining:

- Almost every part must be made in stages, often requiring multiple passes for material removal and setups.
- All machining is performed from an approach direction (sometimes referred to as 2.5D rather than fully 3D manufacture). This requires that the stock material be held in a particular orientation and that not all the material can be accessible at any one stage in the process.

NC machining, therefore, only requires surface modeling software. All early CAM systems were based on surface modeling CAD. AM technology was the first automated computer-aided manufacturing process that truly required 3D solid modeling CAD. It was necessary to have a fully enclosed surface to generate the driving coordinates for AM. This can be achieved using surface modeling systems, but because surfaces are described by boundary curves it is often difficult to precisely and seamlessly connect these together. Even if the gaps are imperceptible, the resulting models may be difficult to build using AM. At the very least, any inaccuracies in the 3D model would be passed on to the AM part that was constructed. Early AM applications often displayed difficulties because of associated problems with surface modeling software.

Since it is important for AM systems to have accurate models that are fully enclosed, the preference is for solid modeling CAD. Solid modeling CAD ensures that all models made have a volume and, therefore, by definition are fully enclosed surfaces. While surface modeling can be used in part construction, we cannot always be sure that the final model is faithfully represented as a solid. Such models are generally necessary for Computer-Aided Engineering (CAE) tools like Finite Element Analysis (FEA), but are also very important for other CAM processes.

Most CAD systems can now quite readily run on PCs. This is generally a result of the improvements in computer technology mentioned earlier, but is also a result in improvements in the way CAD data is presented, manipulated, and stored. Most CAD systems these days utilize Non-Uniform Rational Basis-Splines, or NURBS [6]. NURBS are an excellent way of precisely defining the curves and surfaces that correspond to the outer shell of a CAD model. Since model definitions can include free form surfaces as well as simple geometric shapes, the representation must accommodate this and splines are complex enough to represent such shapes without making the files too large and unwieldy. They are also easy to manipulate to modify the resulting shape.

CAD technology has rapidly improved along the following lines:

- *Realism:* With lighting and shading effects, ray tracing and other photorealistic imaging techniques, it is becoming possible to generate images of the CAD models that are difficult to distinguish from actual photographs. In some ways, this reduces the requirements on AM models for visualization purposes.
- *Usability and user interface:* Early CAD software required the input of text-based instructions through a dialog box. Development of Windows-based GUIs has led to graphics-based dialogs and even direct manipulation of models within virtual 3D environments. Instructions are issued through the use of drop-down menu systems and context-related commands. To suit different user preferences and styles, it is often possible to execute the same instruction in different ways. Keyboards are still necessary for input of specific measurements, but the usability of CAD systems has improved dramatically. There is still some way to go to make CAD systems available to those without engineering knowledge or without training, however.
- *Engineering content:* Since CAD is almost an essential part of a modern engineer's training, it is vital that the software includes as much engineering content as possible. With solid modeling CAD it is possible to calculate the volumes and masses of models, investigate fits and clearances according to tolerance variations, and to export files with mesh data for FEA. FEA is often even possible without having to leave the CAD system.
- *Speed:* As mentioned previously, the use of NURBS assists in optimizing CAD data manipulation. CAD systems are constantly being optimized in various ways, mainly by exploiting the hardware developments of computers.
- *Accuracy:* If high tolerances are expected for a design then it is important that calculations are precise. High precision can make heavy demands on processing time and memory.
- *Complexity:* All of the above characteristics can lead to extremely complex systems. It is a challenge to software vendors to incorporate these features without making them unwieldy and unworkable.
- *Usability:* Recent developments in CAD technology have focused on making the systems available to a wider range of users. In particular the aim has been to allow untrained users to be able to design complex geometry parts for themselves. There are now 3D solid modeling CAD systems that run entirely within a web browser with similar capabilities of workstation systems of only 10 years ago.

Many CAD software vendors are focusing on producing highly integrated design environments that allow designers to work as teams and to share designs across different platforms and for different departments. Industrial designers must work with sales and marketing, engineering designers, analysts, manufacturing engineers, and many other branches of an organization to bring a design to fruition as a product. Such branches may even be in different regions of the world and may be part of the same organization or acting as subcontractors. The Internet must therefore also be integrated with these software systems, with appropriate measures for fast and accurate transmission and protection of intellectual property.

It is quite possible to directly manipulate the CAD file to generate the slice data that will drive an AM machine, and this is commonly referred to as direct slicing [7]. However, this would mean every CAD system must have a direct slicing algorithm that would have to be compatible with all the different types of AM technology. Alternatively, each AM system vendor would have to write a routine for every CAD system. Both of these approaches are impractical. The solution is to use a generic format that is specific to the technology. This generic format was developed by 3D Systems, USA, who was the first company to commercialize AM technology and called the file format "STL" after their stereolithography technology (an example of which is shown in Fig. 2.2).

The STL file format was made public domain to allow all CAD vendors to access it easily and hopefully integrate it into their systems. This strategy has been successful and STL is now a standard output for nearly all solid modeling CAD systems and has also been adopted by AM system vendors [8]. STL uses triangles to describe the surfaces to be built. Each triangle is described as three points and a facet normal vector indicating the outward side of the triangle, in a manner similar to the following:

facet normal −4.470293E−02 7.003503E−01−7.123981E-01
outer loop
vertex −2.812284E+00 2.298693E+01 0.000000E+00
vertex −2.812284E+00 2.296699E+01−1.960784E−02
vertex −3.124760E+00 2.296699E+01 0.000000E+00
endloop
endfacet

Fig. 2.2 A CAD model on the *left* converted into STL format on the *right*

The demands on CAD technology in the future are set to change with respect to AM. As we move toward more and more functionality in the parts produced by AM, we must understand that the CAD system must include rules associated with AM. To date, the focus has been on the external geometry. In the future, we may need to know rules associated with how the AM systems function so that the output can be optimized.

2.4 Other Associated Technologies

Aside from computer technology there are a number of other technologies that have developed along with AM that are worthy of note here since they have served to contribute to further improvement of AM systems.

2.4.1 Lasers

Many of the earliest AM systems were based on laser technology. The reasons are that lasers provide a high intensity and highly collimated beam of energy that can be moved very quickly in a controlled manner with the use of directional mirrors. Since AM requires the material in each layer to be solidified or joined in a selective manner, lasers are ideal candidates for use, provided the laser energy is compatible with the material transformation mechanisms. There are two kinds of laser processing used in AM; curing and heating. With photopolymer resins the requirement is for laser energy of a specific frequency that will cause the liquid resin to solidify, or "cure." Usually this laser is in the ultraviolet range but other frequencies can be used. For heating, the requirement is for the laser to carry sufficient thermal energy to cut through a layer of solid material, to cause powder to melt, or to cause sheets of material to fuse. For powder processes, for example, the key is to melt the material in a controlled fashion without creating too great a build-up of heat, so that when the laser energy is removed, the molten material rapidly solidifies again. For cutting, the intention is to separate a region of material from another in the form of laser cutting. Earlier AM machines used tube lasers to provide the required energy but many manufacturers have more recently switched to solid-state technology, which provides greater efficiency, lifetime, and reliability.

2.4.2 Printing Technologies

Ink-jet or droplet printing technology has rapidly developed in recent years. Improvements in resolution and reduction in costs has meant that high-resolution printing, often with multiple colors, is available as part of our everyday lives. Such improvement in resolution has also been supported by improvement in material handling capacity and reliability. Initially, colored inks were low in viscosity and fed into the print heads at ambient temperatures. Now it is possible to generate

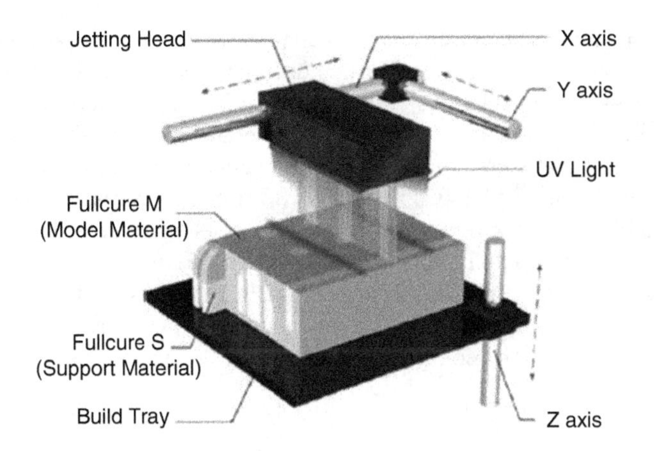

Fig. 2.3 Printer technology used on an AM machine (photo courtesy of Stratasys)

much higher pressures within the droplet formation chamber so that materials with much higher viscosity and even molten materials can be printed. This means that droplet deposition can now be used to print photocurable and molten resins as well as binders for powder systems. Since print heads are relatively compact devices with all the droplet control technology highly integrated into these heads (like the one shown in Fig. 2.3), it is possible to produce low-cost, high-resolution, high-throughput AM technology. In the same way that other AM technologies have applied the mass-produced laser technology, other technologies have piggy-backed upon the larger printing industry.

2.4.3 Programmable Logic Controllers

The input CAD models for AM are large data files generated using standard computer technology. Once they are on the AM machine, however, these files are reduced to a series of process stages that require sensor input and signaling of actuators. This is process and machine control that often is best carried out using microcontroller systems rather than microprocessor systems. Industrial microcontroller systems form the basis of Programmable Logic Controllers (PLCs), which are used to reliably control industrial processes. Designing and building industrial machinery, like AM machines, is much easier using building blocks based around modern PLCs for coordinating and controlling the various steps in the machine process.

2.4.4 Materials

Earlier AM technologies were built around materials that were already available and that had been developed to suit other processes. However, the AM processes are somewhat unique and these original materials were far from ideal for these new applications. For example, the early photocurable resins resulted in models that were brittle and that warped easily. Powders used in laser-based powder bed fusion processes degraded quickly within the machine and many of the materials used resulted in parts that were quite weak. As we came to understand the technology better, materials were developed specifically to suit AM processes. Materials have been tuned to suit more closely the operating parameters of the different processes and to provide better output parts. As a result, parts are now much more accurate, stronger, and longer lasting and it is even possible to process metals with some AM technologies. In turn, these new materials have resulted in the processes being tuned to produce higher temperature materials, smaller feature sizes, and faster throughput.

2.4.5 Computer Numerically Controlled Machining

One of the reasons AM technology was originally developed was because CNC technology was not able to produce satisfactory output within the required time frames. CNC machining was slow, cumbersome, and difficult to operate. AM technology on the other hand was quite easy to set up with quick results, but had poor accuracy and limited material capability. As improvements in AM technologies came about, vendors of CNC machining technology realized that there was now growing competition. CNC machining has dramatically improved, just as AM technologies have matured. It could be argued that high-speed CNC would have developed anyway, but some have argued that the perceived threat from AM technology caused CNC machining vendors to rethink how their machines were made. The development of hybrid prototyping technologies, such as Space Puzzle Molding that use both high-speed machining and additive techniques for making large, complex and durable molds and components, as shown in Fig. 2.4 [9], illustrate how the two can be used interchangeably to take advantage of the benefits of both technologies. For geometries that can be machined using a single set-up orientation, CNC machining is often the fastest, most cost-effective method. For parts with complex geometries or parts which require a large proportion of the overall material volume to be machined away as scrap, AM can be used to more quickly and economically produce the part than when using CNC.

2.5 The Use of Layers

A key enabling principle of AM part manufacture is the use of layers as finite 2D cross-sections of the 3D model. Almost every AM technology builds parts using layers of material added together; and certainly all commercial systems work that way, primarily due to the simplification of building 3D objects. Using 2D representations to represent cross-sections of a more complex 3D feature has been common in many applications outside AM. The most obvious example of

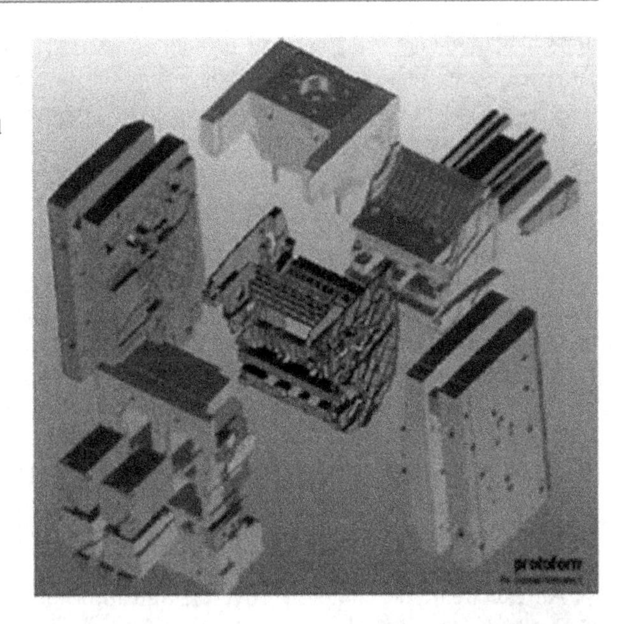

Fig. 2.4 Space puzzle molding, where molds are constructed in segments for fast and easy fabrication and assembly (photo courtesy of Protoform, Germany)

Fig. 2.5 An architectural landscape model, illustrating the use of layers (photo courtesy of LiD)

this is how cartographers use a line of constant height to represent hills and other geographical reliefs. These contour lines, or iso-heights, can be used as plates that can be stacked to form representations of geographical regions. The gaps between these 2D cross sections cannot be precisely represented and are therefore approximated, or interpolated, in the form of continuity curves connecting these layers. Such techniques can also be used to provide a 3D representation of other physical properties, like isobars or isotherms on weather maps.

Architects have also used such methods to represent landscapes of actual or planned areas, like that used by an architect firm in Fig. 2.5 [10]. The concept is particularly logical to manufacturers of buildings who also use an additive approach, albeit not using layers. Consider how the pyramids in Egypt and in South America were created. Notwithstanding how they were fabricated, it's clear that they were created using a layered approach, adding material as they went.

2.6 Classification of AM Processes

There are numerous ways to classify AM technologies. A popular approach is to classify according to baseline technology, like whether the process uses lasers, printer technology, extrusion technology, etc. [11, 12]. Another approach is to collect processes together according to the type of raw material input [13]. The problem with these classification methods is that some processes get lumped together in what seems to be odd combinations (like Selective Laser Sintering (SLS) being grouped together with 3D Printing) or that some processes that may appear to produce similar results end up being separated (like Stereolithography and material jetting with photopolymers). It is probably inappropriate, therefore, to use a single classification approach.

An excellent and comprehensive classification method is described by Pham [14], which uses a two-dimensional classification method as shown in Fig. 2.6. The first dimension relates to the method by which the layers are constructed. Earlier technologies used a single point source to draw across the surface of the base material. Later systems increased the number of sources to increase the throughput, which was made possible with the use of droplet deposition technology, for example, which can be constructed into a one dimensional array of deposition

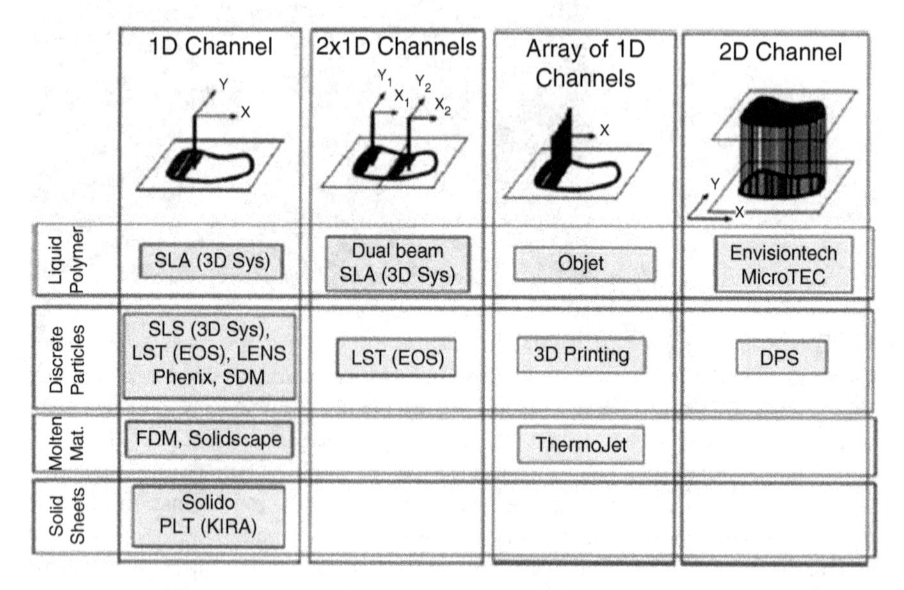

Fig. 2.6 Layered manufacturing (LM) processes as classified by Pham (note that this diagram has been amended to include some recent AM technologies)

heads. Further throughput improvements are possible with the use of 2D array technology using the likes of Digital Micro-mirror Devices (DMDs) and high-resolution display technology, capable of exposing an entire surface in a single pass. However, just using this classification results in the previously mentioned anomalies where numerous dissimilar processes are grouped together. This is solved by introducing a second dimension of raw material to the classification. Pham uses four separate material classifications; liquid polymer, discrete particles, molten material, and laminated sheets. Some more exotic systems mentioned in this book may not fit directly into this classification. An example is the possible deposition of composite material using an extrusion-based technology. This fits well as a 1D channel but the material is not explicitly listed, although it could be argued that the composite is extruded as a molten material. Furthermore, future systems may be developed that use 3D holography to project and fabricate complete objects in a single pass. As with many classifications, there can sometimes be processes or systems that lie outside them. If there are sufficient systems to warrant an extension to this classification, then it should not be a problem.

It should be noted that, in particular 1D and 2 × 1D channel systems combine both vector- and raster-based scanning methods. Often, the outline of a layer is traced first before being filled in with regular or irregular scanning patterns. The outline is generally referred to as vector scanned while the fill pattern can often be generalized as a raster pattern. The array methods tend not to separate the outline and the fill.

Most AM technology started using a 1D channel approach, although one of the earliest and now obsolete technologies, Solid Ground Curing from Cubital, used liquid photopolymers and essentially (although perhaps arguably) a 2D channel method. As technology developed, so more of the boxes in the classification array began to be filled. The empty boxes in this array may serve as a guide to researchers and developers for further technological advances. Most of the 1D array methods use at least 2 × 1D lines. This is similar to conventional 2D printing where each line deposits a different material in different locations within a layer. The recent Connex process using the Polyjet technology from Stratasys is a good example of this where it is now possible to create parts with different material properties in a single step using this approach. Color 3D Printing is possible using multiple 1D arrays with ink or separately colored material in each. Note however that the part coloration in the sheet laminating Mcor process [15] is separated from the layer formation process, which is why it is defined as a 1D channel approach.

2.6.1 Liquid Polymer Systems

As can be seen from Fig. 2.5 liquid polymers appear to be a popular material. The first commercial system was the 3D Systems Stereolithography process based on liquid photopolymers. A large portion of systems in use today are, in fact, not just liquid polymer systems but more specifically liquid photopolymer systems. However, this classification should not be restricted to just photopolymers, since a number of experimental systems are using hydrogels that would also fit into this category. Furthermore, the Fab@home system developed at Cornell University in the USA and the open source RepRap systems originating from Bath University in the UK also use liquid polymers with curing techniques other than UV or other wavelength optical curing methods [16, 17].

Using this material and a 1D channel or 2 × 1D channel scanning method, the best option is to use a laser like in the Stereolithography process. Droplet deposition of polymers using an array of 1D channels can simplify the curing process to a floodlight (for photopolymers) or similar method. This approach is used with machines made by the Israeli company Objet (now part of Stratasys) who uses printer technology to print fine droplets of photopolymer "ink" [18]. One unique feature of the Objet system is the ability to vary the material properties within a single part. Parts can for example have soft-feel, rubber-like features combined with more solid resins to achieve a result similar to an overmolding effect.

Controlling the area to be exposed using DMDs or other high-resolution display technology obviates the need for any scanning at all, thus increasing throughput and reducing the number of moving parts. DMDs are generally applied to micron-scale additive approaches, like those used by Microtec in Germany [19]. For normal-scale systems Envisiontec uses high-resolution DMD displays to cure photopolymer resin in their low-cost AM machines. The 3D Systems V-Flash process is also a variation on this approach, exposing thin sheets of polymer spread onto a build surface.

2.6.2 Discrete Particle Systems

Discrete particles are normally powders that are generally graded into a relatively uniform particle size and shape and narrow size distribution. The finer the particles the better, but there will be problems if the dimensions get too small in terms of controlling the distribution and dispersion. Again, the conventional 1D channel approach is to use a laser, this time to produce thermal energy in a controlled manner and, therefore, raise the temperature sufficiently to melt the powder. Polymer powders must therefore exhibit thermoplastic behavior so that they can be melted and re-melted to permit bonding of one layer to another. There are a wide variety of such systems that generally differ in terms of the material that can be processed. The two main polymer-based systems commercially available are the SLS technology marketed by 3D Systems [20] and the EOSint processes developed by the German company EOS [21].

Application of printer technology to powder beds resulted in the (original) 3D Printing (3DP) process. This technique was developed by researchers at MIT in the USA [22]. Droplet printing technology is used to print a binder, or glue, onto a powder bed. The glue sticks the powder particles together to form a 3D structure. This basic technique has been developed for different applications dependent on the type of powder and binder combination. The most successful approaches use low-cost, starch- and plaster-based powders with inexpensive glues, as commercialized by ZCorp, USA [23], which is now part of 3D Systems. Ceramic powders and appropriate binders as similarly used in the Direct Shell Production Casting (DSPC) process used by Soligen [24] as part of a service to create shells for casting of metal parts. Alternatively, if the binder were to contain an amount of drug, 3DP can be used to create controlled delivery-rate drugs like in the process developed by the US company Therics. Neither of these last two processes has proven to be as successful as that licensed by ZCorp/3D Systems. One particular advantage of the former ZCorp technology is that the binders can be jetted from multinozzle print heads. Binders coming from different nozzles can be different and, therefore, subtle variations can be incorporated into the resulting part. The most obvious of these is the color that can be incorporated into parts.

2.6.3 Molten Material Systems

Molten material systems are characterized by a pre-heating chamber that raises the material temperature to melting point so that it can flow through a delivery system. The most well-known method for doing this is the Fused Deposition Modeling (FDM) material extrusion technology developed by the US company Stratasys [25]. This approach extrudes the material through a nozzle in a controlled manner. Two extrusion heads are often used so that support structures can be fabricated from a different material to facilitate part cleanup and removal. It should be noted that there are now a huge number of variants of this technology due to the lapse of key FDM patents, with the number of companies making these perhaps even into three figures. This competition has driven the price of these machines down to such a level that individual buyers can afford to have their own machines at home.

Printer technology has also been adapted to suit this material delivery approach. One technique, developed initially as the Sanders prototyping machine, that later became Solidscape, USA [26] and which is now part of Stratasys, is a 1D channel system. A single jet piezoelectric deposition head lays down wax material. Another head lays down a second wax material with a lower melting temperature that is used for support structures. The droplets from these print heads are very small so the resulting parts are fine in detail. To further maintain the part precision, a planar cutting process is used to level each layer once the printing has been completed. Supports are removed by inserting the complete part into a temperature-controlled bath that melts the support material away, leaving the part material intact. The use of wax along with the precision of Solidscape machines makes this approach ideal for precision casting applications like jewelry, medical devices, and dental castings. Few machines are sold outside of these niche areas.

The 1D channel approach, however, is very slow in comparison with other methods and applying a parallel element does significantly improve throughput. The Thermojet technology from 3D Systems also deposits a wax material through droplet-based printing heads. The use of parallel print heads as an array of 1D channels effectively multiplies the deposition rate. The Thermojet approach, however, is not widely used because wax materials are difficult and fragile when handled. Thermojet machines are no longer being made, although existing machines are commonly used for investment casting patterns.

2.6.4 Solid Sheet Systems

One of the earliest AM technologies was the Laminated Object Manufacturing (LOM) system from Helisys, USA. This technology used a laser to cut out profiles from sheet paper, supplied from a continuous roll, which formed the layers of the final part. Layers were bonded together using a heat-activated resin that was coated on one surface of the paper. Once all the layers were bonded together the result was very much like a wooden block. A hatch pattern cut into the excess material allowed the user to separate away waste material and reveal the part.

A similar approach was used by the Japanese company Kira, in their Solid Center machine [27], and by the Israeli company Solidimension with their Solido machine. The major difference is that both these machines cut out the part profile using a blade similar to those found in vinyl sign-making machines, driven using a 2D plotter drive. The Kira machine used a heat-activated adhesive applied using laser printing technology to bond the paper layers together. Both the Solido and Kira machines have been discontinued for reasons like poor reliability material wastage and the need for excessive amounts of manual post-processing. Recently, however, Mcor Technologies have produced a modern version of this technology, using low-cost color printing to make it possible to laminate color parts in a single process [28].

2.6.5 New AM Classification Schemes

In this book, we use a version of Pham's classification introduced in Fig. 2.6. Instead of using the 1D and 2 × 1D channel terminology, we will typically use the terminology "point" or "point-wise" systems. For arrays of 1D channels, such as when using ink-jet print heads, we refer to this as "line" processing. 2D Channel technologies will be referred to as "layer" processing. Last, although no current commercialized processes are capable of this, holographic-like techniques are considered "volume" processing.

The technology-specific descriptions starting in Chap. 4 are based, in part, upon a separation of technologies into groups where processes which use a common type of machine architecture and similar materials transformation physics are grouped together. This grouping is a refinement of an approach introduced by Stucker and Janaki Ram in the CRC Materials Processing Handbook [29]. In this grouping scheme, for example, processes which use a common machine architecture developed for stacking layers of powdered material and a materials transformation mechanism using heat to fuse those powders together are all discussed in the Powder Bed Fusion chapter. These are grouped together even though these processes encompass polymer, metal, ceramic, and composite materials, multiple types of energy sources (e.g., lasers, and infrared heaters), and point-wise and layer processing approaches. Using this classification scheme, all AM processes fall into one of seven categories. An understanding of these seven categories should enable a person familiar with the concepts introduced in this book to quickly grasp and understand an unfamiliar AM process by comparing its similarities, benefits, drawbacks, and processing characteristics to the other processes in the grouping into which it falls.

This classification scheme from the first edition of this textbook had an important impact on the development and adoption of ASTM/ISO standard terminology. The authors were involved in these consensus standards activities and we have agreed to adopt the modified terminology from ASTM F42 and ISO TC 261 in the second edition. Of course, in the future, we will continue to support the ASTM/ISO standardization efforts and keep the textbook up to date.

The seven process categories are presented here. Chapters 4–10 cover each one in detail:

- Vat photopolymerization: processes that utilize a liquid photopolymer that is contained in a vat and processed by selectively delivering energy to cure specific regions of a part cross-section.
- Powder bed fusion: processes that utilize a container filled with powder that is processed selectively using an energy source, most commonly a scanning laser or electron beam.
- Material extrusion: processes that deposit a material by extruding it through a nozzle, typically while scanning the nozzle in a pattern that produces a part cross-section.
- Material jetting: ink-jet printing processes.
- Binder jetting: processes where a binder is printed into a powder bed in order to form part cross-sections.
- Sheet lamination: processes that deposit a layer of material at a time, where the material is in sheet form.
- Directed energy deposition: processes that simultaneously deposit a material (usually powder or wire) and provide energy to process that material through a single deposition device.

2.7 Metal Systems

One of the most important recent developments in AM has been the proliferation of direct metal processes. Machines like the EOSint M [21] and Laser Engineered Net Shaping (LENS) have been around for a number of years [30, 31]. Recent additions from other companies and improvements in laser technology, machine accuracy, speed, and cost have opened up this market.

Most direct metal systems work using a point-wise method and nearly all of them utilize metal powders as input. The main exception to this approach is the sheet lamination processes, particularly the Ultrasonic Consolidation process from the Solidica, USA, which uses sheet metal laminates that are ultrasonically welded together [32]. Of the powder systems, almost every newer machine uses a powder spreading approach similar to the SLS process, followed by melting using an energy beam. This energy is normally a high-power laser, except in the case of the Electron Beam Melting (EBM) process by the Swedish company Arcam [33]. Another approach is the LENS powder delivery system used by Optomec [31]. This machine employs powder delivery through a nozzle placed above the part. The powder is melted where the material converges with the laser and the substrate. This approach allows the process to be used to add material to an existing part, which means it can be used for repair of expensive metal components that may have been damaged, like chipped turbine blades and injection mold tool inserts.

2.8 Hybrid Systems

Some of the machines described above are, in fact, hybrid additive/subtractive processes rather than purely additive. Including a subtractive component can assist in making the process more precise. An example is the use of planar milling at the end of each additive layer in the Sanders and Objet machines. This stage makes for a smooth planar surface onto which the next layer can be added, negating cumulative effects from errors in droplet deposition height.

It should be noted that when subtractive methods are used, waste will be generated. Machining processes require removal of material that in general cannot easily be recycled. Similarly, many additive processes require the use of support structures and these too must be removed or "subtracted."

It can be said that with the Objet process, for instance, the additive element is dominant and that the subtractive component is important but relatively insignificant. There have been a number of attempts to merge subtractive and additive technologies together where the subtractive component is the dominant element. An excellent example of this is the Stratoconception approach [34], where the original CAD models are divided into thick machinable layers. Once these layers are machined, they are bonded together to form the complete solid part. This approach works very well for very large parts that may have features that would be difficult to machine using a multi-axis machining center due to the accessibility of the tool. This approach can be applied to foam and wood-based materials or to metals. For structural components it is important to consider the bonding methods. For high strength metal parts diffusion bonding may be an alternative.

A lower cost solution that works in a similar way is Subtractive RP (SRP) from Roland [35], who is also famous for plotter technology. SRP makes use of Roland desktop milling machines to machine sheets of material that can be sandwiched together, similar to Stratoconception. The key is to use the exterior material as a frame that can be used to register each slice to others and to hold the part in place. With this method not all the material is machined away and a web of connecting spars are used to maintain this registration.

Another variation of this method that was never commercialized was Shaped Deposition Manufacturing (SDM), developed mainly at Stanford and Carnegie-Mellon Universities in the USA [36]. With SDM, the geometry of the part is devolved into a sequence of easier to manufacture parts that can in some way be combined together. A decision is made concerning whether each subpart should be manufactured using additive or subtractive technology dependent on such factors as the accuracy, material, geometrical features, functional requirements, etc. Furthermore, parts can be made from multiple materials, combined together using a variety of processes, including the use of plastics, metals and even ceramics. Some of the materials can also be used in a sacrificial way to create cavities and clearances. Additionally, the "layers" are not necessarily planar, nor constant in thickness. Such a system would be unwieldy and difficult to realize commercially, but the ideas generated during this research have influenced many studies and systems thereafter.

In this book, for technologies where both additive and subtractive approaches are used, these technologies are discussed in the chapter where their additive approach best fits.

2.9 Milestones in AM Development

We can look at the historical development of AM in a variety of different ways. The origins may be difficult to properly define and there was certainly quite a lot of activity in the 1950s and 1960s, but development of the associated technology (computers, lasers, controllers, etc.) caught up with the concept in the early 1980s. Interestingly, parallel patents were filed in 1984 in Japan (Murutani), France (Andre et al.) and in the USA (Masters in July and Hull in August). All of these patents described a similar concept of fabricating a 3D object by selectively adding material layer by layer. While earlier work in Japan is quite well-documented, proving that this concept could be realized, it was the patent by Charles Hull that is generally recognized as the most influential since it gave rise to 3D Systems. This was the first company to commercialize AM technology with the Stereolithography apparatus (Fig. 2.7).

Further patents came along in 1986, resulting in three more companies, Helisys (Laminated Object Manufacture or LOM), Cubital (with Solid Ground Curing, SGC), and DTM with their SLS process. It is interesting to note neither Helisys nor Cubital exist anymore, and only SLS remains as a commercial process with DTM merging with 3D Systems in 2001. In 1989, Scott Crump patented the FDM process, forming the Stratasys Company. Also in 1989 a group from MIT patented the 3D Printing (3DP) process. These processes from 1989 are heavily used today, with FDM variants currently being the most successful. Rather than forming a company, the MIT group licensed the 3DP technology to a number of different companies, who applied it in different ways to form the basis for different applications of their AM technology. The most successful of these was ZCorp, which focused mainly on low-cost technology.

Fig. 2.7 The first AM technology from Hull, who founded 3D systems (photo courtesy of 3D Systems)

Ink-jet technology has become employed to deposit droplets of material directly onto a substrate, where that material hardens and becomes the part itself rather than just as a binder. Sanders developed this process in 1994 and the Objet Company also used this technique to print photocurable resins in droplet form in 2001.

There have been numerous failures and successes in AM history, with the previous paragraphs mentioning only a small number. However, it is interesting to note that some technology may have failed because of poor business models or by poor timing rather than having a poor process. Helisys appears to have failed with their LOM machine, but there have been at least five variants from Singapore, China, Japan, Israel, and Ireland. The most recent Mcor process laminates colored sheets together rather than the monochrome paper sheets used in the original LOM machine. Perhaps this is a better application and perhaps the technology is in a better position to become successful now compared with the original machines that are over 20 years old. Another example may be the defunct Ballistic Particle Manufacturing process, which used a 5-axis mechanism to direct wax droplets onto a substrate. Although no company currently uses such an approach for polymers, similar 5-axis deposition schemes are being used for depositing metal and composites.

Another important trend that is impacting the development of AM technology is the expiration of many of the foundational patents for key AM processes. Already, we are seeing an explosion of material extrusion vendors and systems since the first FDM patents expired in the early 2010s. Patents in the stereolithography, laser sintering, and LOM areas are expiring (or have already expired) and may lead to a proliferation of technologies, processes, machines, and companies.

2.10 AM Around the World

As was already mentioned, early patents were filed in Europe (France), USA, and Asia (Japan) almost simultaneously. In early years, most pioneering and commercially successful systems came out of the USA. Companies like Stratasys, 3D Systems, and ZCorp have spearheaded the way forward. These companies have generally strengthened over the years, but most new companies have come from outside the USA.

In Europe, the primary company with a world-wide impact in AM is EOS, Germany. EOS stopped making SL machines following settlement of disputes with 3D Systems but continues to make powder bed fusion systems which use lasers to melt polymers, binder-coated sand, and metals. Companies from France, The Netherlands, Sweden, and other parts of Europe are smaller, but are competitive in their respective marketplaces. Examples of these companies include Phenix [37] (now part of 3D Systems), Arcam, Strataconception, and Materialise. The last of these, Materialise from Belgium [38], has seen considerable success in developing software tools to support AM technology.

In the early 1980s and 1990s, a number of Japanese companies focused on AM technology. This included startup companies like Autostrade (which no longer appears to be operating). Large companies like Sony and Kira, who established subsidiaries to build AM technology, also became involved. Much of the Japanese technology was based around the photopolymer curing processes. With 3D Systems dominant in much of the rest of the world, these Japanese companies struggled to find market and many of them failed to become commercially viable, even though their technology showed some initial promise. Some of this demise may have resulted in the unusually slow uptake of CAD technology within Japanese industry in general. Although the Japanese company CMET [39] still seems to be doing quite well, you will likely find more non-Japanese made machines in Japan than home-grown ones. There is some indication however that this is starting to change.

AM technology has also been developed in other parts of Asia, most notably in Korea and China. Korean AM companies are relatively new and it remains to be seen whether they will make an impact. There are, however, quite a few Chinese manufacturers who have been active for a number of years. Patent conflicts with the earlier USA, Japanese, and European designs have meant that not many of these machines can be found outside of China. Earlier Chinese machines were also thought to be of questionable quality, but more recent machines have markedly improved performance (like the machine shown in Fig. 2.8). Chinese machines primarily reflect the SL, FDM, and SLS technologies found elsewhere in the world.

A particular country of interest in terms of AM technology development is Israel. One of the earliest AM machines was developed by the Israeli company Cubital. Although this technology was not a commercial success, in spite of early installations around the world, they demonstrated a number of innovations not found in other machines, including layer processing through a mask, removable secondary support materials and milling after each layer to maintain a constant layer thickness. Some of the concepts used in Cubital can be found in Sanders

Fig. 2.8 AM technology from Beijing Yinhua Co. Ltd., China

machines as well as machines from another Israeli company, Objet. Although one of the newer companies, Objet (now Stratasys) is successfully using droplet deposition technology to deposit photocurable resins.

2.11 The Future? Rapid Prototyping Develops into Direct Digital Manufacturing

How might the future of AM look? The ability to "grow" parts may form the core to the answer to that question. The true benefit behind AM is the fact that we do not really need to design the part according to how it is to be manufactured. We would prefer to design the part to perform a particular function. Avoiding the need to consider how the part can be manufactured certainly simplifies the process of design and allows the designer to focus more on the intended application. The design flexibility of AM is making this more and more possible.

An example of geometric flexibility is customization of a product. If a product is specifically designed to suit the needs of a unique individual then it can truly be said to be customized. Imagine being able to modify your mobile phone so that it fits snugly into your hand based on the dimensions gathered directly from your hand. Imagine a hearing aid that can fit precisely inside your ear because it was made from an impression of your ear canal (like those shown in Fig. 2.9). Such things are possible using AM because it has the capacity to make one-off parts, directly from digital models that may not only include geometric features but may also include biometric data gathered from a specific individual.

With improvements in AM technology the speed, quality, accuracy, and material properties have all developed to the extent that parts can be made for final use and not just for prototyping. The terms Rapid Manufacturing and Direct Digital

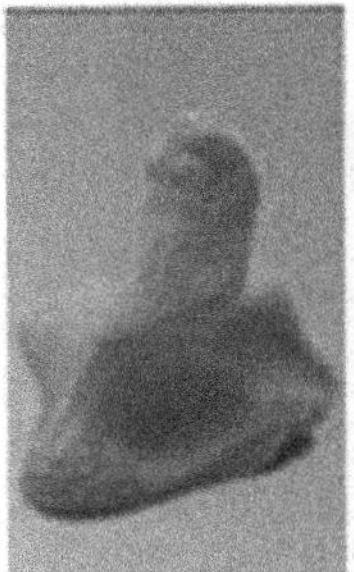

Fig. 2.9 RM of custom hearing aids, from a wax ear impression, on to the machine to the final product (photo courtesy of Phonak)

Manufacturing (RM and DDM) have gained popularity to represent the use of AM to produce parts which will be used as an end-product. Certainly we will continue to use this technology for prototyping for years to come, but we are already entering a time when it is commonplace to manufacture products in low volumes or unique products using AM. Eventually we may see these machines being used as home fabrication devices.

2.12 Exercises

1. (a) Based upon an Internet search, describe the Solid Ground Curing process developed by Cubital. (b) Solid Ground Curing has been described as a 2D channel (layer) technique. Could it also be described in another category? Why?
2. Make a list of the different metal AM technologies that are currently available on the market today. How can you distinguish between the different systems? What different materials can be processed in these machines?
3. NC machining is often referred to as a 2.5D process. What does this mean? Why might it not be regarded as fully 3D?
4. Provide three instances where a layer-based approach has been used in fabrication, other than AM.
5. Find five countries where AM technology has been developed commercially and describe the machines.
6. Consider what a fabrication system in the home might look like, with the ability to manufacture many of the products around the house. How do you think this could be implemented?

References

1. Zuse Z3 Computer. http://www.zib.de/zuse
2. Goldstine HH, Goldstine A (1946) The electronic numerical integrator and computer (ENIAC). Math Tables Other Aids Comput 2(15):97–110

3. Wilkes MV, Renwick W (1949) The EDSAC – an electronic calculating machine. J Sci Instrum 26:385–391
4. Waterloo Computer Science Club, talk by Bill Gates. http://csclub.uwaterloo.ca/media
5. Gatlin J (1999) Bill Gates – the path to the future. Quill, New York, p 39
6. Piegl L, Tiller W (1997) The NURBS book, 2nd edn. Springer, New York
7. Jamieson R, Hacker H (1995) Direct slicing of CAD models for rapid prototyping. Rapid Prototyping J 1(2):4–12
8. 3D Systems Inc. (1989) Stereolithography interface specification. 3D Systems, Valencia
9. Protoform. Space Puzzle Moulding. http://www.protoform.de
10. LiD, Architects. http://www.lid-architecture.net
11. Kruth JP, Leu MC, Nakagawa T (1998) Progress in additive manufacturing and rapid prototyping. Ann CIRP 47(2):525–540
12. Burns M (1993) Automated fabrication: improving productivity in manufacturing. Prentice Hall, Englewood Cliffs
13. Chua CK, Leong KF (1998) Rapid prototyping: principles and applications in manufacturing. Wiley, New York
14. Pham DT, Gault RS (1998) A comparison of rapid prototyping technologies. Int J Mach Tools Manuf 38(10–11):1257–1287
15. MCor. http://www.mcortechnologies.com
16. Fab@home. http://www.fabathome.org
17. Reprap. http://www.reprap.org
18. Objet technologies. http://www.objet.com
19. Microtec. http://www.microtec-d.com
20. 3D Systems. Stereolithography and selective laser sintering machines. http://www.3dsystems.com
21. EOS. http://www.eos.info
22. Sachs EM, Cima MJ, Williams P, Brancazio D, Cornie J (1992) Three dimensional printing: rapid tooling and prototypes directly from a CAD model. J Eng Ind 114(4):481–488
23. ZCorp. http://www.zcorp.com
24. Soligen. http://www.soligen.com
25. Stratasys. http://www.stratasys.com
26. Solidscape. http://www.solid-scape.com
27. Kira. Solid Center machine. www.kiracorp.co.jp/EG/pro/rp/top.html
28. MCor Technologies. http://www.mcortechnologies.com
29. Stucker BE, Janaki Ram GD (2007) Layer-based additive manufacturing technologies. In: Groza J et al (eds) CRC materials processing handbook. CRC, Boca Raton, pp 26.1–26.31
30. Atwood C, Ensz M, Greene D et al (1998) Laser engineered net shaping (LENS(TM)): a tool for direct fabrication of metal parts. Paper presented at the 17th International Congress on Applications of Lasers and Electro-Optics, Orlando, 16–19 November 1998. http://www.osti.gov/energycitations
31. Optomec. LENS process. http://www.optomec.com
32. White D (2003) Ultrasonic object consolidation. US Patent 6,519,500 B1
33. Arcam. Electron Beam Melting. http://www.arcam.com
34. Stratoconception. Thick layer hybrid AM. http://www.stratoconception.com
35. Roland. SRP technology. http://www.rolanddga.com/solutions/rapidprototyping/
36. Weiss L, Prinz F (1998) Novel applications and implementations of shape deposition manufacturing. Naval research reviews, vol L(3). Office of Naval Research, Arlington
37. Phenix. Metal RP technology. http://www.phenix-systems.com
38. Materialise. AM software systems and service provider. http://www.materialise.com
39. CMET. Stereolithography technology. http://www.cmet.co.jp

Generalized Additive Manufacturing Process Chain

3

3.1 Introduction

Every product development process involving an additive manufacturing machine requires the operator to go through a set sequence of tasks. Easy-to-use "personal" 3D printing machines emphasize the simplicity of this task sequence. These desktop-sized machines are characterized by their low cost, simplicity of use, and ability to be placed in a home or office environment. The larger and more "industrial" AM machines are more capable of being tuned to suit different user requirements and therefore require more expertise to operate, but with a wider variety of possible results and effects that may be put to good use by an experienced operator. Such machines also usually require more careful installation in industrial environments.

This chapter will take the reader through the different stages of the process that were described in Chap. 1. Where possible, the different steps in the process will be described with reference to different processes and machines. The objective is to allow the reader to understand how these machines may differ and also to see how each task works and how it may be exploited to the benefit of higher quality results. As mentioned before, we will refer to eight key steps in the process sequence:

- Conceptualization and CAD
- Conversion to STL/AMF
- Transfer and manipulation of STL/AMF file on AM machine
- Machine setup
- Build
- Part removal and cleanup
- Post-processing of part
- Application

There are other ways to breakdown this process flow, depending on your perspective and equipment familiarity. For example, if you are a designer, you

I. Gibson et al., *Additive Manufacturing Technologies*,
DOI 10.1007/978-1-4939-2113-3_3

may see more stages in the early product design aspects. Model makers may see more steps in the post-build part of the process. Different AM technologies handle this process sequence differently, so this chapter will also discuss how choice of machine affects the generic process.

The use of AM in place of conventional manufacturing processes, such as machining and injection molding, enables designers to ignore some of the constraints of conventional manufacturing. However, conventional manufacturing will remain core to how many products are manufactured. Thus, we must also understand how conventional technologies, such as machining, integrate with AM. This may be particularly relevant to the increasingly popular metal AM processes. Thus, we will discuss how to deal with metal systems in detail.

3.2 The Eight Steps in Additive Manufacture

The above-mentioned sequence of steps is generally appropriate to all AM technologies. There will be some variations dependent on which technology is being used and also on the design of the particular part. Some steps can be quite involved for some machines but may be trivial for others.

3.2.1 Step 1: Conceptualization and CAD

The first step in any product development process is to come up with an idea for how the product will look and function. Conceptualization can take many forms, from textual and narrative descriptions to sketches and representative models. If AM is to be used, the product description must be in a digital form that allows a physical model to be made. It may be that AM technology will be used to prototype and not build the final product, but in either case, there are many stages in a product development process where digital models are required.

AM technology would not exist if it were not for 3D CAD. Only after we gained the ability to represent solid objects in computers were we able to develop technology to physically reproduce such objects. Initially, this was the principle surrounding CNC machining technology in general. AM can thus be described as a direct or streamlined Computer Aided Design to Computer Aided Manufacturing (CAD/CAM) process. Unlike most other CAD/CAM technologies, there is little or no intervention between the design and manufacturing stages for AM.

The generic AM process must therefore start with 3D CAD information, as shown in Fig. 3.1. There may be a variety of ways for how the 3D source data can be created. This model description could be generated by a design expert via a user-interface, by software as part of an automated optimization algorithm, by 3D scanning of an existing physical part, or some combination of all of these. Most 3D CAD systems are solid modeling systems with surface modeling components; solid models are often constructed by combining surfaces together or by adding thickness to a surface. In the past, 3D CAD modeling software had difficulty

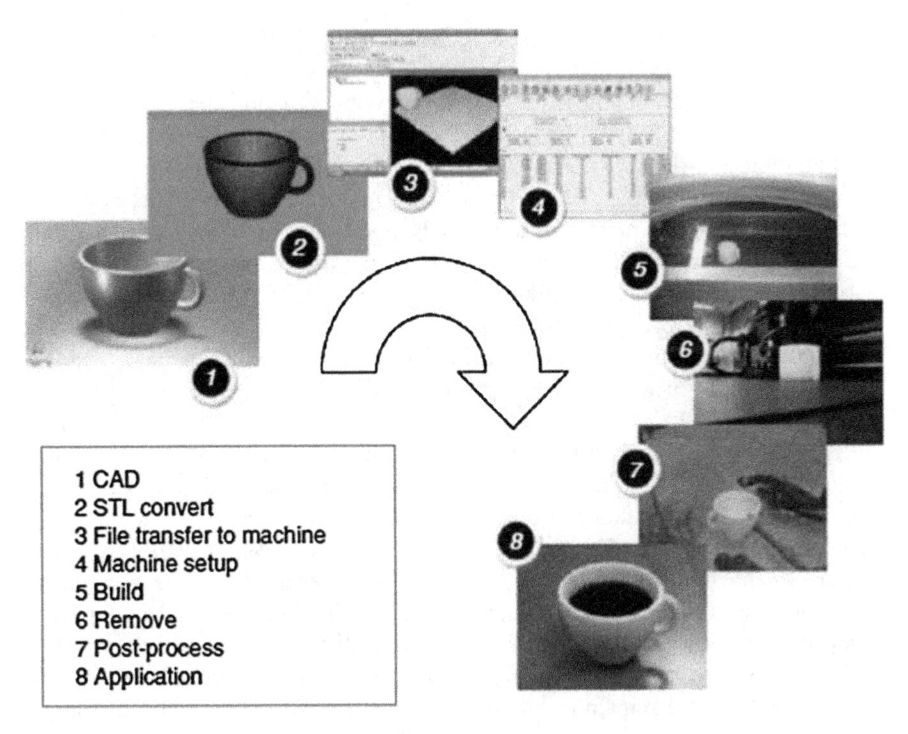

Fig. 3.1 The eight stages of the AM process

creating fully enclosed solid models, and often models would appear to the casual observer to be enclosed but in fact were not mathematically closed. Such models could result in unpredictable output from AM machines, with different AM technologies treating gaps in different ways.

Most modern solid modeling CAD tools can now create files without gaps (e.g., "water tight"), resulting in geometrically unambiguous representations of a part. Most CAD packages treat surfaces as construction tools that are used to act on solid models, and this has the effect of maintaining the integrity of the solid data. Provided it can fit inside the machine, typically any CAD model can be made using AM technology without too many difficulties. However, there still remain some older or poorly developed 3D CAD software that may result in solids that are not fully enclosed and produce unreliable AM output. Problems of this manner are normally detected once the CAD model has been converted into the STL format for building using AM technology.

3.2.2 Step 2: Conversion to STL/AMF

Nearly every AM technology uses the STL file format. The term STL was derived from STereoLithograhy, which was the first commercial AM technology from 3D

Systems in the 1990s. Considered a de facto standard, STL is a simple way of describing a CAD model in terms of its geometry alone. It works by removing any construction data, modeling history, etc., and approximating the surfaces of the model with a series of triangular facets. The minimum size of these triangles can be set within most CAD software and the objective is to ensure the models created do not show any obvious triangles on the surface. The triangle size is in fact calculated in terms of the minimum distance between the plane represented by the triangle and the surface it is supposed to represent. In other words, a basic rule of thumb is to ensure that the minimum triangle offset is smaller than the resolution of the AM machine. The process of converting to STL is automatic within most CAD systems, but there is a possibility of errors occurring during this phase. There have therefore been a number of software tools developed to detect such errors and to rectify them if possible.

STL files are an unordered collection of triangle vertices and surface normal vectors. As such, an STL file has no units, color, material, or other feature information. These limitations of an STL file have led to the recent adoption of a new "AMF" file format. This format is now an international ASTM/ISO standard format which extends the STL format to include dimensions, color, material, and many other useful features. As of the writing of this book, several major CAD companies and AM hardware vendors had publically announced that they will be supporting AMF in their next generation software. Thus, although the term STL is used throughout the remainder of this textbook, the AMF file could be simply substituted wherever STL appears, as the AMF format has all of the benefits of the STL file format with many fewer limitations.

STL file repair software, like the MAGICS software from the Belgian company Materialise [1], is used when there are problems with the STL file that may prevent the part from being built correctly. With complex geometries, it may be difficult for a human to detect such problems when inspecting the CAD or the subsequently generated STL data. If the errors are small then they may even go unnoticed until after the part has been built. Such software may therefore be applied as a checking stage to ensure that there are no problems with the STL file data before the build is performed.

Since STL is essentially a surface description, the corresponding triangles in the files must be pointing in the correct direction; in other words, the surface normal vector associated with the triangle must indicate which side of the triangle is outside vs. inside the part. The cross-section that corresponds to the part layers of a region near an inverted normal vector may therefore be the inversion of what is desired. Additionally, complex and highly discontinuous geometry may result in triangle vertices that do not align correctly. This may result in gaps in the surface. Various AM technologies may react to these problems in different ways. Some machines may process the STL data in such a way that the gaps are bridged. This bridge may not represent the desired surface, however, and it may be possible that additional, unwanted material may be included in the part.

While most errors can be detected and rectified automatically, there may also be a requirement for manual intervention. Software should therefore highlight the problem, indicating what is thought to be inverted triangles for instance. Since geometries can become very complex, it may be difficult for the software to establish whether the result is in fact an error or something that was part of the original design intent.

3.2.3 Step 3: Transfer to AM Machine and STL File Manipulation

Once the STL file has been created and repaired, it can be sent directly to the target AM machine. Ideally, it should be possible to press a "print" button and the machine should build the part straight away. This is not usually the case however and there may be a number of actions required prior to building the part.

The first task would be to verify that the part is correct. AM system software normally has a visualization tool that allows the user to view and manipulate the part. The user may wish to reposition the part or even change the orientation to allow it to be built at a specific location within the machine. It is quite common to build more than one part in an AM machine at a time. This may be multiples of the same part (thus requiring a copy function) or completely different STL files. STL files can be linearly scaled quite easily. Some applications may require the AM part to be slightly larger or slightly smaller than the original to account for process shrinkage or coatings; and so scaling may be required prior to building. Applications may also require that the part be identified in some way and some software tools have been developed to add text and simple features to STL formatted data for this purpose. This would be done in the form of adding 3D embossed characters. More unusual cases may even require segmentation of STL files (e.g., for parts that may be too large) or even merging of multiple STL files. It should be noted that not all AM machines will have all the functions mentioned here, but numerous STL file manipulation software tools are available for purchase or, in some cases, for free download to perform these functions prior to sending the file to a machine.

3.2.4 Step 4: Machine Setup

All AM machines will have at least some setup parameters that are specific to that machine or process. Some machines are only designed to run a few specific materials and give the user few options to vary layer thickness or other build parameters. These types of machines will have very few setup changes to make from build to build. Other machines are designed to run with a variety of materials and may also have some parameters that require optimization to suit the type of part that is to be built, or permit parts to be built quicker but with poorer resolution. Such machines can have numerous setup options available. It is common in the more complex cases to have default settings or save files from previously defined setups to help speed up the machine setup process and to prevent mistakes being made.

Normally, an incorrect setup procedure will still result in a part being built. The final quality of that part may, however, be unacceptable.

In addition to setting up machine software parameters, most machines must be physically prepared for a build. The operator must check to make sure sufficient build material is loaded into the machine to complete the build. For machines which use powder, the powder is often sifted and subsequently loaded and leveled in the machine as part of the setup operation. For processes which utilize a build plate, the plate must be inserted and leveled with respect to the machine axes. Some of these machine setup operations are automated as part of the start-up of a build, but for most machines these operations are done manually by a trained operator.

3.2.5 Step 5: Build

Although benefitting from the assistance of computers, the first few stages of the AM process are semiautomated tasks that may require considerable manual control, interaction, and decision making. Once these steps are completed, the process switches to the computer-controlled building phase. This is where the previously mentioned layer-based manufacturing takes place. All AM machines will have a similar sequence of layering, including a height adjustable platform or deposition head, material deposition/spreading mechanisms, and layer cross-section formation. Some machines will combine the material deposition and layer formation simultaneously while others will separate them. As long as no errors are detected during the build, AM machines will repeat the layering process until the build is complete.

3.2.6 Step 6: Removal and Cleanup

Ideally, the output from the AM machine should be ready for use with minimal manual intervention. While sometimes this may be the case, more often than not, parts will require a significant amount of post-processing before they are ready for use. In all cases, the part must be either separated from a build platform on which the part was produced or removed from excess build material surrounding the part. Some AM processes use additional material other than that used to make the part itself (secondary support materials). Later chapters describe how various AM processes need these support structures to help keep the part from collapsing or warping during the build process. At this stage, it is not necessary to understand exactly how support structures work, but it is necessary to know that they need to be dealt with. While some processes have been developed to produce easy-to-remove supports, there is often a significant amount of manual work required at this stage. For metal supports, a wire EDM machine, bandsaw, and/or milling equipment may be required to remove the part from the baseplate and the supports from the part. There is a degree of operator skill required in part removal, since mishandling of parts and poor technique can result in damage to the part. Different AM parts have different cleanup requirements, but suffice it to say that all processes have some requirement at this stage. The cleanup stage may also be considered as the initial part of the post-processing stage.

3.2.7 Step 7: Post-Processing

Post-processing refers to the (usually manual) stages of finishing the parts for application purposes. This may involve abrasive finishing, like polishing and sandpapering, or application of coatings. This stage in the process is very application specific. Some applications may only require a minimum of post-processing. Other applications may require very careful handling of the parts to maintain good precision and finish. Some post-processing may involve chemical or thermal treatment of the part to achieve final part properties. Different AM processes have different results in terms of accuracy, and thus machining to final dimensions may be required. Some processes produce relatively fragile components that may require the use of infiltration and/or surface coatings to strengthen the final part. As already stated, this is often a manually intensive task due to the complexity of most AM parts. However, some of the tasks can benefit from the use of power tools, CNC milling, and additional equipment, like polishing tubs or drying and baking ovens.

3.2.8 Step 8: Application

Following post-processing, parts are ready for use. It should be noted that, although parts may be made from similar materials to those available from other manufacturing processes (like molding and casting), parts may not behave according to standard material specifications. Some AM processes inherently create parts with small voids trapped inside them, which could be the source for part failure under mechanical stress. In addition, some processes may cause the material to degrade during build or for materials not to bond, link, or crystallize in an optimum way. In almost every case, the properties are anisotropic (different properties in different direction). For most metal AM processes, rapid cooling results in different microstructures than those from conventional manufacturing. As a result, AM produced parts behave differently than parts made using a more conventional manufacturing approach. This behavior may be better or worse for a particular application, and thus a designer should be aware of these differences and take them into account during the design stage. AM materials and processes are improving rapidly, and thus designers must be aware of recent advancements in materials and processes to best determine how to use AM for their needs.

3.3 Variations from One AM Machine to Another

The above generic process steps can be applied to every commercial AM technology. As has been noted, different technologies may require more or less attention for a number of these stages. Here we discuss the implications of these variations, not only from process to process but also in some cases within a specific technology.

The nominal layer thickness for most machines is around 0.1 mm. However, it should be noted that this is just a rule of thumb. For example, the layer thickness for some material extrusion machines is 0.254 mm, whereas layer thicknesses between 0.05 and 0.1 mm are commonly used for vat photopolymerization processes, and small intricate parts made for investment casting using material jetting technology may have layer thicknesses of 0.01 mm. Many technologies have the capacity to vary the layer thickness. The reasoning is that thicker layer parts are quicker to build but are less precise. This may not be a problem for some applications where it may be more important to make the parts as quickly as possible.

Fine detail in a design may cause problems with some AM technologies, such as wall thickness, particularly if there is no choice but to build the part vertically. This is because even though positioning within the machine may be very precise, there is a finite dimension to the droplet size, laser diameter, or extrusion head that essentially defines the finest detail or thinnest wall that can be fabricated.

There are other factors that may not only affect the choice of process but also influence some of the steps in the process chain. In particular, the use of different materials even within the same process may affect the time, resources, and skill required to carry out a stage. For example, the use of water soluble supports in material extrusion processes may require specialist equipment but will also provide better finish to parts with less hand finishing required than when using conventional supports. Alternatively, some polymers require special attention, like the use (or avoidance) of particular solvents or infiltration compounds. A number of processes benefit from application of sealants or even infiltration of liquid polymers. These materials must be compatible with the part material both chemically and mechanically. Post-processing that involves heat must include awareness of the heat resistance or melting temperature of the materials involved. Abrasive or machining-based processing must also require knowledge of the mechanical properties of the materials involved. If considerable finishing is required, it may also be necessary to include an allowance in the part geometry, perhaps by using scaling of the STL file or offsetting of the part's surfaces, so that the part does not become worn away too much.

Variations between AM technologies will become clarified further in the following chapters, but a general understanding can be achieved by considering whether the build material is processed as a powder, molten material, solid sheet, vat of liquid photopolymer, or ink-jet deposited photopolymer.

3.3.1 Photopolymer-Based Systems

It is quite easy to set up systems which utilize photopolymers as the build material. Photopolymer-based systems, however, require files to be created which represent the support structures. All liquid vat systems must use supports from essentially the same material as that used for the part. For material jetting systems it is possible to use a secondary support material from parallel ink-jet print heads so that the supports will come off easier. An advantage of photopolymer systems is that accuracy is generally very good, with thin layers and fine precision where required compared with other systems. Photopolymers have historically had poor material properties when compared with many other AM materials, however newer resins have been developed that offer improved temperature resistance, strength, and ductility. The main drawback of photopolymer materials is that degradation can occur quite rapidly if UV protective coatings are not applied.

3.3.2 Powder-Based Systems

There is no need to use supports for powder systems which deposit a bed of powder layer-by-layer (with the exception of supports for metal systems, as addressed below). Thus, powder bed-based systems are among the easiest to set up for a simple build. Parts made using binder jetting into a powder bed can be colored by using colored binder material. If color is used then coding the file may take a longer time, as standard STL data does not include color. There are, however, other file formats based around VRML that allow colored geometries to be built, in addition to AMF. Powder bed fusion processes have a significant amount of unused powder in every build that has been subjected to some level of thermal history. This thermal history may cause changes in the powder. Thus, a well-designed recycling strategy based upon one of several proven methods can help ensure that the material being used is within appropriate limits to guarantee good builds [2].

It is also important to understand the way powders behave inside a machine. For example, some machines use powder feed chambers at either side of the build platform. The powder at the top of these chambers is likely to be less dense than the powder at the bottom, which will have been compressed under the weight of the powder on top. This in turn may affect the amount of material deposited at each layer and density of the final part built in the machine. For very tall builds, this may be a particular problem that can be solved by carefully compacting the powder in the feed chambers before starting the machine and also by adjusting temperatures and powder feed settings during the build.

3.3.3 Molten Material Systems

Systems which melt and deposit material in a molten state require support structures. For droplet-based systems like with the Thermojet process these supports are automatically generated; but with material extrusion processes or directed energy deposition systems supports can either be generated automatically or the user can use some flexibility to change how supports are made. With water soluble supports it is not too important where the supports go, but with breakaway support systems made from the same material as the build material, it is worthwhile to check where the supports go, as surface damage to the part will occur to some extent where these supports were attached before breaking them away. Also, fill patterns for material extrusion may require some attention, based upon the design intent. Parts can be easily made using default settings, but there may be some benefit in changing aspects of the build sequence if a part or region of a part requires specific characteristics. For example, there are typically small voids in FDM parts that can be minimized by increasing the amount of material extruded in a particular region. This will minimize voids, but at the expenses of part accuracy. Although wax parts made using material jetting are good for reproducing fine features, they are difficult to handle because of their low strength and brittleness. ABS parts made using material extrusion, on the other hand, are among the strongest AM polymer parts available, but when they are desired as a functional end-use part, this may mean they need substantial finishing compared with other processes as they exhibit lower accuracy than some other AM technologies.

3.3.4 Solid Sheets

With sheet lamination methods where the sheets are first placed and then cut, there is no need for supports. Instead, there is a need to process the waste material in such a way that it can be removed from the part. This is generally a straightforward automated process but there may be a need for close attention to fine detail within a part. Cleaning up the parts can be the most laborious process and there is a general need to know exactly what the final part is supposed to look like so that damage is not caused to the part during the waste removal stage. The paper-based systems experienced problems with handling should they not be carefully and comprehensively finished using sealants and coatings. For polymer sheet lamination, the parts are typically not as sensitive to damage. For metal sheet lamination processes, typically the sheets are cut first and then stacked to form the 3D shape, and thus support removal becomes unnecessary.

3.4 Metal Systems

As previously mentioned, operation of metal-based AM systems is conceptually similar to polymer systems. However, the following points are worth considering.

3.4.1 The Use of Substrates

Most metal systems make use of a base platform or substrate onto which parts are built and from which they must be removed using machining, wire cutting, or a similar method. The need to attach the parts to a base platform is mainly because of the high-temperature gradients between the temporarily molten material and its surroundings, resulting in large residual stress. If the material was not rigidly attached to a solid platform then there would be a tendency for the part to warp as it cools, which means further layers of powder could not be spread evenly over top. Therefore, even though these processes may build within a powder bed, there is still a need for supports.

3.4.2 Energy Density

The energy density required to melt metals is obviously much higher than for melting polymers. The high temperatures achieved during metal melting may require more stringent heat shielding, insulation, temperature control, and atmospheric control than for polymer systems.

3.4.3 Weight

Metal powder systems may process lightweight titanium powders but they also process high-density tool steels. The powder handling technology must be capable of withstanding the mass of these materials. This means that power requirements for positioning and handling equipment must be quite substantial or gear ratios must be high (and corresponding travel speeds lower) to deal with these tasks.

3.4.4 Accuracy

Metal powder systems are generally at least as accurate as corresponding polymer powder systems. Surface finish is characteristically grainy but part density and part accuracy are very good. Surface roughness is on the order of a few tens to a few hundreds of microns depending on the process, and can be likened in general appearance to precision casting technology. For metal parts, this is often not satisfactory and at least some shot-peening is required to smooth the surface. Key mating features on metal parts often require surface machining or grinding. The part density will be high (generally over 99 %), although some voids may still be seen.

3.4.5 Speed

Since there are heavy requirements on the amount of energy to melt the powder particles and to handle the powders within the machine, the build speed of metal systems is generally slower than a comparable polymer system. Laser powers are usually just a few 100 W (polymer systems start at around 50 W of laser power). This means that the laser scanning speed is lower than for polymer systems, to ensure enough energy is delivered to the powder.

3.5 Maintenance of Equipment

While numerous stages in the AM process have been discussed, it is important to realize that many machines require careful maintenance. Some machines use sensitive laser or printer technology that must be carefully monitored and that should preferably not be used in a dirty or noisy (both electrical noise and mechanical vibration) environment. Similarly, many of the feed materials require careful handling and should be used in low humidity conditions. While machines are designed to operate unattended, it is important to include regular checks in the maintenance schedule. Many machine vendors recommend and provide test patterns that should be used periodically to confirm that the machines are operating within acceptable limits.

Laser-based systems are generally expensive because of the cost of the laser and scanner system. Furthermore, maintenance of a laser can be very expensive, particularly for lasers with limited lifetimes. Printheads are also components that have finite lifetimes for material jetting and binder jetting systems. The fine nozzle dimensions and the use of relatively high viscosity fluids mean they are prone to clogging and contamination effects. Replacement costs are, however, generally quite low.

3.6 Materials Handling Issues

In addition to the machinery, AM materials often require careful handling. The raw materials used in some AM processes have limited shelf-life and must also be kept in conditions that prevent them from chemical reaction or degradation. Exposure to moisture and to excess light should be avoided. Most processes use materials that can be used for more than one build. However, it may be that this could degrade the material if used many times over and therefore a procedure for maintaining consistent material quality through recycling should also be observed.

While there are some health concerns with extended exposure to some photopolymer resins, most AM polymer raw materials are safe to handle. Powder materials may in general be medically inert, but excess amounts of powder can make the workplace slippery, contaminate mechanisms, and create a breathing hazard. In addition, reactive powders can be a fire hazard. These issues may cause problems if machines are to be used in a design center environment rather than in a workshop. AM system vendors have spent considerable effort to simplify and facilitate material handling. Loading new materials is often a procedure that can be done offline or with minimal changeover time so that machines can run continuously. Software systems are often tuned to the materials so that they can recognize different materials and adjust build parameters accordingly.

Many materials are carefully tuned to work with a specific AM technology. There are often warranty issues surrounding the use of third party materials that users should be aware of. For example, some polymer laser sintering powders may have additives that prevent degradation due to oxidation since they are kept at elevated temperatures for long periods of time. Also, material extrusion filaments need a very tight diametric tolerance not normally available from conventional extruders. Since a material extrusion drive pushes the filament through the machine, variations in diameter may cause slippage. Furthermore, build parameters are designed around the standard materials used. Since there are huge numbers of material formulations, changing one material for another, even though they appear to be the same, may require careful build setup and process parameter optimization.

Some machines allow the user to recycle some or all of the material used in a machine but not consumed during the build of a prior part. This is particularly true with the powder-based systems. Also photopolymer resins can be reused. However, there may be artifacts and other contaminants in the recycled materials and it is important to carefully inspect, sift, or sieve the material before returning it to the machine. Many laser sintering builds have been spoiled, for example, by hairs that have come off a paintbrush used to clean the parts from a previous build.

3.7 Design for AM

Designers and operators should consider a number of build-related factors when considering the setup of an AM machine, including the following sections. This is a brief introduction, but more information can be found in Chap.17.

3.7.1 Part Orientation

If a cylinder was built on its end, then it would consist of a series of circular layers built on top of each other. Although layer edges may not be precisely vertical for all AM processes, the result would normally be a very well-defined cylinder with a relatively smooth edge. The same cylinder built on its side will have distinct layer stair-step patterning on the sides. This will result in less accurate reproduction of the original CAD data with a poorer aesthetic appearance. Additionally, as the layering process for most AM machines takes additional time, a long cylinder built vertically will take more time to build than if it is laid horizontally. For material extrusion processes, however, the time to build a part is solely a factor of the total build volume (including supports) and thus a cylinder should always be built vertically if possible.

Orientation of the part within the machine can affect part accuracy. Since many parts will have complex features along multiple axes, there may not be an ideal orientation for a particular part. Furthermore, it may be more important to maintain the geometry of some features when compared with others, so correct orientation may be a judgment call. This judgment may also be in contrast with other factors like the time it takes to build a part (e.g., taller builds take longer than shorter ones so high aspect ratio parts may be better built lying down), whether a certain orientation will generate more supports, or whether certain surfaces should be built face-up to ensure good surface finish in areas that are not in contact with support structures.

In general upward-facing features in AM have the best quality. The reason for this depends upon the process. For instance, upward-facing features are not in contact with the supports required for many processes. For powder beds, the upward-facing features are smooth since they solidify against air, whereas downward-facing and sideways-facing features solidify against powder and thus have a powdery texture. For extrusion processes, upward-facing surfaces are smoothed by the extrusion tip. Thus, this upward-facing feature quality rule is one of the few rules-of-thumb that are generically applicable to every AM process.

3.7.2 Removal of Supports

For those technologies that require supports, it is a good idea to try and minimize the amount. Wherever the supports meet the part there will be small marks and reducing the amount of supports would reduce the amount of part cleanup and post-process finishing. However, as mentioned above, some surfaces may not be as important as others and so positioning of the part must be weighed against the relative importance of an affected surface. In addition, removal of too many supports may mean that the part becomes detached from the baseplate and will move around during subsequent layering. If distortion causes a part to extend in the z direction enough that it hits the layering mechanism (such as a powder-spreading blade) then the build will fail.

Parts that require supports may also require planning for their removal. Supports may be located in difficult-to-reach regions within the part. For example, a hollow cylinder with end caps built vertically will require supports for the top surface. However, if there is no access hole then these supports cannot be removed. Inclusion of access holes (which could be plugged later) is a possible solution to this, as may be breaking up the part so the supports can be removed before reassembly. Similarly, parts made using vat photopolymerization processes may require drain holes for any trapped liquid resin.

3.7.3 Hollowing Out Parts

Parts that have thick walls may be designed to include hollow features if this does not reduce the part's functionality. The main benefits of doing this are the reduced build time, the reduced cost from the use of less material, and the reduced mass in the final component. As mentioned previously, some liquid-based resin systems would require drain holes to remove excess resin from inside the part, and the same is true for powder. A honeycomb- or truss-like internal structure can assist in providing support and strength within a part, while reducing its overall mass and volume. All of these approaches must be balanced against the additional time that it would take to design such a part. However, there are software systems that would allow this to be done automatically for certain types of parts.

3.7.4 Inclusion of Undercuts and Other Manufacturing Constraining Features

AM models can be used at various stages of the product development process. When evaluating initial designs, focus may be on the aesthetics or ultimate functionality of the part. Consideration of how to include manufacturing-related features would have lower priority at this stage. Conventional manufacturing would require considerable planning to ensure that a part is fabricated correctly. Undercuts, draft angles, holes, pockets, etc. must be created in a specific order when using multiple-stage conventional processes. While this can be ignored when designing the part for AM, it is important not to forget them if AM is being used just as a prototype process. AM can be used in the design process to help determine where and what type of rib, boss, and other strengthening approaches should be used on the final part. If the final part is to be injection molded, the AM part can be used to determine the best location for the parting lines in the mold.

3.7.5 Interlocking Features

AM machines have a finite build volume and large parts may not be capable of being built inside them. A solution may be to break the design up into segments that can fit into the machine and manually assemble them together later. The designer must therefore consider the best way to break up the parts. The regions where the breaks are made can be designed in such a way as to facilitate reassembly. Techniques can include incorporation of interlocking features and maximizing surface area so that adhesives can be most effective. Such regions should also be in easy-to-reach but difficult-to-observe locations.

This approach of breaking parts up may be helpful even when they can still fit inside the machine. Consider the design shown in Fig. 3.2. If it was built as a single part, it would take a long time and may require a significant amount of supports (as shown in the left-hand figure). If the part were built as two separate pieces the

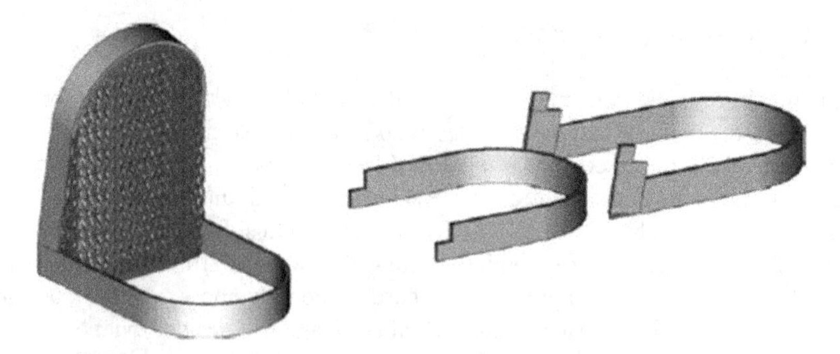

Fig. 3.2 The build on the *left* (shown with support materials within the arch) can be broken into the two parts on the *right*, which may be stronger and can be glued together later. Note the reduction in the amount of supports and the reduced build height

resulting height would be significantly reduced and there would be few supports. The part could be glued together later. This glued region may be slightly weakened, but the individual segments may be stronger. Since the example has a thin wall section, the top of the upright band shown in the left side of the figure will exhibit stair-stepping and may also be weaker than the rest of the part, whereas the part build lying down would typically be stronger. For the bonded region, it is possible to include large overlapping regions that will enable more effective bonding.

3.7.6 Reduction of Part Count in an Assembly

There are numerous sections in this book that discuss the use of AM for direct manufacture of parts for end-use applications. The AM process is therefore toward the end of the product development process and the design does not need to consider alternative manufacturing processes. This in turn means that if part assembly can be simplified using AM, then this should be done. For example, it is possible to build fully assembled hinge structures by providing clearance around the moving features. In addition, complex assemblies made up of multiple injection molded parts, for instance, could be built as a single component. Thus, when producing components with AM, designers should always look for ways to consolidate multiple parts into a single part and to include additional part complexity where it can improve system performance. Several of the parts in Fig. 1.4 provide good examples of these concepts.

3.7.7 Identification Markings/Numbers

Although AM parts are often unique, it may be difficult for a company to keep track of them when they are possibly building hundreds of parts per week. It is a straightforward process to include identifying features on the parts. This can be done when designing the CAD model but that may not be possible since the models may come from a third party. There are a number of software systems that provide tools for labeling parts by embossing alphanumeric characters onto them as 3D models. In addition, some service providers build all the parts ordered by a particular customer (or small parts which might otherwise get lost) within a mesh box so that they are easy to find and identify during part cleanup.

3.8 Application Areas That Don't Involve Conventional CAD Modeling

Additive manufacturing technology opens up opportunities for many applications that do not take the standard product development route. The capability of integrating AM with customizing data or data from unusual sources makes for rapid response and an economical solution. The following sections are examples where nonstandard approaches are applicable.

3.8.1 Medical Modeling

AM is increasingly used to make parts based on an individual person's medical data. Such data are based on 3D scanning obtained from systems like Computerized Tomography (CT), Magnetic Resonance Imaging (MRI), 3D ultrasound, etc. These datasets often need considerable processing to extract the relevant sections before it can be built as a model or further incorporated into a product design. There are a few software systems that can process medical data in a suitable way, and a range of applications have emerged. For example, Materialise [1] developed software used in the production of hearing aids. AM technology helps in customizing these hearing aids from data that are collected from the ear canals of individual patients.

3.8.2 Reverse Engineering Data

Medical data from patients is just one application that benefits from being able to collect and process complex surface information. For nonmedical data collection, the more common approach is to use laser scanning technology. Such technology has the ability to faithfully collect surface data from many types of surfaces that are difficult to model because they cannot be easily defined geometrically. Similar to medical data, although the models can just be reproduced within the AM machine (like a kind of 3D copy machine), the typical intent is to merge this data into product design. Interestingly, laser scanners for reverse engineering and inspection run the gamut from very expensive, very high-quality systems (e.g., from Leica and Steinbichler) to mid-range systems (from Faro and Creaform) to Microsoft Kinect™ controllers.

3.8.3 Architectural Modeling

Architectural models are usually created to emphasize certain features within a building design and so designs are modified to show textures, colors, and shapes that may not be exact reproductions of the final design. Therefore, architectural packages may require features that are tuned to the AM technology.

3.9 Further Discussion

AM technologies are beginning to move beyond a common set of basic process steps. In the future we will likely see more processes using variations of the conventional AM approach, and combinations of AM with conventional manufacturing operations. Some technologies are being developed to process regions rather than layers of a part. As a result, more intelligent and complex software systems will be required to effectively deal with segmentation.

We can expect processes to become more complex within a single machine. We already see numerous additive processes combined with subtractive elements. As technology develops further, we may see commercialization of hybrid technologies that include additive, subtractive, and even robotic handling phases in a complex coordinated and controlled fashion. This will require much more attention to software descriptions, but may also lead to highly optimized parts with multiple functionality and vastly improved quality with very little manual intervention during the actual build process.

Another trend we are likely to see is the development of customized AM systems. Presently, AM machines are designed to produce as wide a variety of possible part geometries with as wide a range of materials as possible. Reduction of these variables may result in machines that are designed only to build a subset of parts or materials very efficiently or inexpensively. This has already started with the proliferation of "personal" versus "industrial" material extrusion systems. In addition, many machines are being targeted for the dental or hearing aid markets, and system manufacturers have redesigned their basic machine architectures and/or software tools to enable rapid setup, building, and post-processing of patient-specific small parts.

Software is increasingly being optimized specifically for AM processing. Special software has been designed to increase the efficiency of hearing aid design and manufacture. There is also special software designed to convert the designs of World of Warcraft models into "FigurePrints" (see Fig. 3.3) as well as specially designed post-processing techniques [3]. As Direct Digital Manufacturing becomes more common, we will see the need to develop standardized software processes based around AM, so that we can better control, track, regulate, and predict the manufacturing process.

Fig. 3.3 FigurePrints model, post-processed for output to an AM machine

3.9.1 Exercises

1. Investigate some of the web sites associated with different AM technologies. Find out information on how to handle the processes and resulting parts according to the eight stages mentioned in this chapter. What are four different tasks that you would need to carry out using a vat photopolymerization process that you wouldn't have to do using a binder jetting technology and vice versa?
2. Explain why surface modeling software is not ideal for describing models that are to be made using AM, even though the STL file format is itself a surface approximation. What kind of problems may occur when using surface modeling only?
3. What is the VRML file format like? How is it more suitable for specifying color models to be built using Color ZCorp machines than the STL standard? How does it compare with the AMF format?
4. What extra considerations might you need to give when producing medical models using AM instead of conventionally engineered products?
5. Consider the FigurePrints part shown in Fig. 3.3, which is made using a color binder jetting process. What finishing methods would you use for this application?

References

1. Materialise, AM software systems and service provider. www.materialise.com
2. Choren J, Gervasi V, Herman T et al (2001) SLS powder life study. Solid Freeform Fabrication Proceedings, pp 39–45
3. Figureprints, 3DP models from World of Warcraft figures. www.figureprints.com

4 Vat Photopolymerization Processes

Abstract

Photopolymerization processes make use of liquid, radiation-curable resins, or photopolymers, as their primary materials. Most photopolymers react to radiation in the ultraviolet (UV) range of wavelengths, but some visible light systems are used as well. Upon irradiation, these materials undergo a chemical reaction to become solid. This reaction is called photopolymerization, and is typically complex, involving many chemical participants.

Photopolymers were developed in the late 1960s and soon became widely applied in several commercial areas, most notably the coating and printing industry. Many of the glossy coatings on paper and cardboard, for example, are photopolymers. Additionally, photo-curable resins are used in dentistry, such as for sealing the top surfaces of teeth to fill in deep grooves and prevent cavities. In these applications, coatings are cured by radiation that blankets the resin without the need for patterning either the material or the radiation. This changed with the introduction of stereolithography, the first vat photopolymerization process.

4.1 Introduction

Photopolymerization processes make use of liquid, radiation-curable resins, or photopolymers, as their primary materials. Most photopolymers react to radiation in the ultraviolet (UV) range of wavelengths, but some visible light systems are used as well. Upon irradiation, these materials undergo a chemical reaction to become solid. This reaction is called photopolymerization, and is typically complex, involving many chemical participants.

Photopolymers were developed in the late 1960s and soon became widely applied in several commercial areas, most notably the coating and printing industry. Many of the glossy coatings on paper and cardboard, for example, are photopolymers. Additionally, photo-curable resins are used in dentistry, such as

I. Gibson et al., *Additive Manufacturing Technologies*,
DOI 10.1007/978-1-4939-2113-3_4

for sealing the top surfaces of teeth to fill in deep grooves and prevent cavities. In these applications, coatings are cured by radiation that blankets the resin without the need for patterning either the material or the radiation. This changed with the introduction of stereolithography.

In the mid-1980s, Charles (Chuck) Hull was experimenting with UV-curable materials by exposing them to a scanning laser, similar to the system found in laser printers. He discovered that solid polymer patterns could be produced. By curing one layer over a previous layer, he could fabricate a solid 3D part. This was the beginning of stereolithography (SL) technology. The company 3D Systems was created shortly thereafter to market SL machines as "rapid prototyping" machines to the product development industry. Since then, a wide variety of SL-related processes and technologies has been developed. The term "vat photopolymerization" is a general term that encompasses SL and these related processes. SL will be used to refer specifically to macroscale, laser scan vat photopolymerization; otherwise, the term vat polymerization will be used and will be abbreviated as VP.

Various types of radiation may be used to cure commercial photopolymers, including gamma rays, X-rays, electron beams, UV, and in some cases visible light. In VP systems, UV and visible light radiation are used most commonly. In the microelectronics industry, photomask materials are often photopolymers and are typically irradiated using far UV and electron beams. In contrast, the field of dentistry uses visible light predominantly.

Two primary configurations were developed for photopolymerization processes in a vat, plus one additional configuration that has seen some research interest. Although photopolymers are also used in some ink-jet printing processes, this method of line-wise processing is not covered in this chapter, as the basic processing steps are more similar to the printing processes covered in Chap. 7. The configurations discussed in this chapter include:

- *Vector scan*, or point-wise, approaches typical of commercial SL machines
- *Mask projection*, or layer-wise, approaches, that irradiate entire layers at one time, and
- *Two-photon* approaches that are essentially high-resolution point-by-point approaches

These three configurations are shown schematically in Fig. 4.1. Note that in the vector scan and two-photon approaches, scanning laser beams are needed, while the mask projection approach utilizes a large radiation beam that is patterned by another device, in this case a Digital Micromirror Device™ (DMD). In the two-photon case, photopolymerization occurs at the intersection of two scanning laser beams, although other configurations use a single laser and different photoinitiator chemistries. Another distinction is the need to recoat, or apply a new layer of resin, in the vector scan and mask projection approaches, while in the two-photon approach, the part is fabricated below the resin surface, making

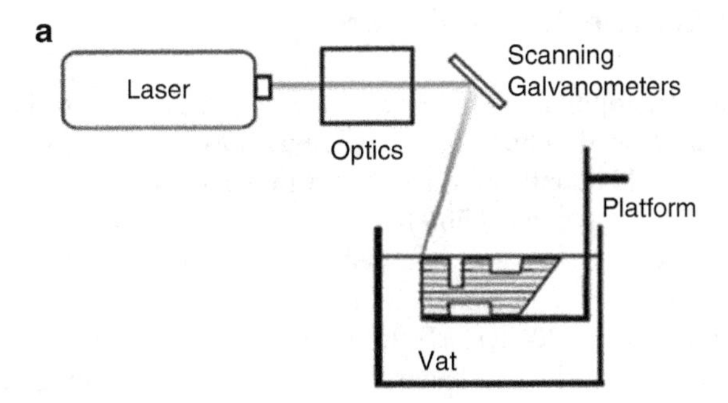

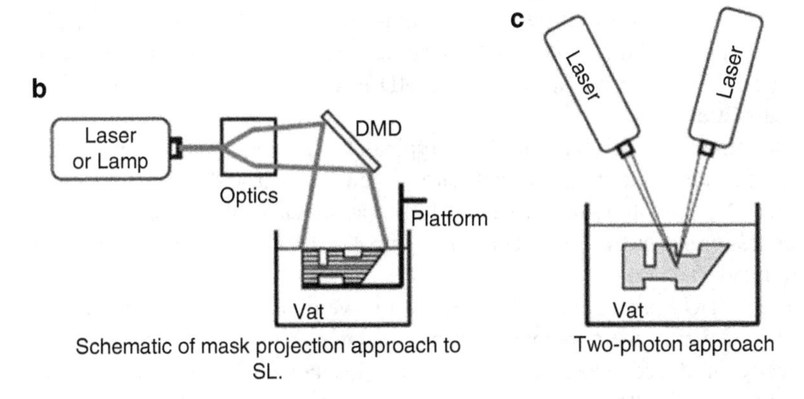

Fig. 4.1 Schematic diagrams of three approaches to photopolymerization processes

recoating unnecessary. Approaches that avoid recoating are faster and less complicated.

In this chapter, we first introduce photopolymer materials, then present the vector scan SL machines, technologies, and processes. Mask projection approaches are presented and contrasted with the vector scan approach. Additional configurations, along with their applications, are presented at the end of the chapter. Advantages, disadvantages, and uniquenesses of each approach and technology are highlighted.

4.2 Vat Photopolymerization Materials

Some background of UV photopolymers will be presented in this section that is common to all configurations of photopolymerization processes. Two subsections on reaction rates and characterization methods conclude this section. Much of this material is from the Jacobs book [1] and from a Master's thesis from the early 2000s [2].

4.2.1 UV-Curable Photopolymers

As mentioned, photopolymers were developed in the late 1960s. In addition to the applications mentioned in Sect. 4.1, they are used as photoresists in the microelectronics industry. This application has had a major impact on the development of epoxy-based photopolymers. Photoresists are essentially one-layer SL, but with critical requirements on accuracy and feature resolution.

Various types of radiation may be used to cure commercial photopolymers, including gamma rays, X-rays, electron beams, UV, and in some cases visible light, although UV and electron beam are the most prevalent. In AM, many of these radiation sources have been utilized in research, however only UV and visible light systems are utilized in commercial systems. In SL systems, for example, UV radiation is used exclusively although, in principle, other types could be used. In the SLA-250 from 3D Systems, a helium–cadmium (HeCd) laser is used with a wavelength of 325 nm. In contrast, the solid-state lasers used in the other SL models are Nd-YVO_4. In mask projection DMD-based systems, UV and visible light radiation are used.

Thermoplastic polymers that are typically injection molded have a linear or branched molecular structure that allows them to melt and solidify repeatedly. In contrast, VP photopolymers are cross-linked and, as a result, do not melt and exhibit much less creep and stress relaxation. Figure 4.2 shows the three polymer structures mentioned [3].

The first US patents describing SL resins were published in 1989 and 1990 [4, 5]. These resins were prepared from acrylates, which had high reactivity but typically produced weak parts due to the inaccuracy caused by shrinkage and curling. The acrylate-based resins typically could only be cured to 46 % completion when the image was transferred through the laser [6]. When a fresh coating was put on the exposed layer, some radiation went through the new coating and initiated new photochemical reactions in the layer that was already partially cured. This layer was less susceptible to oxygen inhibition after it had been coated. The additional cross-linking on this layer caused extra shrinkage, which increased stresses in the layer, and caused curling that was observed either during or after the part fabrication process [7].

The first patents that prepared an epoxide composition for SL resins appeared in 1988 [8, 9] (Japanese). The epoxy resins produced more accurate, harder, and stronger parts than the acrylate resins. While the polymerization of acrylate compositions leads to 5–20 % shrinkage, the ring-opening polymerization of epoxy compositions only leads to a shrinkage of 1–2 % [10]. This low level of shrinkage associated with epoxy chemistry contributes to excellent adhesion and reduced tendency for flexible substrates to curl during cure. Furthermore, the polymerization of the epoxy-based resins is not inhibited by atmospheric oxygen. This enables low-photoinitiator concentrations, giving lower residual odor than acrylic formulations [11].

However, the epoxy resins have disadvantages of slow photospeed and brittleness of the cured parts. The addition of some acrylate to epoxy resins is required to

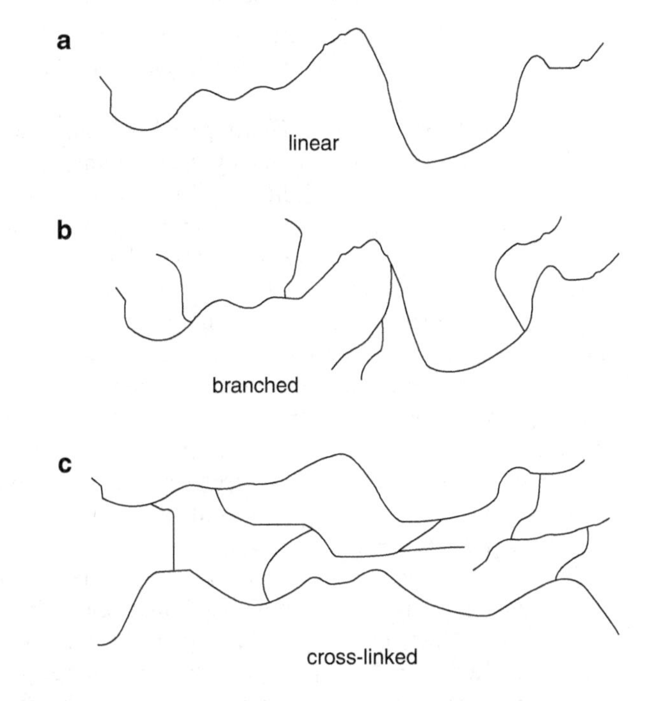

Fig. 4.2 Schematics of polymer types

rapidly build part strength so that they will have enough integrity to be handled without distortion during fabrication. The acrylates are also useful to reduce the brittleness of the epoxy parts [7]. Another disadvantage of epoxy resins is their sensitivity to humidity, which can inhibit polymerization [11].

As a result, most SL resins commercially available today are epoxides with some acrylate content. It is necessary to have both materials present in the same formulation to combine the advantages of both curing types. The improvement in accuracy resulting from the use of hybrid resins has given SL a tremendous boost.

4.2.2 Overview of Photopolymer Chemistry

VP photopolymers are composed of several types of ingredients: photoinitiators, reactive diluents, flexibilizers, stabilizers, and liquid monomers. Broadly speaking, when UV radiation impinges on VP resin, the photoinitiators undergo a chemical transformation and become "reactive" with the liquid monomers. A "reactive" photoinitiator reacts with a monomer molecule to start a polymer chain. Subsequent reactions occur to build polymer chains and then to cross-link—creation of strong covalent bonds between polymer chains. Polymerization is the term used to describe the process of linking small molecules (monomers) into larger molecules (polymers) composed of many monomer units [1]. Two main types of photopolymer chemistry are commercially evident:

- Free-radical photopolymerization—acrylate
- Cationic photopolymerization—epoxy and vinylether

The molecular structures of these types of photopolymers are shown in Fig. 4.5. Symbols C and H denote carbon and hydrogen atoms, respectively, while R denotes a molecular group which typically consists of one or more vinyl groups. A vinyl group is a molecular structure with a carbon–carbon double bond. It is these vinyl groups in the R structures that enable photopolymers to become cross-linked.

Free-radical photopolymerization was the first type that was commercially developed. Such SL resins were acrylates. Acrylates form long polymer chains once the photoinitiator becomes "reactive," building the molecule linearly by adding monomer segments. Cross-linking typically happens after the polymer chains grow enough so that they become close to one another. Acrylate photopolymers exhibit high photospeed (react quickly when exposed to UV radiation), but have a number of disadvantages including significant shrinkage and a tendency to warp and curl. As a result, they are rarely used now without epoxy or other photopolymer elements.

The most common cationic photopolymers are epoxies, although vinylethers are also commercially available. Epoxy monomers have rings, as shown in Fig. 4.3. When reacted, these rings open, resulting in sites for other chemical bonds. Ring-opening is known to impart minimal volume change on reaction, because the number and types of chemical bonds are essentially identical before and after reaction [12]. As a result, epoxy SL resins typically have much smaller shrinkages and much less tendency to warp and curl. Almost all commercially available SL resins have significant amounts of epoxies.

Polymerization of VP monomers is an exothermic reaction, with heats of reaction around 85 kJ/mol for an example acrylate monomer. Despite high heats of reaction, a catalyst is necessary to initiate the reaction. As described earlier, a photoinitiator acts as the catalyst.

Schematically, the free radical-initiated polymerization process can be illustrated as shown in Fig. 4.4 [1]. On average, for every two photons (from the laser), one radical will be produced. That radical can easily lead to the polymerization of over 1,000 monomers, as shown in the intermediate steps of the process, called propagation. In general, longer polymer molecules are preferred, yielding higher molecular weights. This indicates a more complete reaction. In Fig. 4.4, the *P–I* term indicates a photoinitiator, the $-I^{\bullet}$ symbol is a free radical, and *M* in a monomer.

Polymerization terminates from one of three causes, recombination, disproportionation, or occlusion. Recombination occurs when two polymer chains merge by joining two radicals. Disproportionation involves essentially the cancelation of one radical by another, without joining. Occlusion occurs when free radicals become "trapped" within a solidified polymer, meaning that reaction sites remain available, but are prevented from reacting with other monomers or polymers by the limited mobility within the polymer network. These occluded sites will most certainly react eventually, but not with another polymer chain or monomer. Instead, they will react

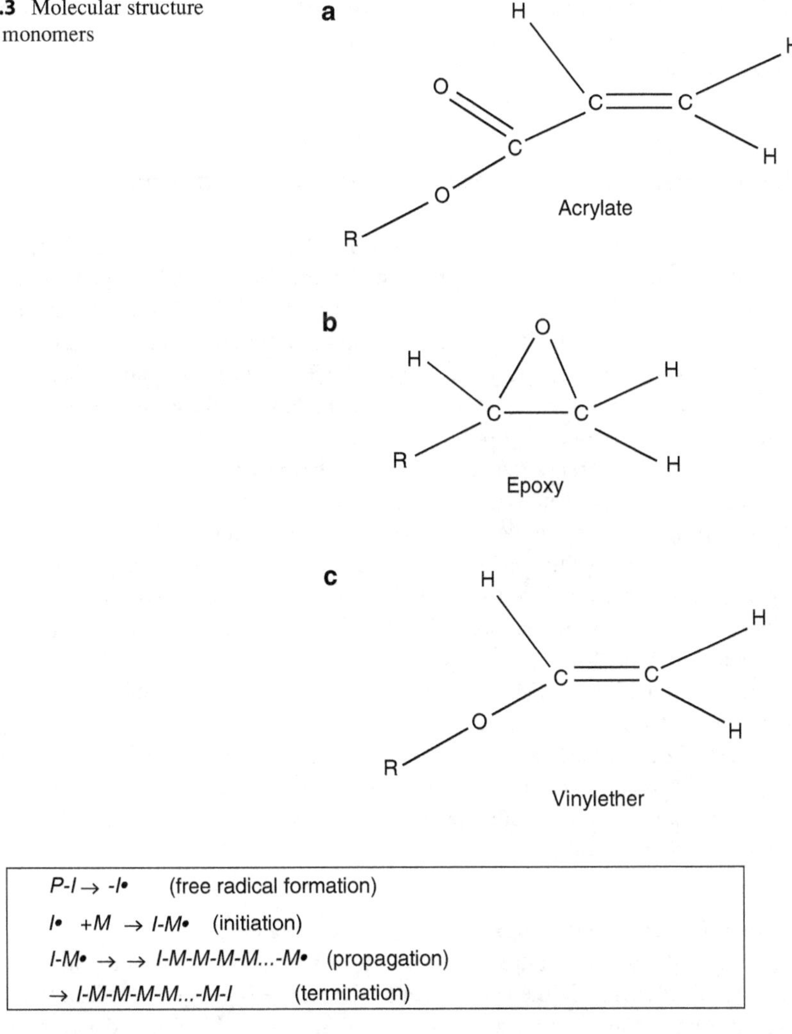

Fig. 4.3 Molecular structure of VP monomers

Fig. 4.4 Free-radical polymerization process

with oxygen or another reactive species that diffuses into the occluded region. This may be a cause of aging or other changes in mechanical properties of cured parts, which should be a topic of future research.

Cationic photopolymerization shares the same broad structure as free-radical polymerization, where a photoinitiator generates a cation as a result of laser energy, the cation reacts with a monomer, propagation occurs to generate a polymer, and a termination process completes the reaction. A typical catalyst for a cationic polymerization is a Lewis Acid, such as BF_3 [13]. Initially, cationic photopolymerization received little attention, but that has changed during the 1990s due to advances in the microelectronics industry, as well as interest in SL technology. We will not investigate the specifics of cationic reactions here, but will note that the ring-opening reaction mechanism of epoxy monomers is similar to radical propagation in acrylates.

4.2.3 Resin Formulations and Reaction Mechanisms

Basic raw materials such as polyols, epoxides, (meth) acrylic acids and their esters, and diisocyanates are used to produce the monomers and oligomers used for radiation curing. Most of the monomers are multifunctional monomers (MFM) or polyol polyacrylates which give a cross-linking polymerization. The main chemical families of oligomers are polyester acrylate (PEA), epoxy acrylates (EA), urethane acrylates (UA), amino acrylates (used as photoaccelerator in the photoinitiator system), and cycloaliphatic epoxies [11].

Resin suppliers create ready-to-use formulations by mixing the oligomers and monomers with a photoinitiator, as well as other materials to affect reaction rates and part properties. In practice, photosensitizers are often used in combination with the photoinitiator to shift the absorption towards longer wavelengths. In addition, supporting materials may be mixed with the initiator to achieve improved solubility in the formulation. Furthermore, mixtures of different types of photoinitiators may also be employed for a given application. Thus, photoinitiating systems are, in practice, often highly elaborate mixtures of various compounds which provide optimum performance for specific applications [10].

Other additives facilitate the application process and achieve products of good properties. A reactive diluent, for example, is usually added to adjust the viscosity of the mixtures to an acceptable level for application [14]; it also participates in the polymerization reaction.

4.2.3.1 Photoinitiator System

The role of the photoinitiator is to convert the physical energy of the incident light into chemical energy in the form of reactive intermediates. The photoinitiator must exhibit a strong absorption at the laser emission wavelength, and undergo a fast photolysis to generate the initiating species with a great quantum yield [15]. The reactive intermediates are either radicals capable of adding to vinylic or acrylic double bonds, thereby initiating radical polymerization, or reactive cationic species which can initiate polymerization reactions among epoxy molecules [10].

The free-radical polymerization process was outlined in Fig. 4.4, with the formation of free radicals as the first step. In the typical case in VP, radical photoinitiator systems include compounds that undergo unimolecular bond cleavage upon irradiation. This class includes aromatic carbonyl compounds that are known to undergo a homolytic C–C bond scission upon UV exposure [16]. The benzoyl radical is the major initiating species, while the other fragment may, in some cases, also contribute to the initiation. The most efficient photoinitiators include benzoin ether derivatives, benzyl ketals, hydroxyalkylphenones, α-amino ketones, and acylphosphine oxides [16, 17]. The Irgacure family of radical photoinitiators from Ciba Specialty Chemicals is commonly used in VP.

While photoinitiated free-radical polymerizations have been investigated for more than 60 years, the corresponding photoinduced cationic polymerizations have received much less attention. The main reason for the slow development in this area was the lack of suitable photoinitiators capable of efficiently inducing cationic polymerization [18]. Beginning in 1965, with the earliest work on diazonium salt initiators, this situation has markedly changed. The discovery in the 1970s of onium salts or organometallic compounds with excellent photoresponse and high efficiency has initiated the very rapid and promising development of cationic photopolymerization, and made possible the concurrent radical and cationic reaction in hybrid systems [19]. Excellent reviews have been published in this field [10, 18, 20–23]. The most important cationic photoinitiators are the onium salts, particularly the triarylsulfonium and diaryliodonium salts. Examples of the cationic photoinitiator are triaryl sulfonium hexafluorophosphate solutions in propylene carbonate such as Degacure KI 85 (Degussa), SP-55 (Asahi Denka), Sarcat KI-85 (Sartomer), and 53,113-8 (Aldrich), or mixtures of sulfonium salts such as SR-1010 (Sartomer, currently unavailable), UVI 6976 (B-V), and UVI 6992 (B-VI) (Dow).

Initiation of cationic polymerization takes place from not only the primary products of the photolysis of triarylsulfonium salts but also from secondary products of the reaction of those reactive species with solvents, monomers, or even other photolysis species. Probably the most ubiquitous species present is the protonic acid derived from the anion of the original salt. Undoubtedly, the largest portion of the initiating activity in cationic polymerization by photolysis of triarylsulfonium salts is due to protonic acids [18].

4.2.3.2 Monomer Formulations

The monomer formulations presented here are from a set of patents from the mid- to late-1990s. Both di-functional and higher functionality monomers are used typically in VP resins. Poly(meth)acrylates may be tri-, pentafunctional monomeric or oligomeric aliphatic, cycloaliphatic or aromatic (meth)acrylates, or polyfunctional urethane (meth)acrylates [24–27]. One specific compound in the Huntsman SL-7510 resin includes the dipentaerythritol monohydroxy penta(meth)acrylates [26], such as Dipentaerythritol Pentaacrylate (SR-399, Sartomer).

The cationically curable epoxy resins may have an aliphatic, aromatic, cycloaliphatic, araliphatic, or heterocyclic structure; they on average possess more than one epoxide group (oxirane ring) in the molecule and comprise epoxide groups as side groups, or those groups form part of an alicyclic or heterocyclic ring system. Examples of epoxy resins of this type are also given by these patents such as polyglycidyl esters or ethers, poly(N or S-glycidyl) compounds, and epoxide compounds in which the epoxide groups form part of an alicyclic or heterocyclic ring system. One specific composition includes at least 50 % by weight of a cycloaliphatic diepoxide [26] such as bis(2,3-epoxycyclopentyl) ether (formula A-I), 3,4-epoxycyclohexyl-methyl 3,4-epoxycyclohexanecarboxylate (A-II), dicyclopentadiene diepoxide (A-III), and bis-(3,4-epoxycyclohexylmethyl) adipate (A-IV).

Additional insight into compositions can be gained by investigating the patent literature further.

4.2.3.3 Interpenetrating Polymer Network Formation

As described earlier, acrylates polymerize radically, while epoxides cationically polymerize to form their respective polymer networks. In the presence of each other during the curing process, an interpenetrating polymer network (IPN) is finally obtained [28, 29]. An IPN can be defined as a combination of two polymers in network form, at least one of which is synthesized and/or cross-linked in the immediate presence of the other [30]. It is therefore a special class of polymer blends in which both polymers generally are in network form [30–32], and which is originally generated by the concurrent reactions instead of by a simple mechanical mixing process. In addition, it is a polymer blend rather than a copolymer that is generated from the hybrid curing [33], which indicates that acrylate and epoxy monomers undergo independent polymerization instead of copolymerization. However, in special cases, copolymerization can occur, thus leading to a chemical bonding of the two networks [34].

It is likely that in typical SL resins, the acrylate and epoxide react independently. Interestingly, however, these two monomers definitely affect each other physically during the curing process. The reaction of acrylate will enhance the photospeed and reduce the energy requirement of the epoxy reaction. Also, the presence of acrylate monomer may decrease the inhibitory effect of humidity on the epoxy polymerization. On the other hand, the epoxy monomer acts as a plasticizer during the early polymerization of the acrylate monomer where the acrylate forms a network while the epoxy is still at liquid stage [31]. This plasticizing effect, by increasing molecular mobility, favors the chain propagation reaction [35]. As a result, the acrylate polymerizes more extensively in the presence of epoxy than in the neat acrylate monomer. Furthermore, the reduced sensitivity of acrylate to oxygen in the hybrid system than in the neat composition may be due to the simultaneous polymerization of the epoxide which makes the viscosity rise, thus slowing down the diffusion of atmospheric oxygen into the coating [31].

In addition, it has been shown [31] that the acrylate/epoxide hybrid system requires a shorter exposure to be cured than either of the two monomers taken separately. It might be due to the plasticizing effect of epoxy monomer and the contribution of acrylate monomer to the photospeed of the epoxy polymerization. The two monomers benefit from each other by a synergistic effect.

It should be noted that if the concentration of the radical photoinitiator was decreased so that the two polymer networks were generated simultaneously, the plasticizing effect of the epoxy monomer would become less pronounced. As a result, it would be more difficult to achieve complete polymerization of the acrylate monomer and thus require longer exposure time.

Although the acrylate/epoxy hybrid system proceeds via a heterogeneous mechanism, the resultant product (IPN) seems to be a uniphase component [36]. The properties appear to be extended rather than compromised [31, 34]. The optimal properties of IPNs for specific applications can be obtained by selecting two appropriate components and adjusting their proportions [34]. For example, increasing the acrylate content increases the cure speed but decreases the adhesion characteristics, while increasing the epoxy content reduces the shrinkage of curing and improves the adhesion, but decreases the cure speed [36].

4.3 Reaction Rates

As is evident, the photopolymerization reaction in VP resins is very complex. To date, no one has published an analytical photopolymerization model that describes reaction results and reaction rates. However, qualitative understanding of reaction rates is straightforward for simple formulations. Broadly speaking, reaction rates for photopolymers are controlled by concentrations of photoinitiators [I] and monomers [M]. The rate of polymerization is the rate of monomer consumption, which can be shown as [3]:

$$R_p = -d[\mathrm{M}]/dt \,\alpha\, [\mathrm{M}] \;\; (k[\mathrm{I}])^{1/2} \tag{4.1}$$

where k = constant that is a function of radical generation efficiency, rate of radical initiation, and rate of radical termination. Hence, the polymerization rate is proportional to the concentration of monomer, but is only proportional to the square root of initiator concentration.

Using similar reasoning, it can be shown that the average molecular weight of polymers is the ratio of the rate of propagation and the rate of initiation. This average weight is called the kinetic average chain length, v_o, and is given in (4.2):

$$v_o = R_p/R_i \;\; \alpha\, [\mathrm{M}]/[\mathrm{I}]^{1/2} \tag{4.2}$$

where R_i is the rate of initiation of macromonomers.

Equations (4.1) and (4.2) have important consequences for the VP process. The higher the rate of polymerization, the faster parts can be built. Since VP resins are predominantly composed of monomers, the monomer concentration cannot be changed much. Hence, the only other direct method for controlling the polymerization rate and the kinetic average chain length is through the concentration of initiator. However, (4.1) and (4.2) indicate a trade-off between these characteristics. Doubling the initiator concentration only increases the polymerization rate by a factor of 1.4, but reduces the molecular weight of resulting polymers by the same amount. Strictly speaking, this analysis is more appropriate for acrylate resins, since epoxies continue to react after laser exposure, so (4.2) does not apply well for epoxies. However, reaction of epoxies is still limited, so it can be concluded that a trade-off does exist between polymerization rate and molecular weight for epoxy resins.

4.4 Laser Scan Vat Photopolymerization

Laser scan VP creates solid parts by selectively solidifying a liquid photopolymer resin using an UV laser. As with many other AM processes, the physical parts are manufactured by fabricating cross-sectional contours, or slices, one on top of another. These slices are created by tracing 2D contours of a CAD model in a vat of photopolymer resin with a laser. The part being built rests on a platform that is dipped into the vat of resin, as shown schematically in Fig. 4.1a. After each slice is created, the platform is lowered, the surface of the vat is recoated, then the laser starts to trace the next slice of the CAD model, building the prototype from the bottom up. A more complete description of the SL process may be found in [12]. The creation of the part requires a number of key steps: input data, part preparation, layer preparation, and finally laser scanning of the two-dimensional cross-sectional slices. The input data consist of an STL file created from a CAD file or reverse engineering data. Part preparation is the phase at which the operator specifies support structures, to hold each cross section in place while the part builds, and provides values for machine parameters. These parameters control how the prototype is fabricated in the VP machine. Layer preparation is the phase in which the STL model is divided into a series of slices, as defined by the part preparation phase, and translated by software algorithms into a machine language. This information is then used to drive the SL machine and fabricate the prototype. The laser scanning of the part is the phase that actually solidifies each slice in the VP machine.

After building the part, the part must be cleaned, post-cured, and finished. During the cleaning and finishing phase, the VP machine operator may remove support structures. During finishing, the operator may spend considerable time sanding and filing the part to provide the desired surface finishes.

4.5 Photopolymerization Process Modeling

Background on SL materials and energy sources enables us to investigate the curing process of photopolymers in SL machines. We will begin with an investigation into the fundamental interactions of laser energy with photopolymer resins. Through the application of the Beer–Lambert law, the theoretical relationship between resin characteristics and exposure can be developed, which can be used to specify laser scan speeds. This understanding can then be applied to investigate mechanical properties of cured resins. From here, we will briefly investigate the ranges of size scales and time scales of relevance to the SL process. Much of this section is adapted from [1].

Nomenclature

C_d = cure depth = depth of resin cure as a result of laser irradiation [mm]

D_p = depth of penetration of laser into a resin until a reduction in irradiance of $1/e$ is reached = key resin characteristic [mm]

E = exposure, possibly as a function of spatial coordinates [energy/unit area] [mJ/mm^2]
E_c = critical exposure = exposure at which resin solidification starts to occur [mJ/mm^2]
E_{max} = peak exposure of laser shining on the resin surface (center of laser spot) [mJ/mm^2]
$H(x, y, z)$ = irradiance (radiant power per unit area) at an arbitrary point in the resin = time derivative of $E(x, y, z)$.[W/mm^2]
P_L = output power of laser [W]
V_s = scan speed of laser [mm/s]
W_0 = radius of laser beam focused on the resin surface [mm]

4.5.1 Irradiance and Exposure

As a laser beam is scanned across the resin surface, it cures a line of resin to a depth that depends on many factors. However, it is also important to consider the width of the cured line as well as its profile. The shape of the cured line depends on resin characteristics, laser energy characteristics, and the scan speed. We will investigate the relationships among all of these factors in this subsection.

The first concept of interest here is *irradiance*, the radiant power of the laser per unit area, $H(x, y, z)$. As the laser scans a line, the radiant power is distributed over a finite area (beam spots are not infinitesimal). Figure 4.5 shows a laser scanning a line along the x-axis at a speed V_s [1]. Consider the z-axis oriented perpendicular to the resin surface and into the resin, and consider the origin such that the point of interest, p', has an x coordinate of 0. The irradiance at any point x, y, z in the resin is related to the irradiance at the surface, assuming that the resin absorbs radiation according to the Beer–Lambert Law. The general form of the irradiance equation for a Gaussian laser beam is given here as (4.3).

$$H(x,y,z) = H(x,y,0)e^{-z/D_p} \tag{4.3}$$

From this relationship, we can understand the meaning of the penetration depth, D_p. Setting $z = D_p$, we get that the irradiance at a depth D_p is about 37 % ($e^{-1} = 0.36788$) of the irradiance at the resin surface. Thus, D_p is the depth into the resin at which the irradiance is 37 % of the irradiance at the surface. Furthermore, since we are assuming the Beer–Lambert Law holds, D_p is only a function of the resin.

Without loss of generality, we will assume that the laser scans along the x-axis from the origin to point b. Then, the irradiance at coordinate x along the scan line is given by

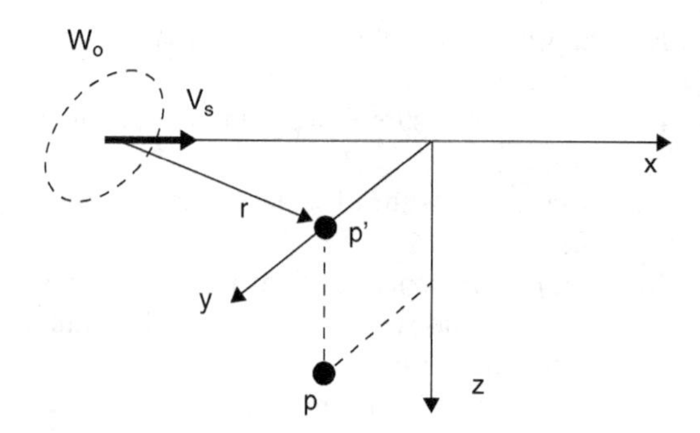

Fig. 4.5 Scan line of Gaussian laser

$$H(x, y, 0) = H(x, y) = H_0 e^{-2x^2/W_0^2} e^{-2y^2/W_0^2} \tag{4.4}$$

where $H_0 = H(0,0)$ when $x = 0$, and W_0 is the $1/e^2$ Gaussian half-width of the beam spot. Note that when $x = W_0$, $H(x,0) = H_0 e^{-2} = 0.13534 H_0$.

The maximum irradiance, H_0, occurs at the center of the beam spot ($x = 0$). H_0 can be determined by integrating the irradiance function over the area covered by the beam at any particular point in time. Changing from Cartesian to polar coordinates, the integral can be set equal to the laser power, P_L, as shown in (4.5).

$$P_L = \int_{r=0}^{r=\infty} H(r, 0)\, dA \tag{4.5}$$

When solved, H_o turns out to be a simple function of laser power and beam half-width, as in (4.6).

$$H_0 = \frac{2P_L}{\pi W_0^2} \tag{4.6}$$

As a result, the irradiance at any point x, y between $x = 0$ and $x = b$ is given by:

$$H(x, y) = \frac{2p_L}{\pi W_0^2} e^{-2x^2/W_0^2} e^{-2y^2/W_0^2} \tag{4.7}$$

However, we are interested in *exposure* at an arbitrary point, p, not irradiance, since exposure controls the extent of resin cure. Exposure is the energy per unit area; when exposure at a point in the resin vat exceeds a critical value, called E_c, we assume that resin cures. Exposure can be determined at point p by appropriately integrating (4.7) along the scan line, from time 0 to time t_b, when the laser reaches point b.

$$E(y, 0) = \int_{t=0}^{t=t_b} H[x(t), 0]dt \tag{4.8}$$

It is far more convenient to integrate over distance than over time. If we assume a constant laser scan velocity, then it is easy to substitute t for x, as in (4.9).

$$E(y, 0) = \frac{2P_{\mathrm{L}}}{\pi V_{\mathrm{s}} W_0^2} \mathrm{e}^{-2y^2/W_0^2} \int_{x=0}^{x=b} \mathrm{e}^{-2x^2/W_0^2} dx \tag{4.9}$$

The exponential term is difficult to integrate directly, so we will change the variable of integration. Define a variable of integration, v, as

$$v^2 \equiv \frac{2x^2}{W_0^2}$$

Then, take the square root of both sides, take the derivative, and rearrange to give

$$dx = \frac{W_0}{\sqrt{2}} dv$$

Due to the change of variables, it is also necessary to convert the integration limit to $b = \sqrt{2}/W_0 x_e$.

Several steps in the derivation will be skipped. After integration, the exposure received at a point x, y between $x = (0, b)$ can be computed as:

$$E(y, 0) = \frac{P_{\mathrm{L}}}{\sqrt{2\pi} W_0 V_{\mathrm{s}}} \mathrm{e}^{-\frac{2y^2}{W_0^2}} [\mathrm{erf}(b)] \tag{4.10}$$

where $erf(x)$ is the error function evaluated at x. $erf(x)$ is close to -1 for negative values of x, is close to 1 for positive values of x, and rapidly transitions from -1 to 1 for values of x close to 0. This behavior localizes the exposure within a narrow range around the scan vector. This makes sense since the laser beam is small and we expect that the energy received from the laser drops off quickly outside of its radius.

Equation (4.10) is not quite as easy to apply as a form of the exposure equation that results from assuming an infinitely long scan vector. If we make this assumption, then (4.10) becomes

$$E(y, 0) = \frac{2P_{\mathrm{L}}}{\pi V_{\mathrm{s}} W_0^2} \mathrm{e}^{-2y^2/W_0^2} \int_{x=\infty}^{x=\infty} \mathrm{e}^{-2x^2/W_0^2} \mathrm{d}x$$

and after integration, exposure is given by

$$E(y,0) = \sqrt{\frac{2}{\pi}} \frac{P_L}{W_0 V_s} e^{-2y^2/W_0^2} \tag{4.11}$$

Combining this with (4.3) yields the fundamental general exposure equation:

$$E(x,y,z) = \sqrt{\frac{2}{\pi}} \frac{P_L}{W_0 V_s} e^{-2y^2/W_0^2} e^{-z/D_p} \tag{4.12}$$

4.5.2 Laser–Resin Interaction

In this subsection, we will utilize the irradiance and exposure relationships to determine the shape of a scanned vector line and its width. As we will see, the cross-sectional shape of a cured line becomes a parabola.

Starting with (4.12), the locus of points in the resin that is just at its gel point, where $E = E_c$, is denoted by y^* and z^*. Equation (4.12) can be rearranged, with y^*, z^*, and E_c substituted to give (4.13).

$$e^{2y^{*2}/W_0^2 + z^*/D_P} = \sqrt{\frac{2}{\pi}} \frac{P_L}{W_0 V_s E_c} \tag{4.13}$$

Taking natural logarithms of both sides yields

$$2\frac{y^{*2}}{W_0^2} + \frac{z^*}{D_p} = \ln\left[\sqrt{\frac{2}{\pi}} \frac{P_L}{W_0 V_s E_c}\right] \tag{4.14}$$

This is the equation of a parabolic cylinder in y^* and z^*, which can be seen more clearly in the following form,

$$ay^{*2} + bz^* = c \tag{4.15}$$

where a, b, and c are constants, immediately derivable from (4.14). Figure 4.6 illustrates the parabolic shape of a cured scan line.

To determine the maximum depth of cure, we can solve (4.14) for z^* and set $y^* = 0$, since the maximum cure depth will occur along the center of the scan vector. Cure depth, C_d, is given by

$$C_d = D_P \ln\left[\sqrt{\frac{2}{\pi}} \frac{P_L}{W_0 V_s E_c}\right] \tag{4.16}$$

As is probably intuitive, the width of a cured line of resin is the maximum at the resin surface; i.e., y_{max} occurs at $z = 0$. To determine line width, we start with the line shape function (4.14). Setting $z = 0$ and letting line width, L_w, equal $2y_{max}$, the line width can be found:

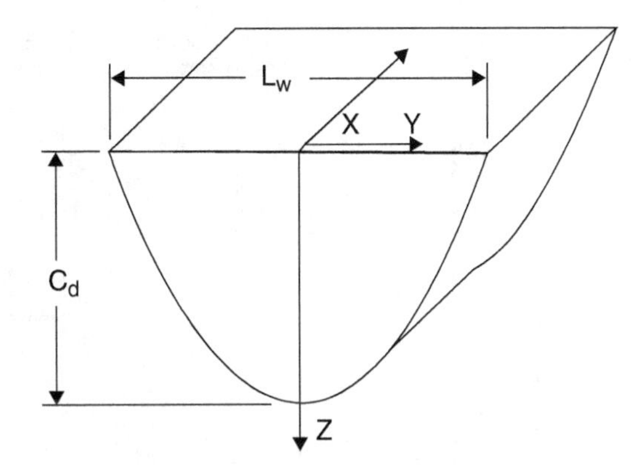

Fig. 4.6 Cured line showing parabolic shape, cure depth, and line width

$$L_W = W_0 \sqrt{2C_d / D_p} \tag{4.17}$$

As a result, two important aspects become clear. First, line width is proportional to the beam spot size. Second, if a greater cure depth is desired, line width must increase, all else remaining the same. This becomes very important when performing line width compensation during process planning.

The final concept to be presented in this subsection is fundamental to commercial SL. It is the *working curve*, which relates exposure to cure depth, and includes the two key resin constants, D_p and E_c. At the resin surface and in the center of the scan line:

$$E(0,0) \equiv E_{max} = \sqrt{\frac{2}{\pi}} \frac{P_L}{W_0 V_s} \tag{4.18}$$

which is most of the expression within the logarithm term in (4.16). Substituting (4.18) into (4.16) yields the working curve equation:

$$C_d = D_P \ln\left(\frac{E_{max}}{E_c}\right) \tag{4.19}$$

In summary, a laser of power P_L scans across the resin surface at some speed V_s solidifying resin to a depth C_d, the cure depth, assuming that the total energy incident along the scan vector exceeds a critical value called the critical exposure, E_c. If the laser scans too quickly, no polymerization reaction takes place; i.e., exposure E is less than E_c. E_c is assumed to be a characteristic quantity of a particular resin.

An example working curve is shown in Fig. 4.7, where measured cure depths at a given exposure are indicated by "*." The working curve equation, (4.19), has several major properties [1]:

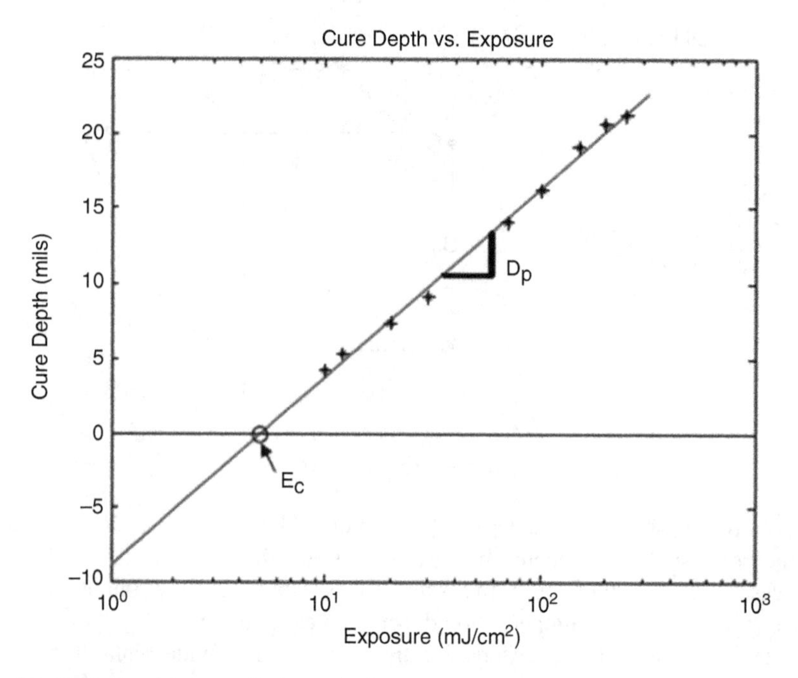

Fig. 4.7 Resin "working curve" of cure depth vs. exposure

1. The cure depth is proportional to the natural logarithm of the maximum exposure on the centerline of a scanned laser beam.
2. A semilog plot of C_d vs. E_{max} should be a straight line. This plot is known as the *working curve* for a given resin.
3. The slope of the working curve is precisely D_p at the laser wavelength being used to generate the working curve.
4. The x-axis intercept of the working curve is E_c, the critical exposure of the resin at that wavelength. Theoretically, the cure depth is 0 at E_c, but this does indicate the gel point of the resin.
5. Since D_p and E_c are purely resin parameters, the slope and intercept of the working curve are independent of laser power.

In practice, various E_{max} values can be generated easily by varying the laser scan speed, as indicated by (4.19).

4.5.3 Photospeed

Photospeed is typically used as an intuitive approximation of SL photosensitivity. But it is useful in that it relates to the speed at which the laser can be scanned across the polymer surface to give a specified cure depth. The faster the laser can be scanned to give a desired cure depth, the higher the photospeed. Photospeed is a characteristic of the resin and does not depend upon the specifics of the laser or optics subsystems. In particular, photospeed is indicated by the resin constants E_c and D_p, where higher levels of D_p and lower values of E_c indicate higher photospeed.

To determine scan velocity for a desired cure depth, it is straightforward to solve (4.16) for V_s. Recall that at the maximum cure depth, the exposure received equals the cure threshold, E_c. Scan velocity is given by (4.20).

$$V_s = \sqrt{\frac{2}{\pi}} \frac{P_L}{W_0 E_c} e^{-C_d/D_p} \tag{4.20}$$

This discussion can be related back to the working curve. Both E_c and D_p must be determined experimentally. 3D Systems has developed a procedure called the WINDOWPANE procedure for finding E_c and D_p values [41]. The cure depth, C_d, can be measured directly from specimens built on an SL machine that are one layer thickness in depth. The WINDOWPANE procedure uses a specific part shape, but the principle is simply to build a part with different amounts of laser exposure in different places in the part. By measuring the part thickness, C_d, and correlating that with the exposure values, a "working curve" can easily be plotted. Note that (4.19) is log-linear. Hence, C_d is plotted linearly vs. the logarithm of exposure to generate a working curve.

So how is exposure varied? Exposure is varied by simply using different scan velocities in different regions of the WINDOWPANE part. The different scan velocities will result in different cure depths. In practice, (4.20) is very useful since we want to directly control cure depth, and want to determine how fast to scan the laser to give that cure depth. Of course, for the WINDOWPANE experiment, it is more useful to use (4.16) or (4.19).

4.5.4 Time Scales

It is interesting to investigate the time scales at which SL operates. On the short end of the time scale, the time it takes for a photon of laser light to traverse a photopolymer layer is about a picosecond (10^{-12} s). Photon absorption by the photoinitiator and the generation of free radicals or cations occur at about the same time frame. A measure of photopolymer reaction speed is the kinetic reaction rates, t_k, which are typically several microseconds.

The time it takes for the laser to scan past a particular point on the resin surface is related to the size of the laser beam. We will call this time the characteristic exposure time, t_e. Values of t_e are typically 50–2,000 μs, depending on the scan speed (500–5,000 mm/s). Laser exposure continues long after the onset of polymerization. Continued exposure generates more free radicals or cations and, presumably, generates these at points deeper in the photopolymer. During and after the laser beam traverses the point of interest, cross-linking occurs in the photopolymer.

The onset of measurable shrinkage, $t_{s,o}$, lags exposure by several orders of magnitude. This appears to be due to the rate of cross-linking, but for the epoxy-based resins, may have more complicated characteristics. Time at corresponding completion of shrinkage is denoted $t_{s,c}$. For the acrylate-based resins of the early 1990s, times for the onset and completion of shrinkage were typically 0.4–1 and 4–10 s, respectively. Recall that epoxies can take hours or days to polymerize. Since shrinkage lags exposure, this is clearly a phenomenon that complicates the VP process. Shrinkage leads directly to accuracy problems, including deviation from nominal dimensions, warpage, and curl.

The final time dimension is that of scan time for a layer, denoted t_d, which typically spans 10–300 s. The time scales can be summarized as

$$t_t \ll t_k \ll t_e \ll t_{s,o} < t_{s,c} \ll t_d \tag{4.21}$$

As a result, characteristic times for the VP process span about 14 orders of magnitude.

4.6 Vector Scan VP Machines

At present (2014), 3D Systems is the predominant manufacturer of laser scanning VP machines in the world, although several other companies in Japan and elsewhere in Asia also market VP machines. Fockele & Schwarze in Germany produces a micro-VP technology, although they only sell design and manufacturing services. Several Japanese companies produce or produced machines, including Denken Engineering, CMET (Mitsubishi), Sony, Meiko Corp., Mitsui Zosen, and Teijin Seiki (license from Dupont). Formlabs is a start-up company, funded in part by a Kickstarter campaign, markets a small, high-resolution SL machine.

A schematic of a typical VP machine was illustrated in Fig. 4.1a, which shows the main subsystems, including the laser and optics, the platform and elevator, the vat and resin-handling subsystem, and the recoater. The machine subsystem hierarchy is given in Fig. 4.8. Note that the five main subsystems are: recoating system, platform system, vat system, laser and optics system, and control system.

Typically, recoating is done using a shallow dip and recoater blade sweeping. Recoating issues are discussed in [37]. The process can be described as follows:

- After a layer has been cured, the platform dips down by a layer thickness.
- The recoater blade slides over the whole build depositing a new layer of resin and smoothing the surface of the vat.

A common recoater blade type is the zephyr blade, which is a hollow blade that is filled with resin. A vacuum system pulls resin into the blade from the vat. As the blade translates over the vat to perform recoating, resin is deposited in regions where the previous part cross section was built. When the blade encounters a region in the vat without resin, the resin falls into this region since its weight is stronger

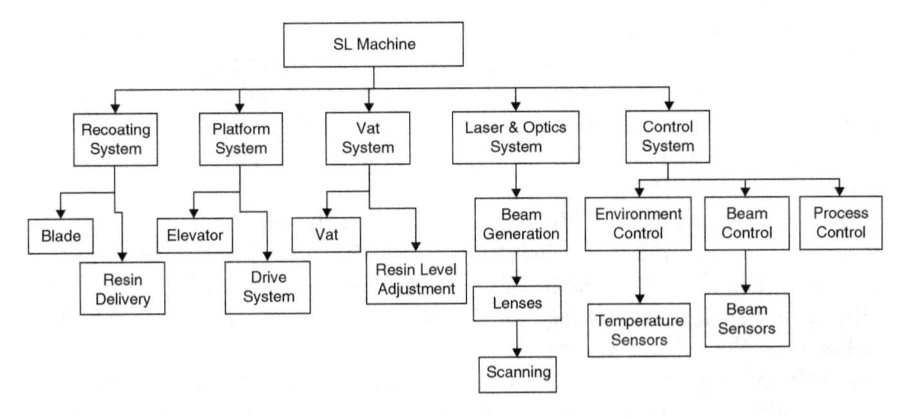

Fig. 4.8 Subsystems for SL technology

than the vacuum force. Blade alignment is critical to avoid "blade crashes," when the blade hits the part being built and often delaminates the previous layer. The blade gap (distance between the bottom of the blade and the resin surface) and speed are important variables under user control.

The platform system consists of a build platform that supports the part being built and an elevator that lowers and raises the platform. The elevator is driven by a lead-screw. The vat system is simply the vat that holds the resin, combined with a level adjustment device, and usually an automated refill capability.

The optics system includes a laser, focusing and adjustment optics, and two galvanometers that scan the laser beam across the surface of the vat. Modern VP machines have solid-state lasers that have more stable characteristics than their predecessors, various gas lasers. SL machines from 3D Systems have Nd-YVO_4 lasers that output radiation at about 1,062 nm wavelength (near infrared). Additional optical devices triple the frequency to 354 nm, in the UV range. These lasers have relatively low power, in the range of 0.1–1 W, compared with lasers used in other AM and material processing applications.

The control system consists of three main subsystems. First, a process controller controls the sequence of machine operations. Typically, this involves executing the sequence of operations that are described in the build file that was prepared for a specific part or set of parts. Commands are sent to the various subsystems to actuate the recoating blade, to adjust resin level or changing the vat height, or to activate the beam controller. Sensors are used to detect resin height and to detect forces on the recoater blade to detect blade crashes. Second, the beam controller converts operation descriptions into actions that adjust beam spot size, focus depth, and scan speed, with some sensors providing feedback. Third, the environment controller adjusts resin vat temperature and, depending on machine model, adjusts environment temperature and humidity.

Two of the main advantages of VP technology over other AM technologies are part accuracy and surface finish, in combination with moderate mechanical properties. These characteristics led to the widespread usage of VP parts as form, fit, and, to a lesser extent, functional prototypes. Typical dimensional accuracies for VP machines are often quoted as a ratio of an error per unit length. For example, accuracy of an SLA-250 is typically quoted as 0.002 in./in. [38]. Modern VP machines are somewhat more accurate. Surface finish of SL parts ranges from submicron Ra for upfacing surfaces to over 100 μm Ra for surfaces at slanted angles [39].

The current commercial VP product line from 3D Systems consists of two families of models: the SLA Viper Si2, and the iPro SLA Centers (iPro 9000XL, iPro 9000, and iPro 8000). Some of these machines are summarized in Table 4.1 [40]. Both the Viper Si2 and the iPro models have dual laser spot size capabilities. In the Viper Si2, a "high-resolution" mode is available that provides a spot size of about 80 μm in diameter, useful for building small parts with fine features. In the iPro machines, in contrast, the machine automatically switches between the "normal" beam of 0.13 mm diameter for borders and fills and the "wide" beam of 0.76 mm diameter for hatch vectors (filling in large areas). The wide beam enables much faster builds. The iPro line replaces other machines, including the popular SLA-3500, SLA-5000, and SLA-7000 machines, as well as the SLA Viper Pro. Additionally, the SLA-250 was a very popular model that was discontinued in 2001 with the introduction of the Viper Si2 model.

4.7 Scan Patterns

4.7.1 Layer-Based Build Phenomena and Errors

Several phenomena should be noted since they are common to all radiation and layer-based AM processes. The most obvious phenomenon is discretization, e.g., a stack of layers causes "stair steps" on slanted or curved surfaces. So, the layer-wise nature of most AM processes causes edges of layers to be visible. Conventionally, commercial AM processes build parts in a "material safe" mode, meaning that the stair steps are on the outside of the CAD part surfaces. Technicians can sand or finish parts; the material they remove is outside of the desired part geometry. Other discretization examples are the set of laser scans or the pixels of a DMD. In most processes, individual laser scans or pixels are not visible on part surfaces, but in other processes such as material extrusion, the individual filaments can be noticeable.

As a laser scans a cross section, or a lamp illuminates a layer, the material solidifies and, as a result, shrinks. When resins photopolymerize, they shrink since the volume occupied by monomer molecules is larger than that of reacted polymer. Similarly, after powder melts, it cools and freezes, which reduces the volume of the material. When the current layer is processed, its shrinkage pulls on the previous layers, causing stresses to build up in the part. Typically, those stresses remain and are called residual stresses. Also, those stresses can cause part edges to curl upwards. Other warpage or part deformations can occur due to these residual stresses, as well.

Table 4.1 Selected SL systems (photos courtesy of 3D Systems, Inc.)

iPro 9000XL SLA center	
Laser type	Solid-state frequency tripled Nd:$YV0_4$
Wavelength	354.7 nm
Power at vat @ 5,000 h	1,450 mW
Recoating system	
Process	Zephyr™ Recoater
Layer thickness min	0.05 mm (0.002 in)
Layer thickness max	0.15 mm (0.006 in)
Optical and scanning	
Beam diameter (@ 1/e2)	0.13 mm (borders)
	0.76 (large hatch)
Drawing speed	3.5 m/s (borders)
	25 m/s (hatch)
Maximum part weight	150 kg (330 lb)
Vat: Max. build envelope, capacity	650 × 350 × 300 (39.1 gal)
	650 × 750 × 50 (25.1 gal)
	650 × 750 × 275 (71.9 gal)
	650 × 750 × 550 (109 gal)
	1,500 × 750 × 550
iPro 8000 SLA center	
Specifications are the same as the iPro 9000XL, except the following	
Maximum part weight	75 kg (165 lb)
Vat: Max. build envelope, capacity	650 × 350 × 300 (39.1 gal)
	650 × 750 × 50 (25.1 gal)
	650 × 750 × 275 (71.9 gal)
	650 × 750 × 550 (109 gal)
SLA Viper Si2	
Laser type	Solid-State Nd:$YV0_4$
Wavelength	354.7 nm
Power at vat	100 mW

(continued)

Table 4.1 (continued)

Recoating System: process	Zephyr recoater	
Typical	0.1 mm (0.004 in) app.	
Minimum	0.05 mm (0.002 in) app.	
Optical and scanning		
Beam diameter (@ 1/e2): standard mode	0.25 ± 0.025 mm (0.01 ± 0.001 in)	
High-resolution	0.075 ± 0.015 mm (0.003 ± 0.0005 in)	
Part drawing speed	5 mm/s (0.2 in/s)	
Maximum part weight	9.1 kg (20 lb)	
Vat capacity	Volume	
Maximum build envelope	250 × 250 × 200 mm XYZ (10 × ×10 × 10 in)	
	32.2 L (8.5 U.S. gal)	
High-res. build envelope	125 × 125 × 250 mm	
	(5 × 5 × 10 in)	

The last phenomenon to be discussed is that of print-through errors. In photopolymerization processes, it is necessary to have the current layer cure into the previous layer. In powder bed fusion processes, the current layer needs to melt into the previous layer so that one solid part results, instead of a stack of disconnected solid layers. The extra energy that extends below the current layer results in thicker part sections. This extra thickness is called print-through error in SL and "bonus Z" in laser sintering. Most process planning systems compensate for print-through by giving users the option of skipping the first few layers of a part, which works well unless important features are contained within those layers.

These phenomena will be illustrated in this section through an investigation of scan patterns in SL.

4.7.2 WEAVE

Prior to the development of WEAVE, scan patterns were largely an ad hoc development. As a result, post-cure curl distortion was the major accuracy problem. The WEAVE scan pattern became available for use in late 1990 [1].

The development of WEAVE began with the observation that distortion in post-cured parts was proportional to the percent of uncured resin after removal from the vat. Another motivating factor was the observation that shrinkage lags exposure and that this time lag must be considered when planning the pattern of laser scans. The key idea in WEAVE development was to separate the curing of the majority of a layer from the adherence of that layer to the previous layer. Additionally, to prevent laser scan lines from interfering with one another while each is shrinking, parallel scans were separated from one another by more than a line width.

The WEAVE style consists of two sets of parallel laser scans:

- First, parallel to the x-axis, spaced 1 mil (1 mil = 0.001 in. = 0.0254 mm, which historically is a standard unit of measure in SL) apart, with a cure depth of 1 mil less than the layer thickness.
- Second, parallel to the y-axis, spaced 1 mil apart, again with a cure depth of 1 mil less than the layer thickness.

However, it is important to understand the relationships between cure depth and exposure. On the first pass, a certain cure depth is achieved, C_{d1}, based on an amount of exposure, E_{max1}. On the second pass, the same amount of exposure is provided and the cure depth increases to C_{d2}. A simple relationship can be derived among these quantities, as shown in (4.21).

$$C_{d2} = D_p \ln(2E_{max\,1}/E_c) = D_p \ln(2) + D_p \ln(E_{max\,1}/E_c) \quad (4.21)$$

$$C_{d2} = C_{d1} + D_p \ln(2) \quad (4.22)$$

It is the second pass that provides enough exposure to adhere the current layer to the previous one. The incremental cure depth caused by the second pass is just $\ln(2)D_p$, or about $0.6931D_p$. This distance is always greater than 1 mil.

As mentioned, a major cause of post-cure distortion was the amount of uncured resin after scanning. The WEAVE build style cures about 99 % of the resin at the vat surface and about 96 % of the resin volume through the layer thickness. Compared with previous build styles, WEAVE provided far superior results in terms of eliminating curl and warpage. Figure 4.9 shows a typical WEAVE pattern, illustrating how WEAVE gets its name.

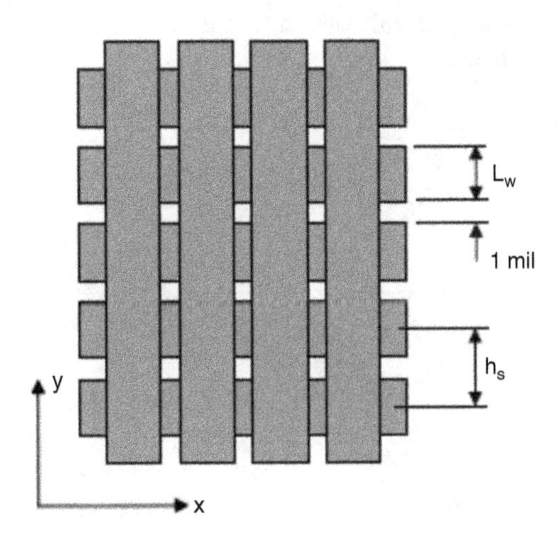

Fig. 4.9 WEAVE scan pattern

Even though WEAVE was a tremendous improvement, several flaws were observed with its usage. Corners were distorted on large flat surfaces, one of these corners always exhibited larger distortion, and it was always the same corner. Some microfissures occurred; on a flat plate with a hole, a macrofissure tangent to the hole would appear.

It was concluded that significant internal stresses developed within parts during part building, not only post-cure. As a result, improvements to WEAVE were investigated, leading to the development of STAR-WEAVE.

4.7.3 STAR-WEAVE

STAR-WEAVE was released in October 1991, roughly 1 year after WEAVE [1]. STAR-WEAVE addressed all of the known deficiencies of WEAVE and worked very well with the resins available at the time. WEAVE's deficiencies were traced to the consequences of two related phenomena: the presence of shrinkage and the lag of shrinkage relative to exposure. These phenomena led directly to the presence of large internal stresses in parts. STAR-WEAVE gets its name from the three main improvements from WEAVE:

1. Staggered hatch
2. Alternating sequence
3. Retracted hatch

Staggered hatch directly addresses the observed microfissures. Consider Fig. 4.10 which shows a cross-sectional view of the hatch vectors from two layers. In Fig. 4.10a, the hatch vectors in WEAVE form vertical "walls" that do not directly touch. In STAR-WEAVE, Fig. 4.10b, the hatch vectors are staggered such that they directly adhere to the layer below. This resulting overlap from one layer to the next eliminated microfissures and eliminated stress concentrations in the regions between vectors.

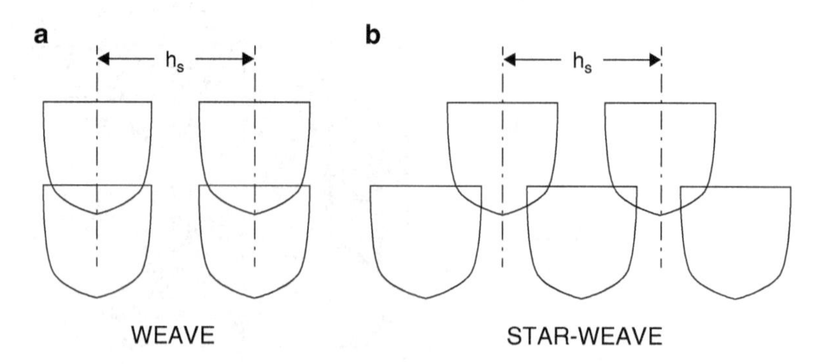

Fig. 4.10 Cross-sectional view of WEAVE and STAR-WEAVE patterns

Fig. 4.11 WEAVE problem example

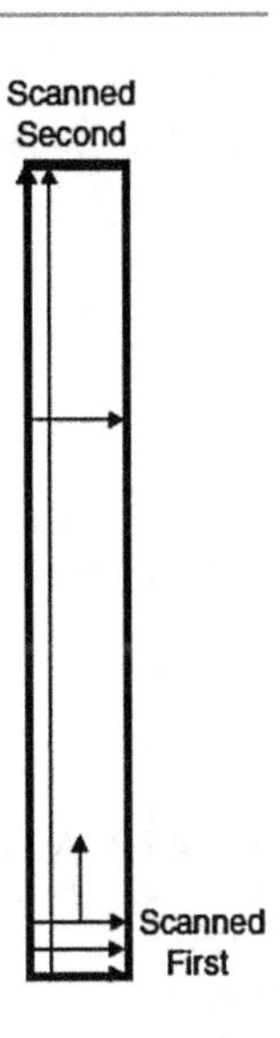

Upon close inspection, it became clear why the WEAVE scan pattern tended to cause internal stresses, particularly if a part had a large cross section. Consider a thin cross section, as shown in Fig. 4.11. The WEAVE pattern was set up to always proceed in a certain manner. First, the x-axis vectors were drawn left to right, and front to back. Then, the y-axis vectors were drawn front to back and left to right. Consider what happens as the y-axis vectors are drawn and the fact that shrinkage lags exposure. As successive vectors are drawn, previous vectors are shrinking, but these vectors have adhered to the x-axis vectors and to the previous layer. In effect, the successive shrinkage of y-axis vectors causes a "wave" of shrinkage from left to right, effectively setting up significant internal stresses. These stresses cause curl.

Given this behavior, it is clear that square cross sections will have internal stresses, possibly without visible curl. However, if the part cannot curl, the stresses will remain and may result in warpage or other form errors.

With a better understanding of curing and shrinking behavior, the Alternating Sequence enhancement to building styles was introduced. This behavior can be alleviated to a large extent simply by varying the x and y scan patterns. There are two vector types: x and y. These types can be drawn left to right, right to left, front to back, and back to front. Looking at all combinations, eight different scan sequences are possible. As a part is being built, these eight scan sequences alternate, so that eight consecutive layers have different patterns, and this pattern is repeated every eight layers.

The good news is that internal stresses were reduced and the macrofissures disappeared. However, internal stresses were still evident. To alleviate the internal stresses to a greater extent, the final improvement in STAR-WEAVE was introduced, that of Retracted Hatch. It is important to realize that the border of a cross section is scanned first, then the hatch is scanned. As a result, the x-axis vectors adhere to both the left and right border vectors. When they shrink, they pull

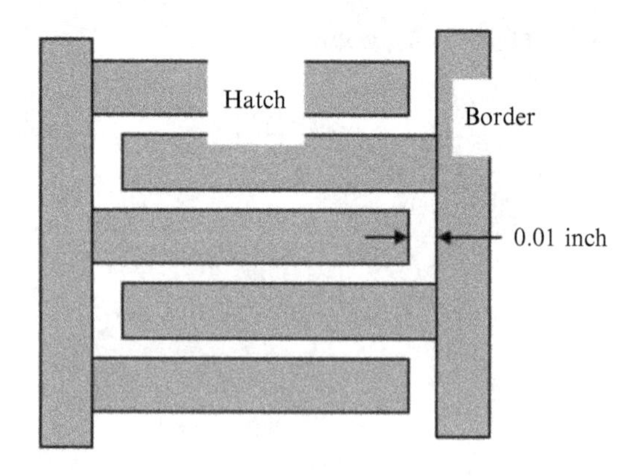

Fig. 4.12 Retracted hatch of the STAR-WEAVE pattern

Table 4.2 ACES process variables for the SLA-250

Variable	Range
Layer thickness	0.002–0.008 in.
Hatch spacing	0.006–0.012 in.
Hatch overcure	(−0.003)–(+0.001) in.
Fill overcure	0.006–0.012 in.
Blade gap %	100–200
Sweep period	5–15 s
Z-wait	0–20 s
Pre-dip delay	0–20 s

on the borders, bending them towards one another, causing internal stresses. To alleviate this, alternating hatch vectors are retracted from the border, as shown in Fig. 4.12. This retracted hatch is performed for both the x and y vectors.

4.7.4 ACES Scan Pattern

With the development of epoxy-based photopolymers in 1992–1993, new scan patterns were needed to best adopt to their curing characteristics. ACES (Accurate, Clear, Epoxy, Solid) was the answer to these needs. ACES is not just a scan pattern, but is a family of build styles. The operative word in the ACES acronym is Accurate. ACES was mainly developed to provide yet another leap in part accuracy by overcoming deficiencies in STAR-WEAVE, most particularly, in percent of resin cured in the vat. Rather than achieving 96 % solidification, ACES is typically capable of 98 %, further reducing post-cure shrinkage and the associated internal stresses, curl, and warpage [12].

Machine operators have a lot of control over the particular scan pattern used, along with several other process variables. For example, while WEAVE and STAR-WEAVE utilized 0.001 in. spacings between solidified lines, ACES allows the user to specify hatch spacing. Table 4.2 shows many of the process variables for the SLA-250 along with typical ranges of variable settings.

In Table 4.2, the first four variables are called scan variables since they control the scan pattern, while the remaining variables are recoat variables since they control how the vat and part are recoated. With this set of variables, the machine operator has a tremendous amount of control over the process; however, the number of variables can cause a lot of confusion since it is difficult to predict exactly how the part will behave as a result of changing a variable's value. To address this issue, 3D Systems provides nominal values for many of the variables as a function of layer thickness.

The fundamental premise behind ACES is that of curing more resin in a layer before bonding that layer to the previous one. This is accomplished by overlapping hatch vectors, rather than providing 0.001 in. spacing between hatch vectors. As a result, each point in a layer is exposed to laser radiation from multiple scans. Hence, it is necessary to consider these multiple scans when determining cure depth for a layer. ACES also makes use of two passes of scan vectors, one parallel to the x-axis and one parallel to the y-axis. In the first pass, the resin is cured to a depth 1 mil less than the desired layer thickness. Then on the second pass, the remaining resin is cured and the layer is bonded to the previous one.

As might be imagined, more scan vectors are necessary using the ACES scan pattern, compared with WEAVE and STAR-WEAVE.

The remaining presentation in this section is on the mathematical model of cure depth as a function of hatch spacing to provide insight into the cure behavior of ACES.

Consider Fig. 4.13 that shows multiple, overlapping scan lines with hatch spacing h_s. Also shown is the cure depth of each line, C_{d0}, and the cure depth,

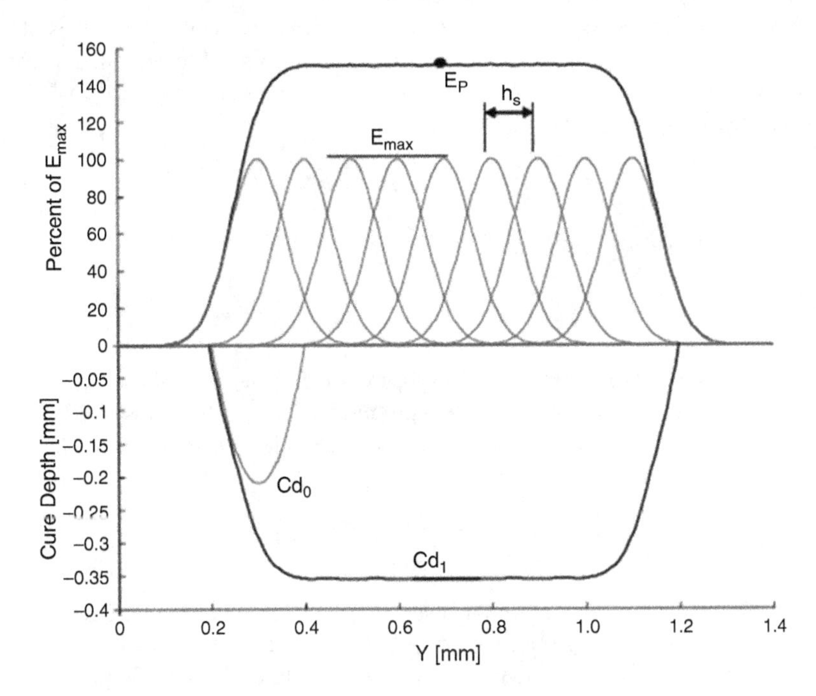

Fig. 4.13 Cure depth and exposure for the ACES scan pattern

C_{d1}, of the entire scan pass. As we know from earlier, the relationship between exposure and cure depth is given by (4.23).

$$C_{d_0} = D_\mathrm{p}\ln(E_\mathrm{max}/E_\mathrm{c}) \tag{4.23}$$

The challenge is to find an expression for cure depth of a scan pass when the scan vectors overlap. This can be accomplished by starting from the relationship describing the spatial distribution of exposure. From earlier, we know that:

$$E(y,0) = E_\mathrm{max}e^{-\left(2y^2/W_0^2\right)} \tag{4.24}$$

Consider the progression of curing that results from many more scans in Fig. 4.13. If we consider a point P in the region of the central scan, we need to determine the number of scan vectors that provide significant exposure to P. Since the region of influence is proportional to beam spot size, the number of scans depends upon the beam size and the hatch spacing. Considering that the ratio of hatch spacing to beam half-width, W_0, is rarely less than 0.5 (i.e., $h_s/W_0 \geq 0.5$), then we can determine that point P receives about 99 % of its exposure from a distance of 4 h_s or less. In other words, if we start at the center of a scan vector, at most, we need to consider 4 scans to the left and 4 scans to the right when determining cure depth.

In this case, we are only concerned with the variation of exposure with y, the dimension perpendicular to the scan direction.

Given that it is necessary to consider 9 scans, we know the various values of y in (4.24). We can consider that $y = nh_s$, and let n range from −4 to +4. Then, the total exposure received at a point P is the sum of the exposures received over those 9 scans, as shown in (4.25) and (4.26).

$$E_\mathrm{P} = E_0 + 2E_1 + 2E_2 + 2E_3 + 2E_4 \tag{4.25}$$

where $E_n \equiv E(n\,h_s, 0) = E_\mathrm{max}\mathrm{e}^{-2(n\,h_s/W_0)^2}$

$$E_\mathrm{P} = E_\mathrm{max}\left[1 + 2\mathrm{e}^{-2(h_s/W_0)^2} + 2\mathrm{e}^{-8(h_s/W_0)^2} + 2\mathrm{e}^{-18(h_s/W_0)^2} + 2\mathrm{e}^{-32(h_s/W_0)^2}\right] \tag{4.26}$$

It is convenient to parameterize exposure vs. E_max against the ratio of hatch spacing vs. beam half-width. A simple rearrangement of (4.26) yields (4.27). A plot of (4.27) over the typical range of size ratios (h_s/W_0) is shown in Fig. 4.14.

$$\frac{E_\mathrm{p}}{E_\mathrm{max}} = 1 + \sum_{n=1}^{4}\mathrm{e}^{-2(nh_s/W_0)^2} \tag{4.27}$$

We can now return to our initial objective of determining the cure depth for a single pass of overlapping scan vectors. Further, we can determine the increase in cure depth from a single scan to the entire layer. A cure depth for a single pass, C_{d1},

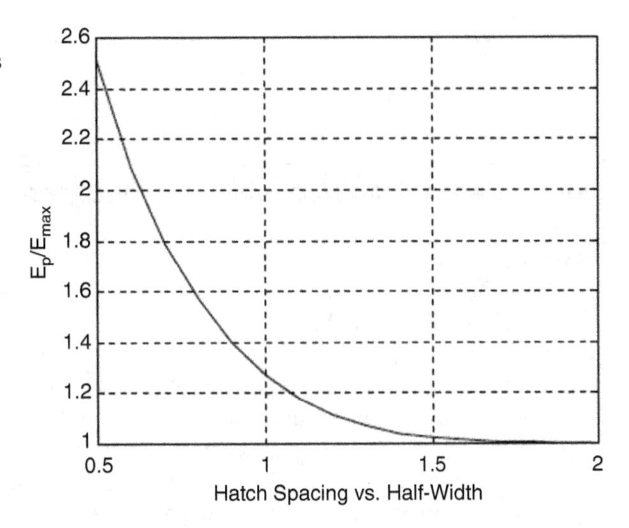

Fig. 4.14 Plot of (4.27): exposure ratios vs. size ratios

with overlapping scans is a function of the total exposure given in (4.26). C_{d1} is determined using (4.28).

$$C_{d_1} = D_P \ln(E_p/E_c) \tag{4.28}$$

The cure depth increase is given by $C_{d_1} - C_{d_0}$ and can be determined using (4.29).

$$C_{d_1} - C_{d_0} = D_P \ln(E_p/E_{max}) \tag{4.29}$$

As an example, consider that we desire a layer thickness to be 4 mils using a resin with a D_p of 5.8 mils. Assume further that the desired hatch spacing is 6 mils and the beam half-width is 5 mils, giving a size ratio of $h_s/W_0 = 1.2$. On the first pass, the cure depth, C_{d1}, should be $4 - 1 = 3$ mils. From (4.27), the exposure ratio can be determined to be 1.1123 (or see Fig. 4.14). The cure depth for a single scan vector can be determined by rearranging (4.29) to solve for C_{d_0}.

$$\begin{aligned} C_{d_0} &= C_{d_1} - D_P \ln(E_P/E_{max}) \\ &= 3 \; - \; 5.8 * \; \ln(1.1123) \\ &= 2.383 \text{ mils} \end{aligned}$$

From this calculation, it is evident that the cure depth of a single scan vector is 1.6 mils less than the desired layer thickness. Rounding up from 1.6 mils, we say that the hatch overcure of this situation is −2 mils. Recall that the hatch overcure is one of the variables that can be adjusted by the SL machine operator.

This concludes the presentation of traditional vector scan VP. We now proceed to discuss micro-vat photopolymerization and mask projection-based systems, where areas of the vat surface are illuminated simultaneously to define a part cross section.

4.8 Vector Scan Micro-Vat Photopolymerization

Several processes were developed exclusively for microfabrication applications based on photopolymerization principles using both lasers and X-rays as the energy source. These processes build complex shaped parts that are typically less than 1 mm in size. They are referred to as Microstereolithography (MSL), Integrated Hardened Stereolithography (IH), LIGA [42], Deep X-ray Lithography (DXRL), and other names. In this section, we will focus on those processes that utilize UV radiation to directly process photopolymer materials.

In contrast to convention VP, vector scan technologies for the microscale typically have moved the vat in x, y, and z directions, rather than scanning the laser beam. To focus a typical laser to spot sizes less than 20 μm requires the laser's focal length to be very short, causing difficulties for scanning the laser. For an SLA-250 with a 325 nm wavelength HeCd laser, the beam has a diameter of 0.33 mm and a divergence of 1.25 mrad as it exits the laser. It propagates 280 mm then encounters a diverging lens (focal length −25 mm) and a converging lens (focal length 100 mm) which is 85 mm away. Using simple thin-lens approximations, the distance from the converging lens to the focal point, where the laser reaches a spot size of 0.2 mm is 940 mm and its Rayleigh range is 72 mm. Hence, the focused laser spot is a long distance from the focusing optics and the Rayleigh range is long enough to enable a wide scanning region and a large build area.

In contrast, a typical calculation is presented here for a high-resolution micro-SL system with a laser spot size of 10 μm. A 325 nm wavelength HeCd laser used in SL is included here to give the reader an idea of the challenge. The beam, as it exits the laser, has a diameter of 0.33 mm and a divergence of 1.25 mrad. It propagates 280 mm then encounters a diverging lens (focal length −25 mm) and a converging lens (focal length 36.55 mm). The distance from the converging lens to the focal point is 54.3 mm and its Rayleigh range is only 0.24 mm. It would be very difficult to scan this laser beam across a vat without severe spot distortions.

Scanning micro-VP systems have been presented in literature since 1993 with the introduction of the Integrated Hardening method of Ikuta and Hirowatari [43]. They used a laser spot focused to a 5-μm diameter and the resin vat is scanned underneath it to cure a layer. Examples of devices built with this method include tubes, manifolds, and springs and flexible microactuators [44] and fluid channels on silicon [45]. Takagi and Nakajima [46] have demonstrated the use of this technology for connecting MEMS gears together on a substrate. The artifact fabricated using micro-VP can be used as a mold for subsequent electroplating followed by removal of the resin [47]. This method has been able to achieve sub-1 μm minimum feature size.

The following specifications of a typical point-wise microstereolithography process have been presented in [48]:

- 5-μm spot size of the UV beam
- Positional accuracy is 0.25 μm (in the x–y directions) and 1.0 μm in the z-direction
- Minimum size of the unit of hardened polymer is 5 μm × 5 μm × 3 μm (in x, y, z)
- Maximum size of fabrication structure is 10 mm × 10 mm × 10 mm

The capability of building around inserted components has also been proposed for components such as ultrafiltration membranes and electrical conductors. Applications include fluid chips for protein synthesis [49] and bioanalysis [50]. The bioanalysis system was constructed with integrated valves and pumps that include a stacked modular design, 13 × 13 mm^2 and 3 mm thick, each of which has a different fluid function. However, the full extent of integrated processing on silicon has not yet been demonstrated. The benefits of greater design flexibility and lower cost of fabrication may be realized in the future.

4.9 Mask Projection VP Technologies and Processes

Technologies to project bitmaps onto a resin surface to cure a layer at a time were first developed in the early 1990s by researchers who wanted to develop special VP machines to fabricate microscale parts. Several groups in Japan and Europe pursued what was called mask projection stereolithography technology at that time. The main advantage of mask projection methods is speed: since an entire part cross section can be cured at one time, it can be faster than scanning a laser beam. Dynamic masks can be realized by LCD screens, by spatial light modulators, or by DMDs, such as the Digital Light Processing (DLP™) chips manufactured by Texas Instruments [51].

4.9.1 Mask Projection VP Technology

Mask projection VP (MPVP) systems have been realized by several groups around the world. Some of the earlier systems utilized LCD displays as their dynamic mask [52, 53], while another early system used a spatial light modulator [54]. The remaining systems all used DMDs as their dynamic masks [55–58]. These latest systems all use UV lamps as their radiation source, while others have used lamps in the visible range [55] or lasers in the UV. A good overview of micro-VP technology, systems, and applications is the book by Varadan et al. [59].

Microscale VP has been commercialized by MicroTEC GmbH, Germany. Although machines are not for sale, the company offers customer-specific services. The company has developed machines based on point-wise as well as layer-wise photopolymerization principles. Their Rapid Micro Product Development (RMPD)

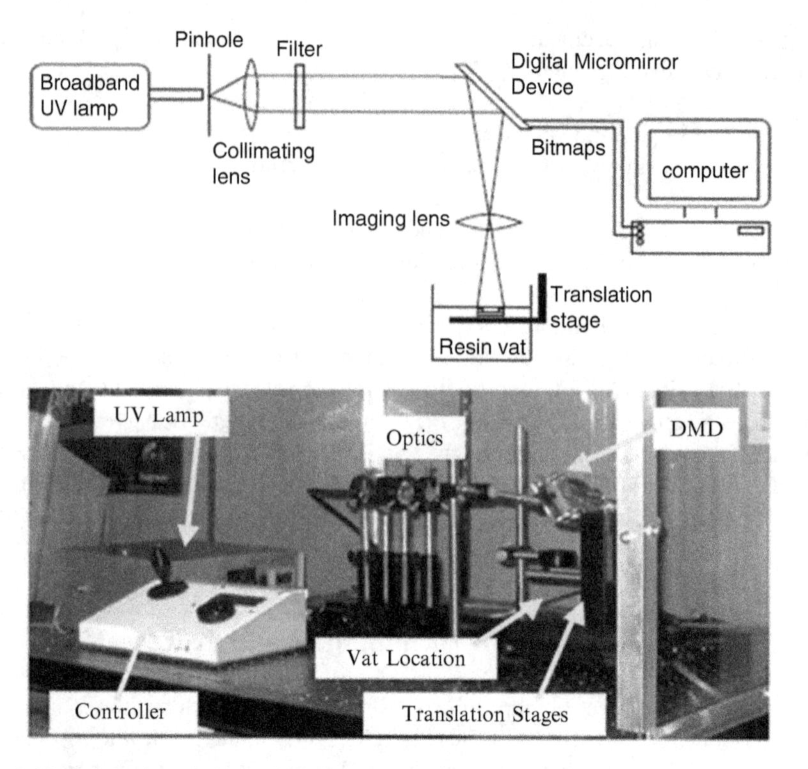

Fig. 4.15 Schematic and photo of mask projection VP machine

machines using a He–Cd laser enable construction of small parts layer-by-layer (as thin as 1 μm) with a high surface quality in the subnanometer range and with a feature definition of <10 μm.

A schematic and photograph of a MPVP system from Georgia Tech is shown in Fig. 4.15. Similar to conventional SL, the MPVP process starts with the CAD model of the part, which is then sliced at various heights. Each resulting slice cross section is stored as bitmaps to be displayed on the dynamic mask. UV radiation reflects off of the "on" micromirrors and is imaged onto the resin surface to cure a layer. In the system at Georgia Tech, a broadband UV lamp is the light source, a DMD is the dynamic mask, and an automated XYZ stage is used to translate the vat of resin in three dimensions. Standard VP resins are typically used, although other research groups formulate their own.

4.9.2 Commercial MPVP Systems

Several companies market VP systems based on mask projection technology, including EnvisionTec and 3D Systems. New companies in Europe and Asia have also started recently to market MPVP systems.

EnvisionTEC first marketed their MPVP systems in 2003. They now have several lines of machines with various build envelopes and resolutions based on the MPVP process, including the Perfactory, Perfactory Desktop, Aureus, Xede/Xtreme, and Ultra. Variants of some of these models are available, including specialized Perfactory machines for dental restorations or for hearing aid shells. A photo of the Perfactory Standard machine is shown in Fig. 4.16 and its technical specifications are listed in Table 4.3.

Schematically, their machines are very similar to the Georgia Tech machine in Fig. 4.15 and utilize a lamp for illuminating the DMD and vat. However, several of their machine models have a very important difference: they build parts upside down and do not use a recoating mechanism. The vat is illuminated vertically upwards through a clear window. After the system irradiates a layer, the cured resin sticks to the window and cures into the previous layer. The build platform pulls away from the window at a slight angle to gently separate the part from the window. The advantage of this approach is threefold. First, no separate recoating mechanism is needed since gravity forces the resin to fill in the region between the cured part and the window. Second, the top vat surface being irradiated is a flat window, not a free surface, enabling more precise layers to be fabricated. Third, they have devised a build process that eliminates a regular vat. Instead, they have a supply-on-demand

Fig. 4.16 EnvisionTEC Perfactory model

Table 4.3 Specifications on EnvisionTEC Perfactory Standard Zoom machine

Lens system		$f = 25$–45 mm
Build envelope	Standard	190 × 142 × 230 mm
	High-resolution	120 × 90 × 230 mm
Pixel size	Standard	86–136 μm
	High-resolution	43–68 μm
Layer thickness	25–150 mm	

material feed system. The disadvantage is that small or fine features may be damaged when the cured layer is separated from the window.

3D Systems introduced their V-Flash machine in 2008. It utilizes MPVP technology and a novel material handling approach [60]. The V-Flash was intended to be an inexpensive prototyping machine (under $10,000) that was as easy to use as a typical home ink-jet printer. Its build envelope was 230 × 170 × 200 mm (9 × 7 × 8 in.). During operation, parts were built upside down. For each layer, a blade coated a layer of resin onto a film that spanned the build chamber. The build platform slid down until the platform or the in-process part contacted the resin layer and film. A cartridge provided a supply of unused film for each layer. That layer was cured by the machine's "UV Imager," which consisted of the MPVP technology. Some rinsing of the part was required, similar to SL, and support structures may have to be removed during the post-processing phase of part fabrication.

4.9.3 MPVP Modeling

Most of the research presented on MPVP technology is experimental. As in SL, it is possible to develop good predictive models of curing for MPVP systems. Broadly speaking, models of the MPVP process can be described by a model that determines the irradiation of the vat surface and its propagation into the resin, followed by a model that determines how the resin reacts to that irradiation. Schematically, the MPVP model can be given by Fig. 4.17, showing an Irradiance Model and a Cure Model.

As a given bitmap pattern is displayed, the resin imaged by the "on" mirrors is irradiated. The exposure received by the resin is simply the product of the irradiance and the time of exposure. The dimensional accuracy of an imaged part cross

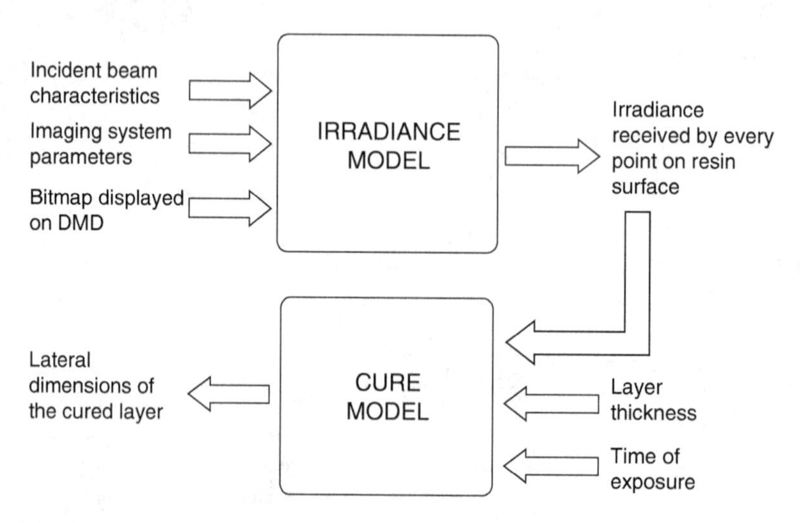

Fig. 4.17 Model of the MPVP process

section is a function of the radiation uniformity across the DMD, the collimation of the beam, and the capability of the optics system in delivering an undistorted image.

If the MPVP machine's optical system produces a plane wave that is neither converging nor diverging, then it is easy to project rays from the DMD to the resin surface. The irradiance model in this case is very straightforward. However, in most practical cases, it is necessary to model the cone of rays that project from each micromirror on the DMD to the resin. As a result, a point on the resin may receive radiation from several micromirrors. Standard ray-tracing methods can be used to compute the irradiance field that results from a bitmap [61].

After computing the irradiance distribution on the vat surface, the cured shape can be predicted. The depth of cure can be computed in a manner similar to that used in Sect. 4.5. Cure depth is computed as the product of the resin's D_p value and the exponential of the exposure received divided by the resin's E_c value, as in (4.15). The exposure received is simply the product of the irradiance at a point and the time of exposure, T.

$$C_d = D_p e^{-E/E_c} = D_p e^{-H \cdot T/E_c} \tag{4.30}$$

In the build direction, overcure and print-through errors are evident, as in SL. In principle, however, it is easier to correct for these errors than in point-wise SL systems. A method called the "Compensation Zone" approach was developed to compensate for this unwanted curing [61]. A tailored volume (Compensation Zone) is subtracted from underneath the CAD model to compensate for the increase in the Z dimension that would occur due to print-through. Using this method, more accurate parts and better surface finish can be achieved.

4.10 Two-Photon Vat Photopolymerization

In the two-photon vat photopolymerization (2p-VP) process, the photoinitiator requires two photons to strike it before it decomposes to form a free radical that can initiate polymerization. The effect of this two-photon requirement is to greatly increase the resolution of photopolymerization processes. This is true since only near the center of the laser is the irradiance high enough to provide the photon density necessary to ensure that two photons will strike the same photoinitiator molecule. Feature sizes of 0.2 μm or smaller have been achieved using 2p-VP.

2p-VP was first invented in the 1970s for the purposes of fabricated three-dimensional parts [62]. Interestingly, this predates the development of stereolithography by over 10 years. In this approach, two lasers were used to irradiate points in a vat of photopolymer. When the focused laser spots intersected, the photon density was high enough for photopolymerization.

More recently, 2p-VP received research attention in the late 1990s. A schematic of a typical research setup for this process is shown in Fig. 4.18 [63]. In this system, they used a high-power Ti:Sapphire laser, with wavelength 790 nm, pulse-width 200 fs, and peak power 50 kW. The objective lens had an NA = 0.85. Similarly to

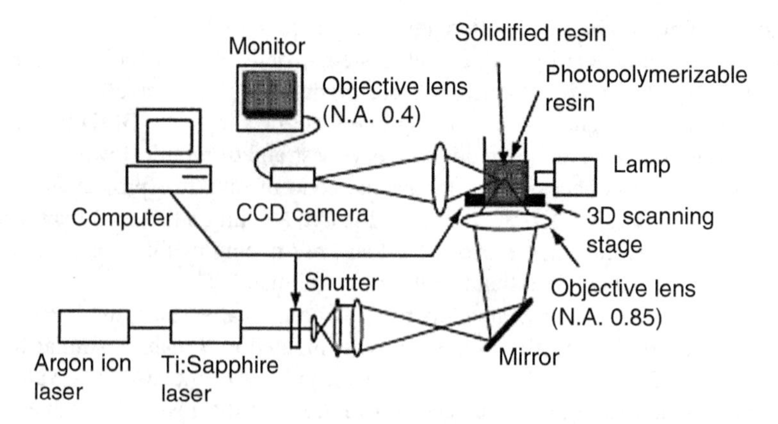

Fig. 4.18 Schematic of typical two-photon equipment

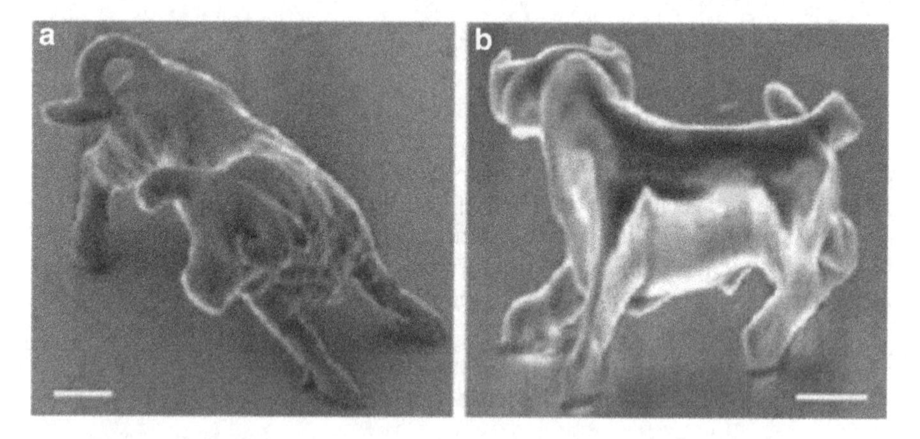

Fig. 4.19 Bull model fabricated by 2p-VP. The size scale bar is 1 μm

other micro-VP approaches, the vat was scanned by a 3D scanning stage, not the laser beam. Parts were built from the bottom up. The viscosity of the resin was enough to prevent the micropart being cured from floating away. Complicated parts have been produced quickly by various research groups. For example, the micro-bull in Fig. 4.19 was produced in 13 min [64]. The shell of the micro-bull was cured by 2p-VP, while the interior was cured by flood exposure to UV light.

Typical photopolymer materials can be used in 2p-VP machines [64–66]. The most commonly used resin was SCR500 from Japan Synthetic Rubber Company, which was a common SL resin in Japan, where this research started during the 1990s. SCR500 is a mixture of urethane acrylate oligomers/monomers and common free-radical generating photoinitiators. The absorption spectrum of the resin shows that it is transparent beyond 550 nm, which is a significant advantage since photons can penetrate the resin to a great depth (D_p is very large). One implication is that parts can be built inside the resin vat, not just at the vat surface, which eliminates the need for recoating.

Photosensitivity of a 2p-VP resin is measured in terms of the two-photon absorption cross section (Δ) of the initiator molecule corresponding to the wavelength used to irradiate it. The larger the value of Δ, the more sensitive is the resin to two-photon polymerization, possibly enabling lower power lasers.

Acrylate photopolymer systems exhibit low photosensitivity as the initiators have small two-photon absorption cross sections. Consequently, these initiators require high-laser power and longer exposure times. Other materials have been investigated for 2p-VP, specifically using initiators with larger Δ. New types of photoinitiators tend to be long molecules with certain patterns that make them particularly good candidates for decomposing into free radicals if two photons strike it in rapid succession [67]. By tuning the design of the photoinitiators, large absorption cross sections and low-polymerization threshold energies can be achieved [68].

4.11 Process Benefits and Drawbacks

Two of the main advantages of vat photopolymerization technology over other AM technologies are part accuracy and surface finish. These characteristics led to the widespread usage of vector scan stereolithography parts as form, fit, and, to a lesser extent, functional prototypes as the rapid prototyping field developed. Typical dimensional accuracies for SL machines are often quoted as a ratio of an error per unit length. For example, accuracy of an SLA-250 is typically quoted as 0.002 in./in. Modern SL machines are somewhat more accurate. Surface finish of SL parts ranges from submicron Ra for upfacing surfaces to over 100 μm Ra for surfaces at slanted angles.

Another advantage of VP technologies is their flexibility, supporting many different machine configurations and size scales. Different light sources can be used, including lasers, lamps, or LEDs, as well as different pattern generators, such as scanning galvanometers or DMDs. The size range that has been demonstrated with VP technologies is vast: from the 1.5 m vat in the iPro 9000XL SLA Center to the 100 nm features possible with 2-photon photopolymerization.

Mask projection VP technologies have an inherent speed advantage over laser scan SL. By utilizing a mask, an entire part cross section can be projected, rather than having to sequentially scan the vector pattern for the cross section. There is a trade-off between resolution and the size of the pattern (and size of the solidified cross section) due to the mask resolution. For example, typical DMDs have $1,024 \times 780$ or $1,280 \times 720$ resolution, although newer HDTV DMDs have resolutions of $1,920 \times 1,080$. Nonetheless, for fabricated part resolution of 50 μm or better, the projected area can be a maximum of 96×54 mm.

A drawback of VP processes is their usage of photopolymers, since the chemistries are limited to acrylates and epoxies for commercial materials. Although quite a few other material systems are photopolymerizable, none have emerged as commercial successes to displace the current chemistries. Generally, the current SL materials do not have the impact strength and durability of good quality injection molded thermoplastics. Additionally, they are known to age, resulting in degraded mechanical properties over time. These limitations prevent SL processes from being used for many production applications.

4.12 Summary

Photopolymerization processes make use of liquid, radiation-curable resins called photopolymers to fabricate parts. Upon irradiation, these materials undergo a chemical reaction to become solid. Several methods of illuminating photopolymers for part fabrication were presented, including vector scan point-wise processing, mask projection layer-wise processing, and two-photon approaches. The vector scan approach is used with UV lasers in the VP process, while DLP micromirror array chips are commonly used for mask projection technologies. Two-photon approaches, which have the highest resolution, remain of research interest only. Advantages, disadvantages, and unique characteristics of these approaches were summarized.

Photopolymerization processes lend themselves to accurate analytical modeling due to the well-defined interactions between radiation and photopolymers. An extensive model for laser scan VP was presented, while a simpler one for MPVP was summarized. Discretization errors and scan patterns were covered in this chapter to convey a better understanding of these concepts as they apply to photopolymerization processes, as well as many of the processes still to be presented in this book.

4.13 Exercises

1. Explain why VP is a good process to use to fabricate patterns for investment casting of metal parts (0.5 page+).
2. Explain why two photoinitiators are needed in most commercial VP resins. Explain what these photoinitiators do.
3. Assume you are building with the STAR-WEAVE build style under the following conditions: layer thickness = 0.006″, $D_p = 6.7$ mil, $E_c = 9.9$ mJ/cm^2 (SL-5240), machine = SLA-250/50.
   (a) Determine the cure depths C_{d1} and C_{d2} needed.
   (b) Compute the laser scan speeds required for C_{d1} and C_{d2}.
   (c) Determine laser scan speeds required C_{d1} and C_{d2} when building along an edge of the vat.

4. Assume you are building with the ACES build style under the following conditions: layer thickness $= 0.004''$, $D_p = 4.1$ mil, $E_c = 11.4$ mJ/cm^2 (SL-5510), machine = SLA-Viper Si2.
   (a) Determine the cure depths C_{d1} and C_{d2} needed.
   (b) Compute the laser scan speeds required for C_{d1} and C_{d2}.
   (c) Determine laser scan speeds required for C_{d1} and C_{d2} when building along an edge of the vat, taking into account the laser beam angle.
5. In the derivation of exposure (4.9) for a scan from 0 to $x = b$, several steps were skipped.
   (a) Complete the derivation of (4.9). Note that the integral of e^{-v^2} from 0 to b is $\int_0^b e^{-v^2} dv = \frac{\sqrt{\pi}}{2} \operatorname{erf}(v) \Big|_0^b$, where erf($v$) is the error function of variable v (see Matlab or other math source for explanation of erf(v)).
   (b) Compute the exposure received from this scan at the origin, at $x = 10$ mm, and at $b = 20$ mm using the conditions in Prob. 3b, where laser power is 60 mW.
   (c) Now, let $b = 0.05$ mm and recompute the exposure received at the origin and point b. Compare with results of part (b). Explain the differences observed.
6. Consider a tall thin rib that consists of a stack of 10 vector scans. That is, the rib consists of 10 layers and on each layer, only 1 vector scan is drawn.
   (a) Derive an expression for the width of the rib at any point z along its height.
   (b) Develop a computer program to solve your rib width equation.
   (c) Using your program, compute the rib widths along the height of the rib and plot a graph of rib width. Use the conditions of Prob. 4 and a scan speed of 1,000 mm/s.
   (d) Repeat part (c) using a scan speed of 5,000 mm/s. Note the differences between your graphs from (c) and (d).

References

1. Jacobs PF (1992) Rapid prototyping & manufacturing, fundamentals of stereolithography. SME, New York
2. Tang, Y (2002) Stereolithography (SL) cure modeling. Masters Thesis, School of Chemical Engineering, Georgia Institute of Technology
3. Beaman JJ, Barlow JW, Bourell DL, Crawford RH, Marcus HL, McAlea KP (1997) Solid freeform fabrication: a new direction in manufacturing. Kluwer Academic, Boston
4. Hull CW (1990) Method for production of three-dimensional objects by stereolithography, 3D Systems, Inc. US Patent 4,929,402, 29 May 1990
5. Murphy EJ, Ansel RE, Krajewski JJ (1989) Investment casting utilizing patterns produced by stereolithography, DeSoto, Inc. US Patent 4,844,144, 4 July 1989
6. Wohlers T (1991) Rapid prototyping: an update on RP applications, technology improvements, and developments in the industry. Wohlers, Fort Collins
7. Lu L, Fuh JYH, Nee AYC, Kang ET, Miyazawa T, Cheah CM (1995) Origin of shrinkage, distortion and fracture of photopolymerized material. Mater Res Bull 30(12):1561–1569
8. Asahi Denka JP (1988) Patent 2,138,471, filed Feb 1988

9. Asahi Denka JP (1988) Patent 2,590,216, filed Jul 1988
10. Crivello JV, Dietliker K (1998) Photoinitiators for free radical, cationic & anionic photopolymerisation. In: Bradley G (ed) Chemistry & technology of UV & EB formulation for coatings, inks & paints, vol III, 2nd edn. Wiley, Chichester
11. Dufour P (1993) State-of-the-art and trends in radiation curing. In: Fouassier JP, Rabek JF (eds) Radiation curing in polymer science and technology, vol I, Fundamentals and methods. Elsevier Applied Science, London, p P1
12. Jacobs PF (1996) Stereolithography and other RP & M technologies. SME, Dearborn
13. Wilson JE (1974) Radiation chemistry of monomers, polymers, and plastics. Marcel Dekker, New York
14. Fouassier JP (1993) An introduction to the basic principles in UV curing. In: Fouassier JP, Rabek JF (eds) Radiation curing in polymer science and technology, vol I, Fundamentals and methods. Elsevier Applied Science, London, p P49
15. Decker C, Elazouk B (1995) Laser curing of photopolymers. In: Allen NS et al (eds) Current trends in polymer photochemistry. Ellis Horwood, New York, p P130
16. Andrzejewska E (2001) Photopolymerization kinetics of multifunctional monomers. Prog Polym Sci 26:605
17. Hageman HJ (1989) Photoinitiators and photoinitiation mechanisms of free-radical polymerization processes. In: Allen NS (ed) Photopolymerization and photoimaging science and technology. Elsevier Science, London, p P1
18. Crivello JV (1993) Latest developments in the chemistry of onium salts. In: Fouassier JP, Rabek JF (eds) Radiation curing in polymer science and technology, vol II, Photoinitiated systems. Elsevier Applied Science, London, p P435
19. Bassi GL (1993) Formulation of UV-curable coatings—how to design specific properties. In: Fouassier JP, Rabek JF (eds) Radiation curing in polymer science and technology, vol II, Photoinitiated systems. Elsevier Applied Science, London, p P239
20. Crivello JV (1984) Cationic polymerization—iodonium and sulfonium salt photoinitiators. Adv Polym Sci 62:1
21. Crivello JV, Lee JL (1988) Method for making polymeric photoactive aryl iodonium salts, products obtained therefrom, and use. General Electric Company, US Patent 4,780,511, 25 Oct 1988
22. Crivello JV, Lee JL (1989) Alkoxy-substituted diaryliodonium salt cationic photoinitiators. J Polym Sci Pol Chem 27:3951–3968
23. Crivello JV, Lee JL (1990) Synthesis, characterization, and photoinitiated cationic polymerization of silicon-containing epoxy resins. J Polym Sci Pol Chem 28:479–503
24. Melisaris AP, Renyi W, Pang TH (2000) Liquid, radiation-curable composition, especially for producing flexible cured articles by stereolithography. Vantico Inc., US Patent 6,136,497, 24 Oct 2000
25. Pang TH, Melisaris AP, Renyi W, Fong JW (2000) Liquid radiation-curable composition especially for producing cured articles by stereolithography having high heat deflection temperatures. Ciba Specialty Chemicals Corp., US Patent 6,100,007, 8 Aug 2000
26. Steinmann B, Wolf JP, Schulthess A, Hunziker M (1995) Photosensitive compositions. Ciba-Geigy Corporation, US Patent 5,476,748, 19 Dec 1995
27. Steinmann B, Schulthess A (1999) Liquid, radiation-curable composition, especially for stereolithography. Ciba Specialty Chemicals Corp., US Patent 5,972,563, 26 Oct 1999
28. Crivello JV, Lee JL, Conlon DA (1983) Photoinitiated cationic polymerization with multifunctional vinyl ether monomers. J Radiat Cur 10(1):6–13
29. Decker C, Decker D (1994) Kinetic and mechanistic study of the UV-curing of vinyl ether based systems. Proc Rad Tech Conf, Orlando, vol I, p 602
30. Sperling LH (1981) Interpenetrating polymer networks and related materials. Plenum, New York
31. Decker C, Viet TNT, Decker D, Weber-Koehl E (2001) UV-radiation curing of acrylate/epoxide systems. Polymer 42:5531–5541

32. Sperling LH, Mishra V (1996) In: Salomone JC (ed) Polymer materials encyclopedia, vol 5. CRC, New York, p P3292
33. Chen M, Chen Q, Xiao S, Hong X (2001) Mechanism and application of hybrid UV curing system. Photogr Sci Photochem 19(3):208–216
34. Decker C (1996) Photoinitiated crosslinking polymerization. Prog Polym Sci 21:593–650
35. Decker C, Xuan HL, Viet TNT (1996) Photocrosslinking of functionalized rubber. III. Polymerization of multifunctional monomers in epoxidized liquid natural rubber. J Polym Sci Pol Chem 34:1771–1781
36. Perkins WC (1981) New developments in photo-induced cationic polymerization. J Radiat Cur 8(1):16
37. Renap K, Kruth JP (1995) Recoating issues in stereolithography. Rapid Prototyping J 1(3):4–16
38. Lynn-Charney CM, Rosen DW (2000) Accuracy models and their use in stereolithography process planning. Rapid Prototyping J 6(2):77–86
39. West AP (1999) A decision support system for fabrication process planning in stereolithography. Masters Thesis, Georgia Institute of Technology
40. 3D Systems web page: http://www.3dsystems.com
41. 3D Systems, Inc, (1996) AccuMax™ toolkit user guide. 3D Systems, Valencia
42. Yi F, Wu J, Xian D (1993) LIGA technique for microstructure fabrication. Microfabrication Technol 4:1
43. Ikuta K, Hirowatari K (1993) Real three dimensional microfabrication using stereolithography and metal molding. Proc. IEEE MEMS, Fort Lauderdale, pp 42–47, 7–10 Feb 1993
44. Suzumori K, Koga A, Haneda R (1994) Microfabrication of integrated FMA's using stereo lithography. Proc. MEMS, Oiso, Japan, 25–28 Jan 1994, pp 136–141
45. Ikuta K, Hirowatari K, Ogata T (1994) Three dimensional micro integrated fluid systems fabricated by micro stereolithography. Proc. IEEE MEMS, Oiso, Japan, 25–28 Jan 1994, pp 1–6,
46. Takagi T, Nakajima N (1994) Architecture combination by microphotoforming process. Proc. IEEE MEMS, pp 211–216
47. Ikuta K, Maruo S, Fujisawa T, Yamada A (1999) Micro concentrator with opto-sense micro reactor for biochemical IC chip family. Proc. MEMS, Orlando, 17–21 Jan 1999, pp 376–380
48. Gardner J, Varadan V, Awadelkarim O (2001) Microsensors MEMS and smart devices. Wiley, New York
49. Ikuta K, Ogata T, Tsubio M, Kojima S (1996) Development of mass productive microstereolithography. Proc. MEMS, San Diego, pp 301–305, 11–15 Feb 1996
50. Ikuta K, Maruo S, Fujisawa T, Fukaya Y (1998) Chemical IC chip for dynamical control of protein synthesis. Proc Int Symp Micromechatronics and Human Science, Nagoya, Japan, 25–28 Nov 1998, pp 249–254
51. Dudley D, Duncan W, Slaughter J (2003) Emerging Digital Micromirror Device (DMD) applications. Proc. SPIE, Vol. 4985, San Jose, pp 14–25, 28–29 Jan 2003
52. Bertsch A, Zissi S, Jezequel J, Corbel S, Andre J (1997) Microstereolithography using liquid crystal display as dynamic mask-generator. Microsyst Technol 3(2):42–47
53. Monneret S, Loubere V, Corbel S (1999) Microstereolithography using dynamic mask generator and a non-coherent visible light source. Proc SPIE 3680:553–561
54. Chatwin C, Farsari M, Huang S, Heywood M, Birch P, Young R, Richardson J (1998) UV microstereolithography system that uses spatial light modulator technology. Appl Opt 37 (32):7514–7522
55. Bertsch A, Bernhard P, Vogt C, Renaud P (2000) Rapid prototyping of small size objects. Rapid Prototyping J 6(4):259–266
56. Hadipoespito G, Yang Y, Choi H, Ning G, Li X (2003) Digital Micromirror device based microstereolithography for micro structures of transparent photopolymer and nanocomposites. Proceedings of the 14th Solid Freeform Fabrication Symposium, Austin, pp 13–24

57. Limaye A, Rosen DW (2004) Quantifying dimensional accuracy of a mask projection micro stereolithography system. Proc. Solid Freeform Fabrication Symposium, Austin, Aug 2–4
58. Sun C, Fang N, Wu D, Zhang X (2005) Projection micro-stereolithography using digital micro-mirror dynamic mask. Sensor Actuat A-Phys 121:113–120
59. Varadan VK, Jiang S, Varadan VV (2001) Microstereolithography and other fabrication techniques for 3D MEMS. Wiley, Chichester
60. V-Flash Modeler, www.modelin3d.com
61. Limaye A, Rosen DW (2006) Compensation zone approach to avoid Z errors in Mask Projection Stereolithography builds. Rapid Prototyping J 12(5):283–291
62. Swanson WK, Kremer SD (1975) Three dimensional systems. US Patent 4078229, filed 27 Jan 1975
63. Maruo S, Nakamura O, Kawata S (1997) Three-dimensional microfabrication with two-photon-absorbed photopolymerization. Opt Lett 22(2):132–134
64. Kawata S, Sun H, Tanaka T, Takada K (2001) Finer features for functional microdevices. Nature 412:697–698
65. Miwa M, Juodkazis S, Kawakami T, Matsuo S, Misawa H (2001) Femtosecond two-photon stereolithography. Appl Phys A-Mater 73:561–566
66. Sun H, Kawakami T, Xu Y, Ye J, Matsuo S, Misawa H, Miwa M, Kaneko R (2000) Real three-dimensional microstructures fabricated by photopolymerization of resins through two-photon absorption. Opt Lett 25(15):1110–1112
67. Albota M (1998) Design of organic molecules with large two-photon absorption cross sections. Science 281:1653–1656
68. Cumpston B, Ananthavel S, Barlow S, Dyer D, Ehrlich J, Erskine L, Heikal A, Kuebler S, Lee I, Mc-Cord Maughon D, Qin J, Rockel H, Rumi M, Wu X, Marder S, Perry J (1999) Two photon polymerization for three dimensional optical data storage and microfabrication. Nature 398:51–54

Powder Bed Fusion Processes

5

5.1 Introduction

Powder bed fusion (PBF) processes were among the first commercialized AM processes. Developed at the University of Texas at Austin, USA, selective laser sintering (SLS) was the first commercialized PBF process. Its basic method of operation is schematically shown in Fig. 5.1, and all other PBF processes modify this basic approach in one or more ways to enhance machine productivity, enable different materials to be processed, and/or to avoid specific patented features.

All PBF processes share a basic set of characteristics. These include one or more thermal sources for inducing fusion between powder particles, a method for controlling powder fusion to a prescribed region of each layer, and mechanisms for adding and smoothing powder layers. The most common thermal sources for PBF are lasers. PBF processes which utilize lasers are known as laser sintering (LS) machines. Since polymer laser sintering (pLS) machines and metal laser sintering (mLS) machines are significantly different from each other, we will address each separately. In addition, as electron beam and other thermal sources require significantly different machine architectures than laser sintering machines, non-laser thermal sources will be addressed separately from laser sources at the end of the chapter.

LS processes were originally developed to produce plastic prototypes using a point-wise laser scanning technique. This approach was subsequently extended to metal and ceramic powders; additional thermal sources are now utilized; and variants for layer-wise fusion of powdered materials are being commercially introduced. As a result, PBF processes are widely used worldwide, have a broad range of materials (including polymers, metals, ceramics, and composites) which can be utilized, and are increasingly being used for direct manufacturing of end-use products, as the material properties are comparable to many engineering-grade polymers, metals, and ceramics.

In order to provide a baseline description of powder fusion processes, pLS will be described as the paradigm approach to which the other PBF processes will be

I. Gibson et al., *Additive Manufacturing Technologies*,
DOI 10.1007/978-1-4939-2113-3_5

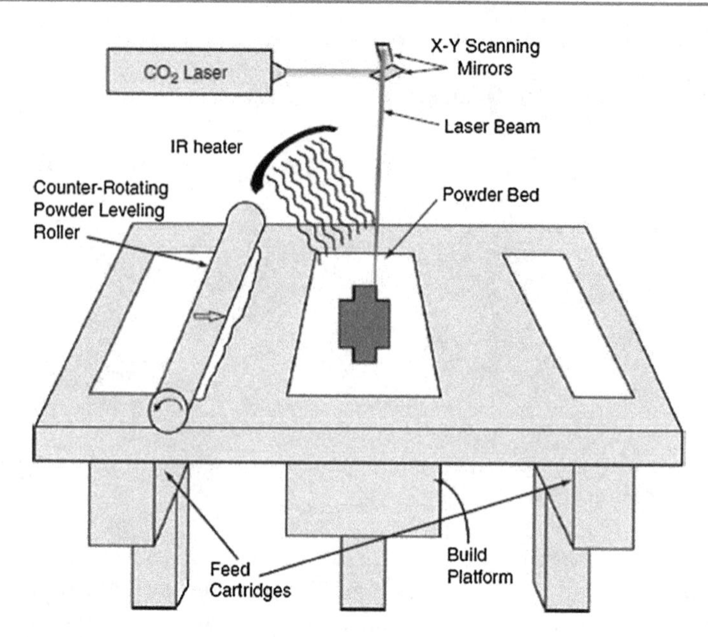

Fig. 5.1 Schematic of the selective laser sintering process

compared. As shown in Fig. 5.1, pLS fuses thin layers of powder (typically 0.075–0.1 mm thick) which have been spread across the build area using a counter-rotating powder leveling roller. The part building process takes place inside an enclosed chamber filled with nitrogen gas to minimize oxidation and degradation of the powdered material. The powder in the build platform is maintained at an elevated temperature just below the melting point and/or glass transition temperature of the powdered material. Infrared heaters are placed above the build platform to maintain an elevated temperature around the part being formed, as well as above the feed cartridges to preheat the powder prior to spreading over the build area. In some cases, the build platform is also heated using resistive heaters around the build platform. This preheating of powder and maintenance of an elevated, uniform temperature within the build platform is necessary to minimize the laser power requirements of the process (with preheating, less laser energy is required for fusion) and to prevent warping of the part during the build due to nonuniform thermal expansion and contraction (resulting in curling).

Once an appropriate powder layer has been formed and preheated, a focused CO_2 laser beam is directed onto the powder bed and is moved using galvanometers in such a way that it thermally fuses the material to form the slice cross section. Surrounding powder remains loose and serves as support for subsequent layers, thus eliminating the need for the secondary supports which are necessary for vat photopolymerization processes. After completing a layer, the build platform is lowered by one layer thickness and a new layer of powder is laid and leveled using the counter-rotating roller. The beam scans the subsequent slice cross section. This process repeats until the complete part is built. A cool-down period is typically required to allow the parts to uniformly come to a low-enough temperature that they can be handled and exposed to ambient temperature and atmosphere. If the parts and/or powder bed are prematurely exposed to ambient temperature and atmosphere, the powders may degrade in the presence of oxygen and parts may warp due to uneven thermal contraction. Finally, the parts are removed from the powder bed, loose powder is cleaned off the parts, and further finishing operations, if necessary, are performed.

5.2 Materials

In principle, all materials that can be melted and resolidified can be used in PBF processes. A brief survey of materials processed using PBF processes will be given here. More details can be found in subsequent sections.

5.2.1 Polymers and Composites

Thermoplastic materials are well-suited for powder bed processing because of their relatively low melting temperatures, low thermal conductivities, and low tendency for balling. Polymers in general can be classified as either a thermoplastic or a thermoset polymer. Thermoset polymers are typically not processed using PBF into parts, since PBF typically operates by melting particles to fabricate part cross sections, but thermosets degrade, but do not melt, as their temperature is increased. Thermoplastics can be classified further in terms of their crystallinity. Amorphous polymers have a random molecular structure, with polymer chains randomly intertwined. In contrast, crystalline polymers have a regular molecular structure, but this is uncommon. Much more common are semi-crystalline polymers which have regions of regular structure, called crystallites. Amorphous polymers melt over a fairly wide range of temperatures. As the crystallinity of a polymer increases, however, its melting characteristics tend to become more centered around a well defined melting point.

At present, the most common material used in PBF is polyamide, a thermoplastic polymer, commonly known in the US as nylon. Most polyamides have fairly high crystallinity and are classified as semi-crystalline materials. They have distinct melting points that enable them to be processed reliably. A given amount of laser energy will melt a certain amount of powder; the melted powder fuses and cools quickly, forming part of a cross section. In contrast, amorphous polymers tend to soften and melt over a broad temperature range and not form well defined solidified features. In pLS, amorphous polymers tend to sinter into highly porous shapes, whereas crystalline polymers are typically processed using full melting, which result in higher densities. Polyamide 11 and polyamide 12 are commercially available, where the number designates the number of carbon atoms that are provided by one of the monomers that is reacted to produce polyamide. However, crystalline polymers exhibit greater shrinkage compared to amorphous materials and are more susceptible to curling and distortion and thus require more uniform temperature control. Mechanical properties of pLS parts produced using polyamide powders approach those of injection molded thermoplastic parts, but with significantly reduced elongation and unique microstructures.

Polystyrene-based materials with low residual ash content are particularly suitable for making sacrificial patterns for investment casting using pLS. Interestingly, polystyrene is an amorphous polymer, but is a successful example material due to its intended application. Porosity in an investment casting pattern aids in melting out the pattern after the ceramic shell is created. Polystyrene parts intended for precision investment casting applications should be sealed to prevent ceramic material seeping in and to achieve a smooth surface finish.

Elastomeric thermoplastic polymers are available for producing highly flexible parts with rubber-like characteristics. These elastomers have good resistance to degradation at elevated temperatures and are resistant to chemicals like gasoline and automotive coolants. Elastomeric materials can be used to produce gaskets, industrial seals, shoe soles, and other components.

Additional polymers that are commercially available include flame-retardant polyamide and polyaryletherketone (known as PAEK or PEEK). Both 3D Systems and EOS GmbH offer most of the materials listed in this section.

Researchers have investigated quite a few polymers for biomedical applications. Several types of biocompatible and biodegradable polymers have been processed using pLS, including polycaprolactone (PCL), polylactide (PLA), and poly-L-lactide (PLLA). Composite materials consisting of PCL and ceramic particles, including hydroxyapatite and calcium silicate, have also been investigated for the fabrication of bone replacement tissue scaffolds.

In addition to neat polymers, polymers in PBF can have fillers that enhance their mechanical properties. For example, the Duraform material from 3D Systems is offered as Duraform PA, which is polyamide 12, as well as Duraform GF, which is polyamide 12 filled with small glass beads. The glass additive enhances the material's stiffness significantly, but also causes its ductility to be reduced, compared to polyamide materials without fillers. Additionally, EOS GmbH offers aluminum particle, carbon fiber, and their own glass bead filled polyamide materials.

5.2.2 Metals and Composites

A wide range of metals has been processed using PBF. Generally, any metal that can be welded is considered to be a good candidate for PBF processing. Several types of steels, typically stainless and tool steels, titanium and its alloys, nickel-base alloys, some aluminum alloys, and cobalt-chrome have been processed and are commercially available in some form. Additionally, some companies now offer PBF of precious metals, such as silver and gold.

Historically, a number of proprietary metal powders (either thermoplastic binder-coated or binder mixed) were developed before modern mLS machines were available. RapidSteel was one of the first metal/binder systems, developed by DTM Corp. The first version of RapidSteel was available in 1996 and consisted of a thermoplastic binder coated 1080 carbon steel powder with copper as the infiltrant. Parts produced using RapidSteel were debinded (350–450 °C), sintered (around 1,000 °C), and finally infiltrated with Cu (1,120 °C) to produce a final part with approximately 60 % low carbon steel and 40 % Cu. This is an example of liquid phase sintering which will be described in the next section. Subsequently, RapidSteel 2.0 powder was introduced in 1998 for producing functional tooling, parts, and mold inserts for injection molding. It was a dry blend of 316 stainless steel powder impact milled with thermoplastic and thermoset organic binders with an average particle size of 33 μm. After green part fabrication, the part was debinded and sintered in a hydrogen-rich atmosphere. The bronze infiltrant was introduced in a separate furnace run to produce a 50 % steel and 50 % bronze composite. RapidSteel 2.0 was structurally more stable than the original RapidSteel material because the bronze infiltration temperature was less than the sintering temperature of the stainless steel powder. A subsequent material development was LaserForm ST-100, which had a broader particle size range, with fine particles not being screened out. These fine particles allowed ST-100 particles to be furnace sintered at a lower temperature than RapidSteel 2.0, making it possible to carry out sintering and infiltration in a single furnace run. In addition to the above, H13 and A6 tool steel powders with a polymer binder can also be used for tooling applications. The furnace processing operations (sintering and infiltration) must be carefully designed with appropriate choices of temperature, heating and cooling rates, furnace atmosphere pressure, amount of infiltrant, and other factors, to prevent excessive part distortion. After infiltration, the part is finish machined as needed. These issues are further explored in the post-processing chapter.

Several proprietary metal powders were marketed by EOS for their M250 Xtended metal platforms, prior to the introduction of modern mLS machines. These included liquid-phase sintered bronze-based powders, and steel-based powders and other proprietary alloys (all without polymer binders). These were suitable for producing tools and inserts for injection molding of plastics. Parts made from these powders were often infiltrated with epoxy to improve the surface finish and seal porosity in the parts. Proprietary nickel-based powders for direct tooling applications and Cu-based powders for parts requiring high thermal and electrical conductivities were also available. All of these materials have been successfully used by many organizations; however, the more recent introduction of mLS and electron beam melting (EBM) technology has made these alloys obsolete, as engineering-grade alloys are now able to be processed using a number of manufacturers' machines.

As mentioned, titanium alloys, numerous steel alloys, nickel-based super alloys, CoCrMo, and more are widely available from numerous manufacturers. It should be noted that alloys that crack under high solidification rates are not good candidates for mLS. Due to the high solidification rates in mLS, the crystal structures produced and mechanical properties are different than those for other manufacturing processes. These structures may be metastable, and the heat treatment recipes needed to produce standard microstructures may be different. As mLS and EBM processes advance, the types of metal alloys which are commonly utilized will grow and new alloys specifically tailored for PBF production will be developed.

5.2.3 Ceramics and Ceramic Composites

Ceramic materials are generally described as compounds that consist of metal-oxides, carbides, and nitrides and their combinations. Several ceramic materials are available commercially including aluminum oxide and titanium oxide. Commercial machines were developed by a company called Phenix Systems in France, which was acquired by 3D Systems in 2013. 3D Systems also says it offers cermets, which are metal-ceramic composites.

Ceramics and metal-ceramic composites have been demonstrated in research. Typically, ceramic precipitates form through reactions occurring during the sintering process. One example is the processing of aluminum in a nitrogen atmosphere, which forms an aluminum matrix with small regions of AlN interspersed throughout. This process is called chemically induced sintering and is described further in the next section.

Biocompatible materials have been developed for specific applications. For example, calcium hydroxyapatite, a material very similar to human bone, has been processed using pLS for medical applications.

5.3 Powder Fusion Mechanisms

Since the introduction of LS, each new PBF technology developer has introduced competing terminology to describe the mechanism by which fusion occurs, with variants of "sintering" and "melting" being the most popular. However, the use of a single word to describe the powder fusion mechanism is inherently problematic as multiple mechanisms are possible. There are four different fusion mechanisms which are present in PBF processes [1]. These include solid-state sintering, chemically induced binding, liquid-phase sintering (LPS), and full melting. Most commercial processes utilize primarily LPS and melting. A brief description of each of these mechanisms and their relevance to AM follows.

5.3.1 Solid-State Sintering

The use of the word sintering to describe powder fusion as a result of thermal processing predates the advent of AM. Sintering, in its classical sense, indicates the fusion of powder particles without melting (i.e., in their "solid state") at elevated temperatures. This occurs at temperatures between one half of the absolute melting temperature and the melting temperature. The driving force for solid-state sintering

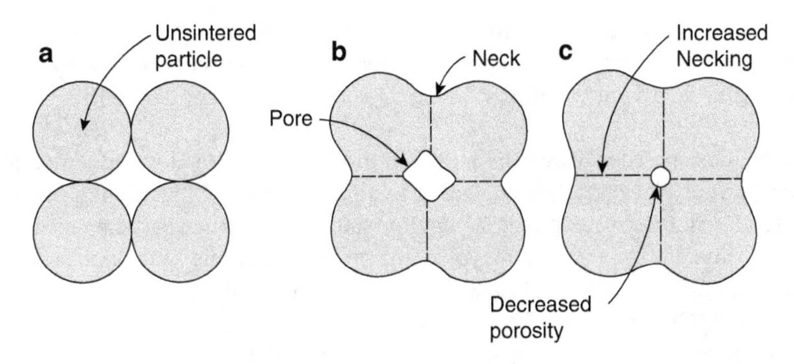

Fig. 5.2 Solid-state sintering. (**a**) Closely packed particles prior to sintering. (**b**) Particles agglomerate at temperatures above one half of the absolute melting temperature, as they seek to minimize free energy by decreasing surface area. (**c**) As sintering progresses, neck size increases and pore size decreases

is the minimization of total free energy, E_s, of the powder particles. The mechanism for sintering is primarily diffusion between powder particles.

Surface energy E_s is proportional to total particle surface area S_A, through the equation $E_s = \gamma_s \times S_A$ (where γ_s is the surface energy per unit area for a particular material, atmosphere, and temperature). When particles fuse at elevated temperatures (see Fig. 5.2), the total surface area decreases, and thus surface energy decreases.

As the total surface area of the powder bed decreases, the rate of sintering slows. To achieve very low porosity levels, long sintering times or high sintering temperatures are required. The use of external pressure, as is done with hot isostatic pressing, increases the rate of sintering.

As total surface area in a powder bed is a function of particle size, the driving force for sintering is directly related to the surface area to volume ratio for a set of particles. The larger the surface area to volume ratio, the greater the free energy driving force. Thus, smaller particles experience a greater driving force for necking and consolidation, and hence, smaller particles sinter more rapidly and initiate sintering at lower temperature than larger particles.

As diffusion rates exponentially increase with temperature, sintering becomes increasingly rapid as temperatures approach the melting temperature, which can be modeled using a form of the Arrhenius equation. However, even at temperatures approaching the melting temperature, diffusion-induced solid-state sintering is the slowest mechanism for selectively fusing regions of powder within a PBF process.

For AM, the shorter the time it takes to form a layer, the more economically competitive the process becomes. Thus, the heat source which induces fusion should move rapidly and/or induce fusion quickly to increase build rates. Since the time it takes for fusion by sintering is typically much longer than for fusion by melting, few AM processes use sintering as a primary fusion mechanism.

Sintering, however, is still important in most thermal powder processes, even if sintering is not the primary fusion mechanism. There are three secondary ways in which sintering affects a build.

1. If the loose powder within the build platform is held at an elevated temperature, the powder bed particles will begin to sinter to one another. This is typically considered a negative effect, as agglomeration of powder particles means that each time the powder is recycled the average particle size increases. This changes the spreading and melting characteristics of the powder each time it is recycled. One positive effect of loose powder sintering, however, is that the powder bed will gain a degree of tensile and compressive strength, thus helping to minimize part curling.
2. As a part is being formed in the build platform, thermally induced fusing of the desired cross-sectional geometry causes that region of the powder bed to become much hotter than the surrounding loose powder. If melting is the dominant fusion mechanism (as is typically the case) then the just-formed part cross section will be quite hot. As a result, the loose powder bed immediately surrounding the fused region heats up considerably, due to conduction from the part being formed. This region of powder may remain at an elevated temperature for a long time (many hours) depending upon the size of the part being built, the heater and temperature settings in the process, and the thermal conductivity of the powder bed. Thus, there is sufficient time and energy for the powder immediately next to the part being built to fuse significantly due to solid-state sintering, both to itself and to the part. This results in "part growth," where the originally scanned part grows a "skin" of increasing thickness the longer the powder bed is maintained at an elevated temperature. This phenomenon can be seen in Fig. 5.3 as unmolten particles fused to the edge of a part. For many materials, the skin formed on the part goes from high density, low porosity near the originally scanned region to lower density, higher porosity further from the part. This part growth can be compensated in the build planning stage by offsetting the laser beam to compensate for part growth or by offsetting the surface of the STL model. In addition, different post-processing methods will remove this skin to a different degree. Thus, the dimensional repeatability of the final part is highly dependent upon effectively compensating for and controlling this part growth. Performing repeatable post-processing to remove the same amount of the skin for every part is thus quite important.
3. Rapid fusion of a powder bed using a laser or other heat source makes it difficult to achieve 100 % dense, porosity-free parts. Thus, a feature of many parts built using PBF techniques (especially for polymers) is distributed porosity throughout the part. This is typically detrimental to the intended part properties. However, if the part is held at an elevated temperature after scanning, solid-state sintering combined with other high-temperature phenomena (such as grain growth in metals) causes the % porosity in the part to decrease. Since lower layers are maintained at an elevated temperature while additional layers are added, this can result in lower regions of a part being denser than upper regions

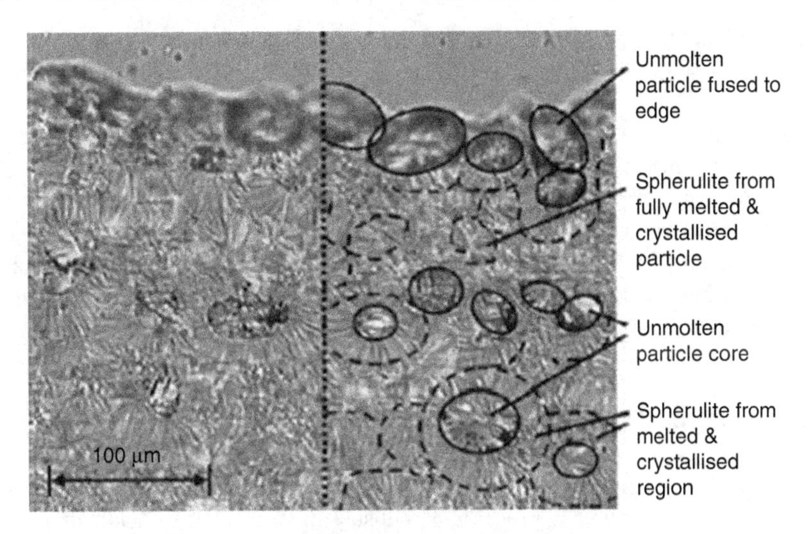

Fig. 5.3 Typical pLS microstructure for nylon polyamide (Materials Science & Engineering. A. Structural Materials: Properties, Microstructure and Processing by Zarringhalam, H., Hopkinson, N., Kamperman, N.F., de Vlieger, J.J. Copyright 2006 by Elsevier Science & Technology Journals. Reproduced with permission of Elsevier Science & Technology Journals in the format Textbook via Copyright Clearance Center) [5]

of a part. This uneven porosity can be controlled, to some extent, by carefully controlling the part bed temperature, cooling rate and other parameters. EBM, in particular, often makes use of the positive aspects of elevated-temperature solid-state sintering and grain growth by purposefully maintaining the metal parts that are being built at a high enough temperature that diffusion and grain growth cause the parts being built to reach 100 % density.

5.3.2 Chemically Induced Sintering

Chemically induced sintering involves the use of thermally activated chemical reactions between two types of powders or between powders and atmospheric gases to form a by-product which binds the powders together. This fusion mechanism is primarily utilized for ceramic materials. Examples of reactions between powders and atmospheric gases include: laser processing of SiC in the presence of oxygen, whereby SiO_2 forms and binds together a composite of SiC and SiO_2; laser processing of ZrB_2 in the presence of oxygen, whereby ZrO_2 forms and binds together a composite of ZrB_2 and ZrO_2; and laser processing of Al in the presence of N_2, whereby AlN forms and binds together the Al and AlN particles.

For chemically induced sintering between powders, various research groups have demonstrated that mixtures of high-temperature structural ceramic and/or intermetallic precursor materials can be made to react using a laser. In this case, raw materials which exothermically react to form the desired by-product are pre-mixed and heated using a laser. By adding chemical reaction energy to the laser energy, high-melting-temperature structures can be created at relatively low laser energies.

One common characteristic of chemically induced sintering is part porosity. As a result, post-process infiltration or high-temperature furnace sintering to higher densities is often needed to achieve properties that are useful for most applications. This post-process infiltration may involve other reactive elements, forming new chemical compounds after infiltration. The cost and time associated with post-processing have limited the adoption of chemically induced sintering in commercial machines.

5.3.3 LPS and Partial Melting

LPS is arguably the most versatile mechanism for PBF. LPS is a term used extensively in the powder processing industry to refer to the fusion of powder particles when a portion of constituents within a collection of powder particles become molten, while other portions remain solid. In LPS, the molten constituents act as the glue which binds the solid particles together. As a result, high-temperature particles can be bound together without needing to melt or sinter those particles directly. LPS is used in traditional powder metallurgy to form, for instance, cemented carbide cutting tools where Co is used as the lower-melting-point constituent to glue together particles of WC.

There are many ways in which LPS can be utilized as a fusion mechanism in AM processes. For purposes of clarity, the classification proposed by Kruth et al. [1] has formed the basis for the distinctions discussed in the following section and shown in Fig. 5.4.

5.3.3.1 Distinct Binder and Structural Materials

In many LPS situations, there is a clear distinction between the binding material and the structural material. The binding and structural material can be combined in three different ways: as separate particles, as composite particles, or as coated particles.

Separate Particles

A simple, well-mixed combination of binder and structural powder particles is sufficient in many cases for LPS. In cases where the structural material has the dominant properties desired in the final structure, it is advantageous for the binder material to be smaller in particle size than the structural material. This enables more efficient packing in the powder bed and less shrinkage and lower porosity after binding. The dispersion of smaller-particle-size binder particles around structural particles also helps the binder flow into the gaps between the structural particles more effectively, thus resulting in better binding of the structural particles. This is often true when, for instance, LS is used to process steel powder with a polymer binder (as discussed more fully in Sect. 5.3.5). This is also true when metal-metal

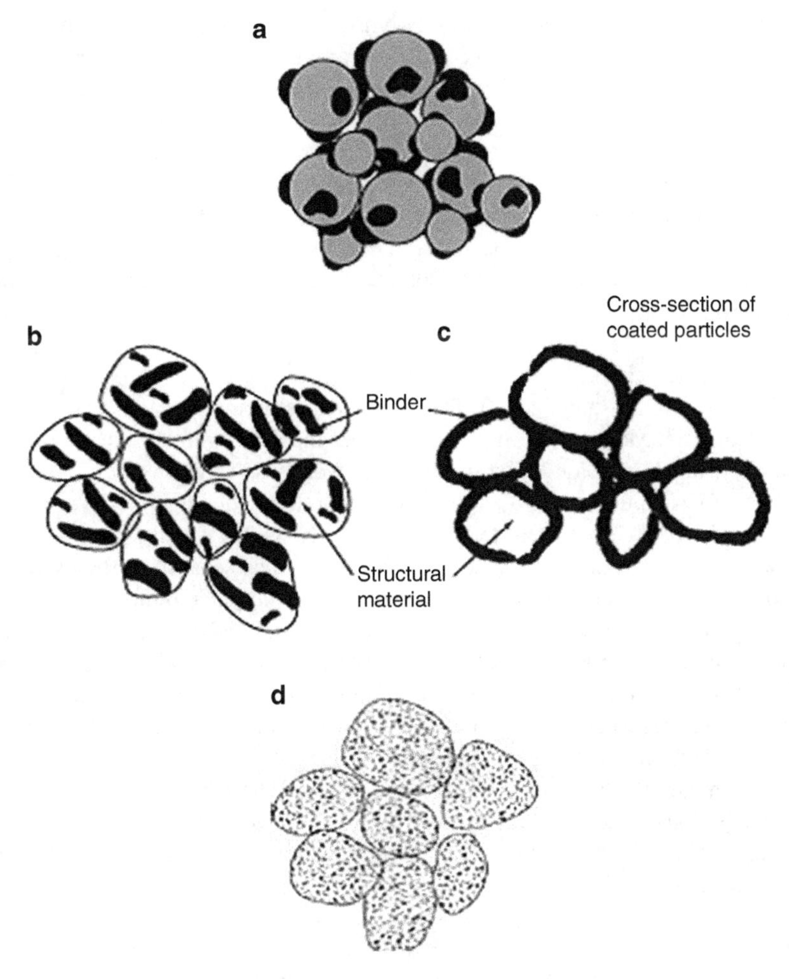

Fig. 5.4 Liquid phase sintering variations used in powder bed fusion processing: (**a**) separate particles, (**b**) composite particles, (**c**) coated particles, and (**d**) indistinct mixtures. *Darker* regions represent the lower-melting-temperature binder material. *Lighter* regions represent the high-melting-temperature structural material. For indistinct mixtures, microstructural alloying eliminates distinct binder and structural regions

mixtures and metal-ceramic mixtures are directly processed without the use of a polymer binder.

In the case of LPS of separate particles, the heat source passes by quickly, and there is typically insufficient time for the molten binder to flow and surface tension to draw the particles together prior to resolidification of the binder unless the binder has a particularly low viscosity. Thus, composite structures formed from separate particles typically are quite porous. This is often the intent for parts made from separate particles, which are then post-processed in a furnace to achieve the final part properties. Parts held together by polymer binders which require further post-processing (e.g., to lower or fill the porosity) are termed as "green" parts.

In some cases, the density of the binder and structural material are quite different. As a result, the binder and structural material may separate during handling. In addition, some powdered materials are most economically manufactured at particle sizes that are too small for effective powder dispensing and leveling (see Sect. 5.5). In either case, it may be beneficial for the structural and/or binder particles to be bound together into larger particle agglomerates. By doing so, composite powder particles made up of both binder and structural material are formed.

Composite Particles

Composite particles contain both the binder and structural material within each powder particle. Mechanical alloying of binder and structural particles or grinding of cast, extruded or molded mixtures into a powder results in powder particles that are made up of binder and structural materials agglomerated together. The benefits of composite particles are that they typically form higher density green parts and typically have better surface finish after processing than separate particles [1].

Composite particles can consist of mixtures of polymer binders with higher melting point polymer, metal, or ceramic structural materials; or metal binders with higher melting point metal or ceramic structural materials. In all cases, the binder and structural portions of each particle, if viewed under a microscope, are distinct from each other and clearly discernable. The most common commercially available composite particle used in PBF processes is glass-filled nylon. In this case, the structural material (glass beads) is used to enhance the properties of the binding material (nylon) rather than the typical use of LPS where the binder is simply a necessary glue to help hold the structural material together in a useful geometric form.

Coated Particles

In some cases, a composite formed by coating structural particles with a binder material is more effective than random agglomerations of binder and structural materials. These coated particles can have several advantages; including better absorption of laser energy, more effective binding of the structural particles, and better flow properties.

When composite particles or separate particles are processed, the random distribution of the constituents means that impinging heat energy, such as laser radiation, will be absorbed by whichever constituent has the highest absorptivity and/or most direct "line-of-sight" to the impinging energy. If the structural materials have a higher absorptivity, a greater amount of energy will be absorbed in the structural particles. If the rate of heating of the structural particles significantly exceeds the rate of conduction to the binder particles, the higher-melting-temperature structural materials may melt prior to the lower-melting-temperature binder materials. As a result, the anticipated microstructure of the processed material will differ significantly from one where the binder had melted and the structural material had remained solid. This may, in some instances, be desirable, but is typically not the intent when formulating a binder/structural material combination. Coated particles can help overcome the structural material heating problem associated with random constituent mixtures and agglomerates. If a structural particle is coated with the binder material then the impinging energy must first pass through the coating before affecting the structural material. As melting of the binder and not the structural material is the objective of LPS, this helps ensure that the proper constituent melts.

Other benefits of coated particles exist. Since there is a direct correlation between the speed of the impinging energy in AM processing and the build rate, it is desirable for the binder to be molten for only a very short period of time. If the binder is present at the surfaces of the structural material, this is the most effective location for gluing adjacent particles together. If the binder is randomly mixed with the structural materials, and/or the binder's viscosity is too high to flow to the contact points during the short time it is molten, then the binder will not be as effective. As a result, the binder % content required for effective fusion of coated particles is usually less than the binder content required for effective fusion of randomly mixed particles.

Many structural metal powders are spherical. Spherical powders are easier to deposit and smooth using powder spreading techniques. Coated particles retain the spherical nature of the underlying particle shape, and thus can be easier to handle and spread.

5.3.3.2 Indistinct Binder and Structural Materials

In polymers, due to their low thermal conductivity, it is possible to melt smaller powder particles and the outer regions of larger powder particles without melting the entire structure (see Fig. 5.3). Whether to more properly label this phenomenon LPS or just "partial melting" is a matter of debate. Also with polymers, fusion can occur between polymer particles above their glass transition temperature, but below their melting temperature. Similarly, amorphous polymers have no distinct melting point, becoming less viscous the higher the temperature goes above the glass transition temperature. As a result, in each of these cases, there can be fusion between polymer powder particles in cases where there is partial but not full melting, which falls within the historical scope of the term "liquid phase sintering."

In metals, LPS can occur between particles where no distinct binder or structural materials are present. This is possible during partial melting of a single particle type, or when an alloyed structure has lower-melting-temperature constituents. For noneutectic alloy compositions, melting occurs between the liquidus and solidus temperature of the alloy, where only a portion of the alloy will melt when the temperature is maintained in this range. Regions of the alloy with higher concentrations of the lower-melting-temperature constituent(s) will melt first. As a result, it is commonly observed that many metal alloys can be processed in such a way that only a portion of the alloy melts when an appropriate energy level is applied. This type of LPS of metal alloys was the method used in the early EOS M250 direct metal laser sintering (DMLS) machines. Subsequent mLS commercialized processes are all designed to fully melt the metal alloys they process.

5.3.4 Full Melting

Full melting is the mechanism most commonly associated with PBF processing of engineering metal alloys and semi-crystalline polymers. In these materials, the entire region of material subjected to impinging heat energy is melted to a depth exceeding the layer thickness. Thermal energy of subsequent scans of a laser or electron beam (next to or above the just-scanned area) is typically sufficient to re-melt a portion of the previously solidified solid structure; and thus, this type of full melting is very effective at creating well-bonded, high-density structures from engineering metals and polymers.

The most common material used in PBF processing is nylon polyamide. As a semi-crystalline material, it has a distinct melting point. In order to produce parts with the highest possible strength, these materials should be fully melted during processing. However, elevated temperatures associated with full melting result in part growth and thus, for practical purposes, many accuracy versus strength optimization studies result in parameters which are at the threshold between full melting and LPS, as can be seen from Fig. 5.3.

For metal PBF processes, the engineering alloys that are utilized in these machines (Ti, Stainless Steel, CoCr, etc.) are typically fully melted. The rapid melting and solidification of these metal alloys results in unique properties that are distinct from, and can sometime be more desirable than, cast or wrought parts made from identical alloys.

Figure 5.5 summarizes the various binding mechanisms which are utilized in PBF processes. Regardless of whether a technology is known as "Selective Laser Sintering," "Selective Laser Melting," "Direct Metal Laser Sintering," "Laser Cusing," "Electron Beam Melting," or some other name, it is possible for any of

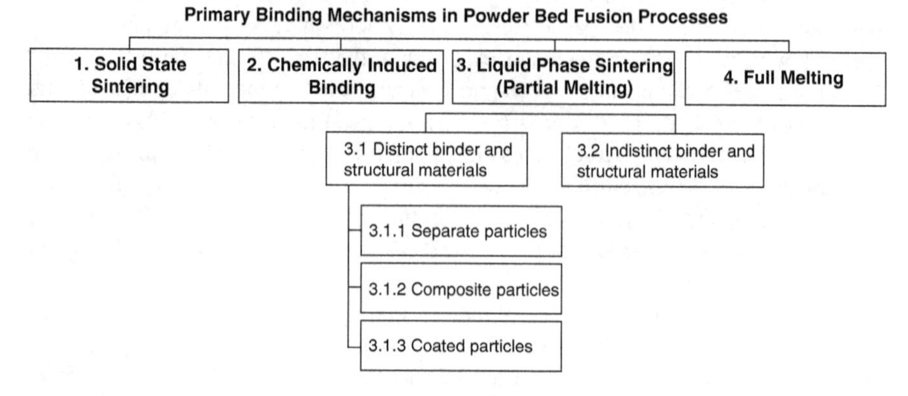

Fig. 5.5 Primary binding mechanisms in powder bed fusion processes (adapted from [1])

these mechanisms to be utilized (and, in fact, often more than one is present) depending upon the powder particle combinations, and energy input utilized to form a part.

5.3.5 Part Fabrication

5.3.5.1 Metal Parts

There are four common approaches for using PBF processes in the creation of complex metal components: full melting, LPS, indirect processing, and pattern methods. In the full melting and LPS (with metal powders) approaches, a metal part is typically usable in the state in which it comes out of the machine, after separation from a build plate.

In indirect processing, a polymer coated metallic powder or a mixture of metallic and polymer powders are used for part construction. Figure 5.6 shows the steps involved in indirect processing of metal powders. During indirect processing, the polymer binder is melted and binds the particles together, and the metal powder remains solid. The metallic powder particles remain largely unaffected by the heat of the laser. The parts produced are generally porous (sometimes exceeding 50 vol. % porosity). The polymer-bound green parts are subsequently furnace processed. Furnace processing occurs in two stages: (1) debinding and (2) infiltration or consolidation. During debinding, the polymer binder is vaporized to remove it from the green part. Typically, the temperature is also raised to the extent that a small degree of necking (sintering) occurs between the metal particles. Subsequently, the remaining porosity is either filled by infiltration of a lower melting point metal to produce a fully dense metallic part, or by further sintering and densification to reduce the part porosity. Infiltration is easier to control, dimensionally, as the overall shrinkage is much less than during consolidation. However, infiltrated structures are always composite in nature whereas consolidated structures can be made up of a single material type.

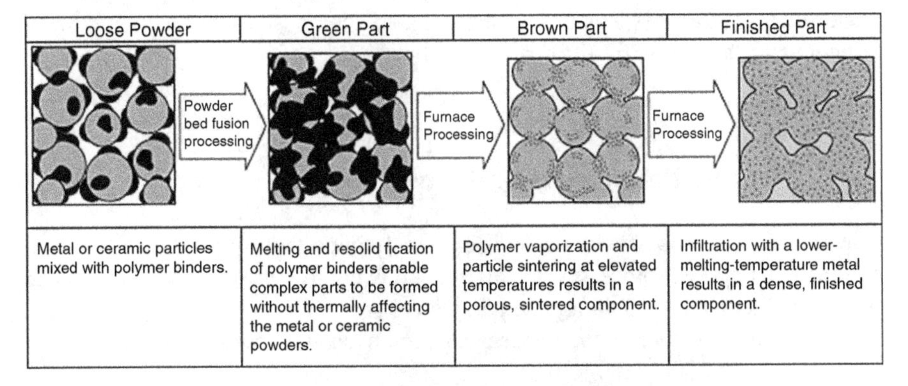

Fig. 5.6 Indirect processing of metal and ceramic powders using PBF

The last approach to metal part creation using PBF is the pattern approach. For the previous three approaches, metal powder is utilized in the PBF process; but in this final approach, the part created in the PBF process is a pattern used to create the metal part. The two most common ways PBF-created parts are utilized as patterns for metal part creation are as investment casting patterns or as sand-casting molds. In the case of investment casting, polystyrene or wax-based powders are used in the machine, and subsequently invested in ceramic during post-processing, and melted out during casting. In the case of sand-casting molds, mixtures of sand and a thermosetting binder are directly processed in the machine to form a sand-casting core, cavity or insert. These molds are then assembled and molten metal is cast into the mold, creating a metal part. Both indirect and pattern-based processes are further discussed in Chap. 18.

5.3.5.2 Ceramic Parts

Similar to metal parts, there are a number of ways that PBF processes are utilized to create ceramic parts. These include direct sintering, chemically induced sintering, indirect processing, and pattern methods. In direct sintering, a high-temperature is maintained in the powder bed and a laser is utilized to accelerate sintering of the powder bed in the prescribed location of each layer. The resultant ceramic parts will be quite porous and thus are often post-processed in a furnace to achieve higher density. This high porosity is also seen in chemically induced sintering of ceramics, as described earlier.

Indirect processing of ceramic powders is identical to indirect processing of metal powders (Fig. 5.6). After debinding, the ceramic brown part is consolidated to reduce porosity or is infiltrated. In the case of infiltration, when metal powders are used as the infiltrant then a ceramic/metal composite structure can be formed. In some cases, such as when creating SiC structures, a polymer binder can be selected, which leaves behind a significant amount of carbon residue within the brown part. Infiltration with molten Si will result in a reaction between the molten Si and the remaining carbon to produce more SiC, thus increasing the overall SiC content and reducing the fraction of metal Si in the final part. These and related approaches have been used to form interesting ceramic-matrix composites and ceramic-metal structures for a number of different applications.

5.4 Process Parameters and Modeling

Use of optimum process parameters is extremely important for producing satisfactory parts using PBF processes. In this section, we will discuss “laser” processing and parameters, but by analogy the parameters and models discussed below could also be applied to other thermal energy sources, such as electron beams or infrared heaters.

5.4.1 Process Parameters

In PBF, process parameters can be lumped into four categories: (1) laser-related parameters (laser power, spot size, pulse duration, pulse frequency, etc.), (2) scan-related parameters (scan speed, scan spacing, and scan pattern), (3) powder-related parameters (particle shape, size, and distribution, powder bed density, layer thickness, material properties, etc.), and (4) temperature-related parameters (powder bed temperature, powder feeder temperature, temperature uniformity, etc.). It should be noted that most of these parameters are strongly interdependent and are mutually interacting. The required laser power, for instance, typically increases with melting point of the material and lower powder bed temperature, and also varies depending upon the absorptivity characteristics of the powder bed, which is influenced by material type and powder shape, size, and packing density.

A typical PBF machine includes two galvanometers (one for the x-axis and one for the y-axis motion). Similar to stereolilthography, scanning often occurs in two modes, contour mode and fill mode, as shown in Fig. 5.7. In contour mode, the outline of the part cross section for a particular layer is scanned. This is typically done for accuracy and surface finish reasons around the perimeter. The rest of the cross section is then scanned using a fill pattern. A common fill pattern is a rastering technique whereby one axis is incrementally moved a laser scan width, and the other axis is continuously swept back and forth across the part being formed. In some cases the fill section is subdivided into strips (where each strip is scanned sequentially and the strip angle is rotated every layer) or squares (with each square being processed separately and randomly). Randomized scanning is sometimes utilized so that there is no preferential direction for residual stresses induced by the scanning. The use of strips or a square-based strategy is primarily for metal parts, whereas a simple raster pattern for the entire part (without subdividing into strips or squares) is typically used for polymers and other low-temperature processing.

In addition to melt pool characteristics, scan pattern and scan strategy can have a profound impact on residual stress accumulation within a part. For instance, if a part

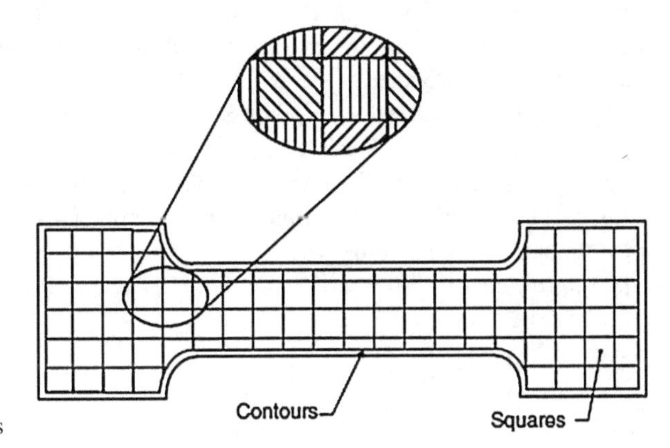

Fig. 5.7 Scan strategies employed in PBF techniques

is moved from one location to another within a machine, the exact laser paths to build the part may change. These laser path changes may cause the part to distort more in one location than another. Thus it is possible for a part to build successfully in one location but not in another location in the same machine due simply to how the scan strategy is applied in different locations.

Powder shape, size, and distribution strongly influence laser absorption characteristics as well as powder bed density, powder bed thermal conductivity, and powder spreading. Finer particles provide greater surface area and absorb laser energy more efficiently than coarser particles. Powder bed temperature, laser power, scan speed, and scan spacing must be balanced to provide the best trade-off between melt pool size, dimensional accuracy, surface finish, build rate, and mechanical properties. The powder bed temperature should be kept uniform and constant to achieve repeatable results. Generally, high-laser-power/high-bed-temperature combinations produce dense parts, but can result in part growth, poor recyclability, and difficulty cleaning parts. On the other hand, low-laser-power/low-bed-temperature combinations produce better dimensional accuracy, but result in lower density parts and a higher tendency for layer delamination. High-laser-power combined with low-part-bed-temperatures result in an increased tendency for nonuniform shrinkage and the build-up of residual stresses, leading to curling of parts.

Laser power, spot size and scan speed, and bed temperature together determine the energy input needed to fuse the powder into a useable part. The longer the laser dwells in a particular location, the deeper the fusion depth and the larger the melt pool diameter. Typical layer thicknesses range from 0.02 to 0.15 mm. Operating at lower laser powers requires the use of lower scan speeds in order to ensure proper particle fusion. Melt pool size is highly dependent upon settings of laser power, scan speed, spot size, and bed temperature. Scan spacing should be selected to ensure a sufficient degree of melt pool overlap between adjacent lines of fused material to ensure robust mechanical properties.

The powder bed density, as governed by powder shape, size, distribution, and spreading mechanism, can strongly influence the part quality. Powder bed densities typically range between 50 and 60 % for most commercially available powders, but may be as low as 30 % for irregular ceramic powders. Generally the higher the powder packing density, the higher the bed thermal conductivity and the better the part mechanical properties.

Most commercialized PBF processes use continuous-wave (CW) lasers. Laser-processing research with pulsed lasers, however, has demonstrated a number of potential benefits over CW lasers. In particular, the tendency of molten metal to form disconnected balls of molten metal, rather than a flat molten region on a powder bed surface, can be partially overcome by pulsed energy. Thus, it is likely that future PBF machines will be commercialized with both CW and pulsed lasers.

5.4.2 Applied Energy Correlations and Scan Patterns

Many common physics, thermodynamics, and heat transfer models are relevant to PBF techniques. In particular, solutions for stationary and moving point-heat-sources in an infinite media and homogenization equations (to estimate, for instance, powder bed thermo-physical properties based upon powder morphology, packing density, etc.) are commonly utilized. The solidification modeling discussed in the directed energy deposition (DED) chapter (Chap. 10) can also be applied to PBF processes. For the purposes of this chapter, a highly simplified model which estimates the energy-input characteristics of PBF processes is introduced and discussed with respect to process optimization for PBF processes.

Melt pool formation and characteristics are fundamentally determined by the total amount of applied energy which is absorbed by the powder bed as the laser beam passes. Both the melt pool size and melt pool depth are a function of absorbed energy density. A simplified energy density equation has been used by numerous investigators as a simple method for correlating input process parameters to the density and strength of produced parts [2]. In their simplified model, applied energy density E_{A} (also known as the Andrews number) can be found using (5.1):

$$E_{\mathrm{A}} = P/(U \times \mathrm{SP}) \tag{5.1}$$

where P is laser power, U is scan velocity, and SP is the scan spacing between parallel scan lines. In this simplified model, applied energy increases with increasing laser power and decreases with increasing velocity and scan spacing. For pLS, typical scan spacing values are ~100 μm, whereas typical laser spot sizes are ~300 μm. Thus, typically every point is scanned by multiple passes of the laser beam.

Although (5.1) does not include powder absorptivity, heat of fusion, laser spot size, bed temperature, or other important characteristics, it provides the simplest analytical approach for optimizing machine performance for a material. For a given material, laser spot size and machine configuration, a series of experiments can be run to determine the minimum applied energy necessary to achieve adequate material fusion for the desired material properties. Subsequently, build speed can be maximized by utilizing the fastest combination of laser power, scan rate, and scan spacing for a particular machine architecture based upon (5.1).

Optimization of build speed using applied energy is reasonably effective for PBF of polymer materials. However, when a molten pool of metal is present on a powder bed, a phenomenon called balling often occurs. When surface tension forces overcome a combination of dynamic fluid, gravitational and adhesion forces, the molten metal will form a ball. The surface energy driving force for metal powders to limit their surface area to volume ratio (which is minimized as a sphere) is much greater than the driving force for polymers, and thus this phenomenon is unimportant for polymers but critically important for metals. An example of balling tendency at various power, P, and scan speed, U, combinations is shown in

Fig. 5.8 [3]. This figure illustrates five typical types of tracks which are formed at various process parameter combinations.

A process map showing regions of power and scan speed combinations which result in each of these track types is shown in Fig. 5.9. As described by Childs et al. tracks of type A were continuous and flat topped or slightly concave. At slightly higher speeds, type B tracks became rounded and sank into the bed. As the speed increased, type C tracks became occasionally broken, although not with the regularity of type D tracks at higher speeds, whose regularly and frequently broken tracks are perfect examples of the balling effect. At even higher speeds, fragile tracks were formed (type E) where the maximum temperatures exceed the solidus temperature but do not reach the liquidus temperature (i.e., partially melted or

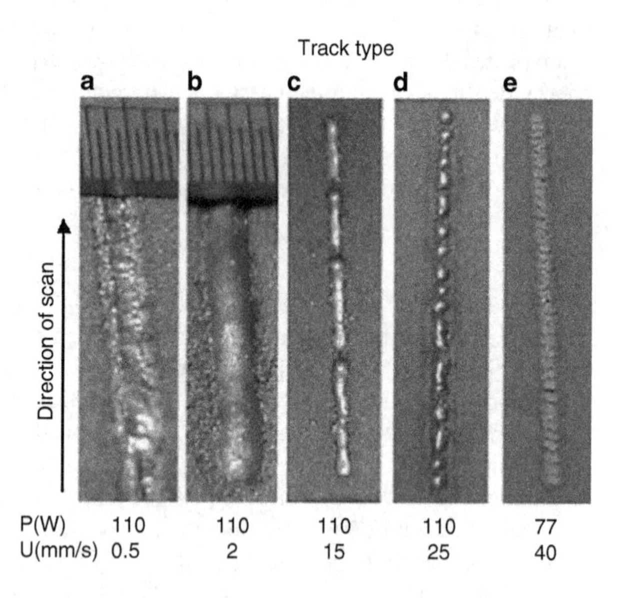

Fig. 5.8 Five examples of test tracks made in −150/+75 μm M2 steel powder in an argon atmosphere with a CO_2 laser beam of 1.1 mm spot size, at similar magnifications (© Professional Engineering Publishing, reproduced from T H C Childs, C Hauser, and M Badrossamay, Proceedings of the Institution of Mechanical Engineers, Part B: Journal of Engineering Manufacture 219 (4), 2005)

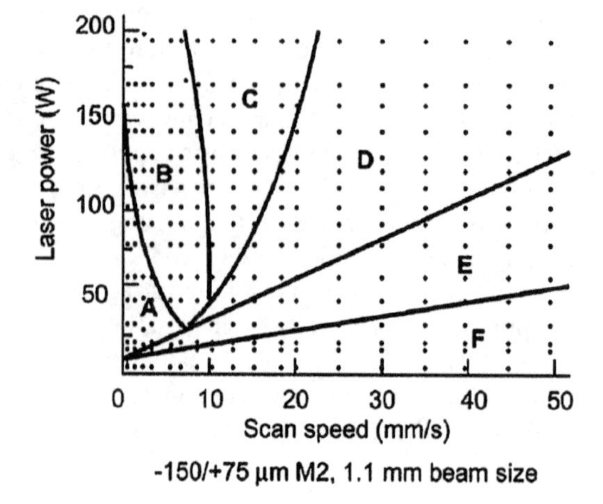

Fig. 5.9 Process map for track types shown in Fig. 5.8 (© Professional Engineering Publishing, reproduced from T H C Childs, C Hauser, and M Badrossamay, Proceedings of the Institution of Mechanical Engineers, Part B: Journal of Engineering Manufacture 219 (4), 2005)

liquid phase sintered tracks). In region F, at the highest speed, lowest power combinations, no melting occurred.

When considering these results, it is clear that build speed optimization for metals is complex, as a simple maximization of scan speed for a particular power and scan spacing based on (5.1) is not possible. However, within process map regions A and B, (5.1) could still be used as a guide for process optimization.

Numerous researchers have investigated residual stresses and distortion in laser PBF processes using analytical and finite element methods. These studies have shown that residual stresses and subsequent part deflection increase with increase in track length. Based on these observations, dividing the scan area into small squares or strips and then scanning each segment with short tracks is highly beneficial. Thus, there are multiple reasons for subdividing the layer cross section into small regions for metals.

Randomization of square scanning (rather than scanning contiguous squares one after the other) and changing the primary scan direction between squares helps alleviate preferential build-up of residual stresses, as shown in Fig. 5.7. In addition, scanning of strips whereby the angle of the strip changes each layer has a positive effect on the build-up of residual stress. As a result, strips and square scan patterns are extensively utilized in PBF processes for metals.

5.5 Powder Handling

5.5.1 Powder Handling Challenges

Several different systems for powder delivery in PBF processes have been developed. The lack of a single solution for powder delivery goes beyond simply avoiding patented embodiments of the counter-rotating roller. The development of other approaches has resulted in a broader range of powder types and morphologies which can be delivered.

Any powder delivery system for PBF must meet at least four characteristics.

1. It must have a powder reservoir of sufficient volume to enable the process to build to the maximum build height without a need to pause the machine to refill the powder reservoir.
2. The correct volume of powder must be transported from the powder reservoir to the build platform that is sufficient to cover the previous layer but without wasteful excess material.
3. The powder must be spread to form a smooth, thin, repeatable layer of powder.
4. The powder spreading must not create excessive shear forces that disturb the previously processed layers.

In addition, any powder delivery system must be able to deal with these universal characteristics of powder feeding.

1. As particle size decreases, interparticle friction and electrostatic forces increase. These result in a situation where powder can lose its flowability. (To illustrate this loss of flowability, compare the flow characteristics of a spoon full of granulated sugar to a spoon full of fine flour. The larger particle size sugar will flow out of the spoon at a relatively shallow angle, whereas the flour will stay in the spoon until the spoon is tipped at a large angle, at which point the flour will fall out as a large clump unless some perturbation (vibration, tapping, etc.) causes it to come out a small amount at a time. Thus, any effective powder delivery system must make the powder flowable for effective delivery to occur.
2. When the surface area to volume ratio of a particle increases, its surface energy increases and becomes more reactive. For certain materials, this means that the powder becomes explosive in the presence of oxygen; or it will burn if there is a spark. As a result, certain powders must be kept in an inert atmosphere while being processed, and powder handling should not result in the generation of sparks.
3. When handled, small particles have a tendency to become airborne and float as a cloud of particles. In PBF machines, airborne particles will settle on surrounding surfaces, which may cloud optics, reduce the sensitivity of sensors, deflect laser beams, and damage moving parts. In addition, airborne particles have an effective surface area greater than packed powders, increasing their tendency to explode or burn. As a result, the powder delivery system should be designed in such a way that it minimizes the creation of airborne particles.
4. Smaller powder particle sizes enable better surface finish, higher accuracy, and thinner layers. However, smaller powder particle sizes exacerbate all the problems just mentioned. As a result, each design for a powder delivery system is inherently a different approach to effectively feed the smallest possible powder particle sizes while minimizing the negative effects of these small powder particles.

5.5.2 Powder Handling Systems

The earliest commercialized LS powder delivery system, illustrated in Fig. 5.1, is one approach to optimizing these powder handling issues. The two feed cartridges represent the powder reservoir with sufficient material to completely fill the build platform to its greatest build height. The correct amount of powder for each layer is provided by accurately incrementing the feed cartridge up a prescribed amount and the build platform down by the layer thickness. The raised powder is then pushed by the counter-rotating roller over the build platform, depositing the powder. As long as the height of the roller remains constant, layers will be created at the thickness with which the build platform moves. The counter-rotating action of the roller creates a "wave" of powder flowing in front of the cylinder. The counter-rotation pushes the powder up, fluidizing the powder being pushed, making it more flowable for a particular particle size and shape. The shear forces on the previously processed

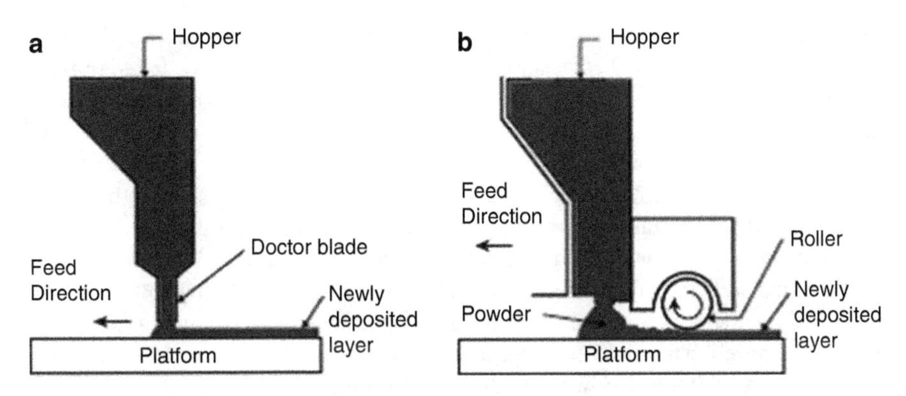

Fig. 5.10 Examples of hopper-based powder delivery systems [6]

layers created by this counter-rotating roller are small, and thus the previously processed layers are relatively undisturbed.

Another commonly utilized solution for powder spreading is a doctor blade. A doctor blade is simply a thin piece of metal that is used to scrape material across the surface of a powder bed. When a doctor blade is used, the powder is not fluidized. Thus, the shear forces applied to the previously deposited layer are greater than for a counter-rotating roller. This increased shear can be reduced if the doctor blade is ultrasonically vibrated, thus partly fluidizing the powder being pushed.

An alternative approach to using a feed cartridge as a powder reservoir is to use a hopper feeding system. A hopper system delivers powder to the powder bed from above rather than beneath. The powder reservoir is typically separate from the build area, and a feeding system is used to fill the hopper. The hopper is then used to deposit powder in front of a roller or doctor blade, or a doctor blade or roller can be integrated with a hopper system for combined feeding and spreading. For both feeding and spreading, ultrasonic vibration can be utilized with any of these approaches to help fluidize the powders. Various types of powder feeding systems are illustrated in Fig. 5.10.

In the case of multimaterial powder bed processing, the only effective method is to use multiple hoppers with separate materials. In a multi-hopper system, the material type can be changed layer-by-layer. Although this has been demonstrated in a research environment, and by some companies for very small parts; to date, all PBF technologies offered for sale commercially utilize a single-material powder feeding system.

5.5.3 Powder Recycling

As mentioned in Sect. 5.3.1, elevated temperature sintering of the powder surrounding a part being built can cause the powder bed to fuse. In addition, elevated temperatures, particularly in the presence of reacting atmospheric gases, will also change the chemical nature of the powder particles. Similarly, holding polymer materials at elevated temperatures can change the molecular weight of the polymer. These combined effects mean that the properties of many different types of powders (particularly polymers) used in PBF processes change their properties when they are recycled and reused. For some materials these changes are small, and thus are considered highly recyclable or infinitely recyclable. In other materials these changes are dramatic, and thus a highly controlled recycling methodology must be used to maintain consistent part properties between builds.

For the most popular PBF polymer material, nylon polyamide, both the effective particle size and molecular weight change during processing. As a result, a number of recycling methodologies have been developed to seek to maintain consistent build properties. The simplest approach to this recycling problem is to mix a specific ratio of unused powder with used powders. An example of a fraction-based mixture might be 1/3 unused powder, 1/3 overflow/feed powder, and 1/3 build platform powder. Overflow/feed and loose part-bed powder are handled separately, as they experience different temperature profiles during the build. The recaptured overflow/feed materials are only slightly modified from the original material as they have been subjected to lower temperatures only in the feed and overflow cartridges; whereas, loose part-bed powder from the build platform has been maintained at an elevated temperature, sometimes for many hours.

Part-bed powder is typically processed using a particle sorting method, most commonly either a vibratory screen-based sifting device or an air classifier, before mixing with other powders. Air classifiers can be better than simple sifting, as they mix the powders together more effectively and help break up agglomerates, thus enabling a larger fraction of material to be recycled. However, air classifiers are more complex and expensive than sifting systems. Regardless of the particle sorting method used, it is critical that the material be well-mixed during recycling; otherwise, parts built from recycled powder will have different properties in different locations.

Although easy to implement, a simple fraction-based recycling approach will always result in some amount of mixing inconsistencies. This is due to the fact that different builds have different part layout characteristics and thus the loose part-bed powder being recycled from one build has a different thermal history than loose part-bed powder being recycled from a different build.

In order to overcome some of the build-to-build inconsistencies inherent in fraction-based mixing, a recycling methodology based upon a powder's melt flow index (MFI) has been developed [4]. MFI is a measure of molten thermoplastic material flow through an extrusion apparatus under prescribed conditions. ASTM and ISO standards, for instance, can be followed to ensure repeatability. When using an MFI-based recycling methodology, a user determines a target MFI, based upon their experience. Used powders (part-bed and overflow/feed materials) are mixed and tested. Unused powder is also tested. The MFI for both is determined, and a well-blended mixture of unused and used powder is created and subsequently tested to achieve the target MFI. This may have to be done iteratively if the target MFI is not reached by the first mixture of unused to used powder. Using this methodology, the closer the target MFI is to the new powder MFI, the higher the new powder fraction, and thus the more expensive the part. The MFI method is generally considered more effective for ensuring consistent build-to-build properties than fractional mixing.

Typically, most users find that they need less of the used build platform powder in their mixture than is produced. Thus, this excess build material becomes scrap. In addition, repeated recycling over a long period of time may result in some powder becoming unusable. As a result, the recyclability of a powder and the target MFI or fractional mixing selected by a user can have a significant effect on part properties and cost.

5.6 PBF Process Variants and Commercial Machines

A large variety of PBF processes has been developed. To understand the practical differences between these processes, it is important to know how the powder delivery method, heating process, energy input type, atmospheric conditions, optics, and other features vary with respect to one another. An overview of commercial processes and a few notable systems under development are discussed in the following section.

5.6.1 Polymer Laser Sintering

Prior to 2014 there were only two major producers of pLS machines, EOS and 3D Systems. The expiration of key patents in 2014 opened the door for many new companies to enter the marketplace. pLS machines are designed for directly processing polymers and for indirect processing of metals and ceramics.

Most commercial polymers were developed for processing via injection molding. The thermal and stress conditions for a material processed via pLS, however, are much different than the thermal and stress conditions for a material processed via injection molding. In injection molding the material is slowly heated under pressure, flows under high shear forces into a mold, and is cooled quickly. In pLS the material is heated very quickly as a laser beam passes, it flows via surface tension under gravitational forces, and it cools slowly over a period of hours to days. Since polymer microstructural features depend upon the time the material is held at elevated temperatures, polymer parts made using LS can have very different properties than polymer parts made using injection molding.

Many polymers which are easy to process using injection molding may not be processable using pLS. Figure 5.11 illustrates a schematic of a differential scanning calorimetry (DSC) curve for the types of melting characteristics which are desirable in a polymer for LS. In order to reduce residual stress induced curling, pLS machines hold the powder bed temperature (T_{bed}) just below the temperature where melting begins ($T_{\text{Melt Onset}}$). When the laser melts a region of the powder bed, it should raise the temperature of the material above the melting temperature, but below the temperature at which the material begins to deteriorate. If there is a

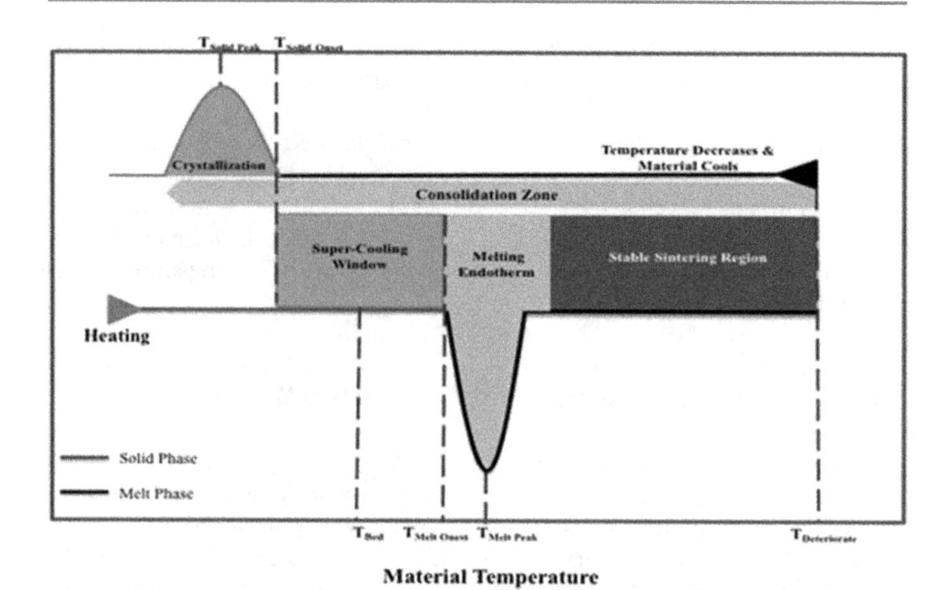

Fig. 5.11 Melting and solidification characteristics for an idealized polymer DSC curve for polymer laser sintering (courtesy: Neil Hopkinson, Sheffield University)

small difference between the melting and deterioration temperatures, then the material will be difficult to successfully process in pLS.

After scanning, the molten cross section will return over a relatively short period of time to the bed temperature (T_{Bed}). If the bed temperature is above the crystallization temperature of the material, then it will remain in a partially molten state for a very long time. This is advantageous for two reasons. First, by keeping the material partially molten, the part will not experience layer-by-layer accumulation of residual stresses and thus will be more accurate. Secondly, by holding the material in a semi-molten state for a long period of time, the part will achieve higher overall density. As a result, parts at the bottom of a build platform (which were built first and experience a longer time at elevated temperature) are denser than the last parts to be built. Thus, a key characteristic of a good polymer for pLS is that it has a broad "Super-Cooling Window" as illustrated in Fig. 5.11. For most commercially available polymers, the melting curve overlaps the crystallization curve and there is no super-cooling window. In addition, for amorphous polymers, there is no sharp onset of melting or crystallization. Thus pLS works best for polymers that are crystalline with a large super-cooling window and a high deterioration temperature.

The SLS Sinterstation 2000 machine was the first commercial PBF system, introduced by the DTM Corporation, USA, in 1992. Subsequently, other variants were commercialized, and these systems are still manufactured and supplied by 3D Systems, USA, which purchased DTM in 2001. Newer machines offer several improvements over previous systems in terms of part accuracy, temperature uniformity, build speed, process repeatability, feature definition, and surface finish, but the basic processing features and system configuration remain unchanged from the description in Sect. 5.1. A typical pLS machine is limited to polymers with a melting temperature below 200 °C, whereas "high-temperature" pLS machines can process polymers with much higher melting temperature. Due to the use of CO_2 lasers and a nitrogen atmosphere with approximately 0.1–3.0 % oxygen, pLS machines are incapable of directly processing pure metals or ceramics. Nylon polyamide materials are the most popular pLS materials, but these processes can also be used for many other types of polymer materials as well as indirect processing of metal and ceramic powders with polymer binders.

EOS GmbH, Germany, introduced its first EOSINT P machine in 1994 for producing plastic prototypes. In 1995, the company introduced its EOSINT M 250 machine for direct manufacture of metal casting molds from foundry sand. In 1998, the EOSINT M 250 Xtended machine was launched for DMLS, which was a LPS approach to processing metallic powders. These early metal machines used a special alloy mixture comprised of bronze and nickel powders developed by Electrolux Rapid Prototyping, and licensed exclusively to EOS. The powder could be processed at low temperatures, required no preheating and exhibited negligible shrinkage during processing; however, the end-product was porous and was not representative of any common engineering metal alloys. Subsequently, EOS introduced many other materials and models, including platforms for foundry sand and full melting of metal powders (which will be discussed in the following section). One unique feature of EOS's large-platform systems for polymers and foundry sand is the use of two laser beams for faster part construction (as illustrated in the 2 × 1D channels example in Fig. 2.6). This multi-machine approach to PBF has made EOS the market leader in this technology segment. A schematic of an EOS machine illustrating their approach to laser sintering powder delivery and processing for foundry sand is shown in Fig. 5.12.

More recent, large-platform pLS systems commonly use a modular design. This modularity can include: removable build platforms so that part cool-down and warm-up can occur outside of the chamber, enabling a fresh build platform to be inserted and used with minimal laser down-time; multiple build platform sizes; automated recycling and feeding of powder using a connected powder handling system; and better thermal control options.

In addition to commercial PBF machines, open-source polymer PBF machines are being developed to mimic the success of the RepRap effort for material extrusion machines. In addition, inventors have applied PBF techniques to nonengineering applications via the CandyFab machine. Sugar is used as the powdered material, and a hot air nozzle is used as the energy source. By scanning the nozzle across the bed in a layer-by-layer fashion, sugar structures are made.

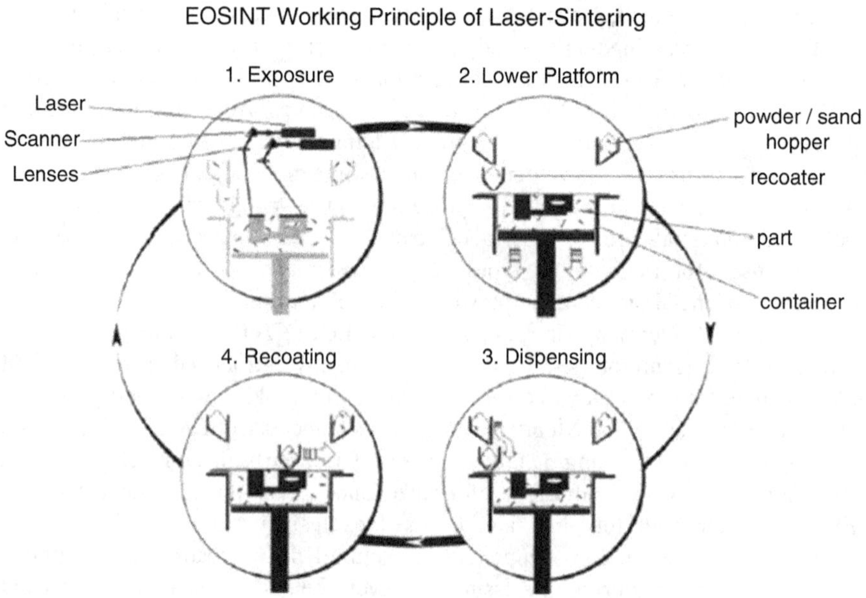

Fig. 5.12 EOSint laser sintering schematic showing the dual-laser system option, hopper powder delivery and a recoater that combines a movable hopper and doctor blade system (courtesy: EOS)

5.6.2 Laser-Based Systems for Metals and Ceramics

There are many companies which make commercially available laser-based systems for direct melting and sintering of metal powders: EOS (Germany), Renishaw (UK), Concept Laser (Germany), Selective Laser Melting (SLM) Solutions (Germany), Realizer (Germany), and 3D Systems (France/USA). There are competing terminologies for these technologies. The term selective laser melting (SLM) is used by numerous companies; however the terms Laser Cusing and DMLS are also used by certain manufacturers. For this discussion, we will use mLS to refer to the technologies in general and not to any particular variant.

mLS research in the late 1980s and early 1990s by various research groups was mostly unsuccessful. Compared to polymers, the high thermal conductivity, propensity to oxidize, high surface tension, and high laser reflectivity of metal powders make them significantly more difficult to process than polymers. Most commercially available mLS systems today are variants of the selective laser powder re-melting (SLPR) approach developed by the Fraunhofer Institute for Laser Technology, Germany. Their research developed the basic processing techniques necessary for successful laser-based, point-wise melting of metals. The use of lasers with wavelengths better tuned to the absorptivity of metal powders was one key for enabling mLS. Fraunhofer used an Nd-YAG laser instead of the CO_2 laser used in pLS, which resulted in a much better absorptivity for metal powders (see Fig. 5.13). Subsequently, almost all mLS machines use fiber lasers, which in general are

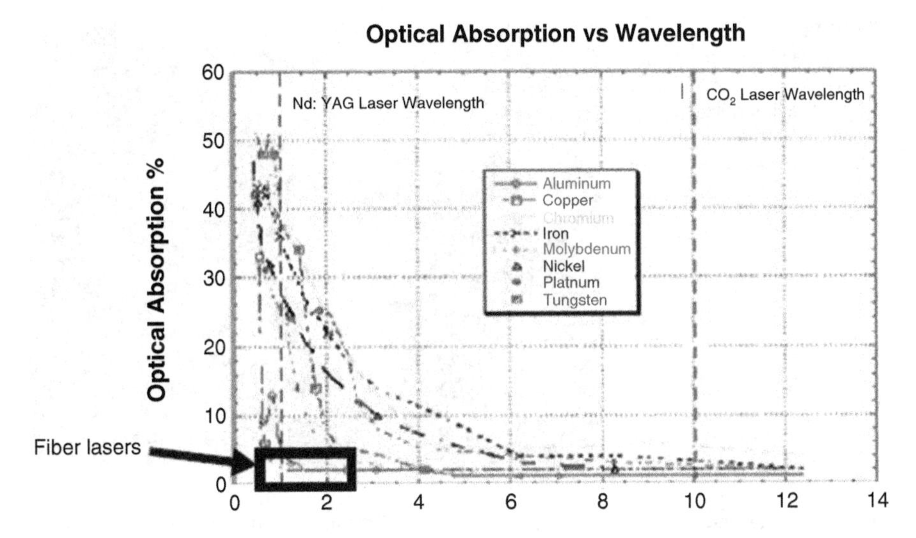

Fig. 5.13 Optical absorption % (absorptivity) of selected metals versus wavelength (units are micrometers) (courtesy: Optomec)

cheaper to purchase and maintain, more compact, energy efficient, and have better beam quality than Nd:YAG lasers. The other key enablers for mLS, compared to pLS, are different laser scan patterns (discussed in the following section), the use of f-theta lenses to minimize beam distortion during scanning, and low oxygen, inert atmosphere control.

One common practice among mLS manufacturers is the rigid attachment of their parts to a base plate at the bottom of the build platform. This is done to keep the metal part being built from distorting due to residual stresses. This means that the design flexibility for parts made from mLS is not quite as broad as the design flexibility for parts made using laser sintering of polymers, due to the need to remove these rigid supports using a machining or cutting operation.

Over the years, various mLS machine manufacturers have sought to differentiate themselves from others by the features they offer. This differentiation includes laser power, number of lasers offered, powder handling systems, scanning strategies offered, maximum build volume, and more. Some machine manufacturers give users more control over the process parameters than other manufacturers, enabling more experimentation by the user, whereas other manufacturers only provide "proven" materials and process parameters. For instance, Renishaw machines have safety features to help minimize the risk of powder fires. EOS, as the world's most successful metal PBF provider, has spent considerable time tuning their machine process parameters and scanning strategies for specific materials which they sell to their customers. Concept Laser has focused on the development of stainless and hot-work steel alloys suitable for injection mold and die cast tooling. 3D Systems (after their acquisition of Phenix Systems) has developed machines which can be held at an elevated temperature, thus enabling efficient sintering of

Fig. 5.14 3D Micromac Powder Feed System. In this picture, only one of the powder feeders (located over the build cylinder) is filled with powder (courtesy: Laserinstitut Mittelsachsen e.V.)

ceramic powders, in addition to melting of metal powders. Another unique characteristic of the 3D Systems machine is its use of a roller to spread and then compact powder, making it the only manufacturer which can directly change the powder bed packing density on-the-fly.

3D-Micromac, Germany, a partner of EOS, produces the only multimaterial, small-scale mLS machine. It has developed small-scale mLS processes with small build cylinders 25 or 50 mm in diameter and 40 mm in height. Their fiber laser is focused to a particularly small spot size, for small feature definition. In order to use the fine powder particle sizes necessary for fine feature reproduction, they have developed a unique two-material powder feeding mechanism, shown in Fig. 5.14. The build platform is located between two powder feed cylinders. When the rotating rocker arm is above a powder feed cylinder, the powder is pushed up into the feeder, thus charging the hopper. When the rocker arm is moved over top of the build platform, it deposits and smoothens the powder, moving away from the build cylinder prior to laser processing. By alternating between feed cylinders, the material being processed can be changed in a layer-by-layer fashion, thus forming multimaterial structures. An example of a small impeller made using aluminum oxide powders is shown in Fig. 5.15.

5.6.3 Electron Beam Melting

Electron beam melting (EBM) has become a successful approach to PBF. In contrast to laser-based systems, EBM uses a high-energy electron beam to induce fusion between metal powder particles. This process was developed at Chalmers University of Technology, Sweden, and was commercialized by Arcam AB, Sweden, in 2001.

Similarly to mLS, in the EBM process, a focused electron beam scans across a thin layer of pre-laid powder, causing localized melting and resolidification per the

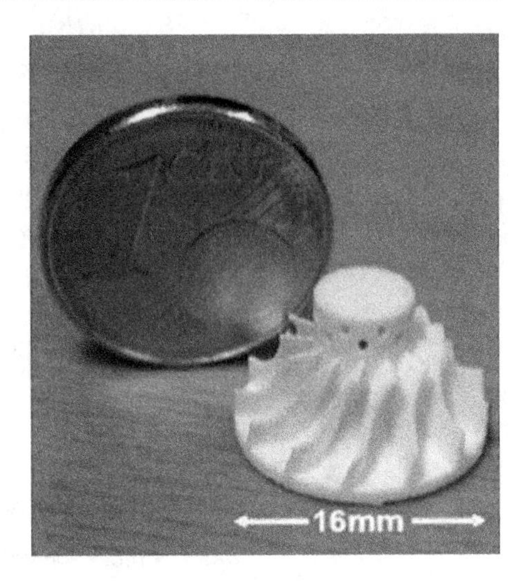

Fig. 5.15 Example 3D Micromac part made from aluminum oxide powders (courtesy: Laserinstitut Mittelsachsen e.V.)

Table 5.1 Differences between EBM and mLS

Characteristic	Electron beam melting	Metal laser sintering
Thermal source	Electron beam	Laser
Atmosphere	Vacuum	Inert gas
Scanning	Deflection coils	Galvanometers
Energy absorption	Conductivity-limited	Absorptivity-limited
Powder preheating	Use electron beam	Use infrared or resistive heaters
Scan speeds	Very fast, magnetically driven	Limited by galvanometer inertia
Energy costs	Moderate	High
Surface finish	Moderate to poor	Excellent to moderate
Feature resolution	Moderate	Excellent
Materials	Metals (conductors)	Polymers, metals and ceramics
Powder particle size	Medium	Fine

slice cross section. There are a number of differences between how mLS and EBM are typically practiced, which are summarized in Table 5.1. Many of these differences are due to EBM having an energy source of electrons, but other differences are due to engineering trade-offs as practiced in EBM and mLS and are not necessarily inherent to the processing. A schematic illustration of an EBM apparatus is shown in Fig. 5.16.

Laser beams heat the powder when photons are absorbed by powder particles. Electron beams, however, heat powder by transfer of kinetic energy from incoming electrons into powder particles. As powder particles absorb electrons they gain an increasingly negative charge. This has two potentially detrimental effects: (1) if the repulsive force of neighboring negatively charged particles overcomes the gravitational and frictional forces holding them in place, there will be a rapid expulsion of powder particles from the powder bed, creating a powder cloud (which is worse for

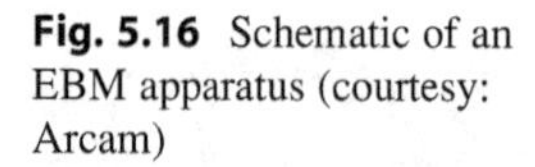
Fig. 5.16 Schematic of an EBM apparatus (courtesy: Arcam)

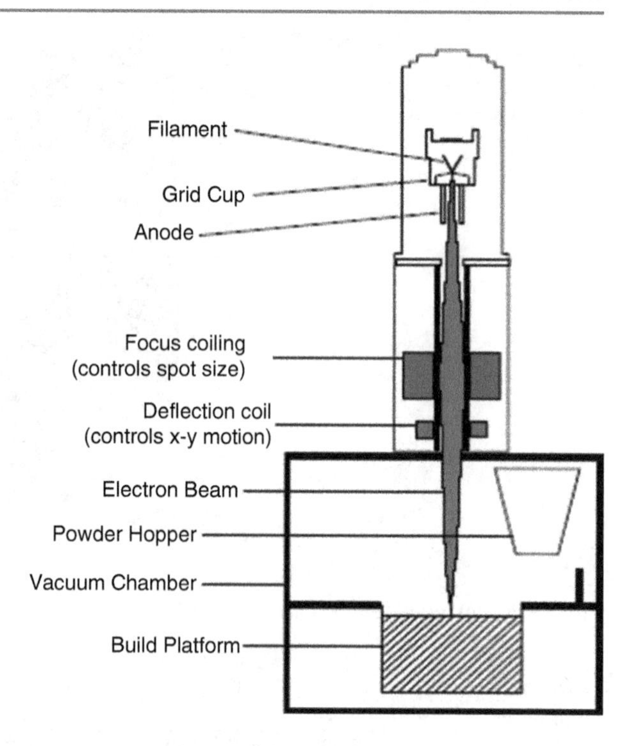

fine powders than coarser powders) and (2) increasing negative charges in the powder particles will tend to repel the incoming negatively charged electrons, thus creating a more diffuse beam. There are no such complimentary phenomena with photons. As a result, the conductivity of the powder bed in EBM must be high enough that powder particles do not become highly negatively charged, and scan strategies must be used to avoid build-up of regions of negatively charged particles. In practice, electron beam energy is more diffuse, in part, so as not to build up too great a negative charge in any one location. As a result, the effective melt pool size increases, creating a larger heat-affected zone. Consequently, the minimum feature size, median powder particle size, layer thickness, resolution, and surface finish of an EBM process are typically larger than for an mLS process.

As mentioned above, in EBM the powder bed must be conductive. Thus, EBM can only be used to process conductive materials (e.g., metals) whereas, lasers can be used with any material that absorbs energy at the laser wavelength (e.g., metals, polymers, and ceramics).

Electron beam generation is typically a much more efficient process than laser beam generation. When a voltage difference is applied to the heated filament in an electron beam system, most of the electrical energy is converted into the electron beam; and higher beam energies (above 1 kW) are available at a moderate cost. In contrast, it is common for only 10–20 % of the total electrical energy input for laser systems to be converted into beam energy, with the remaining energy lost in the form of heat. In addition, lasers with beam energies above 1 kW are typically much more expensive than comparable electron beams with similar energies. Thus, electron beams are a less costly high energy source than laser beams. Newer fiber lasers, however, are more simple in their design, more reliable to maintain, and more efficient to use (with conversion efficiencies reported of 70–80 % for some fiber lasers). Thus, this energy advantage for electron beams may not be a major advantage in the future.

EBM powder beds are maintained at a higher temperature than mLS powder beds. There are several reasons for this. First, the higher energy input of the beam used in the EBM system naturally heats the surrounding loose powder to a higher temperature than the lower energy laser beams. In order to maintain a steady-state uniform temperature throughout the build (rather than having the build become hotter as the build height increases) the EBM process uses the electron beam to heat the metal substrate at the bottom of the build platform before laying a powder bed. By defocusing the electron beam and scanning it very rapidly over the entire surface of the substrate (or the powder bed for subsequently layers) the bed can be preheated rapidly and uniformly to any preset temperature. As a result, the radiative and resistive heaters present in some mLS systems for substrate and powder bed heating are not used in EBM. By maintaining the powder bed at an elevated temperature, however, the resulting microstructure of a typical EBM part is significantly different from a typical mLS part (see Fig. 5.17). In particular, in mLS the individual laser scan lines are typically easily distinguishable, whereas individual scan lines are often indistinguishable in EBM microstructures. Rapid cooling in mLS creates smaller grain sizes and subsequent layer scans only partially re-melt the previously deposited layer. The powder bed is held at a low enough temperature that elevated temperature grain growth does not erase the layering effects. In EBM, the higher temperature of the powder bed, and the larger and more diffuse heat input result in a contiguous grain pattern that is more representative of a cast microstructure, with less porosity than an mLS microstructure.

Although the microstructures presented in Fig. 5.17 are representative of mLS and EBM, it should be noted that the presence of beam traces in the final

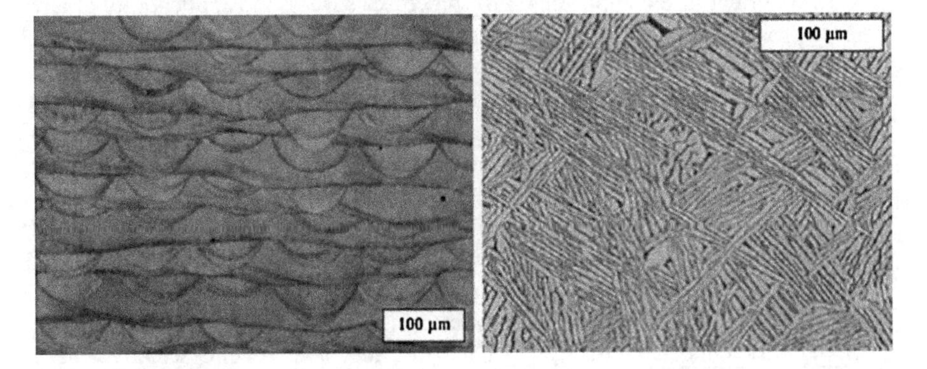

Fig. 5.17 Representative CoCrMo mLS microstructure (*left*, courtesy: EOS), and Ti6Al4V EBM microstructure (*right*, courtesy: Arcam)

microstructure (as seen in the left image of Fig. 5.17) is process parameter and material dependent. For certain alloys, such as titanium, it is not uncommon for contiguous grain growth across layers even for mLS. For other materials, such as those that have a higher melting point, the layering may be more prevalent. In addition, layering is more prevalent for process parameter combinations of lower bed temperature, lower beam energy, faster scan rate, thicker layers, and/or larger scan spacing for both mLS and EBM. The reader is also referred to the presentation of material microstructures and process parameter effects of the DED processes in Sects. 10.6 and 10.7, since the phenomena seen mLS and EBM are similar to those observed in DED processes.

One of the most promising aspects of EBM is the ability to move the beam nearly instantaneously. The current control system for EBM machines makes use of this capability to keep multiple melt pools moving simultaneously for part contour scanning. Future improvements to scanning strategies may dramatically increase the build speed of EBM over mLS, helping to distinguish it even more for certain applications. For instance, when nonsolid cross sections are created, in particular when scanning truss-like structures (with designed internal porosity), nearly instantaneous beam motion from one scan location to another can dramatically speed up the production of the overall product.

In EBM, residual stresses are much lower than for mLS due to the elevated bed temperature. Supports are needed to provide electrical conduction through the powder bed to the base plate, to eliminate electron charging, but the mass of these supports is an order of magnitude less than what is needed for mLS of a similar geometry. Future scan strategies for mLS may help reduce the need for supports to a degree where they can be removed easily, but at present EBM has a clear advantage when it comes to minimizing residual stress and supports.

5.6.4 Line-Wise and Layer-Wise PBF Processes for Polymers

PBF processes have proven to be the most flexible general approach to AM. For production of end-use components, PBF processes surpass the applicability of any other approach. However, the use of expensive lasers in most processes, the fact that these lasers can only process one “point” of material at any instant in time, and the overall cost of the systems means that there is considerable room for improvement. As a result, several organizations are developing ways to fuse lines or layers of polymer material at a time. The potential for polymer processing in a line-wise or layer-wise manner could dramatically increase the build-rate of PBF processes, thus making them more cost-competitive. Three of these processes will be discussed below. All three utilize infrared energy to induce fusion in powder beds; the key differences lay in their approach to controlling which portions of the powder bed fuse and which remain unfused, as illustrated in Fig. 5.18.

Sintermask GmbH, Germany, founded in 2009, sold several selective mask sintering (SMS) machines, based upon technology developed at Speedpart AB. The key characteristics of their technology are exposure of an entire layer at a

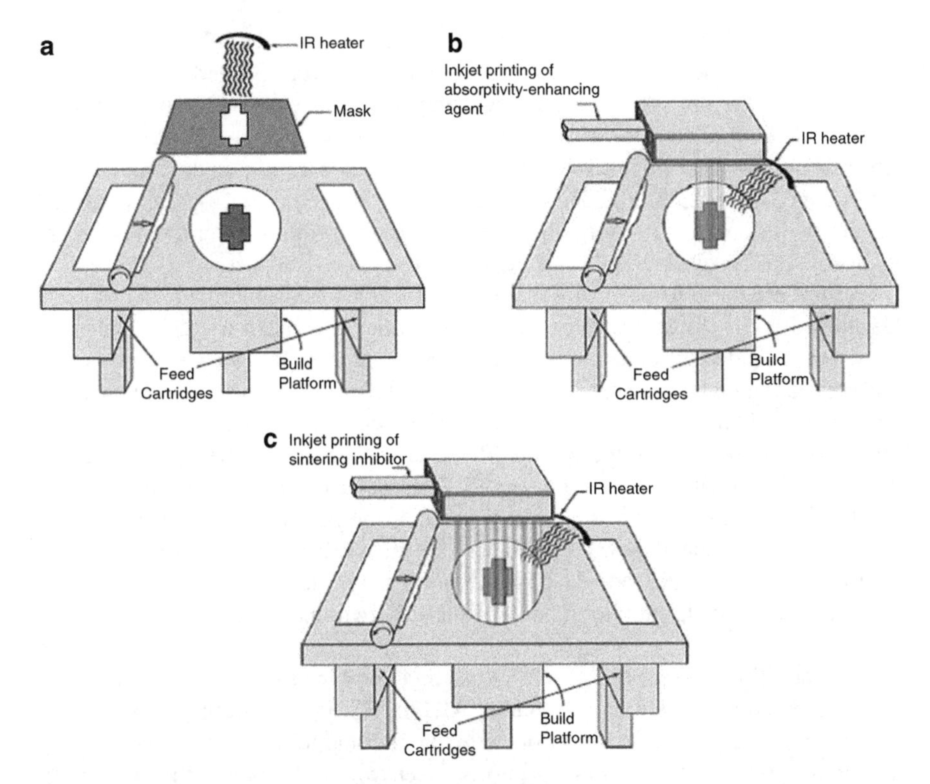

Fig. 5.18 Three different approaches to line- and layer-wise powder bed fusion processing (**a**) mask-based sintering, (**b**) printing of an absorptivity-enhancing agent in the part region, and (**c**) printing of a sintering inhibitor outside the part region

time to infrared thermal energy through a mask, and rapid layering of powdered material. Their powder delivery system can deposit a new layer of powder in 3 s. Heat energy is provided by an infrared heater. A dynamic mask system, similar to those used in a photocopier to transfer ink to paper, is used between the heater and the powder bed. This is a re-birth of an idea conceived by Cubital for layer-wise photopolymerization in the early days of AM, as mentioned in Chap. 2. The SMS mask allows infrared energy to impinge on the powder bed only in the region prescribed by the layer cross section, fusing powder in approximately 1 s. From a materials standpoint, the use of an infrared energy source means that the powder must readily absorb and quickly sinter or melt in the presence of infrared energy. Most materials with this characteristic are dark colored (e.g., gray or black) and thus color-choice limitations may be a factor for some adopters of the technology. It appears that development of this technology is on hold, as of the writing of this book.

High-speed sintering (HSS) is a process developed at Loughborough University and Sheffield University and being commercialized by FACTUM. In HSS, an ink-jet printer is used to deposit ink onto the powder bed, representing a part's cross section for that layer. Inks are specially formulated to significantly enhance infrared absorption compared with the surrounding powder bed. An infrared heater is used to scan the entire powder bed quickly, following ink-jetting. Thus, this process is an example of line-wise processing. The difference between the absorptivity of the unprinted areas compared to the printed areas means that the unprinted areas do not absorb enough energy to sinter, whereas the powder in the printed areas sinters and/or melts. As the distinguishing factor between the fused and unfused region is the enhanced absorption of energy where printing occurs, the inks are typically gray or black and thus affect the color of the final part.

A third approach to rapid PBF is the selective inhibition sintering (SIS) process, developed at the University of Southern California. In contrast to HSS, a sintering inhibitor is printed in regions where fusion is not desired, followed by exposure to infrared radiation. In this case, the inhibitor interferes with diffusion and surface properties to inhibit sintering. In addition, researchers have also utilized movable plates to mask portions of the powder bed where no sintering is desired, in order to minimize the amount of inhibitor required. One benefit of SIS over the previous two are that it does not involve adding an infrared absorption agent into the part itself, and thus the untreated powder becomes the material in the part. However, the unused powder in the powder bed is not easily recyclable, as it has been "contaminated" with inhibitor, and thus, there is significant unrecyclable material created.

Two additional variations of ink-jet printing combined with PBF methodology are also practiced in SIS and by fcubic AB. In SIS, if no sintering is performed during the build (i.e., inhibitor is printed but no thermal infrared energy is scanned) the entire part bed can be moved into an oven where the powder is sintered to achieve fusion within the part, but not in areas where inhibitor has been printed.

fcubic AB, Sweden, uses ink-jet printing plus sintering in a furnace to compete with traditional powder metallurgy for stainless steel components. A sintering aid is printed in the regions representing the part cross section, so that this region will fuse more rapidly in a furnace. A sintering aid is an element or alloy which increases the rate at which solid-state sintering occurs between particles by changing surface characteristics and/or by reacting with the particles. Thus, sintering in the part will occur at lower temperatures and times than for the surrounding powder that has not received a sintering aid.

Both SIS and fcubic are similar to the binder jetting processes described in Chap. 8 (such as practiced by ExOne and Voxeljet) where a binder joins powders in regions of the powder bed where the part is located followed by furnace processing. There is, however, one key aspect of SIS and the fcubic processing which is different than these approaches. In the SIS and fcubic processes, the printed material is a sintering aid or inhibitor rather than a binder, and the part *remains embedded* within the powder bed when sintering in the furnace. Using the ExOne process, for instance, the machine prints a binder to glue powder particles together; and the bound regions are *removed from the powder bed* as a green part before sintering in a furnace (much like the indirect metal processing discussed earlier).

Common to all of the line-wise and layer-wise PBF processes is the need to differentiate between fusion in the part versus the remaining powder. Too low of total energy input will leave the part weak and only partially sintered. Too high of energy levels will result in part growth by sintering of excess surrounding powder to the part and/or degradation of the surrounding powder to the point where it cannot be easily recycled. Most importantly, in all cases it is the *difference* between fusion induced in the part versus fusion induced in the surrounding powder bed that is the key factor to control.

5.7 Process Benefits and Drawbacks

Due to its nature, PBF can process a very wide variety of materials, in contrast to many other AM processes. Although it is easier to control the processing of semi-crystalline polymers, the PBF processing of amorphous polymers has been successful. Many metals can be processed; as mentioned, if a metal can be welded, it is a good candidate for mLS. Some ceramic materials are commercially available, but quite a few others have been demonstrated in research.

During part building, loose powder is a sufficient support material for polymer PBF. This saves significant time during part building and post-processing, and enables advanced geometries that are difficult to post-process when supports are necessary. As a result, internal cooling channels and other complex features that would be impossible to machine are possible in polymer PBF.

Supports, however, are required for most metal PBF processes. The high residual stresses experienced when processing metals means that support structures are typically required to keep the part from excessive warping. This means that post-processing of metal parts after AM can be expensive and time consuming. Small features (including internal cooling channels) can usually be formed without supports; but the part itself is usually constrained to a substrate at the bottom of the build platform to keep it from warping. As a result, the orientation of the part and the location of supports are key factors when setting up a build.

Accuracy and surface finish of powder-based AM processes are typically inferior to liquid-based processes. However, accuracy and surface finish are strongly influenced by the operating conditions and the powder particle size. Finer particle sizes produce smoother, more accurate parts but are difficult to spread and handle. Larger particle sizes facilitate easier powder processing and delivery, but hurt surface finish, minimum feature size and minimum layer thickness. The build materials used in these processes typically exhibit 3–4 % shrinkage, which can lead to part distortion. Materials with low thermal conductivity result in better accuracy as melt pool and solidification are more controllable and part growth is minimized when heat conduction is minimized.

With PBF processes, total part construction time can take longer than other additive manufacturing processes because of the preheat and cool-down cycles involved. However, as is the case with several newer machine designs, removable build platforms enable preheat and cool-down to occur off-line, thus enabling much greater machine productivity. Additionally, the ability to nest polymer parts in three-dimensions, as no support structures are needed, mean that many parts can be produced in a single build, thus dramatically improving the productivity of these processes when compared with processes that require supports.

5.8 Conclusions

PBF processes were one of the earliest AM processes, and continue to be one of the most popular. Polymer-based laser sintering is commonly used for prototyping and end-use applications in many industries, competing with injection molding and other polymer manufacturing processes. PBF processes are particularly competitive for low-to-medium volume geometrically complex parts.

Metal-based processes, including laser and electron beam, are one of the fastest growing areas of AM around the world. Metal PBF processes are becoming increasingly common for aerospace and biomedical applications, due to their inherent geometric complexity benefits and their excellent material properties when compared to traditional metal manufacturing techniques.

As methods for moving from point-wise to line-wise to layer-wise PBF are improved and commercialized, build times and cost will decrease. This will make PBF processing even more competitive. The future for PBF remains bright; and it is likely that PBF processes will remain one of the most common types of AM technologies for the foreseeable future.

5.9 Exercises

1. Find a reference which describes an application of the Arrhenius equation to solid state sintering. If an acceptable level of sintering is achieved within time T_1 at a temperature of 750 K, what temperature would be required to achieve the same level of sintering in half the time?
2. Estimate the energy driving force difference between two different powder beds made up of spherical particles with the same total mass, where the difference in surface area to volume ratio difference between one powder bed and the other is a factor of 2.
3. Explain the pros and cons of the various binder and structural material alternatives in LPS (Sect. 5.3.3.1) for a bone tissue scaffold application, where the binder (matrix material) is PCL and the structural material is hydroxyapatite.
4. Using standard kitchen ingredients, explore the powder characteristics described in Sect. 5.5.1 and powder handling options described in Sect. 5.5.2. Using at least three different ingredients, describe whether or not the issues described are reproducible in your experiments.
5. Using an internet search, find a set of recommended processing parameters for nylon polyamide using laser sintering. Based upon (5.1), are these parameters limited by machine laser power, scan spacing, or scan speed? Why? What machine characteristics could be changed to increase the build rate for this material and machine combination?

6. Using Fig. 5.9 and the explanatory text, estimate the minimum laser dwell time (how long a spot is under the laser as it passes) needed to maintain a type B scan track at 100 W.

References

1. Kruth JP, Mercelis P, Van Vaerenbergh J (2005) Binding mechanisms in selective laser sintering and selective laser melting. Rapid Prototyp J 11(1):26–36
2. Williams JD, Deckard CR (1998) Advances in modeling the effects of selected parameters on the PLS process. Rapid Prototyp J 4(2):90–100
3. THC Childs, Hauser C, Badrossamay M (2005) Selective laser sintering (melting) of stainless and tool steel powders: experiments and modelling. J Proc Inst Mech Eng B J Eng Manuf 219 (4):339–357. doi:10.1007/978-1-4419-1120-9_5, Publisher Professional Engineering Publishing; ISSN 0954-4054
4. Gornet TJ, Davis KR, Starr TL, Mulloy KM (2002) Proceedings of the solid freeform fabrication symposium characterization of selective laser sintering materials to determine process stability, University of Texas at Austin
5. Zarringhalam H, Hopkinson N, Kamperman NF, de Vlieger JJ (2006) Effects of processing on microstructure and properties of PLS Nylon 12. Mater Sci Eng A 435–436:172–180
6. Yan Y, Zhang R, Qingping L, Zhaohui D (1998) Study on multifunctional rapid prototyping manufacturing system. J Integr Manuf Syst 9(4):236–241

Extrusion-Based Systems

6

Abstract
Extrusion-based technology is currently the most popular on the market. Whilst there are other techniques for creating the extrusion, heat is normally used to melt bulk material in a small, portable chamber. The material is pushed through by a tractor-feed system, which creates the pressure to extrude. This chapter deals with AM technologies that use extrusion to form parts. We will cover the basic theory and attempt to explain why it is a leading AM technology.

6.1 Introduction

Material extrusion technologies can be visualized as similar to cake icing, in that material contained in a reservoir is forced out through a nozzle when pressure is applied. If the pressure remains constant, then the resulting extruded material (commonly referred to as "roads") will flow at a constant rate and will remain a constant cross-sectional diameter. This diameter will remain constant if the travel of the nozzle across a depositing surface is also kept at a constant speed that corresponds to the flow rate. The material that is being extruded must be in a semisolid state when it comes out of the nozzle. This material must fully solidify while remaining in that shape. Furthermore, the material must bond to material that has already been extruded so that a solid structure can result.

Since material is extruded, the AM machine must be capable of scanning in a horizontal plane as well as starting and stopping the flow of material while scanning. Once a layer is completed, the machine must index upwards, or move the part downwards, so that a further layer can be produced.

There are two primary approaches when using an extrusion process. The most commonly used approach is to use temperature as a way of controlling the material state. Molten material is liquefied inside a reservoir so that it can flow out through the nozzle and bond with adjacent material before solidifying. This approach is similar to conventional polymer extrusion processes, except the extruder is

I. Gibson et al., *Additive Manufacturing Technologies*,
DOI 10.1007/978-1-4939-2113-3_6

vertically mounted on a plotting system rather than remaining in a fixed horizontal position.

An alternative approach is to use a chemical change to cause solidification. In such cases, a curing agent, residual solvent, reaction with air, or simply drying of a "wet" material permits bonding to occur. Parts may therefore cure or dry out to become fully stable. This approach is can be utilized with paste materials. Additionally, it may be more applicable to biochemical applications where materials must have biocompatibility with living cells and so choice of material is very restricted. However, industrial applications may also exist, perhaps using reaction injection molding-related processes rather than relying entirely on thermal effects.

This chapter will start off by describing the basic principles of extrusion-based additive manufacturing. Following this will be a description of the most widely used extrusion-based technology, developed and commercialized by the Stratasys company. Bioplotting equipment for tissue engineering and scaffold applications commonly use extrusion technology and a discussion on how this differs from the Stratasys approach will ensue. Finally, there have been a number of interesting research projects employing, adapting, and developing this technology, and this will be covered at the end of the chapter.

6.2 Basic Principles

There are a number of key features that are common to any extrusion-based system:

- Loading of material
- Liquification of the material
- Application of pressure to move the material through the nozzle
- Extrusion
- Plotting according to a predefined path and in a controlled manner
- Bonding of the material to itself or secondary build materials to form a coherent solid structure
- Inclusion of support structures to enable complex geometrical features

These will be considered in separate sections to fully understand the intricacies of extrusion-based AM.

A mathematical or physics-based understanding of extrusion processes can quickly become complex, since it can involve many nonlinear terms. The basic science involves extrusion of highly viscous materials through a nozzle. It is reasonable to assume that the material flows as a Newtonian fluid in most cases [1]. Most of the discussion in these sections will assume the extrusion is of molten material and may therefore include temperature terms. For solidification, these temperature terms are generally expressed relative to time; and so temperature could be replaced by other time-dependent factors to describe curing or drying processes.

6.2.1 Material Loading

Since extrusion is used, there must be a chamber from which the material is extruded. This could be preloaded with material, but it would be more useful if there was a continuous supply of material into this chamber. If the material is in a liquid form, then the ideal approach is to pump this material. Most bulk material is, however, supplied as a solid and the most suitable methods of supply are in pellet or powder form, or where the material is fed in as a continuous filament. The chamber itself is therefore the main location for the liquification process. Pellets, granules, or powders are fed through the chamber under gravity or with the aid of a screw or similar propelling process. Materials that are fed through the system under gravity require a plunger or compressed gas to force it through the narrow nozzle. Screw feeding not only pushes the material through to the base of the reservoir but can be sufficient to generate the pressure needed to push it through the nozzle as well. A continuous filament can be pushed into the reservoir chamber, thus providing a mechanism for generating an input pressure for the nozzle.

6.2.2 Liquification

The extrusion method works on the principle that what is held in the chamber will become a liquid that can eventually be pushed through the die or nozzle. As mentioned earlier, this material could be in the form of a solution that will quickly solidify following the extrusion, but more likely this material will be liquid because of heat applied to the chamber. Such heat would normally be applied by heater coils wrapped around the chamber and ideally this heat should be applied to maintain a constant temperature in the melt (see Fig. 6.1). The larger the chamber, the more difficult this can become for numerous reasons related to heat transfer, thermal currents within the melt, change in physical state of the molten material, location of temperature sensors, etc.

The material inside the chamber should be kept in a molten state but care should be taken to maintain it at as low a temperature as possible since some polymers degrade quickly at higher temperatures and could also burn, leaving residue on the inside of the chamber that would be difficult to remove and that would contaminate further melt. A higher temperature inside the chamber also requires additional cooling following extrusion.

6.2.3 Extrusion

The extrusion nozzle determines the shape and size of the extruded filament. A larger nozzle diameter will enable material to flow more rapidly but would result in a part with lower precision compared with the original CAD drawing. The diameter of the nozzle also determines the minimum feature size that can be created. No feature can be smaller than this diameter and in practice features should normally be

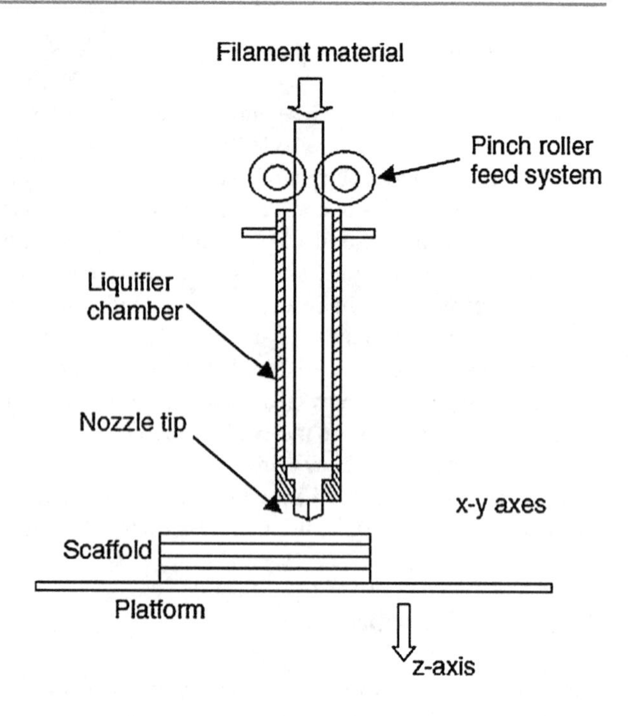

Fig. 6.1 Schematic of extrusion-based systems

large relative to the nozzle diameter to faithfully reproduce them with satisfactory strength. Extrusion-based processes are therefore more suitable for larger parts that have features and wall thicknesses that are at least twice the nominal diameter of the extrusion nozzle used.

Material flow through the nozzle is controlled by the pressure drop between the chamber and the surrounding atmosphere. However, the extrusion process used for AM may not be the same as conventional extrusion. For example, the pressure developed to push the molten material through the nozzle is typically not generated by a screw mechanism. However, to understand the process it may be useful to study a traditional screw-fed extrusion process as described, for example, by Stevens and Covas [2]. Mass flow through a nozzle is related to pressure drop, nozzle geometry, and material viscosity. The viscosity is of course primarily a function of temperature. Since no special dies or material mixing is required for this type of application, it can be said to behave in a similar manner to a single Archimedean screw extruder as shown in Fig. 6.2.

Using simple screw geometry, molten material will gradually move along the screw channel towards the end of the screw where the nozzle is. The velocity W of material flow along the channel will be

$$W = \pi DN \cos \phi \tag{6.1}$$

where D is the screw diameter, N is the screw speed, and ϕ is the screw angle. The velocity of the material U towards the nozzle is therefore

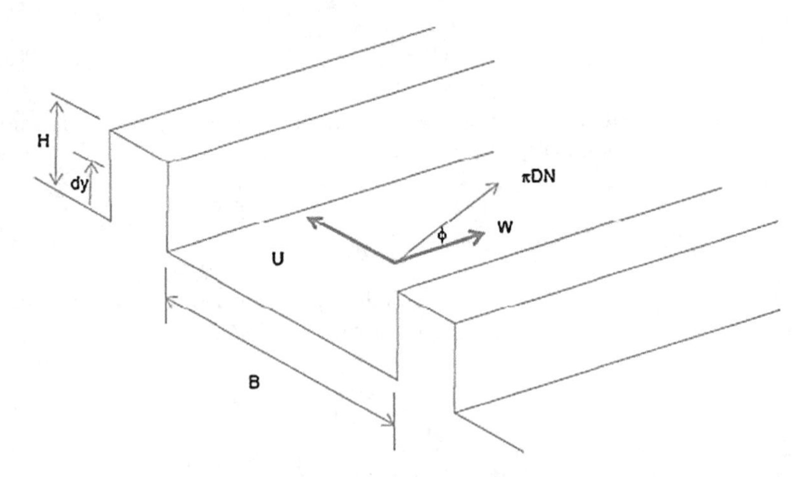

Fig. 6.2 A single Archimedian screw segment. Material flows along the channel in direction *W* and along a fixed barrel in direction *U*

$$U = \pi DN\ \sin\phi \tag{6.2}$$

For a constant helix angle, the volumetric flow caused by the screw in the barrel, known as drag flow Q_D, is proportional to the screw dimensions and speed

$$Q_\mathrm{D}\,\alpha\, D^2NH$$

Since we are operating under drag flow, the relative velocity of the molten material will be W for the material that is in contact with the screw, and 0 for the material in contact with the stationary walls of the barrel. We must therefore integrate over the height of the screw. Generalizing the molten material traveling down this rectangular channel, the along-channel flow Q_D through a channel of B width and dy height can be expressed as

$$\begin{aligned} Q_\mathrm{D} &= \int_0^H \frac{W}{2} B\mathrm{d}y \\ &= \frac{WBH}{2} \end{aligned} \tag{6.3}$$

where H is the screw depth. $W/2$ is defined as the mean down-channel velocity. Substituting for W (6.1) for the screw feed system gives

$$Q_\mathrm{D} = \frac{\pi}{2} DNBH\ \cos\phi$$

We must now consider pressure flow in the channel. Flow through a slit channel, width L, height H, and of infinite length can be derived from the following fundamental equation for shear stress τ

$$\tau(x) = \frac{\Delta P}{L} \cdot x \tag{6.4}$$

where x is perpendicular to the flow direction and ΔP is the pressure change along the channel. For Newtonian flow τ can also be expressed as

$$\tau = -\eta \cdot \frac{\mathrm{d}v_z}{\mathrm{d}x} \tag{6.5}$$

where η is the dynamic viscosity of the molten polymer, defined as a Newtonian fluid. Combining these (6.4) and (6.5), we obtain

$$-\eta \frac{\mathrm{d}v_z}{\mathrm{d}x} = \frac{\Delta P}{L} \mathrm{d}x \tag{6.6}$$

Integration of (6.6) over x, with boundary conditions for $v_z = 0$ when $x = \pm H/2$ (i.e., around the center of the channel and assuming a no-slip boundary) will give the mean velocity for flow of a fluid through a rectangular slit of an infinite length.

$$v_z(x) = \frac{\Delta P}{2\eta L} \left[\frac{H}{2} - x^2 \right]$$

$$\text{Mean velocity } v = \frac{1}{H} \int_{-H/2}^{H/2} v_z(x) \tag{6.7}$$

$$= \frac{\Delta P H^2}{12 \eta L}$$

This velocity can be considered as a result of the back pressure created by the inability for all the molten material to be pushed through an extrusion die (or nozzle) at the end of the channel, which flows opposite the drag flow of the screw. Since the pressure flow rate is volume over time, factoring in B as the screw pitch, or the breadth of the channel and H as the screw depth or the height of the channel:

$$Q_\mathrm{P} = \frac{BH^3}{12\eta} \cdot \frac{\mathrm{d}P}{\mathrm{d}z} \tag{6.8}$$

The pressure calculation for a screw feed is similar to that of flow down a rectangular slit or channel. In order for material to flow down the screw, and therefore material to extrude from the output nozzle, the pressure flow Q_P must exceed the drag flow to give a total flow

$$\begin{aligned} Q_\mathrm{T} &= Q_\mathrm{P} - Q_\mathrm{D} \\ &= \frac{BH^3}{12\eta} \cdot \frac{\mathrm{d}P}{\mathrm{d}z} - \frac{WBH}{2} \end{aligned} \tag{6.9}$$

This provides us with an expression that describes the flow of material back up the rectangular channel as well as down the screw feed, therefore modeling the drag flow generated by the screw and the back flow generated by the pressure differential of the chamber and the output nozzle. A similar pressure flow expression can be formulated for a circular nozzle to model the extrusion process itself. It is assumed that only melt flow exists and that there is a stable and constant temperature within the melt chamber. Both of these are reasonable assumptions.

If a pinch roller feed system is used like in Fig. 6.1, which is in fact the most common approach, one can consider the forces generated by the rollers as the mechanism for generating the extrusion pressure. If the force generated by the rollers exceeds the exit pressure then buckling will occur in the filament feedstock, assuming there is no slippage between the material and the rollers. An excellent model analysis of the pinch roller feed system can be found in Turner et al. [3]. This analysis indicates that the feed forces are related to elastic modulus and that the more brittle the material, the more difficult it is to feed it through the nozzle. This would mean that composite materials that use ceramic fillers for example will require very precise control of feed rates. The increased modulus would lead to higher pressure and thus a higher force generated by the pinch rollers. The greater therefore is the chance of slippage or a mismatch between input and output that would result in non-flow or buckling at the liquifier entry point.

6.2.4 Solidification

Once the material is extruded, it should ideally remain the same shape and size. Gravity and surface tension, however, may cause the material to change shape, while size may vary according to cooling and drying effects. If the material is extruded in the form of a gel, the material may shrink upon drying, as well as possibly becoming porous. If the material is extruded in a molten state, it may also shrink when cooling. The cooling is also very likely to be nonlinear. If this nonlinear effect is significant, then it is possible the resulting part will distort upon cooling. This can be minimized by ensuring the temperature differential between the chamber and the surrounding atmosphere is kept to a minimum (i.e., use of a controlled environmental chamber when building the part) and also by ensuring the cooling process is controlled with a gradual and slow profile.

It is reasonable to assume that an extrusion-based AM system will extrude from a large chamber to a small nozzle through the use of a conical interface. As mentioned before, the melt is generally expected to adhere to the walls of the liquefier and nozzle with zero velocity at these boundaries, subjecting the material to shear deformation during flow. The shear rate $\dot{\gamma}$ can be defined as [1]

$$\dot{\gamma} = -\frac{dv}{dr} \tag{6.10}$$

and the shear stress as

$$\tau = \left(\frac{\dot{\gamma}}{\phi}\right)^{1/m} \tag{6.11}$$

where m represents the flow exponent and ϕ represents the fluidity. The general flow characteristic of a material and its deviation from Newtonian behavior is reflected in the flow exponent m.

6.2.5 Positional Control

Like many AM technologies, extrusion-based systems use a platform that indexes in the vertical direction to allow formation of individual layers. The extrusion head is typically carried on a plotting system that allows movement in the horizontal plane. This plotting must be coordinated with the extrusion rate to ensure smooth and consistent deposition.

Since the plotting head represents a mass and therefore contains an inertial element when moving in a specific direction, any change in direction must result in a deceleration followed by acceleration. The corresponding material flow rate must match this change in speed or else too much or too little material will be deposited in a particular region. For example, if the extrusion head is moving at a velocity v parallel to a nominal x direction and is then required to describe a right angle so that it then moves at the same velocity v in the perpendicular y direction, then, at some point the instantaneous velocity will reach zero. If the extrusion rate is not zero at this point, then excess material will be deposited at the corner of this right angled feature.

Since the requirement is to move a mechanical extrusion head in the horizontal plane then the most appropriate mechanism to use would be a standard planar plotting system. This would involve two orthogonally mounted linear drive mechanisms like belt drives or lead-screws. Such drives need to be powerful enough to move the extrusion chamber at the required velocity and be responsive enough to permit rapid changes in direction without backlash effects. The system must also be sufficiently reliable to permit constant movement over many hours without any loss in calibration. While cheaper systems often make use of belts driven by stepper motors, higher cost systems typically use servo drives with lead-screw technology.

Since rapid changes in direction can make it difficult to control material flow, a common strategy would be to draw the outline of the part to be built using a slower plotting speed to ensure that material flow is maintained at a constant rate. The internal fill pattern can be built more rapidly since the outline represents the external features of the part that corresponds to geometric precision. This outer shell also represents a constraining region that will prevent the filler material from affecting the overall precision. A typical fill pattern can be seen in Fig. 6.3. Determination of the outline and fill patterns will be covered in a later section of this chapter.

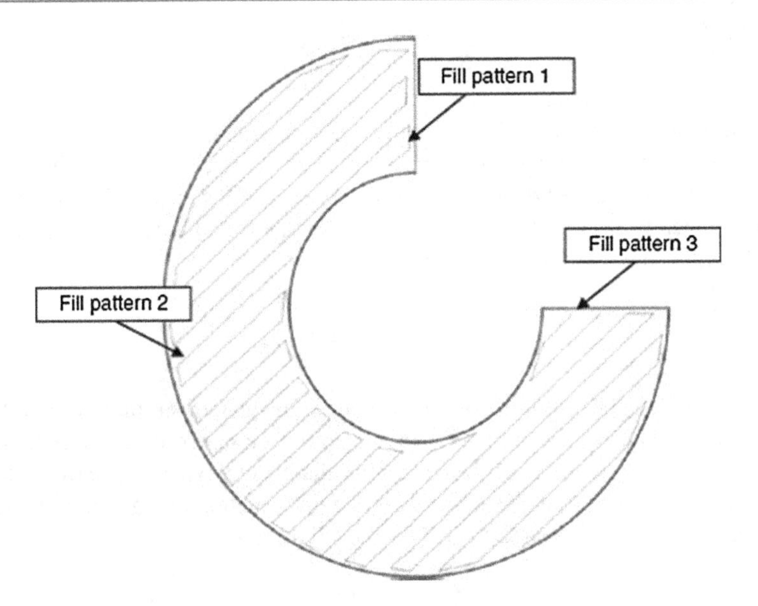

Fig. 6.3 A typical fill pattern using an extrusion-based system, created in three stages (adapted from [4])

6.2.6 Bonding

For heat-based systems there must be sufficient residual heat energy to activate the surfaces of the adjacent regions, causing bonding. Alternatively, gel or paste-based systems must contain residual solvent or wetting agent in the extruded filament to ensure the new material will bond to the adjacent regions that have already been deposited. In both cases, we visualize the process in terms of energy supplied to the material by the extrusion head.

If there is insufficient energy, the regions may adhere, but there would be a distinct boundary between new and previously deposited material. This can represent a fracture surface where the materials can be easily separated. Too much energy may cause the previously deposited material to flow, which in turn may result in a poorly defined part.

Once the material has been extruded, it must solidify and bond with adjacent material. Yardimci defined a set of governing equations that describe the thermal processes at work in a simple extruded road, laid down in a continuous, open-ended fashion along a direction x, based on various material properties [5].

$$\rho\frac{\partial q}{\partial t} = k\frac{\partial^2 T}{\partial x^2} - S_c - S_1 \tag{6.12}$$

where ρ is the material density, q is the specific enthalpy, and k the effective thermal conductivity. T is the cross-sectional average road temperature. The term S_c is a sink term that describes convective losses.

$$S_c = \frac{h}{h_{\text{eff}}}(T - T_\infty) \tag{6.13}$$

h is the convective cooling heat transfer coefficient and h_{eff} is a geometric term representing the ratio of the road element volume to surface for convective cooling. This would be somewhat dependent on the diameter of the nozzle. The temperature T_∞ is the steady-state value of the environment. The term S_l is a sink/source term that describes the thermal interaction between roads:

$$S_l = \frac{k}{\text{Width}^2}\left(T - T_{\text{neigh}}\right) \tag{6.14}$$

where "width" is the width of the road and T_{neigh} is the temperature of the relevant neighboring road. If material is laid adjacent to more material, this sink term will slow down the cooling rate. There is a critical temperature T_c above which a diffusive bonding process is activated and below which bonding is prohibited. On the basis of this, we can state a bonding potential φ as

$$\varphi = \int_0^t (T - T_c)\mathrm{d}\tau \tag{6.15}$$

6.2.7 Support Generation

All AM systems must have a means for supporting free-standing and disconnected features and for keeping all features of a part in place during the fabrication process. With extrusion-based systems such features must be kept in place by the additional fabrication of supports. Supports in such systems take two general forms:

- Similar material supports
- Secondary material supports

If an extrusion-based system is built in the simplest possible way then it will have only one extrusion chamber. If it has only one chamber then supports must be made using the same material as the part. This may require parts and supports to be carefully designed and placed with respect to each other so that they can be separated at a later time. As mentioned earlier, adjustment of the temperature of the part material relative to the adjacent material can result in a fracture surface effect. This fracture surface can be used as a means of separating the supports from the part material. One possible way to achieve this may be to change the layer separation distance when depositing the part material on top of the support material or vice versa. The additional distance can affect the energy transfer sufficiently to result in this fracture phenomenon. Alternatively, adjustment of the chamber or extrusion temperature when extruding supports might be an effective strategy. In all cases however, the support material will be somewhat difficult to separate from the part.

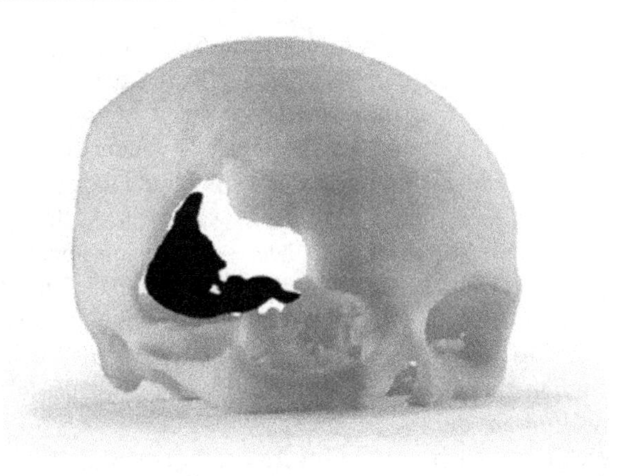

Fig. 6.4 A medical model made using extrusion-based AM technology from two different color materials, highlighting a bone tumor (courtesy of Stratasys)

The most effective way to remove supports from the part is to fabricate them in a different material. The variation in material properties can be exploited so that supports are easily distinguishable from part material, either visually (e.g., using a different color material), mechanically (e.g., using a weaker material for the supports), or chemically (e.g., using a material that can be removed using a solvent without affecting the part material). To do this, the extrusion-based equipment should have a second extruder. In this way, the secondary material can be prepared with the correct build parameters and extruded in parallel with the current layer of build material, without delay. It may be interesting to note that a visually different material, when not used for supports, may also be used to highlight different features within a model, like the bone tumor shown in the medical model of Fig. 6.4.

6.3 Plotting and Path Control

As with nearly all additive manufacturing systems, extrusion-based machines mostly take input from CAD systems using the generic STL file format. This file format enables easy extraction of the slice profile, giving the outline of each slice. As with most systems, the control software must also determine how to fill the material within the outline. This is particularly crucial to this type of system since extrusion heads physically deposit material that fills previously vacant space. There must be clear access for the extrusion head to deposit fill material within the outline without compromising the material that has already been laid down. Additionally, if the material is not laid down close enough to adjacent material, it will not bond effectively. In contrast, laser-based systems can permit, and in fact generally require, a significant amount of overlap from one scan to the next and thus there are no head collision or overfilling-equivalent phenomena.

As mentioned earlier, part accuracy is maintained by plotting the outline material first, which will then act as a constraining region for the fill material. The outline would generally be plotted with a lower speed to ensure consistent material flow. The outline is determined by extracting intersections between a plane (representing the current cross section of the build) and the triangles in the STL file. These intersections are then ordered so that they form a complete, continuous curve for each outline (there may be any number of these curves, either separate or nested inside of each other, depending upon the geometry of that cross section). The only remaining thing for the software to do at this stage is to determine the start location for each outline. Since the extrusion nozzle is a finite diameter, this start location is defined by the center of the nozzle. The stop location will be the final point on this trajectory, located approximately one nozzle diameter from the start location. Since it is better to have a slight overlap than a gap and because it is very difficult to precisely control flow, there is likely to be a slight overfill and thus swelling in this start/stop region. If all the start/stop regions are stacked on top of each other, then there will be a "seam" running down the part. In most cases, it is best to have the start/stop regions randomly or evenly distributed around the part so that this seam is not obvious. However, a counter to this may be that a seam is inevitable and having it in an obvious region will make it more straightforward for removing during the post-processing stage.

Determining the fill pattern for the interior of the outlines is a much more difficult task for the control software. The first consideration is that there must be an offset inside the outline and that the extrusion nozzle must be placed inside this outline with minimal overlap. The software must then establish a start location for the fill and determine the trajectory according to a predefined fill pattern. This fill pattern is similar to those used in CNC planar pocket milling where a set amount of material must be removed with a cylindrical cutter [5]. As with CNC milling, there is no unique solution to achieving the filling pattern. Furthermore, the fill pattern may not be a continuous, unbroken trajectory for a particular shape. It is preferable to have as few individual paths as possible but for complex patterns an optimum value may be difficult to establish. As can be seen with even the relatively simple cross section in Fig. 6.3, start and stop locations can be difficult to determine and are somewhat arbitrary. Even with a simpler geometry, like a circle that could be filled continuously using a spiral fill pattern, it is possible to fill from the outside-in or from the inside-out.

Spiral patterns in CNC are quite common, mainly because it is not quite so important as to how the material is removed from a pocket. However, they are less common as fill patterns for extrusion-based additive manufacturing, primarily for the following reason. Consider the example of building a simple solid cylinder. If a spiral pattern were used, every path on every layer would be directly above each other. This could severely compromise part strength and a weave pattern would be much more preferable. As with composite material weave patterning using material like carbon fiber, for example, it is better to cross the weave over each other at an angle so that there are no weakened regions due to the directionality in the fibers. Placing extrusion paths over each other in a crossing pattern can help to distribute the strength in each part more evenly.

Every additional weave pattern within a specific layer is going to cause a discontinuity that may result in a weakness within the corresponding part. For complex geometries, it is important to minimize the number of fill patterns used in a single layer. As mentioned earlier, and illustrated in Fig. 6.3, it is not possible to ensure that only one continuous fill pattern will successfully fill a single layer. Most outlines can be filled with a theoretically infinite number of fill pattern solutions. It is therefore unlikely that a software solution will provide the best or optimum solution in every case, but an efficient solution methodology should be designed to prevent too many separate patterns from being used in a single layer.

Parts are weakened as a result of gaps between extruded roads. Since weave patterns achieve the best mechanical properties if they are extruded in a continuous path, there are many changes in direction. The curvature in the path for these changes in direction can result in gaps within the part as illustrated in Fig. 6.5. This figure illustrates two different ways to define the toolpath, one that will ensure no additional material will be applied to ensure no part swelling and good part accuracy. The second approach defines an overlap that will cause the material to flow into the void regions, but which may also cause the part to swell. However, in both cases gaps are constrained within the outline material laid down at the perimeter. Additionally, by changing the flow rate at these directional change regions, less or more material can be extruded into these regions to compensate for gaps and swelling. This means that the material flow from the extrusion head should not be directly proportional to the instantaneous velocity of the head when the velocity is low, but rather should be increased or decreased slightly, depending on the toolpath strategy used. Furthermore, if the velocity is zero but the machine is known to be executing a directional change in a weave path, a small amount of flow should ideally be maintained. This will cause the affected region to swell slightly and thus help fill gaps. Obviously, care should be taken to ensure that excess material is not extruded to the extent that part geometry is compromised [6].

It can be seen that precise control of extrusion is a complex trade-off, dependent on a significant number of parameters, including:

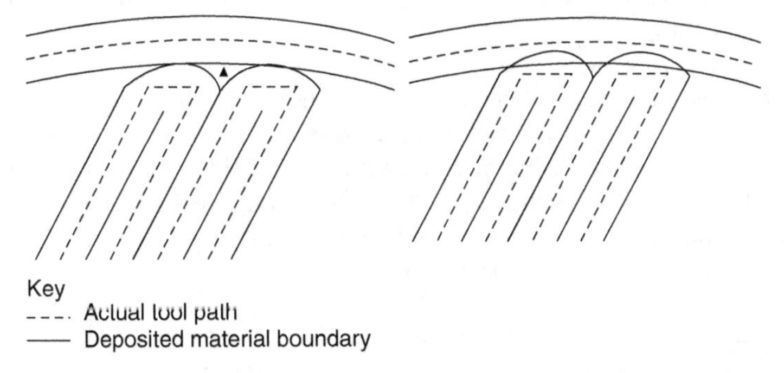

Fig. 6.5 Extrusion of materials to maximize precision (*left*) or material strength (*right*) by controlling voids

- Input pressure: This variable is changed regularly during a build, as it is tightly coupled with other input control parameters. Changing the input pressure (or force applied to the material) results in a corresponding output flow rate change. A number of other parameters, however, also affect the flow to a lesser degree.
- Temperature: Maintaining a constant temperature within the melt inside the chamber would be the ideal situation. However, small fluctuations are inevitable and will cause changes in the flow characteristics. Temperature sensing should be carried out somewhere within the chamber and therefore a loosely coupled parameter can be included in the control model for the input feed pressure to compensate for thermal variations. As the heat builds up, the pressure should drop slightly to maintain the same flow rate.
- Nozzle diameter: This is constant for a particular build, but many extrusion-based systems do allow for interchangeable nozzles that can be used to offset speed against precision.
- Material characteristics: Ideally, control models should include information regarding the materials used. This would include viscosity information that would help in understanding the material flow through the nozzle. Since viscous flow, creep, etc. are very difficult to predict, accurately starting and stopping flow can be difficult.
- Gravity and other factors: If no pressure is applied to the chamber, it is possible that material will still flow due to the mass of the molten material within the chamber causing a pressure head. This may also be exacerbated by gaseous pressure buildup inside the chamber if it is sealed. Surface tension of the melt and drag forces at the internal surfaces of the nozzle may retard this effect.
- Temperature build up within the part: All parts will start to cool down as soon as the material has been extruded. However, different geometries will cool at different rates. Large, massive structures will hold their heat for longer times than smaller, thinner parts, due to the variation in surface to volume ratio. Since this may have an effect on the surrounding environment, it may also affect machine control.

Taking these and other factors into consideration can help one better control the flow of material from the nozzle and the corresponding precision of the final part. However, other uncontrollable or marginally controllable factors may still prove problematic to precisely control flow. Many extrusion-based systems, for instance, resort to periodically cleaning the nozzles from time to time to prevent build up of excess material adhered to the nozzle tip.

6.4 Fused Deposition Modeling from Stratasys

By far the most common extrusion-based AM technology is fused deposition modeling (FDM), produced and developed by Stratasys, USA [7]. FDM uses a heating chamber to liquefy polymer that is fed into the system as a filament. The

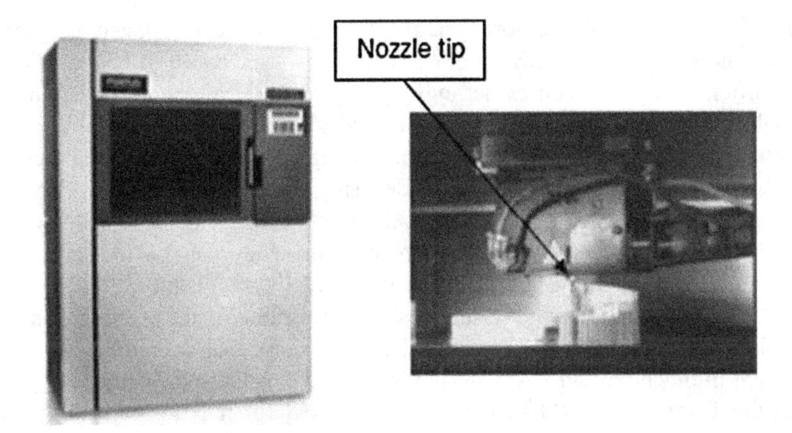

Fig. 6.6 Typical Stratasys machine showing the outside and the extrusion head inside (courtesy of Stratasys)

filament is pushed into the chamber by a tractor wheel arrangement and it is this pushing that generates the extrusion pressure. A typical FDM machine can be seen in Fig. 6.6, along with a picture of an extrusion head.

The initial FDM patent was awarded to Stratasys founder Scott Crump in 1992 and the company has gone from strength to strength to the point where there are more FDM machines than any other AM machine type in the world. The major strength of FDM is in the range of materials and the effective mechanical properties of resulting parts made using this technology. Parts made using FDM are among the strongest for any polymer-based additive manufacturing process.

The main drawback to using this technology is the build speed. As mentioned earlier, the inertia of the plotting heads means that the maximum speeds and accelerations that can be obtained are somewhat smaller than other systems. Furthermore, FDM requires material to be plotted in a point-wise, vector fashion that involves many changes in direction.

6.4.1 FDM Machine Types

The Stratasys FDM machine range is very wide, from low-cost, small-scale, minimal variable machines through to larger, more versatile, and more sophisticated machines that are inevitably more expensive. The company has separated its operations into subsidiaries, each dedicated to different extents of the FDM technology.

The first subsidiary, Dimension, focuses on the low-cost machines currently starting around \$10,000 USD. Each Dimension machine can only process a limited range of materials, with only a few user-controllable parameter options. The Mojo machine is currently the smallest and lowest costing machine, with a maximum part size of $5'' \times 5'' \times 5''$. It has only one layer thickness setting and only one build material, with a soluble support system. There are two further machines that are slightly more expensive, with the uPrint going upwards in size to 10″ × 10″ × 12″ with different layer thickness settings (0.25 and 0.33 mm) and ABS materials available in multiple colors. More expensive variations use the soluble support material while less expensive machines use a single deposition head and breakaway supports. Finer detail parts can be made using the Elite machine, which has a minimum layer thickness of 0.178 mm. All these machines are designed to operate with minimal setup, variation, and intervention. They can be located without special attention to fume extraction and other environmental conditions. This means they can easily be placed in a design office rather than resorting to placing them in a machine shop. Purchasers of Dimension machines would be expected to use them in much the same way as they would an expensive 2D printer.

While Dimension FDM machines can be used for making parts for a wide variety of applications, most parts are likely to be used as concept models by companies investigating the early stages of product development. More demanding applications, like for models for final product approval, functional testing models, and models for direct digital manufacturing, would perhaps require machines that are more versatile, with more control over the settings, more material choices and options that enable the user to correct minor problems in the output model. Higher specification FDM machines are more expensive, not just because of the incorporated technology and increased range of materials, but also because of the sales support, maintenance, and reliability. Stratasys has separated this higher-end technology through the subsidiary named FORTUS, with top-of-the-range models costing around $400,000 USD. The smaller FORTUS 200mc machine starts off roughly where the Dimension machines end, with a slightly smaller build envelope of 8″ × 8″ × 12″ and a similar specification. Further up the range are machines with increases in size, accuracy, range of materials, and range of build speeds. The largest and most sophisticated machine is the FORTUS 900mc, which has the highest accuracy of all Stratasys FDM machines with a layer thickness of 0.076 mm. The build envelope is an impressive 36″ × 24″ × 36″ and there are at least seven different material options.

It should be noted that FDM machines that operate with different layer thicknesses do so because of the use of different nozzle diameters. These nozzles are manually changeable and only one nozzle can be used during a specific build. The nozzle diameter also controls the road width. Obviously, a larger diameter nozzle can extrude more material for a specific plotting speed and thus shorten the build time at the expense of lower precision.

FORTUS software options include the expected file preparation and build setup options. However, there are also software systems that allow the user to remotely monitor the build and schedule builds using a multiple machine setup. Stratasys has many customers who have purchased more than one machine and their software is aimed at ensuring these customers can operate them with a minimum of user intervention. Much of this support was developed because of another Stratasys subsidiary called Redeye, who use a large number of FDM machines as a service bureau for customers. Much of the operation of Redeye is based on customers logging in to an Internet account and uploading STL files. Parts are scheduled for building and sent back to the customer within a few days, depending on part size, amount of finishing required, and order size.

Stratasys also recognizes that many parts coming off their machines will not be immediately suitable for the final application and that there may be an amount of finishing required. To assist in this, Stratasys provides a range of finishing stations that are designed to be compatible with various FDM materials. Finishing can be a mixture of chemically induced smoothing (using solvents that lightly melt the part surface) or burnishing using sodium bicarbonate as a light abrasive cleaning compound. Also, although there is a range of different material colors for the ABS build material, many applications require the application of primers and coatings to achieve the right color and finish on a part.

6.5 Materials

The most popular material is the ABSplus material, which can be used on all current Stratasys FDM machines. This is an updated version of the original ABS (acrylonitrile butadiene styrene) material that was developed for earlier FDM technology. Users interested in a translucent effect may opt for the ABSi material, which has similar properties to other materials in the ABS range. Some machines also have an option for ABS blended with Polycarbonate (PC). Table 6.1 shows properties for various ABS materials and blends.

These properties are quite similar to many commonly used materials. It should be noted, however, that parts made using these materials on FDM machines may exhibit regions of lower strength than shown in this table because of interfacial regions in the layers and possible voids in the parts.

There are three other materials available for FDM technology that may be useful if the ABS materials cannot fulfill the requirements. A material that is predominantly PC-based can provide higher tensile properties, with a flexural strength of

Table 6.1 Variations in properties for the ABS range of FDM materials (compiled from Stratasys data sheets)

Property	ABS	ABSi	ABSplus	ABS/PC
Tensile strength (MPa)	22	37	36	34.8
Tensile modulus (MPa)	1,627	1,915	2,265	1,827
Elongation (%)	6	3.1	4	4.3
Flexural strength (MPa)	41	61	52	50
Flexural modulus (MPa)	1,834	1,820	2,198	1,863
IZOD impact (J/m^2)	106.78	101.4	96	123
Heat deflection at 66 psi (°C)	90	87	96	110
Heat deflection at 264 psi (°C)	76	73	82	96
Thermal expansion (in./in./F)	5.60E – 05	6.7E – 6	4.90E – 05	4.10E – 5
Specific gravity	1.05	1.08	1.04	1.2

104 MPa. A variation of this material is the PC-ISO, which is also PC-based, formulated to ISO 10993-1 and USP Class VI requirements. This material, while weaker than the normal PC with a flexural strength of 90 MPa, is certified for use in food and drug packaging and medical device manufacture. Another material that has been developed to suit industrial standards is the ULTEM 9085 material. This has particularly favorable flame, smoke, and toxicity (FST) ratings that makes it suitable for use in aircraft, marine, and ground vehicles. If applications require improved heat deflection, then an option would be to use the Polyphenylsulfone (PPSF) material that has a heat deflection temperature at 264 psi of 189 °C. It should be noted that these last three materials can only be used in the high-end machines and that they only work with breakaway support system, making their use somewhat difficult and specialized. The fact that they have numerous ASTM and similar standards attached to their materials indicates that Stratasys is seriously targeting final product manufacture (Direct Digital Manufacturing) as a key application for FDM.

Note that FDM works best with polymers that are amorphous in nature rather than the highly crystalline polymers that are more suitable for PBF processes. This is because the polymers that work best are those that are extruded in a viscous paste rather than in a lower viscosity form. As amorphous polymers, there is no distinct melting point and the material increasingly softens and viscosity lowers with increasing temperature. The viscosity at which these amorphous polymers can be extruded under pressure is high enough that their shape will be largely maintained after extrusion, maintaining the extrusion shape and enabling them to solidify quickly and easily. Furthermore, when material is added in an adjacent road or as a new layer, the previously extruded material can easily bond with it. This is different from Selective Laser Sintering, which relies on high crystallinity in the powdered material to ensure that there is a distinct material change from the powder state to a liquid state within a well-defined temperature region.

6.6 Limitations of FDM

FDM machines made by Stratasys are very successful and meet the demands of many industrial users. This is partly because of the material properties and partly because of the low cost of the entry-level machines. There are, however, disadvantages when using this technology, mainly in terms of build speed, accuracy, and material density. As mentioned earlier, they have a layer thickness option of 0.078 mm, but this is only available with the highest-cost machine and use of this level of precision will lead to longer build times. Note also that all nozzles are circular and therefore it is impossible to draw sharp external corners; there will be a radius equivalent to that of the nozzle at any corner or edge. Internal corners and edges will also exhibit rounding. The actual shape produced is dependent on the nozzle, acceleration, and deceleration characteristics, and the viscoelastic behavior of the material as it solidifies.

The speed of an FDM system is reliant on the feed rate and the plotting speed. Feed rate is also dependent on the ability to supply the material and the rate at which the liquefier can melt the material and feed it through the nozzle. If the liquefier were modified to increase the material flow rate, most likely it would result in an increase in mass. This in turn would make it more difficult to move the extrusion head faster. For precise movement, the plotting system is normally constructed using a lead-screw arrangement. Lower cost systems can use belt drives, but flexing in the belts make it less accurate and there is also a lower torque reduction to the drive motor.

One method to improve the speed of motor drive systems is to reduce the corresponding friction. Stratasys used Magnadrive technology to move the plotting head on early Quantum machines. By gliding the head on a cushion of air counterbalanced against magnetic forces attracting the head to a steel platen, friction was significantly reduced, making it easier to move the heads around at a higher speed. The fact that this system was replaced by conventional ball screw drives in the more recent FORTUS 900mc machine indicates that the improvement was not sufficient to balance against the cost.

One method not tried outside the research labs as yet is the use of a particular build strategy that attempts to balance the speed of using thick layers with the precision of using thin layers. The concept here is that thin layers only need to be used on the exterior of a part. The outline of a part can therefore be built using thin layers, but the interior can be built more quickly using thicker layers, similar to the cyclic build styles described in Chap. 4. Since most FDM machines have two extruder heads, it is possible that one head could have a thicker nozzle than the other. This thicker nozzle may be employed to build support structures and to fill in the part interior. However, the difficulty in maintaining a correct registration between the two layer thicknesses has probably prevented this approach from being developed commercially. A compromise on this solution is to use a honeycomb (or similar) fill pattern that uses less material and take less time. This is only appropriate for applications where the reduced mass and strength of such a part is not an issue.

An important design consideration when using FDM is to account for the anisotropic nature of a part's properties. Additionally, different layering strategies result in different strengths. For instance, the right-hand scanning strategy in Fig. 6.5 creates stronger parts than the left-hand scanning strategy. Typically, properties are isotropic in the x–y plane, but if the raster fill pattern is set to preferentially deposit along a particular direction, then the properties in the x–y plane will also be anisotropic. In almost every case, the strength in the z-direction is measurably less than the strength in the x–y plane. Thus, for parts which undergo stress in a particular direction it is best to build the part such that the major stress axes are aligned with the x–y plane rather than in the z-direction.

6.7 Bioextrusion

Extrusion-based technology has a large variety of materials that can be processed. If a material can be presented in a liquid form that can quickly solidify, then it is suitable to this process. As mentioned earlier, the creation of this liquid can be either through thermal processing of the material to create a melt, or by using some form of chemical process where the material is in a gel form that can dry out or chemically harden quickly. These techniques are useful for bioextrusion. Bioextrusion is the process of creating biocompatible and/or biodegradable components that are used to generate frameworks, commonly referred to as "scaffolds," that play host to animal cells for the formation of tissue (tissue engineering). Such scaffolds should be porous, with micropores that allow cell adhesion and macropores that provide space for cells to grow.

There are a few commercial bioextrusion systems, like the modified FDM process used by Osteopore [8] to create scaffolds to assist in primarily head trauma recovery. This machine uses a conventional FDM-like process with settings for a proprietary material, based on the biocompatible polymer, polycaprolactone (PCL). Most tissue engineering is still, however, in research form; investigating many aspects of the process, including material choice, structural strength of scaffolds, coatings, biocompatibility, and effectiveness within various clinical scenarios. Many systems are in fact developed in-house to match the specific interests of the researchers. There are however a small number of systems that are also available commercially to research labs.

6.7.1 Gel Formation

One common method of creating scaffolds is to use hydrogels. These are polymers that are water insoluble but can be dispersed in water. Hydrogels can therefore be extruded in a jelly-like form. Following extrusion, the water can be removed and a solid, porous media remains. Such a media can be very biocompatible and conducive to cell growth with low toxicity levels. Hydrogels can be based on naturally occurring polymers or synthetic polymers. The natural polymers are perhaps more biocompatible whereas the synthetic ones are stronger. Synthetic hydrogels are rarely used in tissue engineering, however, because of the use of toxic reagents. Overall, use of hydrogels results in weak scaffolds that may be useful for soft tissue growth.

6.7.2 Melt Extrusion

Where stronger scaffolds are required, like when used to generate bony tissue, melt extrusion seems to be the process of choice. FDM can be used, but there are some difficulties in using this approach. In particular, FDM is somewhat unsuitable because of the expense of the materials. Biocompatible polymers suitable for tissue

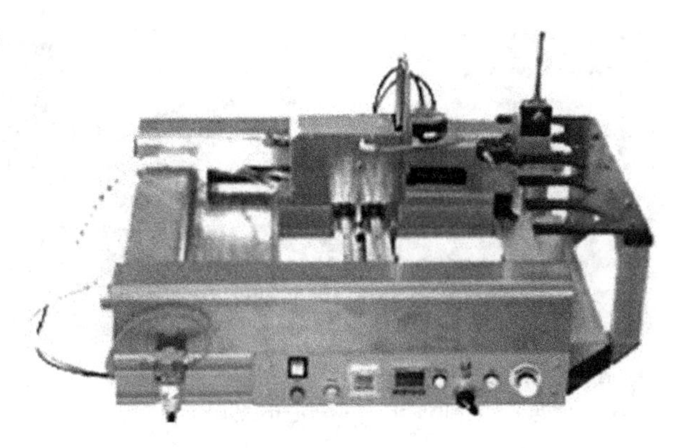

Fig. 6.7 The Envisiontec 3D Bioplotter system (note the multiple head changing system on the *right-hand side*)

engineering are synthesized in relatively small quantities and are therefore only provided at high cost. Furthermore, the polymers often need to be mixed with other materials, like ceramics, that can seriously affect the flow characteristics, causing the material to behave in a non-Newtonian manner. Extrusion using FDM requires the material to be constructed in filament form that is pushed through the system by a pinch roller feed mechanism. This mechanism may not provide sufficient pressure at the nozzle tip, however, and so many of the experimental systems use screw feed, similar to conventional injection molding and extrusion technology. Screw feed systems benefit from being able to feed small amounts of pellet-based feedstock, enabling one to work with a small material volume.

In addition to their layer-wise photopolymerization machines, Envisiontec [9] has also developed the 3D-Bioplotter system (see Fig. 6.7). This system is an extrusion-based, screw feeding technology that is designed specifically for biopolymers. Lower temperature polymers can be extruded using a compressed gas feed, instead of a screw extruder, which results in a much simpler mechanism. Much of the system uses nonreactive stainless steel and the machine itself has a small build envelope and software specifically aimed at scaffold fabrication. The melt chamber is sealed apart from the nozzle, with a compressed air feed to assist the screw extrusion process. The system uses one extrusion head at a time, with a carousel feeder so that extruders can be swapped at any time during the process. This is particularly useful since most tissue engineering research focuses on building scaffolds with different regions made from different materials. Build parameters can be set for a variety of materials with control over the chamber temperature, feed rate, and plotting speed to provide users with a versatile platform for tissue engineering research.

It should be noted that tissue engineering is an extremely complex research area and the construction of physical scaffolds is just the starting point. This approach may result in scaffolds that are comparatively strong compared with hydrogel-based scaffolds, but they may fail in terms of biocompatibility and bio-toxicity. To overcome some of these shortcomings, a significant amount of post-processing is required.

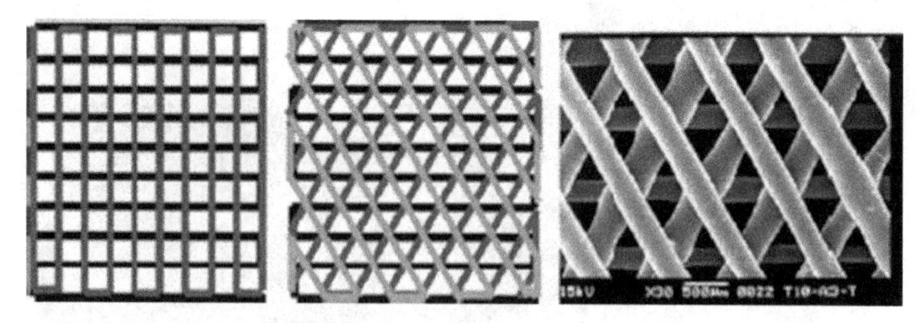

Fig. 6.8 Different scaffold designs showing a porous structure, with an actual image of a scaffold created using a bioextrusion system [10]

6.7.3 Scaffold Architectures

One of the major limitations with extrusion-based systems for conventional manufacturing applications relates to the diameter of the nozzle. For tissue engineering, however, this is not such a limitation. Scaffolds are generally built up so that roads are separated by a set distance so that the scaffold can have a specific macro porosity. In fact, the aim is to produce scaffolds that are as strong as possible but with as much porosity as possible. The greater the porosity, the more space there is for cells to grow. Scaffolds with greater than 66 % porosity are common. Sometimes, therefore, it may be better to have a thicker nozzle to build stronger scaffold struts. The spacing between these struts can be used to determine the scaffold porosity.

The most effective geometry for scaffolds has yet to be determined. For many studies scaffolds with a simple 0° and 90° orthogonal crossover pattern may be sufficient. More complex patterns vary the number of crossovers and their separation. Examples of typical patterns can be seen in Fig. 6.8. Much of the studies involve finding out how cells proliferate in these different scaffold architectures and are usually carried out using bioreactors for in vitro (noninvasive) experiments. As such, samples are usually quite small and often cut from a larger scaffold structure. It is anticipated that it will become commonplace for experiments to be carried out using samples that are as large and complex in shape as the bones they are designed to replace and that are implanted in animal or human subjects. Many more fundamental questions must be answered, however, before this becomes common.

6.8 Other Systems

Although Stratasys owns most of the patents on FDM and similar heat-based extrusion technology, there are a number of other such systems commercially available. The majority of these systems can be purchased only in China, until the expiration of Stratasys' patents. The most successful and well-known system is available from the Beijing Yinhua company. Most of these competing FDM machines utilize a screw extrusion system that are fed using powder or pellet feed rather than continuous filaments.

6.8.1 Contour Crafting

In normal additive manufacturing, layers are considered as 2D shapes extruded linearly in the third dimension. Thicker layers result in lower part precision, particularly where there are slopes or curves in the vertical direction. A major innovative twist on the extrusion-based approach can be found in the Contour Crafting technology developed by Prof. B. Khoshnevis and his team at the University of Southern California [11]. Taking the principle mentioned above that the exterior surface is the most critical in terms of meeting precision requirements, this research team has developed a method to smooth the surface with a scraping tool. This is similar to how artisans shape clay pottery and/or concrete using trowels. By contouring the layers as they are being deposited using the scraping tool to interpolate between these layers, very thick layers can be made that still replicate the intended geometry well.

Using this technique it is conceptually possible to fabricate extremely large objects very quickly compared with other additive processes, since the exterior precision is no longer determined solely by the layer thickness. The scraper tool need not be a straight edge and can indeed be somewhat reconfigurable by positioning different parts of the tool in different regions or by using multiple passes. To illustrate this advantage the team is in fact developing technology that can produce full-sized buildings using a mixture of the Contour Crafting process and robotic assembly (see Fig. 6.9).

6.8.2 Nonplanar Systems

There have been a few attempts at developing AM technology that does not use stratified, planar layers. The most notable projects are Shaped Deposition Manufacture (SDM), Ballistic Particle Manufacture (BPM), and Curved Laminated Object Manufacture (Curved LOM). The Curved LOM [12] process in particular aims at using fiber-reinforced composite materials, sandwiched together for the purposes of making tough-shelled components like nose cones for aircraft using carbon fiber and armored clothing using Kevlar. To work properly, the layers of material must conform to the shape of the part being designed. If edges of laminates are exposed then they can easily come loose by applying shear forces. The Curved LOM process demonstrated feasibility but also quickly became a very complex system that required conformable robotic handling equipment and high powered laser cutting for the laminates.

Fig. 6.9 Contour Crafting technology, developed at USC, showing scraping device and large-scale machine

It is possible to use short fibers mixed with polymer resins in FDM. Fibers can be extruded so long as the diameter and length of the fibers are small enough to prevent clogging of the nozzles. Like Curved LOM, it is somewhat pointless to use such a material in FDM if the layers are aligned with the build plane. However, if the layers were aligned according to the outer layer of the part, then it may be useful. Parts cannot be built using a flat layer approach, in this case, and thus process planning for complex geometries becomes problematic. However, certain parts that require surface toughness can benefit from this nonplanar approach [13].

6.8.3 FDM of Ceramics

Another possible application of FDM is to develop ceramic part fabrication processes. In particular, FDM can be used to extrude ceramic pastes that can quickly solidify. The resulting parts can be fired using a high-temperature furnace to fuse and densify the ceramic particles. Resulting parts can have very good properties with the geometric complexity characteristics of AM processes. Other AM processes have also been used to create ceramic composites, but most work using FDM came out of Rutgers University in the USA [14].

6.8.4 Reprap and Fab@home

The basic FDM process is quite simple; and this can be illustrated by the development of two systems that are extremely low cost and capable of being constructed using minimal tools.

The Reprap project [15] is essentially an experiment in open source technology. The initial idea was developed by a group at the University of Bath in the UK and designs and ideas are being developed by a number of enthusiasts worldwide. One concept being considered is that a machine is capable of producing components for future machines, testing some of the theories of von Neumann on self-replicating machines. A large number of design variants exist, some using cold-cure resins but most using a thermal extrusion head, but all are essentially variants of the FDM process, as illustrated by one of these designs shown in Fig. 6.10.

Another project that aims at low-cost FDM technology is the Fab@home concept. This uses a frame constructed from laser-cut polymer sheets, assembled like a 3D jigsaw. Low-cost stepper motors and drives commonly found in ink-jet printers are used for positioning and the extrusion head is normally a compressed-air-fed syringe that contains a variety of cold-cure materials. The Fab@home designs can be obtained free of charge and kits can be obtained for assembly at a very low cost.

Both of these approaches have inspired a variety of enthusiasts to develop their ideas. Some have focused on improving the designs so that they may be more robust or more versatile. Others have developed software routines that explore things like scanning patterns, more precise control, etc. Yet other enthusiasts have developed new potential applications for this technology, most notably using multiple materials that have unusual chemical or physical behavior. The Fab@home technology has, for example, been used to develop 3D batteries and actuators. Some users have even experimented with chocolate to create edible sculptures.

This development of entry-level AM technology has sparked a 3D Printing revolution that has attracted huge amounts of media attention and brought it into the public domain. This explosion in interest is covered in more detail in a later chapter of this book.

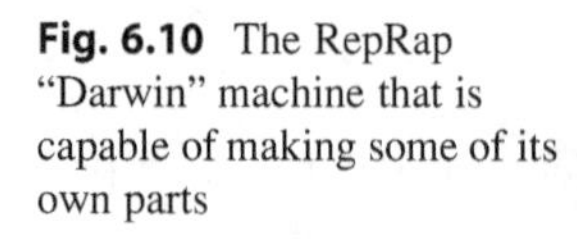

Fig. 6.10 The RepRap "Darwin" machine that is capable of making some of its own parts

6.9 Exercises

1. Derive an expression for Q_T so that we can determine the flow through a circular extrusion nozzle.
2. The expression for total flow does not include a gravity coefficient. Derive an expression for Q_T that includes gravity, assuming there is a constant amount of material in the melt chamber and the nozzle is pointing vertically downwards.
3. The forces generated by the pinch rollers shown in Fig. 6.1 to the elastic modulus.
4. The expressions derived for solidification and bonding assume that a thermal process is being used. What do you think the terms will look like if a curing or drying process were used?
5. Why is extrusion-based AM more suitable for medical scaffold architectures, compared with SLS-fabricated scaffolds made from a similar material?
6. In what ways is extrusion-based AM similar to CNC pocket milling and in what ways is it different?

References

1. Bellini A, Guceri S, Bertoldi M (2004) Liquefier dynamics in fused deposition. J Manuf Sci Eng 126:237–246
2. Stevens MJ, Covas JA (1995) Extruder principles and operation, 2nd edn. Springer, Dordrecht, 494 pp. ISBN: 0412635909, 9780412635908
3. Turner BN, Strong R, Gold SA (2014) A review of melt extrusion additive manufacturing processes. Rapid Prototyp J 20(3):192–204
4. Smid P (2005) CNC programming techniques: an insider's guide to effective methods and applications. Industrial Press Inc., New York, 343 pp. ISBN: 0831131853, 9780831131852
5. Yardimci MA, Guceri S (1996) Conceptual framework for the thermal process of fused deposition. Rapid Prototyp J 2(2):26–31
6. Agarwala MK, Jamalabad VR, Langrana NA et al (1996) Structural quality of parts processed by fused deposition. Rapid Prototyp J 2(4):4–19
7. Stratasys. Fused deposition modelling. http://www.stratasys.com
8. Osteopore. FDM produced tissue scaffolds. http://www.osteopore.com.sg
9. Envisiontec. Bioplotting system. http://www.envisiontec.com
10. Zein I, Hutmacher DW, Tan KC, Teoh SH (2002) Fused deposition modeling of novel scaffold architectures for tissue engineering applications. Biomaterials 23:1169–1185
11. Contour Crafting. large scale extrusion-based process. http://www.contourcrafting.org
12. Klosterman DA, Chartoff RP, Ostborne NR et al (1999) Development of a curved layer LOM process for monolithic ceramics & ceramic matrix composites. Rapid Prototyp J 5(2):61–71
13. Liu Y, Gibson I (2007) A framework for development of a fiber-composite, curved FDM system. Proceedings of the international conference on manufacturing automation, ICMA'07, Singapore, 28–30 May 2007.p 93–102
14. Jafari MA, Han W, Mohammadi F et al (2000) A novel system for fused deposition of advanced multiple ceramics. Rapid Prototyp J 6(3):161–175
15. Reprap. Self replicating AM machine concept. http://www.reprap.org

Material Jetting 7

Abstract

Printing technology has been extensively investigated, with the majority of that investigation historically based upon applications to the two-dimensional printing industry. Recently, however, it has spread to numerous new application areas, including electronics packaging, optics, and additive manufacturing. Some of these applications, in fact, have literally taken the technology into a new dimension. The employment of printing technologies in the creation of three-dimensional products has quickly become an extremely promising manufacturing practice, both widely studied and increasingly widely used.

This chapter will summarize the printing achievements made in the additive manufacturing industry and in academia. The development of printing as a process to fabricate 3D parts is summarized, followed by a survey of commercial polymer printing machines. The focus of this chapter is on material jetting (MJ) in which all of the part material is dispensed from a print head. This is in contrast to binder jetting, where binder or other additive is printed onto a powder bed which forms the bulk of the part. Binder jetting is the subject of Chap. 8. Some of the technical challenges of printing are introduced; material development for printing polymers, metals, and ceramics is investigated in some detail. Models of the material jetting process are introduced that relate pressure required to fluid properties. Additionally, a printing indicator expression is derived and used to analyze printing conditions.

7.1 Evolution of Printing as an Additive Manufacturing Process

Two-dimensional inkjet printing has been in existence since the 1960s, used for decades as a method of printing documents and images from computers and other digital devices. Inkjet printing is now widely used in the desktop printing industry,

I. Gibson et al., *Additive Manufacturing Technologies*,
DOI 10.1007/978-1-4939-2113-3_7

commercialized by companies such as HP and Canon. Le [1] provides a thorough review of the historical development of the inkjet printing industry.

Printing as a three-dimensional building method was first demonstrated in the 1980s with patents related to the development of Ballistic Particle Manufacturing, which involved simple deposition of "particles" of material onto an article [2]. The first commercially successful technology was the ModelMaker from Sanders Prototype (now Solidscape), introduced in 1994, which printed a basic wax material that was heated to liquid state [3]. In 1996, 3D Systems joined the competition with the introduction of the Actua 2100, another wax-based printing machine. The Actua was revised in 1999 and marketed as the ThermoJet [3]. In 2001, Sanders Design International briefly entered the market with its Rapid ToolMaker, but was quickly restrained due to intellectual property conflicts with Solidscape [3]. It is notable that all of these members of the first generation of RP printing machines relied on heated waxy thermoplastics as their build material; they are therefore most appropriate for concept modeling and investment casting patterns.

More recently, the focus of development has been on the deposition of acrylate photopolymer, wherein droplets of liquid monomer are formed and then exposed to ultraviolet light to promote polymerization. The reliance upon photopolymerization is similar to that in stereolithography, but other process challenges are significantly different. The leading edge of this second wave of machines arrived on the market with the Quadra from Objet Geometries of Israel in 2000, followed quickly by the revised QuadraTempo in 2001. Both machines jetted a photopolymer using print heads with over 1,500 nozzles [3]. In 2003, 3D Systems launched a competing technology with its InVision 3D printer. Multi-Jet Modeling, the printing system used in this machine, was actually an extension of the technology developed with the ThermoJet line [3], despite the change in material solidification strategy. The companies continue to innovate, as will be discussed in the next sections.

7.2 Materials for Material Jetting

While industry players have so far introduced printing machines that use waxy polymers and acrylic photopolymers exclusively, research groups around the world have experimented with the potential for printing machines that could build in those and other materials. Among those materials most studied and most promising for future applications are polymers, ceramics, and metals. In addition to the commercially available materials, this section highlights achievements in related research areas.

For common droplet formation methods, the maximum printable viscosity threshold is generally considered to be in the range of 20–40 centipoise (cP) at the printing temperature [4–6]. An equivalent unit of measure is the milli-Pascal-second, denoted mPa s if SI units are preferred. To facilitate jetting, materials that are solid at room temperature must be heated so that they liquefy. For high viscosity fluids, the viscosity of the fluid must be lowered to enable jetting. The most common practices are to use heat or solvents or other low viscosity components in the fluid. In addition to these methods, it is also possible that in some polymer deposition cases shear thinning might occur, dependent upon the material or solution in use; drop-on-demand (DOD) printers are expected to produce strain rates of 10^3–10^4, which should be high enough to produce shear-thinning effects [4, 7]. While other factors such as liquid density or surface tension and print head or nozzle design may affect the results, the limitation on viscosity quickly becomes the most problematic aspect for droplet formation in material jetting.

7.2.1 Polymers

Polymers consist of an enormous class of materials, representing a wide range of mechanical properties and applications. And although polymers are the only material currently used in commercial AM machines, there seems to be relatively little discussion on polymer inkjet production of macro three-dimensional structures in the published scientific literature.

Gao and Sonin [8] present the first notable academic study of the deposition and solidification of groups of molten polymer microdrops. They discuss findings related to three modes of deposition: columnar, sweep (linear), and repeated sweep (vertical walls). The two materials used in their investigations were a candelilla wax and a microcrystalline petroleum wax, deposited in droplets 50 μm in diameter from a print head 3–5 mm from a cooled substrate. The authors first consider the effects of droplet deposition frequency and cooling on columnar formation. As would be expected, if the drops are deposited rapidly ($\geq$50 Hz in this case), the substrate on which they impinge is still at an elevated temperature, reducing the solidification contact angle and resulting in ball-like depositions instead of columns (Fig. 7.1a). Numerical analyses of the relevant characteristic times of cooling are included. Gao and Sonin also consider horizontal deposition of droplets and the subsequent formation of lines. They propose that smooth solid lines will be formed only in a small range of droplet frequencies, dependent upon the sweep speed, droplet size, and solidification contact angle (Fig. 7.1b). Finally, they propose that wall-like deposition will involve a combination of the relevant aspects from each of the above situations.

Reis et al. [9] also provide some discussion on the linear deposition of droplets. They deposited molten Mobilwax paraffin wax with a heated print head from SolidScape. They varied both the print head horizontal speed and the velocity of droplet flight from the nozzle. For low droplet speeds, low sweep speeds created discontinuous deposition and high sweep speeds created continuous lines (Fig. 7.2a–c). High droplet impact speed led to splashing at high sweep speeds and line bulges at low sweep speeds (Fig. 7.2d–f).

From these studies, it is clear that process variables such as print head speed, droplet velocity, and droplet frequency affect the quality of the deposit. These process variables vary depending upon the characteristics of the fluid being printed, so some process development, or fine-tuning, is generally required when trying to print a new material or develop a new printing technology.

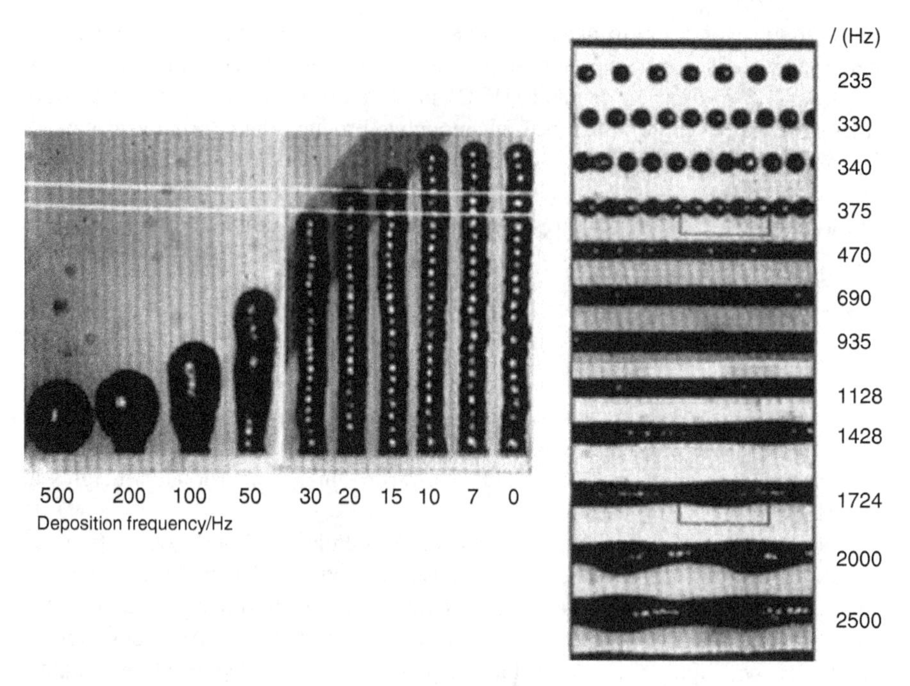

Fig. 7.1 (**a**) Columnar formation and (**b**) line formation as functions of droplet impingement frequency [8]

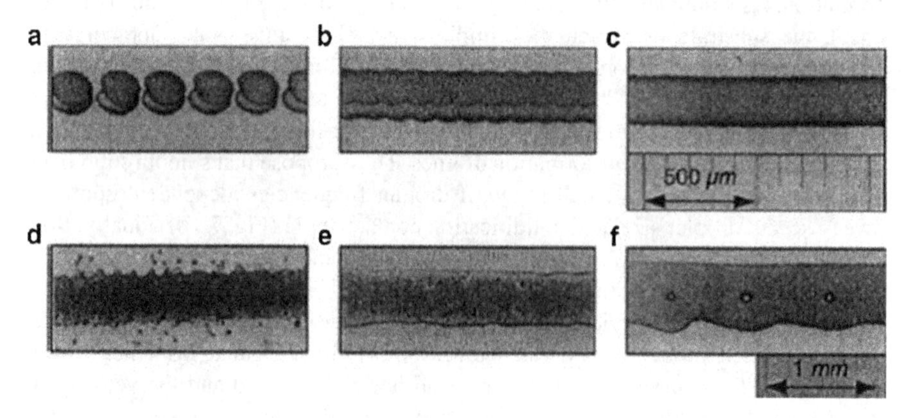

Fig. 7.2 Results of varying sweep and impact speeds [9]

Feng et al. [10] finally present a full system, based on a print head from MicroFab Technologies Inc. that functions similarly to the commercially available machines. It prints a wax material which is heated to 80 °C, more than 10° past its melting point, and deposits it in layers 13–60 μm thick. The deposition pattern is controlled by varying the droplet size and velocity, as well as the pitch and hatch

Fig. 7.3 Wax gear [10]

spacing of the lines produced. An example of the result, a 2.5-dimensional gear, is presented in Fig. 7.3.

The earliest and most often used solution to the problem of high viscosity is to heat the material until its viscosity drops to an acceptable point. As discussed in Sect. 7.5, for example, commercial machines such as 3D Systems' ThermoJet and Solidscape's T66 all print proprietary thermoplastics, which contain mixtures of various waxes and polymers that are solid at ambient temperatures but convert to a liquid phase at elevated printing temperatures [11]. In developing their hot melt materials, for example, 3D Systems investigated various mixtures consisting of a low shrinkage polymer resin, a low viscosity material such as paraffin wax, a microcrystalline wax, a toughening polymer, and a small amount of plasticizer, with the possible additions of antioxidants, coloring agents, or heat dissipating filler [12]. These materials were formulated to have a viscosity of 18–25 cP and a surface tension of 24–29 dyn/cm at the printing temperature of 130 °C. De Gans et al. [13] contend that they have used a micropipette optimized for polymer printing applications that was able to print Newtonian fluids with viscosities up to 160 cP.

The most recent development in addressing the issues of viscosity is the use of prepolymers in the fabrication of polymer parts. This is the method currently employed by the two newest commercially available machine lines, as discussed in Sect. 7.5. For example, 3D Systems investigated a series of UV-curable printing materials, consisting of mixtures of high-molecular weight monomers and oligomers such as urethane acrylate or methacrylate resins, urethane waxes, low molecular weight monomers and oligomers such as acrylates or methacrylates that function as diluents, a small amount of photoinitiators, and other additives such as stabilizers, surfactants, pigments, or fillers [14, 15]. These materials also benefited from the effects of hot melt deposition, as they were printed at a temperature of 70–95 °C, with melting points between 45 and 65 °C. At the printing temperatures, these materials had a viscosity of about 10–16 cP.

One problem encountered, and the reason that the printing temperatures cannot be as high as those used in hot melt deposition, is that when kept in the heated state for extended periods of time, the prepolymers begin to polymerize, raising the viscosity and possibly clogging the nozzles when they are finally printed [15]. Another complication is that the polymerization reaction, which occurs after printing, must be carefully controlled to assure dimensional accuracy.

7.2.2 Ceramics

One significant advance in terms of direct printing for three-dimensional structures has been achieved in the area of ceramic suspensions. As in the case of polymers, studies have been conducted that investigate the basic effects of modifying sweep speed, drop-to-drop spacing, substrate material, line spacing, and simple multilayer forms in the deposition of ceramics [16]. These experiments were conducted with a mixture of zirconia powder, solvent, and other additives, which was printed from a 62 μm nozzle onto substrates 6.5 mm away. The authors found that on substrates that permitted substantial spreading of the deposited materials, neighboring drops would merge to form single, larger shapes, whereas on other substrates the individual dots would remain independent (see Fig. 7.4). In examples where multiple layers were printed, the resulting deposition was uneven, with ridges and valleys throughout.

A sizable body of work has been amassed in which suspensions of alumina particles are printed via a wax carrier [4] which is melted by the print head. Suspensions of up to 40 % solids loading have been successfully deposited at viscosities of 2.9–38.0 cP at a measurement temperature of 100 °C; higher concentrations of the suspended powder have resulted in prohibitively high viscosities. Because this deposition method results in a part with only partial ceramic density, the green part must be burnt out and sintered, resulting in a final product which is 80 % dense but whose dimensions are subject to dramatic shrinkage [17]. A part created in this manner is shown in Fig. 7.5.

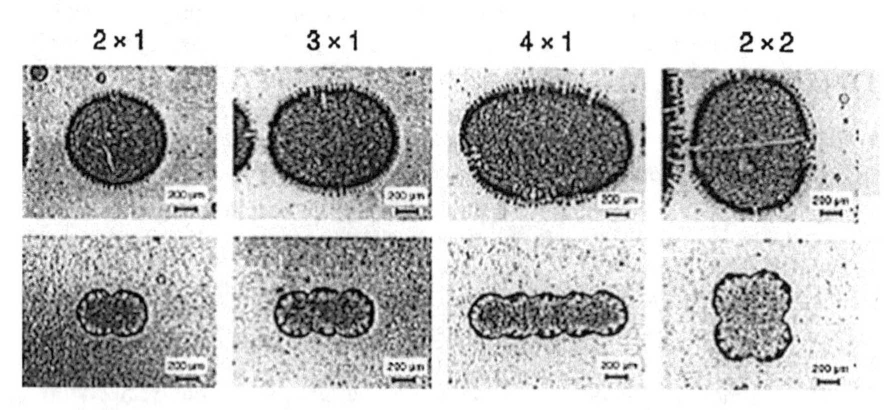

Fig. 7.4 Droplets on two different substrates [16]

Fig. 7.5 Sintered alumina impeller [17]

Fig. 7.6 Sintered zirconia vertical walls [18]

Similar attempts have been made with zirconia powder, using material with 14 % ceramic content by volume [18], with an example shown in Fig. 7.6, as well as with PZT, up to 40 % ceramic particles by volume [19].

7.2.3 Metals

Much of the printing work related to metals has focused upon the use of printing for electronics applications—formation of traces, connections, and soldering. Liu and Orme [20] present an overview of the progress made in solder droplet deposition for the electronics industry. Because solder has a low melting point, it is an obvious choice as a material for printing. They reported use of droplets of 25–500 μm, with

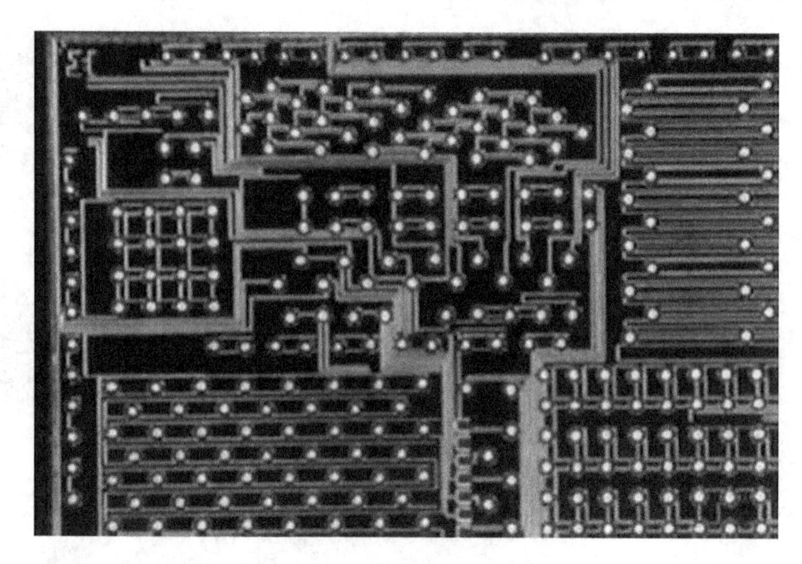

Fig. 7.7 IC test board with solder droplets [20]

results such as the IC test board in Fig. 7.7, which has 70 μm droplets of Sn63/Pb37. In related work, a solder was jetted whose viscosity was approximately 1.3 cP, continuously jetted under a pressure of 138 kPa. Many of the results to which they refer are those of researchers at MicroFab Technologies, who have also produced solder forms such as 25 μm diameter columns.

There is, however, some work in true three-dimensional fabrication with metals. Priest et al. [21] provide a survey of liquid metal printing technologies and history, including alternative technologies employed and ongoing research initiatives. Metals that had been printed included copper, aluminum, tin, various solders, and mercury. One major challenge identified for depositing metals is that the melting point of the material is often high enough to significantly damage components of the printing system.

Orme et al. [22, 23] report on a process that uses droplets of Rose's metal (an alloy of bismuth, lead, and tin). They employ nozzles of diameter 25–150 μm with resulting droplets of 47–283 μm. In specific cases, parts with porosity as low as 0.03 % were formed without post-processing, and the microstructure formed is more uniform than that of standard casting. In discussion of this technology, considerations of jet disturbance, aerodynamic travel, and thermal effects are all presented.

Yamaguchi et al. [24, 25] used a piezoelectrically driven actuator to deposit droplets of an alloy (Bi–Pb–Sn–Cd–In), whose melting point was 47 °C. They heated the material to 55 °C and ejected it from nozzles 200 μm, 50 μm, and less than 8 μm in diameter. As expected, the finer droplets created parts with better resolution. The density, or "packing rate," of some parts reached 98 %. Other examples of fabricated parts are shown in Fig. 7.8.

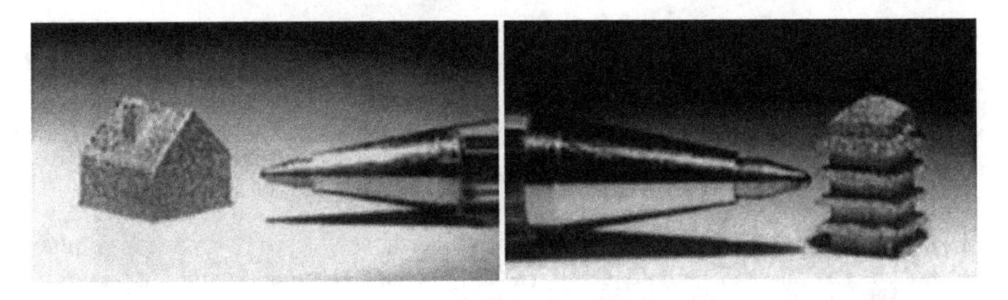

Fig. 7.8 Examples of parts fabricated with metal printing [25]

More recently, several research groups have demonstrated aluminum deposition [26, 27]. In one example, near-net shape components, with fairly simple shapes, have been formed from Al2024 alloy printed from a 100 μm orifice. In another example, pressure pulses of argon gas in the range of 20–100 kPa were used to eject droplets of molten aluminum at the rate of 1–5 drops per second. To achieve this, the aluminum was melted at 750 °C and the substrate to 300 °C. The nozzle orifice used was 0.3 mm in diameter, with a resulting droplet size of 200–500 μm and a deposited line of width 1.00 mm and thickness 0.17 mm. The final product was a near-net shape part of density up to 92 %.

As these examples have shown, printing is well on its way to becoming a viable process for three-dimensional prototyping and manufacturing. While industry has only barely begun to use printing in this arena, the economic and efficiency advantages that printing provide ensure that it will be pursued extensively in the future. Researchers in academia have expanded the use of printing to materials such as ceramics and metals, thus providing additional prospective applications for the technology. Despite its great potential, however, the growth of printing has been hampered significantly by technical challenges inherent to the printing process. These challenges and possible solutions are investigated in subsequent sections.

7.2.4 Solution- and Dispersion-Based Deposition

As hot melt deposition has very specific requirements for the material properties of what is printed, many current applications have turned to solution- or dispersion-based deposition. This allows the delivery of solids or high-molecular weight polymers in a carrier liquid of viscosity low enough to be successfully printed. De Gans et al. [5] provided a review of a number of polymeric applications in which this strategy is employed.

A number of investigators have used solution and dispersion techniques in accurate deposition of very small amounts of polymer in thin layers for mesoscale applications, such as polymer light-emitting displays, electronic components, and surface coatings and masks. For example, Shimoda et al. [28] present a technique to develop light-emitting polymer diode displays using inkjet deposition of conductive polymers. Three different electroluminescent polymers (polyfluorine and two derivatives) were printed in organic solvents at 1–2 wt%. As another example, De Gans et al. [5] report a number of other results related to the creation of polymer light-emitting displays: a precursor of poly(*p*-phenylene vinylene) (PPV) was printed as a 0.3 wt% solution; and PPV derivatives were printed in 0.5–2.0 wt% solutions in solvents such as tetraline, anisole, and o-xylene. Such low weight percentages are typical.

In deposition of ceramics, the use of a low viscosity carrier is also a popular approach. Tay and Edirisinghe [16], for example, used ceramic powder dispersed in industrial methylated spirit with dispersant, binder, and plasticizer additives resulting in a material that was 4.5 % zirconia by volume. The resulting material had a viscosity of 3.0 cP at 20 °C and a shear rate of 1,000 s^{-1}. Zhao et al. [29] tested various combinations of zirconia and wax carried in octane and isopropyl alcohol, with a dispersant added to reduce sedimentation. The viscosities of these materials were 0.6–2.9 cP at 25 °C; the one finally selected was 14.2 % zirconia by volume.

Despite the success of solution and dispersion deposition for these specific applications, however, there are some serious drawbacks, especially in considering the potential for building complex, large, 3D components. The low concentrations of polymer and solid used in the solutions and dispersions will restrict the total amount of material that can be deposited. Additionally, it can be difficult to control the deposition pattern of this material within the area of the droplet's impact. Shimoda et al. [28], among others, report the formation of rings of deposited material around the edge of the droplet. They attribute this to the fact that the contact line of the drying drop is pinned on the substrate. As the liquid evaporates from the edges, it is replenished from the interior, carrying the solutes to the edge. They contend that this effect can be mitigated by control of the droplet drying conditions.

Another difficulty with solutions or dispersions, especially those based on volatile solvents, is that use of these materials can result in precipitations forming in the nozzle after a very short period of time [13], which can clog the nozzle, making deposition unreliable or impossible.

7.3 Material Processing Fundamentals

7.3.1 Technical Challenges of MJ

As evidenced by the industry and research applications discussed in the previous section, material jetting already has a strong foothold in terms of becoming a successful AM technology. There are, however, some serious technical shortcomings that have prevented its development from further growth. To identify and address those problems, the relevant phenomena and strategic approaches taken by its developers must be understood. In the next two sections, the technical challenges of the printing process are outlined, the most important of its limitations relevant to the deposition of functional polymers are identified, and how those limitations are currently addressed is summarized.

Jetting for three-dimensional fabrication is an extremely complex process, with challenging technical issues throughout. The first of these challenges is formulation of the liquid material. If the material is not in liquid form to begin with, this may mean suspending particles in a carrier liquid, dissolving materials in a solvent, melting a solid polymer, or mixing a formulation of monomer or prepolymer with a polymerization initiator. In many cases, other substances such as surfactants are added to the liquid to attain acceptable characteristics. Entire industries are devoted to the mixture of inks for two-dimensional printing, and it is reasonable to assume that in the future this will also be the case for three-dimensional fabrication.

The second hurdle to overcome is droplet formation. To use inkjet deposition methods, the material must be converted from a continuous volume of liquid into a number of small discrete droplets. This function is often dependent upon a finely tuned relationship between the material being printed, the hardware involved, and the process parameters; a number of methods of achieving droplet formation are discussed in this section. Small changes to the material, such as the addition of tiny particles [30], can dramatically change its droplet forming behavior as well, as can changes to the physical setup.

A third challenge is control of the deposition of these droplets; this involves issues of droplet flight path, impact, and substrate wetting or interaction [31–35]. In printing processes, either the print head or the substrate is usually moving, so the calculation of the trajectory of the droplets must take this issue into account. In addition to the location of the droplets' arrival, droplet velocity and size will also affect the deposition characteristics and must be measured and controlled via nozzle design and operation [36]. The quality of the impacted droplet must also be controlled: if smaller droplets, called satellites, break off from the main droplet during flight, then the deposited material will be spread over a larger area than intended and the deposition will not have well-defined boundaries. In the same way, if the droplet splashes on impact, forming what is called a "crown," similar results will occur [37]. All of the effects will negatively impact the print quality of the printed material.

Concurrently, the conversion of the liquid material droplets to solid geometry must be carefully controlled; as discussed in Sect. 7.2, material jetting relies on a phase change of the printed material. Examples of phase change modes employed in existing printing technologies are: solidification of a melted material (e.g., wax, solder), evaporation of the liquid portion of a solution (e.g., some ceramic approaches), and curing of a photopolymer (e.g., Objet, ProJet machines) or other chemical reactions. The phase change must occur either during droplet flight or soon after impact; the time and place of this conversion will also affect the droplet's interaction with the substrate [38, 39] and the final deposition created. To further complicate the matter, drops may solidify nonuniformly, creating warpage and other undesirable results [40].

An additional challenge is to control the deposition of droplets on top of previously deposited layers, rather than only upon the initial substrate

[8, 16]. The droplets will interact differently, for example, with a metal plate substrate than with a surface of previously printed wax droplets. To create substantive three-dimensional parts, each layer deposited must be fully bound to the previous layer to prevent delamination, but must not damage that layer while being created. Commercially available machines tend to approach this problem by employing devices that plane or otherwise smooth the surface periodically [40–42].

Operational considerations also pose a challenge in process planning for MJ. For example, because nozzles are so small, they often clog, preventing droplets from exiting. Much attention has been given to monitoring and maintaining nozzle performance during operation [40]. Most machines currently in use go through purge and cleaning cycles during their builds to keep as many nozzles open as possible; they may also wipe the nozzles periodically [41]. Some machines may also employ complex sensing systems to identify and compensate for malfunctioning or inconsistent nozzles [43, 44]. In addition, many machines, including all commercial AM machines, have replaceable nozzles in case of permanent blockage.

Finally, to achieve the best print resolution, it is advantageous to produce many small droplets very close together. However, this requires high nozzle density in the print head, which is unattainable for many nozzle manufacturing processes. An alternative to nozzle density is to make multiple passes over the same area, effectively using process planning instead of hardware to create the desired effect [41]. Even in cases where high nozzle density is possible, however, problems arise due to crosstalk—basically an "overlapping" of the thermal or pressure differentials used to drive adjacent nozzles.

In approaching a printing process, these numerous challenges must in some sense be addressed sequentially: flight pattern cannot be studied until droplets are formed and layering cannot be investigated until deposition of single droplets is controlled. In terms of functional polymer deposition, the challenge of material preparation has effectively been addressed; numerous polymer resins and mixtures already exist. It is the second challenge—droplet formation—that is therefore the current limiting factor in deposition of these materials. To understand these limitations, Sect. 7.3.2 reviews the dynamic processes that are currently used to form droplets and Sect. 7.2 considered necessary methods of modifying the jetting material for use with those processes.

7.3.2 Droplet Formation Technologies

Over the time that two-dimensional inkjet printing has evolved, a number of methods for creating and expelling droplets have been developed. The main distinction in categorizing the most common of technologies refers to the possible modes of expulsion: continuous stream (CS) and DOD. This distinction refers to the form in which the liquid exits the nozzle—as either a continuous column of liquid or as discrete droplets. Figure 7.9 shows the distinction between continuous (left) and DOD (right) formations.

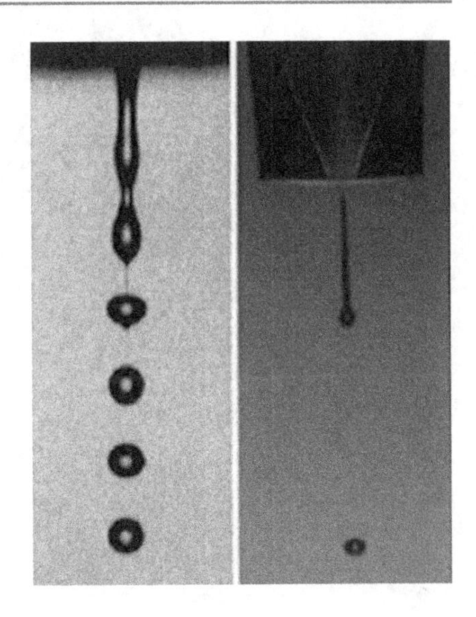

Fig. 7.9 Continuous (*left*) and drop-on-demand (*right*) deposition [46]

7.3.3 Continuous Mode

In CS mode, a steady pressure is applied to the fluid reservoir, causing a pressurized column of fluid to be ejected from the nozzle. After departing the nozzle, this stream breaks into droplets due to Rayleigh instability. The breakup can be made more consistent by vibrating, perturbing, or modulating the jet at a fixed frequency close to the spontaneous droplet formation rate, in which case the droplet formation process is synchronized with the forced vibration, and ink droplets of uniform mass are ejected [45]. Because droplets are produced at constant intervals, their deposition must be controlled after they separate from the jet. To achieve this, they are introduced to a charging field and thus attain an electrostatic charge. These charged particles then pass through a deflection field, which directs the particles to their desired destinations—either a location on the substrate or a container of material to be recycled or disposed. Figure 7.10 shows a schematic of the function of this type of binary deflection continuous system.

An advantage of CS deposition is the high throughput rate; it has therefore seen widespread use in applications such as food and pharmaceutical labeling [5]. Two major constraints related to this method of droplet formation are, however, that the materials must be able to carry a charge and that the fluid deflected into the catcher must be either disposed of or reprocessed, causing problems in cases where the fluid is costly or where waste management is an issue.

In terms of droplets formed, commercially available systems typically generate droplets that are about 150 μm in diameter at a rate of 80–100 kHz, but frequencies of up to 1 MHz and droplet sizes ranging from 6 μm (10 fL) to 1 mm (0.5 μL) have been reported [46]. It has also been shown that, in general, droplets formed from continuous jets are almost twice the diameter of the undisturbed jet [47].

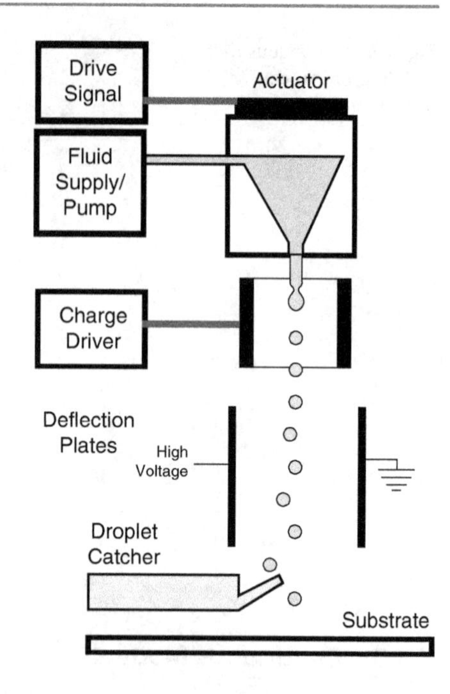

Fig. 7.10 Binary deflection continuous printing [46]

A few investigators of three-dimensional deposition have opted to use continuous printing methods. Blazdell et al. [48] used a continuous printer from Biodot, which was modulated at 66 kHz while ejecting ceramic ink from 50 and 75 μm nozzles. They used 280 kPa of air pressure. Blazdell [49] reports later results in which this Biodot system was modulated at 64 kHz, using a 60 μm nozzle that was also 60 μm in length. For much of the development of the 3D Printing binder jetting process, CS deposition was used. At present, the commercial machines based on 3DP (from 3D Systems and Ex One) use standard DOD print heads. In metal fabrication, Tseng et al. [50] used a continuous jet in depositing their solder alloy, which had a viscosity of about 2 cP at the printing temperature. Orme et al. [22, 23] also report the use of an unspecified continuous system in deposition of solders and metals.

7.3.4 DOD Mode

In DOD mode, in contrast, individual droplets are produced directly from the nozzle. Droplets are formed only when individual pressure pulses in the nozzle cause the fluid to be expelled; these pressure pulses are created at specific times by thermal, electrostatic, piezoelectric, acoustic, or other actuators [1]. Figure 7.11 shows the basic functions of a DOD setup. Liu and Orme [20] assert that DOD methods can deposit droplets of 25–120 μm at a rate of 0–2,000 drops per second.

In the current DOD printing industry, thermal (bubble-jet) and piezoelectric actuator technologies dominate; these are shown in Fig. 7.12. Thermal actuators

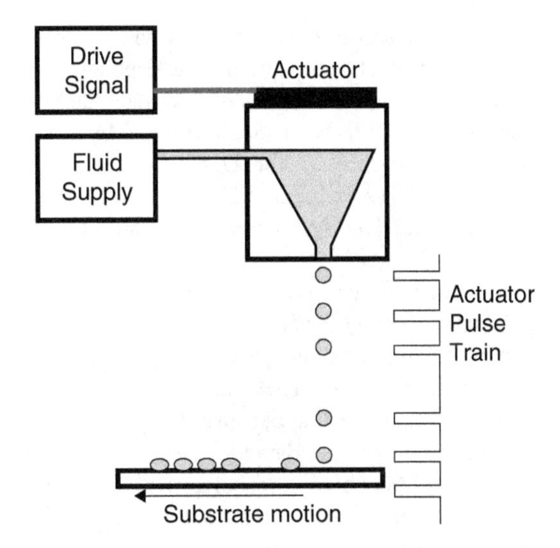

Fig. 7.11 Schematic of drop-on-demand printing system [46]

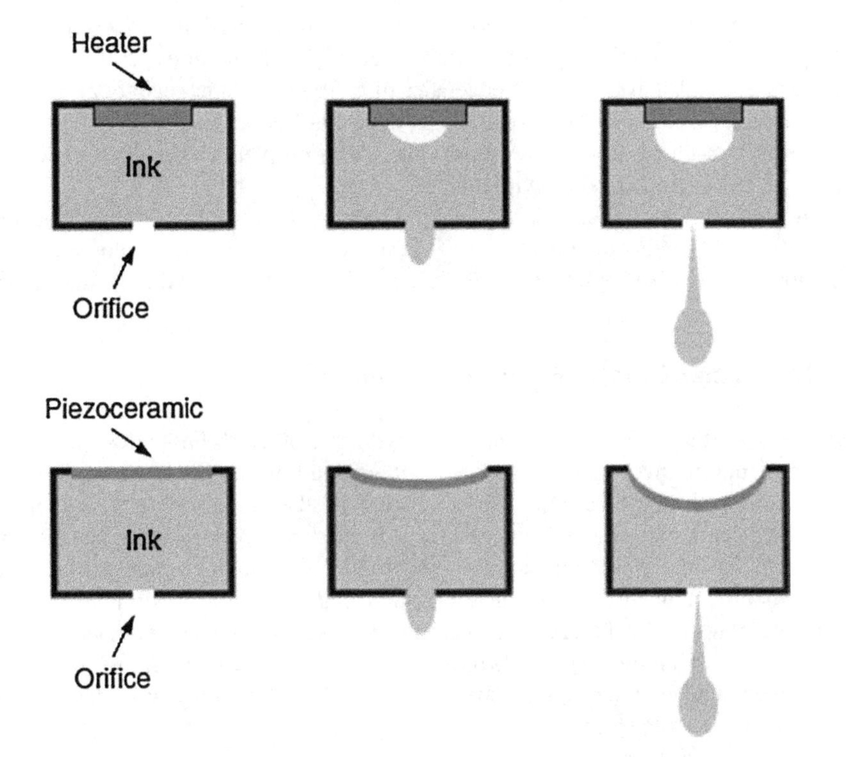

Fig. 7.12 Thermal (*top*) and piezoelectric (*bottom*) DOD ejection

rely on a resistor to heat the liquid within a reservoir until a bubble expands in it, forcing a droplet out of the nozzle. Piezoelectric actuators rely upon the deformation of a piezoelectric element to reduce the volume of the liquid reservoir, which causes a droplet to be ejected. As noted by Basaran [51], the waveforms employed in piezoelectrically driven DOD systems can vary from simple positive square waves to complex negative–positive–negative waves in which the amplitude, duration, and other parameters are carefully modulated to create the droplets as desired.

In their review of polymer deposition, De Gans et al. [5] assert that DOD is the preferable method for all applications that they discuss due to its smaller drop size (often of diameter similar to the orifice) and higher placement accuracy in comparison to CS methods. They further argue that piezoelectric DOD is more widely applicable than thermal because it does not rely on the formation of a vapor bubble or on heating that can damage sensitive materials.

The preference for piezoelectrically driven DOD printing is reflected in the number of investigators who use and study such setups. For example, Gao and Sonin [8] use this technology to deposit 50 μm droplets of two waxes, whose viscosity at 100 °C is about 16 cP. Sirringhaus et al. [52] and Shimoda et al. [28] both use piezoelectric DOD deposition for polymer solutions, as discussed in Sect. 7.2.1. In ceramic deposition, Reis et al. [9] print mixtures with viscosities 6.5 and 14.5 cP at 100 °C and frequencies of 6–20 kHz. Yamaguchi et al. [24, 25] also used a piezoelectrically driven DOD device at frequencies up to 20 Hz in the deposition of metal droplets. Similarly, the solder droplets on the circuit board in Fig. 7.7 were also deposited with a DOD system.

At present, all commercial AM printing machines use DOD print heads, generally from a major manufacturer of printers or printing technologies. Such companies include Hewlett-Packard, Canon, Dimatix, Konica-Minolta, and Xaar.

7.3.5 Other Droplet Formation Methods

Aside from the standard CS and DOD methods, other technologies have been experimentally investigated but have not enjoyed widespread use in industry applications. Liquid spark jetting, a relative of thermal printing, relies on an electrical spark discharge instead of a resistor to form a gas bubble in the reservoir [45, 52]. The electrohydrodynamic inkjet employs an extremely powerful electric field to pull a meniscus and, under very specific conditions, droplets from a pressure-controlled capillary tube; these droplets are significantly smaller than the tube from which they emanate. Electro-rheological fluid jetting uses an ink whose properties change under high electric fields; the fluid flows only when the electric field is turned off [53]. In their flextensional ultrasound droplet ejectors, Percin and Khuri-Yakub [54] demonstrate both DOD and continuous droplet formation with a system in which a plate containing the nozzle orifice acts as the actuator, vibrating at resonant frequencies and forming droplets by creating capillary waves on the liquid surface as well as an increased pressure in the liquid. Focused acoustic beam ejection uses a lens to focus an ultrasound beam onto the free surface of a fluid, using the acoustic pressure transient generated by the focused tone burst to eject a fluid droplet [55]. Meacham et al. extended this work [56] to develop an inexpensive ultrasonic droplet generator and developed a fundamental understanding of its droplet formation mechanisms [57]. These ultrasonic droplet generators show promise in ejecting viscous polymers [58]. Fukumoto et al. [59] present a variant technology in which ultrasonic waves are focused onto the surface of the liquid, forming surface waves that eventually break off into a mist of small droplets. Overviews of these various droplet formation methods are given by Lee [53] and Basaran [51].

Summary: While the general challenges of material jetting for three-dimensional fabrication are identified, there are many aspects that are not well or fully understood. Open research questions abound in almost all stages of the printing process—droplet formation, deposition control, and multilayer accumulation. For the case of polymer jetting, the most appropriate limitation to address is that of droplet formation. Because systems developed for inviscid materials are being used for these applications, numerous accommodations and limitations currently exist; users commonly handle this by modifying the materials to fit the requirements of the existing hardware. However, if the method of droplet formulation could be modified instead, this might allow the deposition of a wider range of materials. A recently developed acoustic focusing ultrasonic droplet generator, under investigation at Georgia Institute of Technology, employs a strategy different from those of existing technologies, which may provide the capabilities to fulfill this need [60].

7.4 MJ Process Modeling

Conservation of energy concepts provides an appropriate context for investigating droplet generation mechanisms for printing. Essentially, the energy imparted by the actuation method to the liquid must be sufficient to balance three requirements: fluid flow losses, surface energy, and kinetic energy. The losses originate from a conversion of kinetic energy to thermal energy due to the viscosity of the fluid within the nozzle; this conversion can be thought of as a result of internal friction of the liquid. The surface energy requirement is the additional energy needed to form the free surface of the droplet or jet. Finally, the resulting droplet or jet must still retain enough kinetic energy to propel the liquid from the nozzle towards the substrate. This energy conservation can be summarized as

$$E_{\text{imparted}} = E_{\text{loss}} + E_{\text{surface}} + E_{\text{kinetic}} \tag{7.1}$$

The conservation law can be considered in the form of actual energy calculations or in the form of pressure, or energy per unit volume, calculations. For example, Sweet used the following approximation for the gauge pressure required in the reservoir of a continuous jetting system [61]:

$$\Delta p = 32\mu d_j^2 v_j \int_{l_1}^{l_2} \frac{dl}{d_n^4} + \frac{2\sigma}{d_j} + \frac{\rho v_j^2}{2} \tag{7.2}$$

where, Δp is the total gauge pressure required, μ is the dynamic viscosity of the liquid, ρ is the liquid's density, σ is the liquid's surface tension, d_j is the diameter of the resultant jet, dn is the inner diameter of the nozzle or supply tubing, v_j is the velocity of the resultant jet, and l is the length of the nozzle or supply tubing. The first term on the right of (7.2) is an approximation of the pressure loss due to viscous friction within the nozzle and supply tubing. The second term is the internal pressure of the jet due to surface tension and the third term is the pressure required to provide the kinetic energy of the droplet or jet.

Energy conservation can also be thought of as a balance among the effects before the fluid crosses a boundary at the orifice of the nozzle and after it crosses that boundary. Before the fluid leaves the nozzle, the positive effect of the driving pressure gradient accelerates it, but energy losses due to viscous flow decelerate it. The kinetic energy with which it leaves the nozzle must be enough to cover the kinetic energy of the traveling fluid as well as the surface energy of the new free surface.

As indicated earlier, actuation energy is typically in the form of heating (bubble-jet) or vibration of a piezoelectric actuator. Various electrical energy waveforms may be used for actuation. In any event, these are standard types of inputs and will not be discussed further.

While the liquid to be ejected travels through the nozzle, before forming droplets, its motion is governed by the standard equations for incompressible, Newtonian fluids, as we are assuming these flows to be. The flow is fully described by the Navier–Stokes and continuity equations; however, these equations are difficult to solve analytically, so we will proceed with a simplification. The first term on the right side of (7.2) takes advantage of one situation for which an analytical solution is possible, that of steady, incompressible, laminar flow through a straight circular tube of constant cross section. The solution is the Hagen–Poiseuille law [62], which reflects the viscous losses due to wall effects:

$$\Delta p = \frac{8Q\mu l}{\pi r^4 \sigma} \tag{7.3}$$

where Q is the flow rate and r is the tube radius. Note that this expression is most applicable when the nozzle is a long, narrow glass tube. However, it can also apply when the fluid is viscous, as we will see shortly.

Another assumption made by using the Hagen–Poiseuille equation is that the flow within the nozzle is fully developed. For the case of laminar flow in a cylindrical pipe, the length of the entry region l_e where flow is not yet fully developed is defined as 0.06 times the diameter of the pipe, multiplied by the Reynolds number [62]:

Table 7.1 Entry lengths for "water" at various viscosities

Viscosity (cP)	Density (kg/m^3)	Entry length (μm)
1	1,000	240
	1,250	300
10	1,000	24
	1,250	30
40	1,000	6
	1,250	7.5
100	1,000	2.4
	1,250	3
200	1,000	1.2
	1,250	1.5

$$l_e = 0.06dRe = \frac{0.06\rho\bar{v}d^2}{\mu} \tag{7.4}$$

where $\bar{v}$ is the average flow velocity across the pipe. To appreciate the magnitude of this effect, consider printing with a 20 μm nozzle in a plate that is 0.1 mm thick, where the droplet ejection speed is 10 m/s. The entry lengths for a fluid with the density of water and varying viscosities are shown in Table 7.1.

Entry lengths are a small fraction of the nozzle length for fluids with viscosities of 40 cP or greater. As a result, we can conclude that flows are fully developed through most of a nozzle for fluids that are at the higher end of the range of printable viscosities.

Most readers will have encountered the primary concepts of fluid mechanics in an undergraduate course and may be familiar with the Navier–Stokes equation, viscosity, surface tension, etc. As a reminder, viscosity is a measure of the resistance of a fluid to being deformed by shear or extensional forces. We will restrict our attention to dynamic, or absolute, viscosity, which has units of pressure-time; in the SI system, units are typically Pa · s or mPa s, for milli-Pascals·seconds. Viscosity is also given in units of poise or centipoises, named after Jean Louis Marie Poiseuille. Centipoise is abbreviated cP, which conveniently has the same magnitude as mPa s. That is, 1 cP is equal to 1 mPa s. Surface tension is given in units of force per length, or energy per unit area; in the SI system, surface tension often has units of N/m or J/m^2.

We can investigate the printing situation further by computing the pressures required for ejection. Equation (7.3) will be used to compute the pressure required to print droplets for various fluid viscosities and nozzle diameters. For many printing situations, wall friction dominates the forces required to print, hence we will only investigate the first term on the right of (7.2) and ignore the second and third terms (which are at least one order of magnitude smaller than wall friction).

Figure 7.13 shows how the pressure required to overcome wall friction varies with fluid viscosity and nozzle diameter. As can be seen, pressure needs to increase sharply as nozzles vary from 0.1 to 0.02 mm in diameter. This could be expected, given the quadratic dependence of pressure on diameter in (7.3). Pressure is seen to

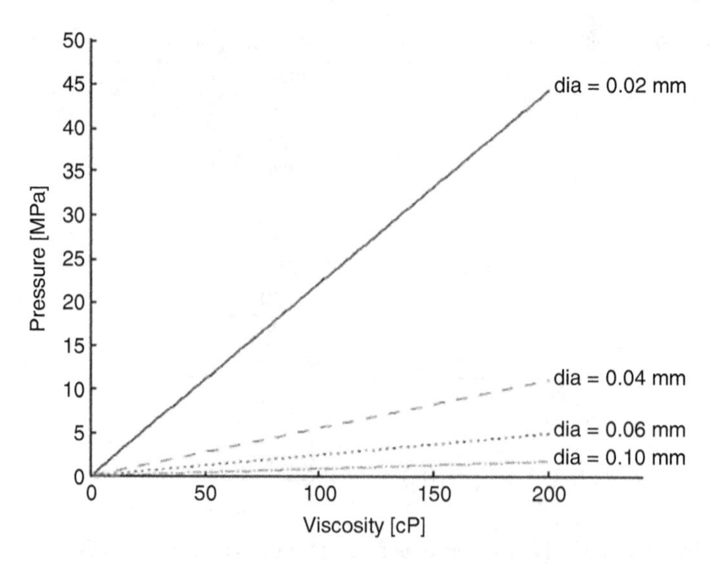

Fig. 7.13 Pressure required to overcome wall friction for printing through nozzles of different diameters

increase linearly with viscosity, which again can be expected from (7.3). As indicated, wall friction dominates for many printing condition. However, as nozzle size increases, the surface tension of the fluid becomes more important. Also, as viscosity increases, viscous losses become important, as viscous fluids can absorb considerable acoustic energy. Regardless, this analysis provides good insight into pressure variations under many typical printing conditions.

Fluid flows when printing are almost always laminar; i.e., the Reynolds number is less than 2,100. As a reminder, the Reynolds number is

$$\mathrm{Re} = \frac{\rho v r}{\mu} \tag{7.5}$$

Another dimensionless number of relevance in printing is the Weber number, which describes the relative importance of a fluid's inertia compared with its surface tension. The expression for the Weber number is:

$$\mathrm{We} = \frac{\rho v^2 r}{\gamma} \tag{7.6}$$

Several research groups have determined that a combination of the Reynolds and Weber numbers is a particularly good indication of the potential for successful printing of a fluid [17]. Specifically, if the ratio of the Reynolds number to the square root of the Weber number has a value between 1 and 10, then it is likely that ejection of the fluid will be successful. This condition will be called the "printing indicator" and is

Table 7.2 Reynolds numbers and printing indicator values for some printing conditions

Nozzle diameter (mm)	Viscosity (cP)	Reynolds no.	Printing indicator
0.02	1	20	37.9
	10	2	3.79
	40	0.5	0.949
	100	0.2	0.379
0.05	1	50	60.0
	10	5	6.00
	40	1.25	1.50
	100	0.5	0.600
0.1	1	100	84.9
	10	10	8.49
	40	2.5	2.12
	100	1	0.849

$$1 \leq \frac{\text{Re}}{\text{We}^{1/2}} = \frac{\sqrt{\rho r \gamma}}{\mu} \leq 10 \tag{7.7}$$

The inverse of the printing indicator is another dimensionless number called the Ohnsorge number, that relates viscous and surface tension forces. Note that values of this ratio that are low indicate that flows are viscosity limited, while large values indicate flows that are dominated by surface tension. The low value of 1 for the printing indicator means that the maximum fluid viscosity should be between 20 and 40 cP.

Some examples of Reynolds numbers and printing indicators are given in Table 7.2. For these results, the surface tension is 0.072 N/m, the density is 1,000 kg/m^3 (same as water at room temperature), and droplet velocity is 1 m/s.

It is important to realize that the printing indicator is a guide, not a law to be followed. Water is usually easy to print through most print heads, regardless of the nozzle size. But the printing indicator predicts that water (with a viscosity of 1 cP) should not be ejectable since its surface tension is too high. We will see in the next section how materials can be modified in order to make printing feasible.

7.5 Material Jetting Machines

The three main companies involved in the development of the RP printing industry are still the main players offering printing-based machines. Solidscape, 3D Systems, and Stratasys (after their merger with Objet Geometries). Solidscape sells the T66 and T612, both descendants of the previous ModelMaker line and based upon the first-generation melted wax technique. Each of these machines employs two single jets—one to deposit a thermoplastic part material and one to deposit a waxy support material—to form layers 0.0005 in. thick [63]. It should be noted that these machines also fly-cut layers after deposition to ensure that the layer

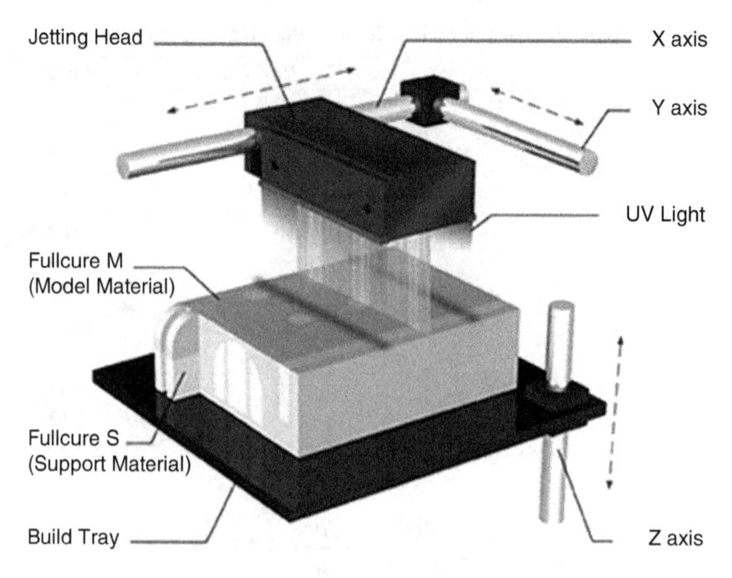

Fig. 7.14 Stratasys Polyjet build process [64]

is flat for the subsequent layer. Because of the slow and accurate build style as well as the waxy materials, these machines are often used to fabricate investment castings for the jewelry and dentistry industries.

3D Systems and Stratasys offer machines using the ability to print and cure acrylic photopolymers. Stratasys markets the Eden, Alaris, and Connex series of printers. These machines print a number of different acrylic-based photopolymer materials in 0.0006 in. layers from heads containing 1,536 individual nozzles, resulting in rapid, line-wise deposition efficiency, as opposed to the slower, point-wise approach used by Solidscape. Each photopolymer layer is cured by ultraviolet light immediately as it is printed, producing fully cured models without post-curing. Support structures are built in a gel-like material, which is removed by hand and water jetting [64]. See Fig. 7.14 for an illustration of Stratasys' Polyjet system, which is employed in all Eden machines. The Connex line of machines provides multimaterial capability. For several years, only two different photopolymers could be printed at one time; however, by automatically adjusting build styles, the machine can print up to 25 different effective materials by varying the relative composition of the two photopolymers. Machines are emerging that print increasing numbers of materials.

In competition with Stratasys, 3D Systems markets the ProJet printers, which print layers 0.0016 in. thick using heads with hundreds of nozzles, half for part material and half for support material [11]. Layers are then flashed with ultraviolet light, which activates the photoinitiated polymerization. The ProJets are the third generation of the Multi-Jet Modeling family from 3D Systems, following the ThermoJet described above and the InVision series. A comparison of the machines currently available is presented in Table 7.3.

Table 7.3 Commercially available printing-based AM machines [11, 63, 64]

Company/product	Cost (1000s)	Build size $X \times Y \times Z$ (mm)	Min. layer (mm)	Resolution X, Y (dpi)	Material	Support
Solidscape						
3Z Studio	$25	152 × 152 × 51	0.01	8,000 × 8,000	Wax-like	Soluble material
3Z Pro	$46	152 × 152 × 102	0.01	8,000 × 8,000	”	”
3D Systems						
ProJet 3510 SD	$60	298 × 185 × 203	0.032	375 × 375	Acrylate photopolymer (AP)	”
ProJet 3510 HD	$78	298 × 185 × 203	0.029	750 × 750	AP	”
ProJet 3500 HD*Max*	$92	298 × 185 × 203	0.016	750 × 750	AP	”
ProJet 3500 CPX*Max*	$99	298 × 185 × 203	0.016	694 × 750	Wax-like	Wax
Stratasys						
Objet 24	$20	240 × 200 × 150	–	–	AP	Gel-like photopolymer
Eden 250	$60	255 × 252 × 200	0.028	600, 300	”	”
Eden 260V	$90	250 × 250 × 200	0.016	600, 600	”	”
Eden 500V	$170	490 × 390 × 200	0.016	600, 600	”	”
Connex500	$240	490 × 390 × 200	0.016	600, 600	”	”
Objet 1000	$500–700	1,000 × 800 × 500	0.016	600, 600	”	”

7.6 Process Benefits and Drawbacks

Each AM process has its advantages and disadvantages. The primary advantages of printing, both direct and binder printing, as an AM process include low cost, high speed, scalability, ease of building parts in multiple materials, and the capability of printing colors. Printing machines are much lower in cost than other AM machines, particularly the ones that use lasers. In general, printing machines can be assembled from standard components (drives, stages, print heads), while other machines have many more machine-specific components. High speed and scalability are related: by using print heads with hundreds or thousands of nozzles, it is possible to deposit a lot of material quickly and over a considerable area. Scalability in this context means that printing speed can be increased by adding another print head to a machine, a relatively easy task, much easier than adding another laser to a SL or SLS machine.

As mentioned, Stratasys markets the Connex machines that print in two or more part materials. One can imagine adding more print heads to increase the capability to many different materials and utilizing dithering deposition patterns raise the number of effective materials into the hundreds. Compatibility and resolution need to be ensured, but it seems that these kinds of improvements should occur in the near future.

Related to multiple materials, colors can be printed by some commercial AM machines (see Sect. 8.3). The capability of printing in color is an important advance in the AM industry; for many years, parts could only be fabricated in one color. The only exception was the selectively colorable SL resins that Huntsman markets for the medical industry, which were developed in the mid-1990s. These resins were capable of only two colors, amber and either blue or red. In contrast, two companies market AM machines that print in high resolution 24-bit color. Several companies are using these machines to produce figurines for video-gamers and other consumers (see Chaps. 3, 8, and 12).

For completeness, a few disadvantages of MJ will provide a more balanced presentation. The choice of materials to date is limited. Only waxes and photopolymers are commercially available. Part accuracy, particularly for large parts, is generally not as good as with some other processes, notably vat photopolymerization and material extrusion. However, accuracies have been improving across the industry and are expected to improve among all processes.

7.7 Summary

Each AM process has its advantages and disadvantages. The primary advantages of printing, both direct and binder printing, as an AM process include low cost, high speed, scalability, ease of building parts in multiple materials, and the capability of printing colors. Printing machines are lower in cost than many other AM machines, particularly the ones that use lasers or electron beams. In general, printing machines can be assembled from standard components (drives, stages, print heads), while other machines have many more machine-specific components. Two primary mechanisms exist for droplet generation, continuous mode and DOD. At present, all commercial MJ machines utilize DOD print heads. High speed and scalability are related: by using print heads with hundreds or thousands of nozzles, it is possible to deposit a lot of material quickly and over a considerable area. Scalability in this context means that printing speed can be increased by adding another print head to a machine, a relatively easy task, much easier than adding another laser to a vat photopolymerization or powder bed fusion machine.

7.8 Exercises

1. List five types of material that can be directly printed.
2. According to the printing indicator (7.7), what is the smallest diameter nozzle that could be used to print a ceramic-wax material that has the following properties:
   (a) Viscosity of 15 cP, density of 1,800 kg/m^3, and surface tension of 0.025 N/m.
   (b) Viscosity of 7 cP, density of 1,500 kg/m^3, and surface tension of 0.025 N/m.
   (c) Viscosity of 38 cP, density of 2,100 kg/m^3, and surface tension of 0.025 N/m.
3. Develop a build time model for a printing machine. Assume that the part platform is to be filled with parts and the platform is L mm long and W mm wide. The print head width is H mm. Assume that a layer requires three passes of the print head, the print head can print in both directions of travel ($+X$ and $-Y$), and the layer thickness is T mm. Figure 7.15 shows a schematic for the problem. Assume that a delay of D seconds is required for cleaning the print heads every K layers. The height of the parts to be printed is P mm.
   (a) Develop a build time model using the variables listed in the problem statement.

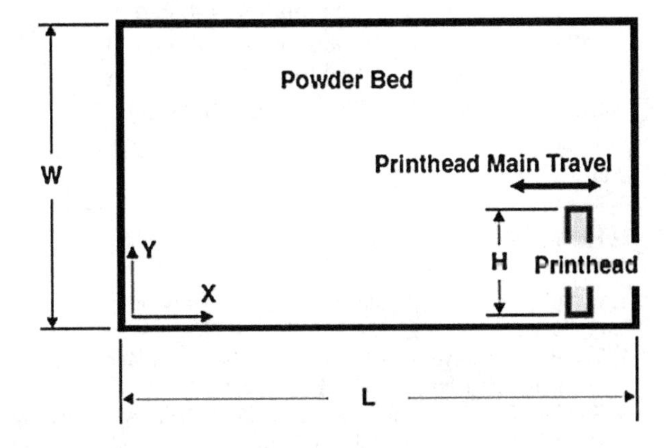

Fig. 7.15 Schematic for problems 4–5

Compute the build time for a layer of parts given the variable values in the following table.

	L	W	H	T	D	K	P
(b)	300	185	50	0.04	10	20	60
(c)	300	185	50	0.028	12	25	85
(d)	260	250	60	0.015	12	25	60
(e)	340	340	60	0.015	12	25	60
(f)	490	390	60	0.015	12	25	80

4. Modify the build time model from Problem 4 for the 3DP process. Assume that the powder bed recoating time is 10 s. Compute build times for a layer of parts using the values in Problem 4, assuming that layer thicknesses are 0.1 mm.
5. The integral in (7.2) can be evaluated analytically for simple nozzle shapes. Assume that the nozzle is conical with the entrance diameter of d_e and the exit diameter d_x.
   (a) evaluate the integral analytically.
   Use your integrated expression to compute pressure drop through the nozzle, instead of (7.3), for the following variable values:

	d_e (mm)	d_x (mm)	l (mm)	μ (cP)	ρ (kg/m^3)	γ (N/m)	v (m/s)
(b)	0.04	0.02	0.1	1	1,000	0.072	10
(c)	0.04	0.02	0.1	40	1,000	0.072	10
(d)	0.04	0.02	1.0	1	1,000	0.072	10
(e)	0.1	0.04	5.0	1	1,000	0.072	10
(f)	0.1	0.04	5.0	40	1,000	0.025	10

6. Using the integral from Problem 6, develop a computer program to compute pressure drop through the nozzle for various nozzle sizes and fluid properties. Compute and plot the pressure drop for the printing conditions of Fig. 7.14, but using nozzles of the following dimensions:
   (a) $l = 0.1$ mm, $d_e = 0.06$ mm, $d_x = 0.02$ mm
   (b) $l = 0.1$ mm, $d_e = 0.08$ mm, $d_x = 0.04$ mm
   (c) $l = 0.1$ mm, $d_e = 0.12$ mm, $d_x = 0.05$ mm
   (d) $l = 5.0$ mm, $d_e = 0.1$ mm, $d_x = 0.05$ mm

References

1. Le HP (1998) Progress and trends in ink-jet printing technology. J Imaging Sci Technol 42 (1):49–62
2. The rapid prototyping patent museum: basic technology patents. http://www.additive3d.com/museum/mus_c.htm. Accessed 17 Aug 2013
3. Wohlers T (2004) Wohlers report 2004. Wohlers Associates, Fort Collins
4. Derby B, Reis N (2003) Inkjet printing of highly loaded particulate suspensions. MRS Bull 28 (11):815–818

5. De Gans BJ, Duineveld PC, Schubert US (2004) Inkjet printing of polymers: state of the art and future developments. Adv Mater 16(3):203–213
6. MicroFab Technologies. MicroFab technote 99-02: fluid properties effects on ink-jet device performance. http://www.microfab.com/equipment/technotes.html
7. Paton A, Kruse J (1995) Reduced nozzle viscous impedance. US Patent 5,463,416
8. Gao F, Sonin AA (1994) Precise deposition of molten microdrops: the physics of digital fabrication. Proc R Soc Lond A 444:533–554
9. Reis N, Seerden KAM, Derby B, Halloran JW, Evans JRG (1999) Direct inkjet deposition of ceramic green bodies: II—jet behaviour and deposit formation. Mater Res Soc Symp Proc 542:147–152
10. Feng W, Fuh J, Wong Y (2006) Development of a drop-on-demand micro dispensing system. Mater Sci Forum 505–507:25–30
11. 3D Systems professional printers. http://www.3dsystems.com/3d-printers/professional/overview. Accessed 17 Aug 2013
12. Leyden R, Hull W (1999) Method for selective deposition modeling. US Patent 5,855,836
13. De Gans BJ, Kazancioglu E, Meyer W, Schubert US (2004) Ink-jet printing polymers and polymer libraries using micropipettes. Macromol Rapid Commun 25:292–296
14. Xu P, Ruatta S, Schmidt K, Doan V (2004) Phase change support material composition. US Patent 7,399,796
15. Schmidt K (2005) Selective deposition modeling with curable phase change materials. US Patent 6,841,116
16. Tay B, Edirisinghe MJ (2001) Investigation of some phenomena occurring during continuous ink-jet printing of ceramics. J Mater Res 16(2):373–384
17. Ainsley C, Reis N, Derby B (2002) Freeform fabrication by controlled droplet deposition of powder filled melts. J Mater Sci 37:3155–3161
18. Zhao X, Evans JRG, Edirisinghe MJ (2002) Direct ink-jet printing of vertical walls. J Am Ceram Soc 85(8):2113–2115
19. Wang T, Derby B (2005) Ink-jet printing and sintering of PZT. J Am Ceram Soc 88 (8):2053–2058
20. Liu Q, Orme M (2001) High precision solder droplet printing technology and the state-of-the-art. J Mater Process Technol 115:271–283
21. Priest JW, Smith C, DuBois P (1997) Liquid metal jetting for printing metal parts. In: Solid freeform fabrication symposium, Austin, TX, 11–13 Aug, pp 1–9
22. Orme M (1993) A novel technique of rapid solidification net-form material synthesis. J Mater Eng Perform 2:399–405
23. Orme M, Huang C, Courter J (1996) Precision droplet-based manufacturing and material synthesis: fluid dynamics and thermal control issues. Atomization Sprays 6:305–329
24. Yamaguchi K (2003) Generation of 3-dimensional microstructure by metal jet. Microsyst Technol 9:215–219
25. Yamaguchi K, Sakai K, Yamanka T, Hirayama T (2000) Generation of three-dimensional micro structure using metal jet. Precis Eng 24:2–8
26. Liu Q, Orme M (2001) On precision droplet-based net-form manufacturing technology. Proc Inst Mech Eng B J Eng Manuf 215:1333–1355
27. Cao W, Miyamoto Y (2006) Freeform fabrication of aluminum parts by direct deposition of molten aluminum. J Mater Process Technol 173:209–212
28. Shimoda T, Morii K, Seki S, Kiguchi H (2003) Inkjet printing of light-emitting polymer displays. MRS Bull 28:821–827
29. Zhao X, Evans JRG, Edirisinghe MJ, Song JH (2001) Ceramic freeforming using an advanced multinozzle ink-jet printer. J Mater Synth Proces 9(6):319–327
30. Furbank RJ, Morris JF (2004) An experimental study of particle effects on drop formation. Phys Fluids 16(5):1777–1790
31. Bechtel SE, Bogy DB, Talke FE (1981) Impact of a liquid drop against a flat surface. IBM J Res Dev 25(6):963–971

32. Pasandideh-Fard M, Qiao Y, Chandra S, Mostaghimi J (1996) Capillary effects during droplet impact on a solid surface. Phys Fluids 8(3):650–659
33. Thoroddsen ST, Sakakibara J (1998) Evolution of the fingering pattern of an impacting drop. Phys Fluids 10(6):1359–1374
34. Bhola R, Chandra S (1999) Parameters controlling solidification of molten wax droplets falling on a solid surface. J Mater Sci 34:4883–4894
35. Attinger D, Zhao Z, Poulikakos D (2000) An experimental study of molten microdroplet surface deposition and solidification: transient behavior and wetting angle dynamics. J Heat Transf 122:544–556
36. Zhou W, Loney D, Fedorov AG, Degertekin FL, Rosen DW (2013) What controls dynamics of droplet shape evolution upon impingement on a solid surface? AIChE J 59(8):3071–3082
37. Bussman M, Chandra S, Mostaghimi J (2000) Modeling the splash of a droplet impacting a solid surface. Phys Fluids 12(12):3121–3132
38. Schiaffino S, Sonin AA (1997) Molten droplet deposition and solidification at low Weber numbers. Phys Fluids 9(11):3172–3187
39. Orme M, Huang C (1997) Phase change manipulation for droplet-based solid freeform fabrication. J Heat Transfer 119:818–823
40. Sanders R, Forsyth L, Philbrook K (1996) 3-D Model maker. US Patent 5,506,706
41. Thayer J, Almquist T, Merot C, Bedal B, Leyden R, Denison K, Stockwell J, Caruso A, Lockard M (2001) Selective deposition modeling system and method. US Patent 6,305,769
42. Gothait H (2005) System and method for three-dimensional model printing. US Patent 6,850,334
43. Gothait H (2001) Apparatus and method for three-dimensional model printing. US Patent 6,259,962
44. Bedal B, Bui V (2002) Method and apparatus for controlling the drop volume in a selective deposition modeling environment. US Patent 6,347,257
45. Tay B, Evans JRG, Edirisinghe MJ (2003) Solid freeform fabrication of ceramics. Int Mater Rev 48(6):341–370
46. MicroFab technote 99-01: background on ink-jet technology. http://www.microfab.com/equipment/technotes/technote99-01.pdf. Accessed 19 Mar 2006
47. Teng W, Edirisinghe MJ (1998) Development of continuous direct ink jet printing of ceramics. Br Ceram Trans 97(4):169–173
48. Blazdell PF, Evans JRG, Edirisinghe MJ, Shaw P, Binstead M (1995) The computer aided manufacture of ceramics using multilayer jet printing. J Mater Sci Lett 54:1562–1565
49. Blazdell PF (2003) Solid free-forming of ceramics using a continuous jet printer. J Mater Process Technol 137:49–54
50. Tseng AA, Lee MH, Zhao B (2001) Design and operation of a droplet deposition system for freeform fabrication of metal parts. J Eng Mater Technol 123:74–84
51. Basaran OA (2002) Small-scale free surface flows with breakup: drop formation and emerging applications. AIChE J 48(9):1842–1848
52. Sirringhaus H, Kawase T, Friend RH, Shimoda T, Inbasekaran M, Wu W, Woo EP (2000) High-resolution inkjet printing of all-polymer transistor circuits. Science 290:2123–2126
53. Lee E (2002) Microdrop generation. CRC Press, Boca Raton
54. Percin G, Khuri-Yakub BT (2002) Piezoelectrically actuated flextensional micromachined ultrasound droplet ejectors. IEEE Trans Ultrason Ferroelectr Freq Control 49(6):756–766
55. Elrod SA, Hadimioglu B, Khuri-Yakub BT, Rawson EG, Richley E, Quate CF (1989) Nozzleless droplet formation with focused acoustic beams. J Appl Phys 65(9):3441–3447
56. Meacham JM, Ejimofor C, Kumar S, Degertekin FL, Fedorov AG (2004) Micromachined ultrasonic droplet generator based on a liquid horn structure. Rev Sci Instrum 75(5):1347–1352
57. Meacham JM, Varady M, Degertekin FL, Fedorov AG (2005) Droplet formation and ejection from a micromachined ultrasonic droplet generator: visualization and scaling. Phys Fluids 17:100605
58. Margolin L (2006) Ultrasonic droplet generation jetting technology for additive manufacturing: an initial investigation. MS Thesis, Georgia Institute of Technology

59. Fukumoto H, Aizawa J, Nakagawa H, Narumiya H (2000) Printing with ink mist ejected by ultrasonic waves. J Imaging Sci Technol 44(5):398–405
60. Meacham JM, O'Rourke A, Yang Y, Fedorov AG, Degertekin FL, Rosen DW (2010) Experimental characterization of high viscosity droplet ejection. J Manuf Sci E-T ASME 132(3):030905
61. Sweet R (1964) High-frequency oscillography with electrostatically deflected ink jets. SEL-64-004, SELTR17221. Stanford Electronics Laboratories, Stanford, CA
62. Munson B, Young D, Okiishi T (1998) Fundamentals of fluid mechanics, 3rd edn. Wiley, New York
63. Solidscape. T66 Benchtop: product description. http://www.solid-scape.com/t66.html. Accessed 31 July 2006
64. Stratasys Polyjet Technology. http://stratasys.com/3d-printers/technology/polyjet-technology. Accessed 17 Aug 2013

Binder Jetting

8

Abstract

Binder jetting methods were developed in the early 1990s, primarily at MIT. They developed what they called the 3D Printing (3DP) process in which a binder is printed onto a powder bed to form part cross sections. This concept can be contrasted with powder bed fusion (PBF), where a laser melts powder particles to define a part cross section. A wide range of polymer composite, metals, and ceramic materials have been demonstrated, but only a subset of these are commercially available. Some binder jetting machines contain nozzles that print color, not binder, enabling the fabrication of parts with many colors. Several companies licensed the 3DP technology from MIT and became successful machine developers, including ExOne and ZCorp (purchased by 3D Systems in 2011). A novel continuous printing technology was been developed recently by Voxeljet that can, in principle, fabricate parts of unlimited length.

8.1 Introduction

The original name for binder jetting was Three-Dimensional Printing (3DP) and it was invented at MIT and has been licensed to more than five companies for commercialization. In contrast to the printing processes described in Chap. 7, binder jetting (BJ) processes print a binder into a powder bed to fabricate a part. Hence, in BJ, only a small portion of the part material is delivered through the print head. Most of the part material is comprised of powder in the powder bed. Typically, binder droplets (80 μm in diameter) form spherical agglomerates of binder liquid and powder particles as well as provide bonding to the previously printed layer. Once a layer is printed, the powder bed is lowered and a new layer of powder is spread onto it (typically via a counter-rotating rolling mechanism) [1], very similar to the recoating methods used in powder bed fusion processes, as presented in Chap. 5. This process (printing binder into bed; recoating bed with new

I. Gibson et al., *Additive Manufacturing Technologies*,
DOI 10.1007/978-1-4939-2113-3_8

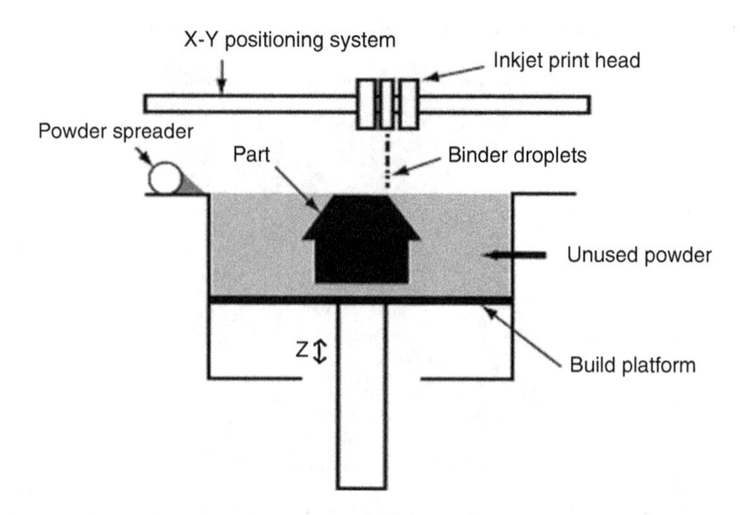

Fig. 8.1 Schematic of the binder jetting process

layer of powder) is repeated until the part, or array of parts, is completed. A schematic of the BJ process is shown in Fig. 8.1.

Because the printer head contains several ejection nozzles, BJ features several parallel one-dimensional avenues for patterning. Since the process can be economically scaled by simply increasing the number of printer nozzles, the process is considered a scalable, line-wise patterning process. Such embodiments typically have a high deposition speed at a relatively low cost (due to the lack of a high-powered energy source) [1], which is the case for BJ machines.

The printed part is typically left in the powder bed after its completion in order for the binder to fully set and for the green part to gain strength. Post-processing involves removing the part from the powder bed, removing unbound powder via pressurized air, and infiltrating the part with an infiltrant to make it stronger and possibly to impart other mechanical properties.

The BJ process shares many of the same advantages of powder bed processes. Parts are self-supporting in the powder bed so that support structures are not needed. Similar to other processes, parts can be arrayed in one layer and stacked in the powder bed to greatly increase the number of parts that can be built at one time. Finally, assemblies of parts and kinematic joints can be fabricated since loose powder can be removed between the parts.

Applications of BJ processes are highly dependent upon the material being processed. Low-cost BJ machines use a plaster-based powder and a water-based binder to fabricate parts. Polymer powders are also available. Some machines have color print heads and can print visually attractive parts. With this capability, a market has developed for colorful figures from various computer games, as well as personal busts or sculptures, with images taken from cameras. Infiltrants are used to strengthen the parts after they are removed from the powder bed. With either the starch or polymer powders, parts are typically considered visual prototypes or light-duty functional prototypes. In some cases, particularly with elastomeric infiltrants, parts can be used for functional purposes. With polymer powders and wax-based infiltrants, parts can be used as patterns for investment casting, since the powder and wax can burn off easily.

For metal powders, parts can be used as functional prototypes or for production purposes, provided that the parts have been designed specifically for the metal alloys available. Molds and cores for sand casting can be fabricated by some BJ machines that use silica or foundry sand as the powder. This is a sizable application in the automotive and heavy equipment industries.

8.2 Materials

8.2.1 Commercially Available Materials

When Z Corporation first started in the mid-1990s, their first material was starch based and used a water-based binder similar to a standard house-hold glue. At present, the commercially available powder from 3D Systems is plaster based (calcium sulfate hemihydrate) and the binder is water based [2]. Printed parts are fairly weak, so they are typically infiltrated with another material. 3D Systems provides three infiltrants, the ColorBond infiltrant, which is acrylate-based and is similar to superglue, StrengthMax infiltrant which is a two-part infiltrant, and Salt Water Cure, an eco-friendly and hazard-free infiltrant. Strength, stiffness, and elongation data are given on 3D Systems' web site for parts fabricated with these infiltrants. In general, parts with any of the infiltrants are much stiffer than typical thermoplastics or VP resins, but are less strong, and have very low elongation at break (0.04–0.23 %).

Voxeljet [3], on the other hand, supplies a PMMA (poly-methyl methacrylate) powder and uses a liquid binder that reacts at room temperature. They recommend that parts stay in the powder bed for several hours to ensure that the binder is completely cured. For investment casting pattern fabrication, they offer a wax-based binder for use with PMMA powder that is somewhat larger in particle size than the powder used for parts. They claim excellent pattern burnout for investment casting.

For materials from both companies, unprinted powders are fully recyclable, meaning that they can be reused in subsequent builds. A desirable characteristic of powders is a high packing density so that printed parts have a high volume fraction of powder and are strong enough to survive depowdering and clean up operations. High packing densities can be achieved by tailoring powder particle shape or by including a range of particle sizes so that small particles fill in gaps between larger particles. In practice, both approaches are used whenever possible.

Quite a few other infiltrant materials have been marketed by ZCorp and 3D Systems and many users have experimented with a variety of materials, so alternatives are possible that can produce parts with a wide range of mechanical properties.

ExOne markets machines that use either metal or sand powders for metal parts or sand-casting molds and cores, respectively [4]. In the metals area, they currently market 3,166 stainless steel and bronze, 420 stainless steel (non-annealed), 420 stainless steel (annealed), bronze, and Inconel 625. For the stainless steel materials, bronze is used as an infiltrant so that parts are virtually fully dense. Polymer binders are used for the metals. In order to fabricate a metal part, the "green" part is removed from the AM machine, then is subject to three furnace cycles. In the first cycle, low temperature is used for several hours to burn off the polymer binder. In the second cycle, high temperature is used to lightly sinter the metal particles together so that the part has decent strength. If this cycle is too long, the metal particles more completely melt, causing the part to lose dimensional accuracy and its desired shape. After this cycle, the part is approximately 60 % dense. In the final cycle, a bronze ingot is placed in the furnace in contact with the part so that bronze infiltrates into the part's pores, resulting in parts that are 90–95 % dense.

An exception to the light sintering and infiltration process is the new Inconel 625 material announced by ExOne in 2014. Although they use a binder, the Inconel material can be sintered to virtually full density (ExOne claims greater than 99 % dense) while maintaining acceptable dimensional accuracy. If this process can be extended to other metals, it could change the economics of metal AM significantly.

Both ExOne and Voxeljet market machines that use sand for the fabrication of molds and cores for sand casting. ExOne offers a silica sand and two-part binder, where one part (binder catalyst) is coated on a layer and the second part is printed onto the layer, causing a polymerization reaction to occur and binding sand particles together. They claim that only standard foundry materials are used so that resulting molds and cores enable easy integration into existing manufacturing and foundry processes. Voxeljet also offers a silica sand with an inorganic binder and claims that their materials also integrate well into existing foundry processes.

Finally, ExOne markets a soda-lime glass material for use in fabricating artwork, jewelry, or other decorative objects. Different colors and finishes are available. An organic binder is used that requires an elevated temperature curing cycle. Then, parts need to be fired at high temperature to sinter the glass particles and impart decent strength and stiffness.

8.2.2 Ceramic Materials in Research

A wide range of materials has been developed for BJ by researchers. Printing into metal and ceramic powder beds was first demonstrated in the early 1990s. Various powder mixes, including compositions and size distributions, have been explored.

Traditional powder-based BJ of ceramics involves the selective printing of a binder over a bed of ceramic powder [5]. Fabrication of ceramic parts follows a very similar process compared with metal parts. Green parts created by this process are subjected to a thermal decomposition prior to sintering to remove the polymer binder. After binder burn off, the furnace temperature is increased until the ceramic's sintering temperature is reached. Sometimes an infiltrant is used that reacts to form a ceramic binder. Another possibility is to infiltrate with a metal to form a ceramic-metal composite. The first report of using BJ for the fabrication of ceramics was in 1993; fired components were reported as typically greater than 99.2 % dense [5]. Alumina, silica, and titanium dioxide have been made with this process [6].

Research involving the BJ of ceramics encountered early setbacks because of the use of dry powders. The fine powders needed for good powder bed density did not generally flow well enough to spread into defect-free layers [5]. Furthermore, since green part density was inadequate with the use of dry powders, isostatic pressing was implemented after the printing process. This extraneous requirement severely limits the types of part shapes capable of being processed.

To counteract the problems encountered with recoating a dry powder bed, research on ceramic BJ has shifted to the use of a slurry-based working material. In this approach, layers are first deposited by ink-jet printing a layer of slurry over the build area. After the slurry dries, binder is selectively printed to define the part shape. This is repeated for each individual layer, at the cost of significantly increased build time. Multiple jets containing different material composition or concentration could be employed to prepare components with composition and density variation on a fine scale (100 μm) [7]. Alumina and silicon nitride have been processed with this technique, improving green part density to 67 %, and utilizing layer thicknesses as small as 10 μm [8].

Recently, a variation of this method was developed to fabricate metal parts starting with metal-oxide powders [9]. The ceramic BJ is used until the furnace sintering step. While in the furnace, a hydrogen atmosphere is introduced, causing a reduction reaction to occur between the hydrogen and the oxygen atoms in the metal-oxide. The reduction reaction converts the oxide to metal. After reduction, the metal particles are sintered to form a metal part. This process has been demonstrated for several material systems, including iron, steels, and copper. Unfortunately, reaction thermodynamics prevent alumina and titanium oxide from being reduced to aluminum and titanium, respectively.

This Metal-Oxide Reduction 3DP (MO3DP) process was demonstrated using a Z405 machine [10]. Metal-oxide powders containing iron oxide, chromium oxide, and a small amount of molybdenum were prepared by spray drying the powder composition with polyvinyl alcohol (PVA) to form clusters of powder particles coated with PVA. Upon reduction, the material composition formed a maraging steel. Water was selectively printed into the powder bed to define part cross sections, since the water will dissolve PVA, causing the clusters to stick together. A variety of shapes (trusses, channels, thin walls) were fabricated using the process to demonstrate the feasibility of producing cellular materials.

The main advantage of BJ, in the context of manufacturing cellular materials, lies in its economic considerations. Simply put, the BJ process does not require high energy, does not involve lasers or any toxic materials, and is relatively inexpensive and fast. Part creation rate is limited to approximately twice the binder flow rate. A typical inkjet nozzle delivers approximately 1 cm^3/min of binder; thus a machine with a 100 nozzle print head could create up to approximately 200 cm^3/min of printed component. Because commercial inkjet printers exist with up to 1,600 nozzles, BJ could be fast enough to be used as a production process.

8.3 Process Variations

Almost all commercially available BJ machines use the architecture shown in Fig. 8.1. An array of print heads is mounted on an XY translation mechanism. If the process is capable of printing colored parts, some print heads are dedicated to printing binder material, while others are dedicated to printing color. Typically, the print heads used are standard, off-the-shelf print heads that are found in machines for 2D printing of posters, banners, and similar applications. Parts are fabricated in batches, just like every other AM process.

Powder handling and recoating systems are similar to those used in powder bed fusion processes. Differences arise when comparing low-cost visual model printers (for plaster or polymer powders) to the metal or sand printers. For the low-cost printers, powder containers (vats) can be hand-carried. In the latter cases, however, powder beds can weigh hundreds or thousands of pounds, necessitating different material handling and powder bed manipulation methods. For the sand printers, the vats utilize a rail system for conveying powder beds to and from depowdering stations and cranes are used for transporting parts or molds.

The capability of continuous printing or of fabricating parts that are larger than the AM machine fabricating them has been discussed in the research community. In recent years, two different approaches have been demonstrated for continuous printing of parts. One approach is being commercialized by Voxeljet in 2013 and is based on linear translation of the part being fabricated. The second approach is called spiral growth manufacturing and was developed by researchers at the University of Liverpool, UK.

The Voxeljet continuous printing process is a novel idea that utilizes an inclined build plane. That is, the build surface of the powder bed is inclined at an angle of 30°, less than the powder's critical angle of repose. Powder recoating and binder jetting are performed on this inclined build surface. The powder bed translates on a conveyor belt from the front towards the back of the machine. In contrast to typical batch fabrication, parts emerge continuously at the back of the machine. In principle, parts could be infinitely long, certainly much longer than the machine. The continuous part fabrication capability could represent an important step in achieving economical manufacture of moderate to high production volumes of parts.

The second continuous fabrication approach, spiral growth manufacturing (SGM) was invented by Chris Sutcliffe at the University of Liverpool in the early 2000s. The patent US2008109102 is a good reference for further information [11]. In the BJ variant of SGM, the powder bed is circular and rotates continuously. Binder printing and recoating are performed continuously also. As the machine operates, the powder bed indexes downwards continuously to accommodate the

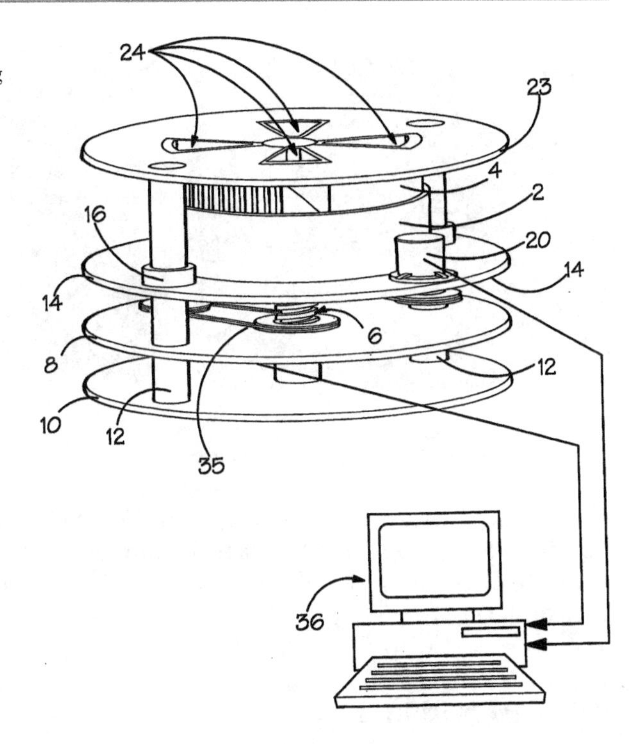

Fig. 8.2 Schematic of a spiral growth manufacturing BJ machine

next layer of powder. As such, the top layer of powder forms a spiral in the powder bed. A machine schematic from the patent is shown in Fig. 8.2.

In the figure, object 2 is the cylindrical build chamber. Plates 10, 8, 14, and 23 do not rotate; plate 14 supports the build chamber and slides up and down on the pillars 12. The build chamber rotates, driven by the lead screw numbered 6. Four powder supply hoppers are shown by objects 24, so this indicates that the machine has four build stations, each with print heads and recoater mechanism. As a consequence, for each rotation of the build chamber, effectively four layers are deposited and processed. So, for example, if the layer thickness is 0.1 mm, each rotation of the build chamber adds 0.4 mm to the powder bed height and plate 14 and the build chamber must translate downward by 0.4 mm to accommodate this increase in bed height.

Each build station typically contains a print head with multiple nozzles. Since the width of the powder bed is typically greater than the print head width, a linear stage must be used to translate the print head across the powder bed. The linear velocity of the outer edge of the build chamber is greater than the linear velocity at the inner edge, which means that the powder along the outer edge passes the print head at a faster speed. This has important consequences for the printing conditions across the width of the chamber: more binder has to be deposited per unit time along the outer edge compared to the inner edge. Effectively this means that the images being printed have to be pre-skewed in order to compensate for the differences in speed. As an example, Fig. 8.3 shows how an image must be skewed so that the

Fig. 8.3 Skewing an image for SGM printing

printed image is fabricated properly [12]. This image was printed on a SGM machine with two print heads, hence the two images on the right. Note that the inner edge is on the left side of the image, which is stretched, while the outer edge is compressed.

8.4 BJ Machines

A wide variety of powder and binder materials can be used which enables significant flexibility in the process. MIT licensed the BJ technology according to the type of material and application that each licensee was allowed to exploit. ZCorp, Inc. was one company that marketed machines that build concept models in starch and plaster powder using a low viscosity glue as binder. At the other end of the spectrum, ExOne markets machines that build in metal powder, with a strong polymer material that is used as the binder, as well as silica sand for sand casting applications. Voxeljet is a relatively new company that markets BJ machines that use polymer and sand powders for concept models, functional models, investment casting patterns, and sand casting applications.

As of 2012, ZCorp was purchased by 3D Systems and their product line was merged into 3D Systems' ProJet line of printers. These printers are now branded as the ProJet X60 line of printers, with the smallest being the ProJet 160 and the largest being the ProJet 860Pro. Specifications for some of these machines are shown in Table 8.1. Machine names consisting only of numbers fabricate parts that are monochrome only, while suffixes of C, Plus, or Pro indicate that parts can be printed in color, up to the full CMYK color model (Cyan, Magenta, Yellow, Key (black)).

Voxeljet sold their first machine in 2005. They now offer a range of machines from the smallest VX200 to the huge VX4000, which has a 4 m long powder bed. The VX4000 processes foundry sand materials for the sand casting industry. They also market the VXC800, which is the infinitely continuous printer that was

Table 8.1 Machine specifications for binder printing machines

Company/ models	Cost (1,000s)	Deposition rate (mm/h)	Build size ($l \times w \times h$) (mm)	Resolution (dpi)	Layer thickness (mm)	Number of nozzles	Materials
3D Systems							
ProJet 160	$16.50	20	236 × 185 × 127	300 × 450	0.1	304	Polymer composite with infiltrants
ProJet 260C	$28.70	20	236 × 185 × 127	300 × 450	0.1	604	
ProJet 860 Pro	$114	15 May	508 × 381 × 229	600 × 540	0.1	1,520	
ExOne							
Lab Platform	$125	6 March	40 × 60 × 35	400 × 400	0.05–0.1		Stainless steel, bronze, ceramics, glass
Flex Platform	$425	12	400 × 250 × 250	400 × 400	0.1		Above and Inconel
Max Platform	>$1,400	20	1,800 × 1,000 × 700	300 × 300	0.28–0.5		Silica sand, ceramics
Voxeljet							
VX200	$159	12	300 × 200 × 150	300 × 300	0.15	256	PMMA, inorganic sand
VXC800	$700	35	850 × 500 × 30°	600 × 600	0.15–0.4	2,656	
VX4000	$1,850	15.4	4,000 × 2,000 × 1,000	600 × 600	0.12–0.3	26,560	

Fig. 8.4 Voxeljet VXC800 machine (Courtesy Voxeljet)

described earlier. A photo of the VXC800 is shown in Fig. 8.4. For this machine, the layer thickness is 150–400 μm and they use 600 dpi print heads. The surface area of the inclined plane is 500 × 850 mm.

The ExOne Corporation markets a line of BJ machines that fabricate metal parts and sand casting molds and cores in foundry sand. Strong polymer binders are required with these heavy powders. As explained earlier, nearly fully dense parts can be fabricated by printing binder into the metal bed, burning off the binder in a furnace at a low temperature, sintering the metal powder during a high temperature furnace cycle, then infiltrating a second metal, such as copper or bronze, at a low temperature. The printed part, when removed from the bed, is a relatively low-density (50–60 %) green part. Other than how the green part is formed, this process is identical to the indirect processing approach for metal and ceramic part fabrication discussed in Chap. 5 and illustrated in Fig. 5.7. Stainless steel-bronze parts have been made with this technology [4]. The process is typically accurate to ±0.125 mm. Several ExOne machine models are also listed in Table 8.1.

Applications for the metal material models include prototypes of metal parts and some low-volume manufacturing, as well as tooling. As parts are fabricated in a powder bed, the surface finish of these parts is comparable to PBF parts. Finish machining is thus required for high tolerance and mating surfaces. ExOne markets another machine that fabricates gold dental restorations, for example, copings for crowns. The materials and binder printing system were developed specifically for this application, since higher resolution is needed.

In the tooling area, ExOne promotes the advantages of conformal cooling in injection molds. In conformal cooling, cooling channels are routed close to the surfaces of the part cavity, particularly where hot spots are predicted. Using conventional machining processes, cooling channels are drilled as straight holes. With AM processes, however, cooling channels of virtually any shape and configuration can be designed into tools. Figure 8.5 illustrates one tool design with conformal cooling channels that was fabricated in an ExOne machine.

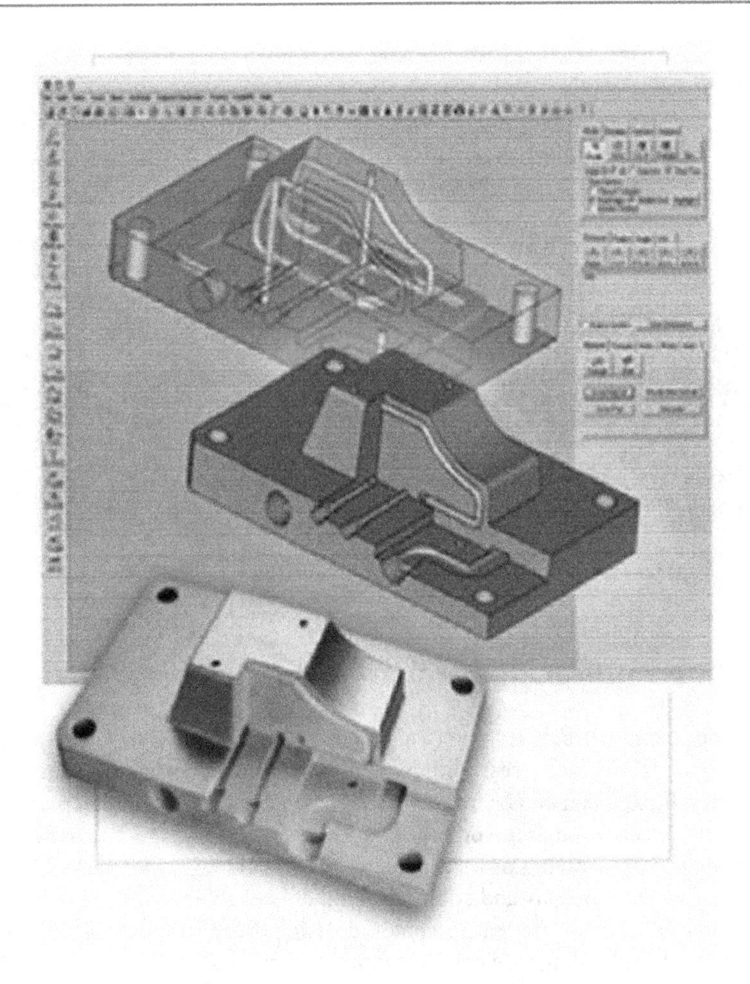

Fig. 8.5 Injection mold with conformal cooling channels fabricated in an ExOne machine (Courtesy ExOne Company)

The largest machine, the S-Max, is intended for companies with large demands for castings, such as the automotive, heavy equipment, and oil & gas industries. A photo of the S-Max is shown in Fig. 8.6. Several dozen of these large machines (S-Max and its predecessor S15) have been installed. The machines print molds and cores for sand casting. Various metals can be cast into the printed molds, including aluminum, zinc, and even magnesium. Special equipment was developed for handling the large volumes of powders and heavy vats, including a silo and powder conveyor, conveyor track for transporting vats of powder and finished molds, and a debinding station. A typical installation with a S-Max or S15 machine occupies a room 40–50 m^2 in size.

Microjet Technology (Taiwan) is another company that currently markets BJ systems.

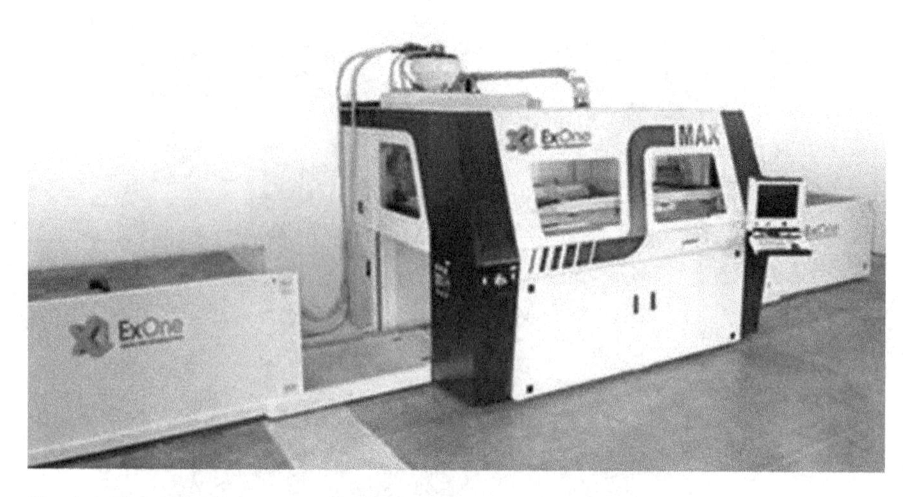

Fig. 8.6 ExOne S-Max system (Courtesy ExOne)

8.5 Process Benefits and Drawbacks

The binder jetting processes share many of the advantages of material jetting relative to other AM processes. With respect to MJ, binder jetting has some distinct advantages. First, it can be faster since only a small fraction of the total part volume must be dispensed through the print heads. However, the need to distribute powder adds an extra step, slowing down binder processes somewhat. Second, the combination of powder materials and additives in binders enables material compositions that are not possible, or not easily achieved, using direct methods. Third, slurries with higher solids loadings are possible with BJ, compared with MJ, enabling better quality ceramic and metal parts to be produced. As mentioned earlier, BJ processes lend themselves readily to printing colors onto parts.

As a general rule, however, parts fabricated using BJ processes tend to have poorer accuracies and surface finishes than parts made with MJ. Infiltration steps are typically needed to fabricate dense parts or to ensure good mechanical properties.

As with any set of manufacturing processes, the choice of manufacturing process and material depends largely on the requirements of the part or device. It is a matter of compromising on the best match between process capabilities and design requirements.

8.6 Summary

The binder jetting processes share many of the advantages of material jetting relative to other AM processes. Compared to MJ, BJ has some distinct advantages. First, it can be faster since only a small fraction of the total part volume must be dispensed through the print heads. However, the need to recoat powder adds an extra step, slowing down binder processes somewhat. Second, the combination of powder materials and additives in binders enables material compositions that are not possible, or not easily achieved, using direct methods. Third, slurries with higher solids loadings are possible with BJ, compared with MJ, enabling better quality ceramic and metal parts to be produced. As mentioned earlier, BJ processes lend themselves readily to printing colors onto parts. Some novel machine architectures have been demonstrated using binder jetting technology that enable continuous printing, including spiral growth manufacturing and an architecture with a slanted build surface.

8.7 Exercises

1. Explain why support structures are not needed in the BJ process.
2. List several characteristics of a good binder material.
3. Identify several methods for achieving a high packing density in the powder bed.
4. Develop a build time model for a conventional binder jetting machine. Assume that the part platform is to be filled with parts and the platform is L mm long and W mm wide. The print head width is H mm. Assume that a layer requires two passes of the print head, the print head can print in both directions of travel ($+X$ and $-Y$), and the layer thickness is T mm. Figure 8.7 shows a schematic for the problem. Assume that a delay of D seconds is required for cleaning the print heads every K layers. The height of the parts to be printed is P mm. Assume that the powder bed recoater moves at 10 cm/s.
   (a) Develop a build time model using the variables listed in the problem statement. Compute the build time for a layer of parts given the variable values in the following table.

	L	W	H	T	D	K	P
(b)	300	185	50	0.04	10	20	60
(c)	300	185	50	0.028	12	25	85
(d)	260	250	60	0.015	12	25	60
(e)	340	340	60	0.015	12	25	60
(f)	490	390	60	0.015	12	25	80

5. Modify the build time model for the continuous printing Voxeljet machine.

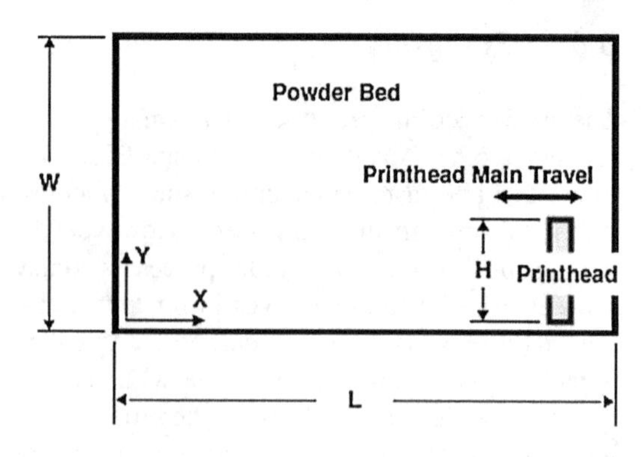

Fig. 8.7 Schematic for problems 4–5

References

1. Sachs EM, Cima MJ, Williams P, Brancazio D, Cornie J (1992) Three-dimensional printing: rapid tooling and prototypes directly from a CAD model. J Eng Ind 114:481–488
2. 3D Systems Professional Printers. http://www.3dsystems.com/3d-printers/professional/overview. Accessed 14 August 2013.
3. Voxeljet Corporation. www.voxeljet.com
4. ExOne. www.exone.com/products. Accessed 10 August 2013
5. Yoo J, Cima MJ, Khanuja S, Sachs EM (1993) Structural ceramic components by 3D printing. Solid freeform fabrication symposium, Austin, TX
6. Uhland S, Holman RK, Morissette S, Cima MJ, Sachs EM (2001) Strength of green ceramics with low binder content. J Am Ceram Soc 84(12):2809–2818
7. Cima MJ, Lauder A, Khanuja S, Sachs E (1992) Microstructural elements of components derived from 3D printing. Solid freeform fabrication symposium, Austin, TX
8. Grau J, Moon J, Uhland S, Cima MJ, Sachs E (1997) High green density ceramic components fabricated by the slurry-based 3DP process. Solid freeform fabrication symposium, Austin, TX
9. Williams CB, Rosen DW (2007) Cellular materials manufactured via 3D printing of metal oxide powders. Solid freeform fabrication symposium, Austin, TX, 6–8 Aug 2007
10. Williams CB (2008) Design and development of a layer-based additive manufacturing process for the realization of metal parts of designed mesostructure. Ph.D. Dissertation, Georgia Institute of Technology
11. European patent, EP1631440 A1, apparatus for manufacturing three dimensional items, Christopher Sutcliffe. US Patent US2008109102
12. Hauser C, Dunschen M, Egan M, Sutcliffe C (2008) Transformation algorithms for image preparation in spiral growth manufacturing (SGM). Rapid Prototyping J 14(4):188–196

Sheet Lamination Processes 9

9.1 Introduction

One of the first commercialized (1991) additive manufacturing techniques was Laminated Object Manufacturing (LOM). LOM involved layer-by-layer lamination of paper material sheets, cut using a CO_2 laser, each sheet representing one cross-sectional layer of the CAD model of the part. In LOM, the portion of the paper sheet which is not contained within the final part is sliced into cubes of material using a crosshatch cutting operation. A schematic of the LOM process can be seen in Fig. 9.1.

A number of other processes have been developed based on sheet lamination involving other build materials and cutting strategies. Because of the construction principle, only the outer contours of the parts are cut, and the sheets can be either cut and then stacked or stacked and then cut. These processes can be further categorized based on the mechanism employed to achieve bonding between layers: (a) gluing or adhesive bonding, (b) thermal bonding, (c) clamping, and (d) ultrasonic welding. As the use of ultrasonic welding involves unique solid state bonding characteristics and can enable a wide range of applications, an extended discussion of this bonding approach is included at the end of this chapter.

9.1.1 Gluing or Adhesive Bonding

The most popular sheet lamination techniques have included a paper build material bonded using a polymer-based adhesive. Initially LOM was developed using adhesive-backed paper similar to the "butcher paper" used to wrap meat. Paper thicknesses range from 0.07 to 0.2 mm. Potentially any sheet material that can be precisely cut using a laser or mechanical cutter and that can be bonded can be utilized for part construction.

A further classification is possible within these processes based upon the order in which they bond and cut the sheet. In some processes the laminate is bonded first to

I. Gibson et al., *Additive Manufacturing Technologies*,
DOI 10.1007/978-1-4939-2113-3_9

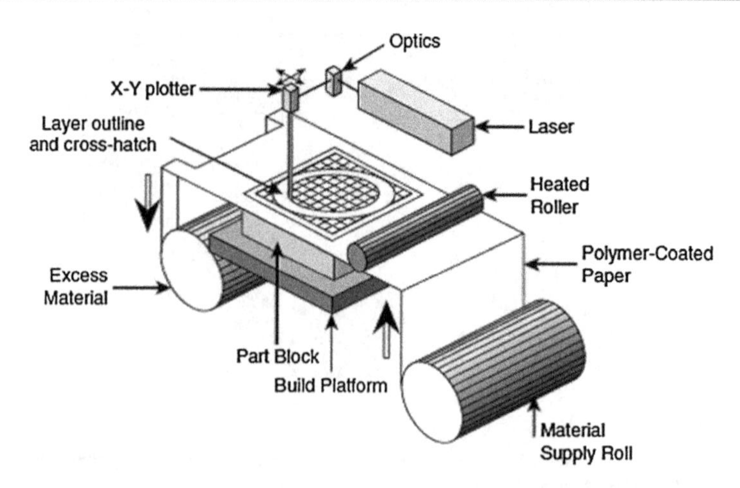

Fig. 9.1 Schematic of the LOM process (based on [1] Journal of Materials Processing Technology by D.I. Wimpenny, B. Bryden, I.R. Pashby. Copyright 2003 by Elsevier Science & Technology Journals. Reproduced with permission of Elsevier Science & Technology Journals in the format Textbook via Copyright Clearance Center.)

the substrate and is then formed into the cross-sectional shape ("bond-then-form" processes). For other processes the laminate is first cut and then bonded to the substrate ("form-then-bond" processes).

9.1.2 Bond-Then-Form Processes

In "bond-then-form" processes, the building process typically consists of three steps in the following sequence: placing the laminate, bonding it to the substrate, and cutting it according to the slice contour. The original LOM machines used this process with adhesive-backed rolls of material. A heated roller passes across the sheet after placing it for each layer, melting the adhesive and producing a bond between layers. A laser (or in some cases a mechanical cutting knife) designed to cut to a depth of one layer thickness cuts the cross-sectional outline based on the slice information. The unused material is left in place as support material and is diced using a crosshatch pattern into small rectangular pieces called "tiles" or "cubes." This process of bonding and cutting is repeated until the complete part is built. After part construction, the part block is taken out and post-processed. The crosshatched pieces of excess material are separated from the part using typical wood carving tools (called decubing). It is relatively difficult to remove the part from the part block when it is cold, therefore, it is often put into an oven for some time before decubing or the part block is processed immediately after part buildup.

Although historically many people continue to associate paper sheet lamination with the LOM machines introduced in 1991 by Helisys Inc., USA and subsequently supported by Cubic Technologies, USA (after Helisys' bankruptcy), new paper-based sheet lamination machines are currently sold by Mcor Technologies (Ireland)

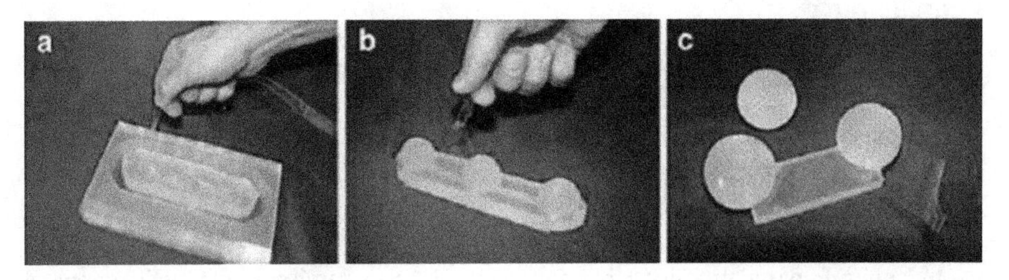

Fig. 9.2 Support material removal for three golf balls made using a Solidimension machine, showing: (**a**) the balls still encased in a central region, being separated from the larger block of bonded material; (**b**) the support material is glued in an accordion-like manner so that the excess material can be pulled out easily as a continuous piece; and (**c**) the balls after complete removal of excess support material (Courtesy 3D Systems)

and Wuhan Binhu (China). These new systems make use of plain paper as the build material, and selectively dispense adhesive only where needed. Because the support material is not adhesively bonded, unlike in LOM, the support removal process is easier. The use of color inkjet printing onto paper by Mcor Technologies enables the production of full-color paper parts directly from a CAD file.

Solidimension (Be'erot, Israel) took the concepts of LOM and further developed them in 1999 into a commercial prototyping system for laminating polyvinyl chloride (PVC) plastic sheets. Solidimension sold its own machines under the Solido name [2] and under other names via resellers. This machine utilized an x–y plotter for cutting the PVC sheets and for writing with "anti glue" pens, which inhibit bonding in prescribed locations. This machine used a unique approach to support material removal. Support material was subdivided into regions, and unique patterns for cutting and bonding the excess material were used to enable easy support material removal. An example of this support material strategy can be seen in Fig. 9.2. Solido machines are no longer offered for sale, however if history is any guide, others may pick up this unique idea and offer similar machines someday.

Bond-then-form sheet lamination principles have also been successfully applied to fabrication of parts from metal, ceramic, and composite materials. In this case, rather than paper or polymer sheets, ceramic or metal-filled tapes are used as the build material to form green parts, and high-temperature furnace post-processing is used to debind and sinter the structure. These tapes are then used for part construction employing a standard sheet lamination process.

Specific advantages of LOM-like bond-then-form adhesive-based processes include: (a) little shrinkage, residual stresses, and distortion problems within the process; (b) when using paper feedstock, the end material is similar to plywood, a typical pattern making material amenable to common finishing operations; (c) large parts can be fabricated rapidly; (d) a variety of build materials can be used, including paper and polymer sheets and metal- or ceramic-filled tapes; (e) nontoxic, stable, and easy-to-handle feedstock; and (f) low material, machine, and process costs relative to other AM systems.

Paper-based sheet lamination has several limitations, including: (a) most paper-based parts require coating to prevent moisture absorption and excessive wear; (b) the control of the parts' accuracy in the Z-dimension is difficult (due to swelling or inconsistent sheet material thickness); (c) mechanical and thermal properties of the parts are inhomogeneous due to the glue used in the laminated structure; and (d) small part feature detail is difficult to maintain due to the manual decubing process.

In general, parts produced by paper-based sheet lamination have been most successfully applied in industries where wooden patterns are often used, or in applications where most features are upward facing. Examples of good applications for paper sheet lamination include patterns for sand casting and 3D topographical maps—where each layer represents a particular elevation of the map.

9.1.3 Form-Then-Bond Processes

In form-then-bond processes, sheet material is cut to shape first and then bonded to the substrate. This approach is popular for construction of parts in metallic or ceramic materials that are thermally bonded (discussed in Sect. 9.3.1) but implementation has primarily been at the research level. One example of a glue-based form-then-bond process is the "Offset Fabbing" system patented by Ennex Corp., USA. In this process, a suitable sheet material with an adhesive backing is placed on a carrier and is cut to the outline of the desired cross section using a two-dimensional plotting knife. Parting lines and outlines of support structures are also cut. The shaped laminate is then placed on top of the previously deposited layers and bonded to it. This process continues until the part is complete. A schematic of the process is shown in Fig. 9.3.

The form-then-bond approach facilitates construction of parts with internal features and channels. Internal features and small channels are difficult or impossible with a bond-then-form approach because the excess material is solid and thus material inside internal features cannot be removed once bonded (unless the part is

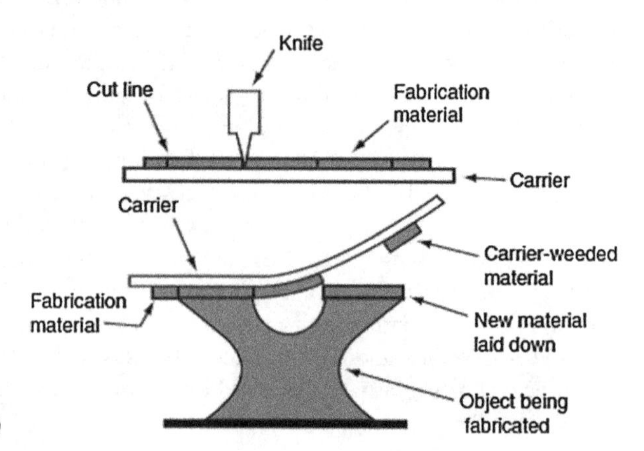

Fig. 9.3 Offset Fabbing system, Ennex Corp (http://www.ennex.com/fab/Offset/)

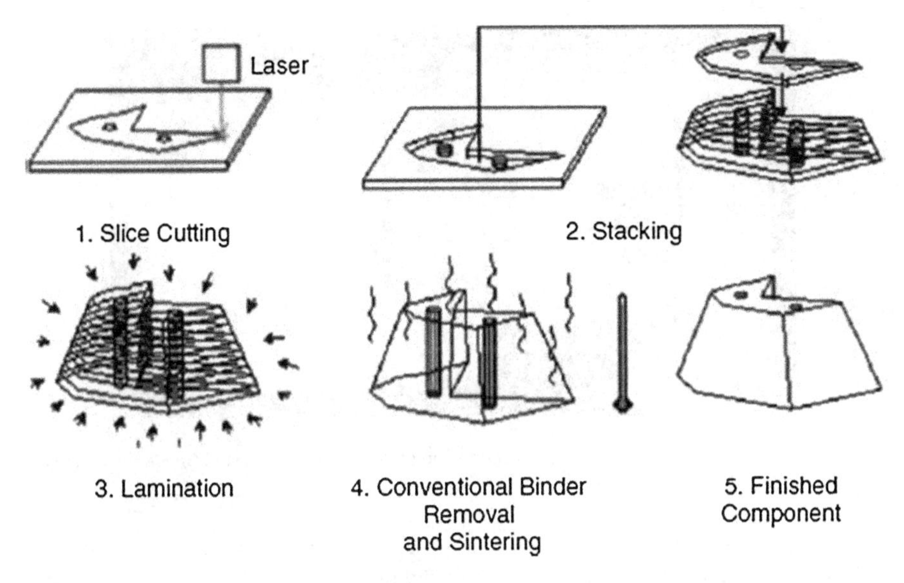

Fig. 9.4 CAM-LEM process (Courtesy CAM-LEM, Inc.)

cut open). Another advantage of form-then-bond approaches is that there is no danger of cutting into the previous layers, unlike in bond-then-form processes where cutting occurs after placing the layer on the previous layer; thus, laser power control or knife pressure is less demanding. Also, the time-consuming and potentially damage-causing decubing step is eliminated. However, these processes require external supports for building overhanging features; some type of tooling or alignment system to ensure a newly bonded layer is registered properly with respect to the previous layers; or a flexible material carrier that can accurately place material regardless of geometry.

Computer-Aided Manufacturing of Laminated Engineering Materials (CAM-LEM, Inc., USA) was developed as a process for fabrication of functional ceramic parts using a form-then-bond method, as shown in Fig. 9.4. In this process, individual slices are laser cut from sheet stock of green ceramic or metal tape. These slices are precisely stacked one over another to create the part. After assembly the layers are bonded using heat and pressure or another adhesive method to ensure intimate contact between layers. The green part is then furnace processed in a manner identical to indirect processing of metal or ceramic green parts made using powder bed fusion, as introduced in Chap. 5. The CL-100 machine produced parts from up to five types of materials, including materials of differing thickness, which were automatically incorporated into a build. One or more of these materials may act as secondary support materials to enable internal voids or channels and overhangs. These support materials were later removed using thermal or chemical means. A key application for this technology is for the fabrication of microfluidic structures (structures with microscale internal cavities and channels). An example microfluidic structure made using CAM-LEM is shown in Fig. 9.5.

Fig. 9.5 A ceramic microfluidic distillation device cutaway view (*left*) and finished part (*right*) (Courtesy CAM-LEM, Inc.)

Another example of a form then bond process is the Stratoconception approach [3], where the model is sliced into thicker layers. These layers are machined and then glued together to form a part. The use of a multiaxis machining center enables the edges of each layer to be contoured to better match the STL file, helping eliminate the stair-step effect that occurs with increasing layer thickness. This and similar cutting techniques have been used by many different researchers to build large structures from foam, wood, and other materials to form statues, large works of art, and other structures.

9.2 Materials

As covered in the previous section, a wide variety of materials has been processed using a variety of sheet lamination processes, including plastics, metals, ceramics, and paper. A brief survey will be offered identifying the materials and their characteristics that facilitate sheet lamination.

Butcher paper was the first material used in the original Helisys LOM process. Butcher paper is coated on one side with a thin layer of a thermoplastic polymer. It is this polymer coating that melts and ensures that one layer of paper bonds to the previous layer. Since butcher paper is fairly strong and heavy, it forms sturdy parts after a suitable thickness has been fabricated (>5–6 mm typically). After part fabrication, parts are finished as if they were wood by sanding, filing, staining, and varnishing or sealing.

The recently developed Mcor Technologies printers use standard copy paper in A4 or US letter sizes with weights of 20 or 43 lb. Either white or colored paper can be used. The water-based glue binds paper sheets and results in fairly rigid parts although, similar to the Helisys process, a minimum thickness of 5 or 6 mm is required to ensure good strength.

In the metals area, both bond-then-form and form-then-bond approaches have been pursued. Perhaps the most conceptually simple fabrication process is the sheet metal clamping approach, where sheet metal is cut to form part cross sections, then simply clamped together. Other processes use several types of bonding methods. Some researchers were interested in demonstrating the feasibility of some metal sheet lamination process advances, rather than fabricating functional devices, and simply used an adhesive to bond sheets together. In other cases, the adhesive bonded structures were meant to be functional prototypes, not just proof-of-concepts. Aluminum and low-carbon steel materials were most commonly used, unless functional molds or dies were desired, in which case tool steels were used. Thermal and diffusion bonding approaches, on the other hand, tend to provide much more strong parts. Thermal bonding, to be discussed in the next section, has been demonstrated with a variety of aluminum and steel sheets and several types of bonding mechanisms, including brazing and welding. Diffusion bonding, to be covered in Sect. 9.4, has also been demonstrated on a variety of metals and is the important joining mechanism for ultrasonic consolidation, where aluminum, titanium, stainless steel, brass, Inconel, and copper materials have been demonstrated.

In sheet lamination processes, ceramic materials are most often fabricated using bond-them-form processes using ceramic-filled tapes. Tape casting methods form sheets of material composed of powdered ceramics, such as SiC, TiC-Ni composite, or alumina, and a polymer binder. Metal powder tapes can also be used to fabricate metal parts. These tapes are then used for part construction employing a standard sheet lamination process. Various SiC, alumina, TiC-Ni composite, and other material tapes have been used to build parts. A challenge with this process is that thermal post-processing to consolidate metal or ceramic powders results in a large amount of shrinkage (12–18 %) which can lead to dimensional inaccuracies and distortion. This is typical of many conventional powder-based processes, such as powder injection molding, and strategies have been developed to address the effects of shrinkage, although limitations exist.

For polymer materials, the Solidimension example is the most well known and used PVC sheets. Foam blocks have also been used in some research machines, as well as by sculptors who create large sculptures by stacking blocks cut by hot wire or CNC milling. Additionally, some research efforts have successfully demonstrated the automated lay-up of polymer composite sheets. The area of polymer sheet lamination is broad and not very well defined, since it stretches from sculpture to composites manufacturing. This is, perhaps, an area that will see significant attention in the near-term due to its potential.

9.3 Material Processing Fundamentals

As indicated, several types of processes are evident under the general category of sheet lamination. Thermal bonding and sheet metal clamping are covered in this section. In the next section, a more in-depth coverage is provided for the ultrasonic consolidation process.

9.3.1 Thermal Bonding

Many organizations around the world have successfully applied thermal bonding to sheet lamination of functional metal parts and tooling. A few examples will be mentioned to demonstrate the flexibility of this approach. Yi et al. [4] have successfully fabricated 3D metallic parts using precut 1-mm thick steel sheets that are then diffusion bonded. They demonstrated continuity in grain structure across sheet interfaces without any physical discontinuities. Himmer et al. [5] produced aluminum injection molding dies with intricate cooling channels using Al 3003 sheets coated with 0.1-mm thick low-melting point Al 4343 (total sheet thickness 2.5 mm). The sheets were laser cut to an approximate, oversized cross section, assembled using mechanical fasteners, bonded together by heating the assembly in a nitrogen atmosphere just above the melting point of the Al 4343 coating material, and then finish machined to the prescribed part dimensions and surface finish. Himmer et al. [6] also demonstrated satisfactory layer bonding using brazing and laser spot welding processes. Obikawa [7] manufactured metal parts employing a similar process from thinner steel sheets (0.2 mm thick), with their top and bottom surface coated with a low-melting-point alloy. Wimpenny et al. [1] produced laminated steel tooling with conformal cooling channels by brazing laser-cut steel sheets. Similarly, Yamasaki [8] manufactured dies for automobile body manufacturing using 0.5-mm thick steel sheets. Each of these, and other investigators, have shown that thermally bonding metal sheets is an effective method for forming complex metal parts and tools, particularly those which have internal cavities and/or cooling channels.

Although extensively studied, sheet metal lamination approaches have gained little traction commercially. This is primarily due to the fact that bond-then-form processes require extensive post-processing to remove support materials, and form-then-bond processes are difficult to automate for arbitrary, complex geometries. In the case of form-then-bond processes, particularly if a cross section has geometry that is disconnected from the remaining geometry, accurate registration of laminates is difficult to achieve and may require a part-specific solution. Thus, upward-facing features where each cross section's geometry is contiguously interconnected are the easiest to handle. Commercial interest in sheet lamination is primarily in the area of inexpensive, full-color paper parts and large tooling, where internal, conformal cooling channels can provide significant benefits over traditional cooling strategies.

Another process that combined sheet lamination with other forms of AM (including beam deposition, extrusion, and subtractive machining) was Shape Deposition Manufacturing (SDM) [9]. With SDM, the geometry of the part is subdivided into nonplanar segments. Each segment is deposited as an over-sized, near-net shape region and then finish machined. Sequential deposition and machining of segments (rather than planar layers) forms the part. A decision is made concerning how each segment should be manufactured dependent on such factors as the accuracy, material, geometrical features, and functional requirements. Secondary support materials were commonly used to enable complex geometry to be made

Fig. 9.6 Profiled edge laminate tool (Courtesy Fraunhofer CCL)

and for clearance between mechanisms that required differential motion after manufacture. A completely automated subdivision routine for arbitrary geometries, however, was never developed and intervention from a human "expert" is required for many types of geometries. As a result, though interesting and useful for certain complex multimaterial structures, such a system was never commercially introduced.

9.3.2 Sheet Metal Clamping

In the case of assembling rigid metal laminates into simple shapes, it may be advantageous to simply clamp the sheets together using bolts and/or a clamping mechanism rather than using an adhesive or thermal bonding method. Clamping is quick and inexpensive and enables the laminates to be disassembled in order to modify a particular laminate's cross section and/or for easy recycling of the materials. In addition, the clamping or bolting mechanism can act as a reference point to register each laminate with respect to one another.

When clamping, it is often advantageous to simply cut a profile into one edge of a laminate, leaving three edges of the rectangular sheet uncut. An example of such a "profiled edge laminate" construction is shown in Fig. 9.6. Of course, this type of profiled edge can also be utilized with adhesive and thermally bonded layers as well. The major benefit of this approach is the ease with which the layers can be clamped (i.e., bolting the laminates together through a set of holes, as could be done using the through-holes visible on the right edge of Fig. 9.6). The drawbacks of a profile approach are that clamping forces for most tools would then be perpendicular to the laminate interface, and the laminates might separate from one another (leaving gaps) under certain conditions, such as when pressurized polymers are injected into a mold made from such a tool.

9.4 Ultrasonic Additive Manufacturing

Ultrasonic Additive Manufacturing (UAM), also known as Ultrasonic Consolidation (UC), is a hybrid sheet lamination process combining ultrasonic metal seam welding and CNC milling, and commercialized by Solidica Inc., USA in 2000, and subsequently licensed to Fabrisonics (USA). In UAM, the object is built up on a rigidly held base plate bolted onto a heated platen, with temperatures ranging from room temperature to approximately 200°C. Parts are built from bottom to top, and each layer is composed of several metal foils laid side by side and then trimmed using CNC milling.

During UAM, a rotating sonotrode travels along the length of a thin metal foil (typically 100–150 μm thick). The foil is held closely in contact with the base plate or previous layer by applying a normal force via the rotating sonotrode, as shown schematically in Fig. 9.7. The sonotrode oscillates transversely to the direction of motion, at a constant 20 kHz frequency and user-set oscillation amplitude. After depositing a foil, another foil is deposited adjacent to it. This procedure is repeated until a complete layer is placed. The next layer is bonded to the previously deposited layer using the same procedure. Typically four layers of deposited metal foils are termed one level in UAM. After deposition of one level, the CNC milling head shapes the deposited foils/layers to their slice contour (the contour does not need to be vertical, but can be a curved or angled surface, based on the local part geometry). This additive-subtractive process continues until the final geometry of the part is achieved. Thus, UAM is a bond-then-form process, where the forming can occur after each layer or after a number of layers, depending on the settings chosen by the user. Additionally, each layer is typically deposited as a

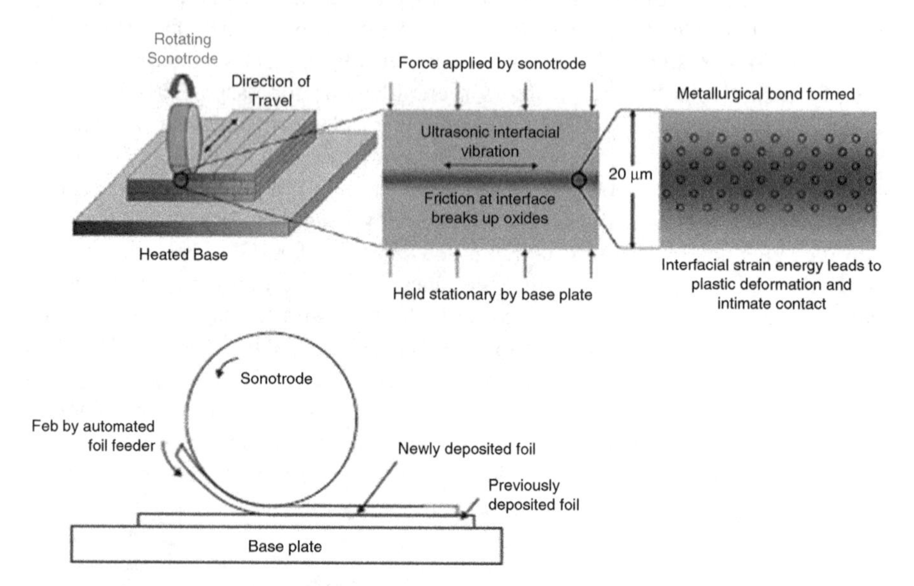

Fig. 9.7 Schematic of ultrasonic consolidation

combination of foils laid side by side rather than a single large sheet, as is typically practiced in sheet lamination processes.

By the introduction of CNC machining, the dimensional accuracy and surface finish of UAM end products is not dependent on the foil thickness, but on the CNC milling approach that is used. This eliminates the stair-stepping effects and layer-thickness-dependent accuracy aspects of other AM processes. Due to the combination of low-temperature ultrasonic bonding, and additive-plus-subtractive processing, the UAM process is capable of creating complex, multifunctional 3D parts, including objects with complex internal features, objects made up of multiple materials, and objects integrated with wiring, fiber optics, sensors, and instruments. The lack of an automated support material in commercial systems, however, means that many types of complex overhanging geometries cannot be built using UAM. However, on-going support material research for UAM will hopefully result in an automated support material approach in the future.

To better illustrate the UAM process, Fig. 9.8a–f illustrates the steps utilized to fabricate a honeycomb panel (270 mm by 240 mm by 10 mm). The cutaway CAD model showing the internal honeycomb features is shown in Fig. 9.8a. The part is fabricated on a 350 mm by 350 mm by 13 mm Al 3003 base plate, which is firmly bolted to a heated platen, as shown in Fig. 9.8b. Metal foils used for this part are Al 3003 foils 25 mm wide and 0.15 mm thick. The first layer of deposited foils is shown in Fig. 9.8c. Since the width of one layer is much larger than the width of the individual metal foils, multiple foils are deposited side by side for one layer. After the deposition of the first layer, a second layer is deposited on the first layer and so on, as seen in Fig. 9.8d. After every four layers of deposition, the UAM machine trims the excess tape ends, and machines internal and external features based on the CAD geometry. After every 40 layers, the machine does a surface machining pass at the exact height of that layer (in this case the z-height of the 40th layer is 0.15 mm per layer times 40 layers, or 6 mm) to compensate for any excess z-height that may occur due to variability in foil thicknesses. A surface machining pass can occur at any point in the process if, for instance, a build interruption or failure occurs (enabling the build to be continued from any user-specific z-height). After a series of repetitive bonding and machining operations the facesheet layers are deposited to enclose the internal features, as shown Fig. 9.8e. Four layers are deposited, and the final panel is shown in Fig. 9.8f.

9.4.1 UAM Bond Quality

There are two widely accepted quality parameters for evaluating UAM-made structures, which are linear welding density (LWD) and part strength. LWD is defined as the percentage of interface which is bonded divided by the total length of the interface between two ultrasonically consolidated foils, determined metallographically. An example of a microstructure sample made from four layers of Al 3003 tapes by UAM is shown in Fig. 9.9. The black areas represent the unbonded

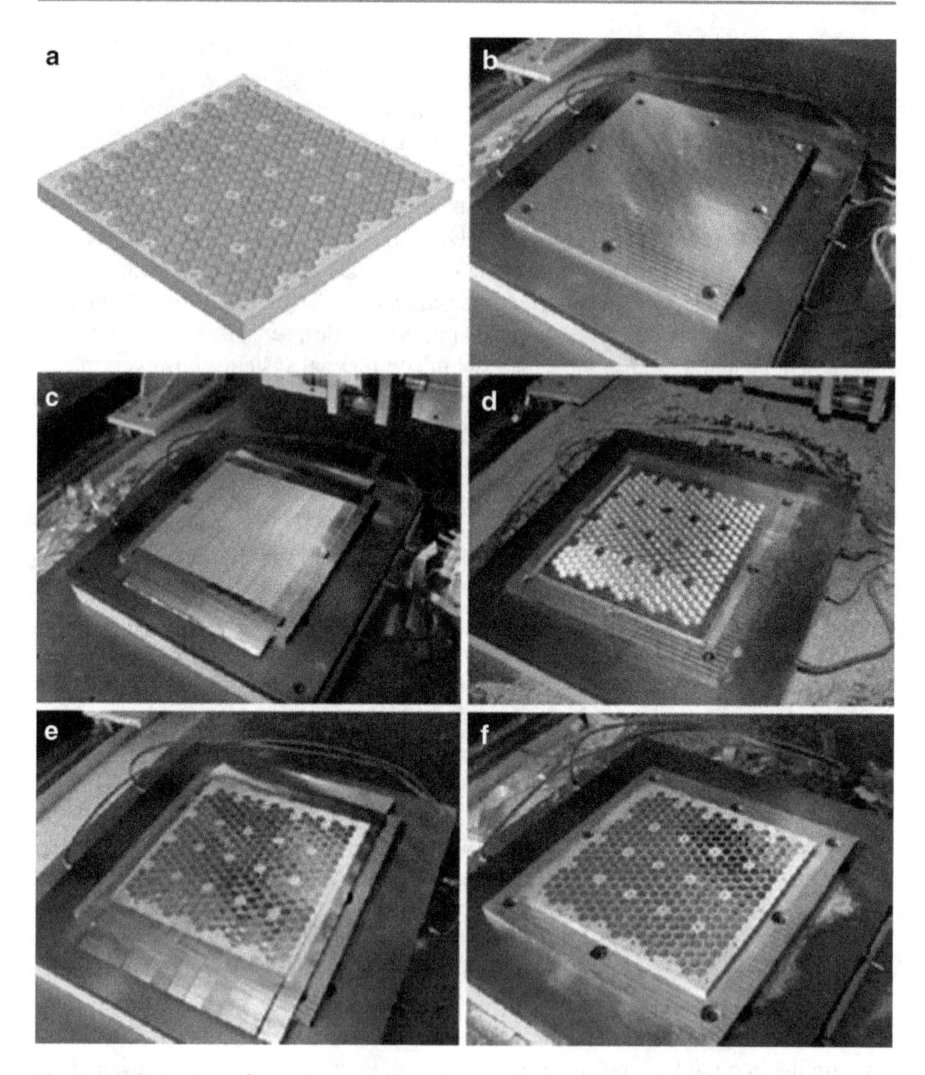

Fig. 9.8 Fabrication procedure for a honeycomb structure using UAM

regions along the interfaces. In this microstructure, a LWD of 100 % occurred only between Layer 1 and the base plate.

9.4.2 Ultrasonic Metal Welding Process Fundamentals

Ultrasonic metal welding (UMW) is a versatile joining technology for various industries, including in electronics, automotive and aerospace. Compared to other metal fusion processes, UMWs solid-state joining approach does not require high

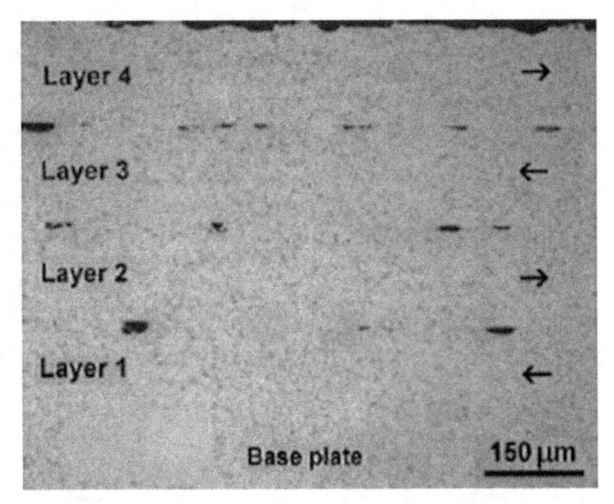

Fig. 9.9 A UAM part made from four layers of Al 3003 foils. LWD is determined by calculating the bonded interface divided by the total interface (*arrows* show the sonotrode traveling direction for each layer). "Effect of Process Parameters on Bond Formation during Ultrasonic Consolidation of Aluminum Alloy 3003," G.D. Janaki Ram, Yanzhe Yang and Brent Stucker, Journal of Manufacturing Systems, 25 (3), pp. 221–238, 2006

temperature diffusion or metal melting; and the maximum processing temperature is generally no higher than 50 % of the melting point of the joined metals. Therefore, thermal residual stresses and thermally induced deformation due to resolidification of molten metals, which are important considerations in thermal welding processes and many AM processes (such as powder bed fusion, beam deposition, and thermal bonding-based sheet lamination processes) are not a major consideration in UAM.

Bonding in UMW can be by (a) mechanical interlocking; (b) melting of interface materials; (c) diffusion bonding; and (d) atomic forces across nascent metal surfaces (e.g., solid-state metallurgical bonding). In UAM, bonding of foils to one another appears to be almost exclusively by nascent metal forces (metallurgical bonding), whereas bonding between foils and embedded structures, such as reinforcement fibers, is primarily by mechanical interlocking. An example of a stainless steel 304 wire mesh embedded between Al 3003 foils using the UAM process is shown as Fig. 9.10. This figure illustrates that the mesh is mechanically interlocked with the Al 3003 matrix, whereas the SS mesh metallurgically bonded to itself and the Al 3003 layers metallurgically bonded to each other. Mechanical interlocking between the Al and SS mesh was due to plastic deformation of Al around and through the mesh. Thus, mechanical interlocking can take place for material combinations between dissimilar metals, or between materials with significant hardness differences. For material combinations of similar materials or materials with similar hardness values, metallurgical bonding appears to be the dominant bond formation mechanism.

Two conditions must be fulfilled for establishment of solid-state bonding during UAM: (a) generation of atomically clean metal surfaces and (b) intimate contact between clean metal surfaces. As all engineering metals contain surface oxides, the oxides must be displaced in order to achieve atomically clean metal surfaces in intimate contact. The ease with which oxide layers can be displaced depends on the

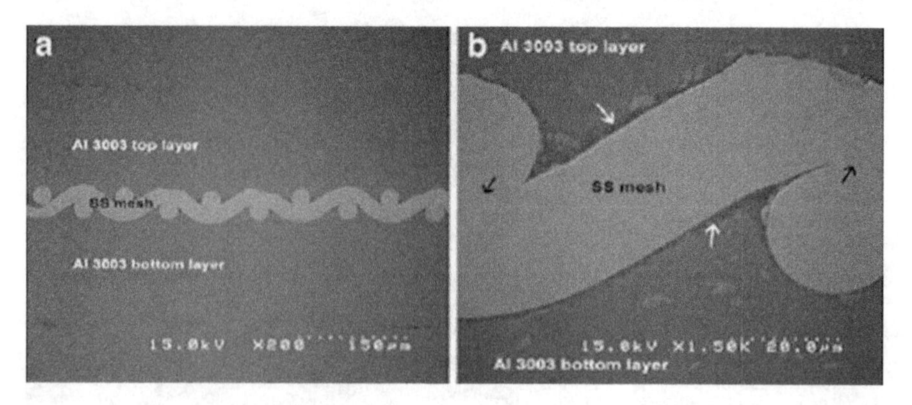

Fig. 9.10 SEM microstructures of Al 3003/SS mesh: (**a**) SS mesh embedded between Al 3003 layers, (**b**) Al 3003/SS mesh interface at a higher magnification. The *white arrows* illustrate the lack of metallurgical bonding between the Al and SS materials. The *black arrows* indicate areas of metallurgical bonding between SS mesh elements. © Emerald Group Publishing Limited, "Use of Ultrasonic Consolidation for Fabrication of Multi-Material Structures," G.D. Janaki Ram, Chris Robinson, Yanzhe Yang and Brent Stucker, Rapid Prototyping Journal, 13 (4), pp. 226–235, 2007

ratio of metal-oxide hardness to base metal hardness, where higher ratios facilitate easier removal. Due to the significant hardness differences between aluminum and aluminum oxide, Al 3003 alloys are one of the best-suited materials for ultrasonic welding. Nonstructural noble metals, such as gold which do not have surface oxide layers, are quite amenable to ultrasonic welding. Materials with difficult-to-remove oxide layers are problematic for ultrasonic welding. However, difficult-to-weld materials have been shown to be UAM compatible when employing chemical or mechanical techniques for removing the surface oxide layers just prior to welding.

Plastic deformation at the foil interfaces is critical for UAM, to break up surface oxides and overcome surface roughness. The magnitude of plastic deformation necessary to achieve bonding can be reduced by decreasing the surface roughness of the interface materials prior to welding, such as by surface machining (which occurred between Layer 1 and the base plate in Fig. 9.9) and/or by removing the surface oxides by chemical stripping or surface finishing. In addition, factors which enhance plastic deformation are also beneficial for bonding, such as using more ductile materials and/or by thermally or acoustically softening the materials during bonding.

Metallic materials experience property changes when subjected to ultrasonic excitations, including acoustic softening, increase in crystallographic defects, and enhanced diffusivities. In particular, metal softening in the presence of ultrasonic excitations, known as the "Blaha effect" or "acoustic softening," means that the magnitude of stresses necessary to initiate plastic deformation are significantly lower [10]. The softening effect of ultrasonic energy on metals is similar to the effect of heating, and can in fact reduce the flow stress of a metallic material more effectively than heating. Thus, acoustic softening results in plastic deformation at strains much less than would otherwise be needed to achieve plastic deformation.

UAM processes also involve metal deformation at high strain rates. High strain rate deformation facilitates formation of vacancies within welded metals, and thus excess vacancy concentration grows rapidly. As a result, the ductility and diffusivity of the metal are enhanced. Both of these characteristics aid in UAM bonding.

9.4.3 UAM Process Parameters and Process Optimization

The important controllable process parameters of UAM are: (a) oscillation amplitude, (b) normal force, (c) travel speed, and (d) temperature. It has been found that the quality of bonding in UAM is significantly affected by each of these process parameters. A brief discussion of each of these parameters and how they affect bonding in UAM follows.

9.4.3.1 Oscillation Amplitude

Energy input directly affects the degree of elastic/plastic deformation between mating metal interfaces, and consequently affects bond formation. Oscillation amplitude and frequency of the sonotrode determine the amount of ultrasonic energy available for bond formation. In commercial UAM machines, the frequency of oscillation is not adjustable, as it is preset based on sonotrode geometry, transducer and booster hardware, and the machine power supply. In UAM, the directly controllable parameter for ultrasonic energy input is oscillation amplitude.

Generally speaking, the higher the oscillation amplitude, the greater the ultrasonic energy delivered. Consequently, for greater energy, more elastic/plastic deformation occurs at the mating metal interface and therefore better welding quality is achieved. However, there is an optimum oscillation amplitude level for a particular foil thickness, geometry, and material combination. A sufficient amount of ultrasonic energy input is needed to achieve plastic deformation, to help fill the voids due to surface roughness that are inherently present at the interface. However, when energy input exceeds a critical level, bonding deteriorates as excess plastic deformation can damage previously formed bonds at the welding interface due to excessive stress and/or fatigue.

9.4.3.2 Normal Force

Normal force is the load applied on the foil by the sonotrode, pressing the layers together. Sufficient normal force is required to ensure that the ultrasonic energy in the sonotrode is delivered to the foils to establish metallurgical bonds across the interface. This process parameter also has an optimized level for best bonding. A normal force higher or lower than the optimum level degrades the quality of bonds and lowers the LWD obtained. When normal force increases beyond the optimum level, the stress condition at the mating interface may be so severe that the formed bonds are damaged, just as it occurs when oscillation amplitude exceeds its optimum level.

9.4.3.3 Sonotrode Travel Speed

Welding exposure time has a direct effect on bond strength during ultrasonic welding. In UAM welding, exposure time is determined by the travel speed of the sonotrode. Higher speeds result in shorter welding exposure times for a given area. Over-input of ultrasonic energy may cause destruction of previously formed metal bonds and metal fatigue. Thus, to avoid bond damage caused by excess ultrasonic energy, an optimum travel speed is important for strong bonds.

9.4.3.4 Preheat Temperature

Metallurgical bonds can be established at ambient temperature during UAM processing. However, for many materials an increased preheat temperature facilitates bond formation. Heating directly benefits bond formation by reducing the flow stress of metals. However, excess heating can have deleterious effects. High levels of metal foil softening can result in pieces of the metal foil sticking to the sonotrode. In addition, in the case of fabrication of structures with embedded electronics, excess temperature may damage embedded electronics. For certain materials, such as Cu, enhanced oxide formation at elevated temperatures will impede oxide removal. Finally, for some materials elevated temperatures cause metallurgical "aging" phenomena such as precipitation hardening, which can embrittle the material and cause premature part failure.

9.4.3.5 Other Parameters

Metal foil thickness is another important factor to be considered in UAM. The most common metal foils used in UAM are on the order of ~150 μm. Generally speaking, bonds are more easily formed between thin metal foils than between thick ones. However, foil damage is a major concern for UAM of thinner metal foils, as they are easily scratched or bent; and thus metal foils between 100 and 200 μm are most often used in UAM.

In addition to material-related constants, process optimization is influenced by the surface condition of the sonotrode, particularly the sonotrode surface roughness. A typical sonotrode in UAM is made of titanium or tool steel. The surface of the sonotrode is EDM roughened to enhance friction between the sonotrode and foil being deposited. However, surface roughness of the sonotrode decreases significantly after extended use. Thus, optimized parameters change along with the condition of the sonotrode surface. Thus it is necessary to practice regular sonotrode roughness measurements and modify process parameters accordingly. Also, the sonotrode surface roughness is imprinted onto the upper-most surface of the just-deposited foil (see upper surface of Fig. 9.9). As a result, this surface roughness must be overcome by plastic deformation during deposition of the next layer. Thus, an optimum surface roughness condition would be one which involves no slip between the sonotrode and the foil being deposited, without significantly increasing the surface roughness of the deposited foil. As slip often increases with decreasing roughness, sonotrode surface roughness is inherently difficult to optimize.

9.4.4 Microstructures and Mechanical Properties of UAM Parts

9.4.4.1 Defects

The most common defects in UAM-made parts are voids. Voids occur either along the interfaces between layers or between the foils that are laid side by side to form each layer. For ease of discussions, defects are classified into three types according to defect origin. Type-1 defects are the voids along layer/layer interfaces due to foil surface roughness and/or insufficient input energy. Type-2 defects are damaged areas, also at the layer/layer interface, that are created when excess energy input during UAM results in the breaking of previously formed bonds. Type-3 defects are found between adjacent foils within a layer.

One can identify defect types by observing the existence of oxide layers on the surfaces of the defects or by looking at the defect morphology. For Type-1 defects, since the metal surfaces have not bonded, oxide layers are not damaged and removed, and can be observed. In addition, Type-1 defects typically have a flat upper surface and a rounded lower surface (where the flat upper surface is the newly deposited, smooth foil and the rounded lower surface is the unbonded upper surface of the previously deposited foil, as seen in Fig. 9.11). For Type-2 defects, since bonding has occurred, oxide layers have been disturbed and are difficult to locate. Type-2 defects thus have a different morphology than Type-1 defects, as they represent voids where the interface has been torn apart after bonding, rather than regions which have never bonded.

Type-3 defects are the physical gaps between adjacent metal foils, as shown in Fig. 9.12. In UAM, the foil width setting within the software determines the offset distance the sonotrode and foil placement mechanism are moved between

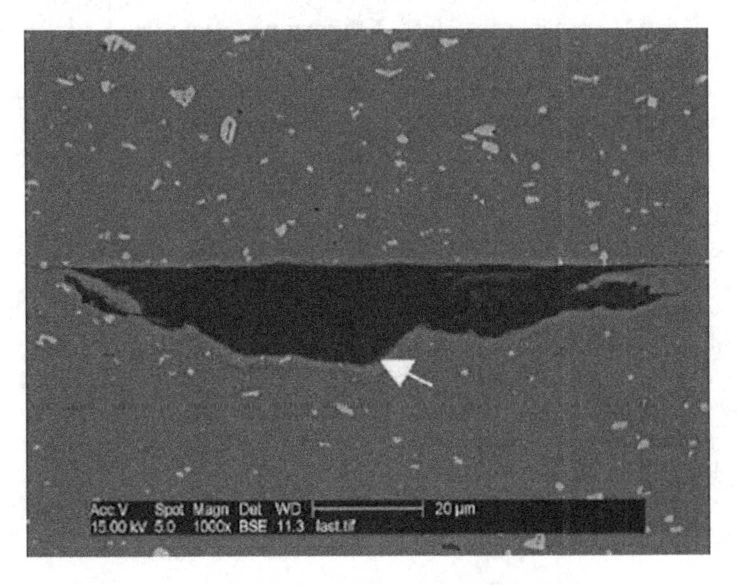

Fig. 9.11 Type-1 UAM defect (*arrow* indicates location of surface oxides)

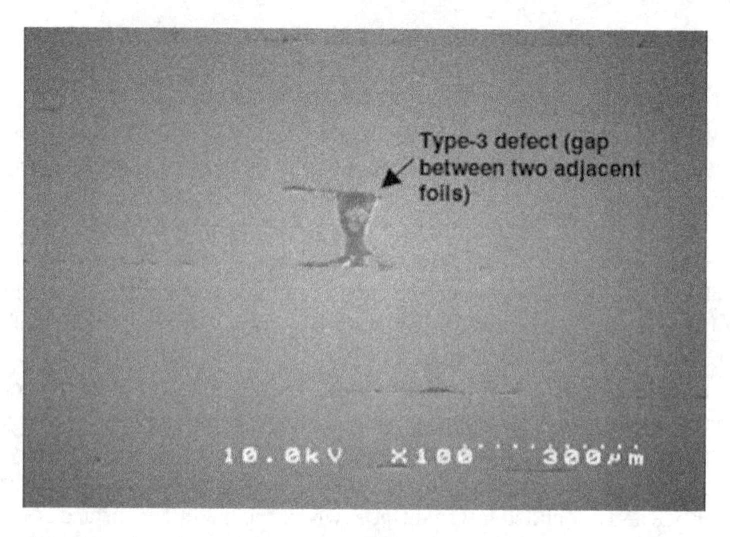

Fig. 9.12 Type-3 defect observed between adjacent foils (Note the morphology of the Type-1 defects between layers indicate that this micrograph is upside-down with respect to build orientation)

depositions of adjacent foils within a layer. If the setting value is larger than the actual metal foil width, there will always be gaps between adjacent foils. The larger the width setting above the foil width, the larger the average physical gap. If the width setting is smaller than the actual width of the foil, gaps will be minimized. However, excess overlap results in surface unevenness at the overlapping areas and difficulty with welding. Thus, positioning inaccuracies of the foil placement mechanism in a UAM machine, combined with improper width settings cause Type-3 defects.

Defects strongly affect the strength of UAM parts. Process parameter optimization (including optimization of width settings) to maximize LWD and minimize Type-3 defects is the most effective means to increase bond strength. With optimized parameters, Type-1 and Type-3 defects are minimized and Type-2- defects do not occur.

Type-1 defects can be reduced by surface machining a small amount of metal (~10 μm, or the largest roughness observed at the upper-most deposited surface, as in Fig. 9.9) after depositing each layer. Post-process heat treatment can also be used to significantly reduce all types of defects.

The degradation of part mechanical properties due to Type-3 defects can be reduced by designed arrangement of successive layers. Successive layers in a UAM part can be arranged so that 50 % overlap across layers is obtained, as shown in Fig. 9.13. Although somewhat counter-intuitive, it has been shown that better tensile properties result from a 50 % overlap than when random foil arrangements are used.

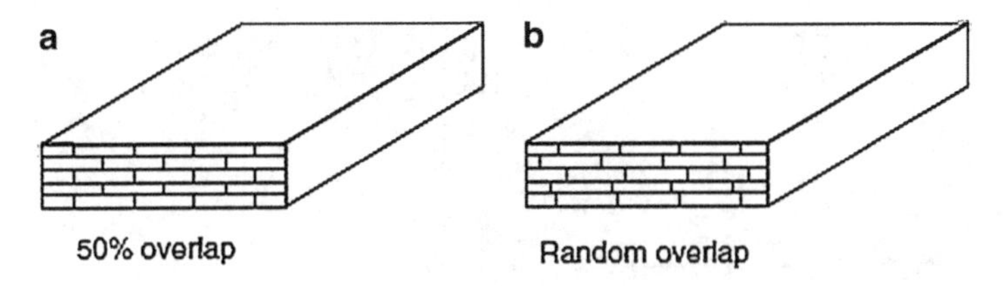

Fig. 9.13 Schematic illustrating (**a**) 50 % foil overlap and (**b**) random foil overlap in UAM

Fig. 9.14 Ultrasonically consolidated Ni 201 foils

9.4.4.2 UAM Microstructures

Typical microstructures from Al 3003 tapes with representative defects were shown in Figs. 9.9 and 9.12. Figure 9.14 shows the microstructure of two Ni 201 foils deposited on an Al 3003 substrate. Plastic deformation of Ni foils near the foil surfaces can be experimentally visualized using orientation imaging microscopy, as shown in Fig. 9.15. Smooth intragrain color transition within a few grains at the surface indicates the foil interfaces undergo some plastic deformation during UAM processing, whereas the absence of intragranular color transitions away from the foil surfaces indicates that the original microstructure is retained in the bulk of the foil.

In addition to UAM of similar materials, UAM of dissimilar materials is quite effective. Many dissimilar metal foils can be bonded with distinct interfaces, with a high degree of LWD and without intermetallic formation [11].

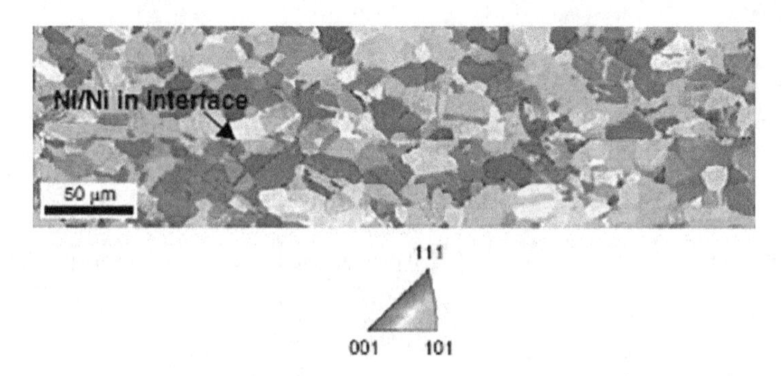

Fig. 9.15 An image of several inverse pole figures of contiguous areas along a well-bonded Ni–Ni interface stitched together. The grains in the image are color coded to reflect their orientation (for color version, see Acta Materialia by Brent L. Adams, Clayton Nylander, Brady Aydelotte, Sadegh Ahmadi, Colin Landon, Brent E. Stucker, G.D. Janaki Ram. Copyright 2008 by Elsevier Science & Technology Journals. Reproduced with permission of Elsevier Science & Technology Journals in the format Textbook via Copyright Clearance Center.)

9.4.4.3 Mechanical Properties

Mechanical properties of UAM parts are highly anisotropic due to the anisotropic properties of metal foils, the presence of defects in particular areas, and the alignment of grain boundaries along the foil-to-foil interfaces. Most metal foils used in UAM are prepared via rolling. Grains within the foils are often elongated along the rolling direction. As a result, foils are typically stronger along the rolling direction, and thus UAM parts are typically stronger in the x-axis than in the y- or z-axes. A typical transverse y-axis strength for a UAM part is about 85 % of the published bulk strength value for a particular material whereas the longitudinal x-axis strength typically exceeds published values for a material. In the *z*-direction, perpendicular to the layer interfaces, UAM parts are much weaker than the x and y properties. This is primarily due to the fact that the bond formed across the foil interfaces, even at 100 % LWD, is not as strong as the more isotropic inter-granular bonding within the foils. Thus, *z*-direction strength values are often 50 % of the published value for a particular material, with very little ductility.

Thus, when considering UAM for part fabrication, it is important to consider the anisotropic aspects of UAM parts with respect to their design. Heat treatment can be used to normalize these properties if this anisotropy results in unacceptable properties.

Another factor which affects mechanical properties is the interfacial plastic deformation which foils undergo during UAM. This plastic deformation increases the hardness of the metal as a result of work hardening effects. Although this work hardening improves the strength, it has a negative effect on ductility.

9.4.5 UAM Applications

UAM provides unique opportunities for manufacture of structures with complex internal geometries, manufacture of structures from multiple materials, fiber embedment during manufacture, and embedding of electronics and other features to form smart structures. Each of these application areas is discussed below.

9.4.5.1 Internal Features

As with other AM techniques, UAM is capable of producing complex internal features within metallic materials. These include honeycomb structures, internal pipes or channels, and enclosed cavities. During UAM, internal geometrical features of a part are fabricated via CNC trimming before depositing the next layer (see Fig. 9.8). Not all internal feature types are possible, and all of the "top" surface of internal features will have a stair-step geometry and not a CNC-milled surface, as the CNC can only mill the upward-facing surfaces of internal geometries. After fabrication of an internal feature is completed, metal foils are placed over the cavities or channels and welded, thus enclosing the internal features.

It has been shown that it becomes quite difficult to bond parts using UAM when their height-to-width ratio is near 1:1 [12]. In order to achieve higher ratios, support materials or other restraints are necessary to make the part rigid enough such that there is differential motion between the existing part and the foils that are being added. The development of an effective support material dispensing system for UAM would dramatically increase its ability to make more free-form shapes and larger internal features. Without support materials, internal features must be designed and oriented in such a way that the sonotrode is always supported by an existing, rigid feature while depositing a subsequent layer. As a result, for instance, internal cooling channels cannot be perpendicular to the sonotrode traveling direction, and honeycomb structures must be small enough that there are always at least two ribs supporting the deposition of the foil face sheets.

9.4.5.2 Material Flexibility

A wide range of metallic materials has been used with UAM. Theoretically, any metal which can be ultrasonically welded is a candidate material for the UAM process. Materials which have been successfully bonded using UAM include Al 3003 (H18 and O condition), Al 6061, Al 2024, Inconel® 600, brass, SS 316, SS 347, Ni 201, and high purity copper. Ultrasonic weldabilities of a number of other metallic materials have been widely demonstrated [11, 13–16]. Thus, there is significant material flexibility for UAM processes. In addition to metal foils, other materials have been used, including MetPreg® (an alumina fiber-reinforced Al matrix composite tape) and prewoven stainless steel AISI 304 wire meshes (see Fig. 9.10), which both have been bonded to Al 3003 using UAM.

By depositing various metal foils at different desired layers or locations during UAM, multimaterial structures or functionally gradient materials can be produced. Composition variation and resultant property changes can be designed to meet various application needs. For instance, by changing materials it is possible to optimize thermal conductivity, wear resistance, strength, ductility, and other properties at specific locations within a part.

9.4.5.3 Fiber Embedment

One of the unique features of UAM is that it enables fiber embedment. As can be seen in Fig. 9.16, bonding near an embedded fiber is much better than bonding away from the fiber for a particular set of process parameter conditions. Plastic flow predicted by modeling done at Sheffield University by Mariani and Ghassemieh (2009) has shown that in some cases there can be one hundred times the degree of interfacial metal flow in the presence of a fiber when compared to bonding of foils without a fiber.

The most commonly embedded fibers are silicon carbide structural fibers within Al matrices (thus forming an Al/SiC metal matrix composite) and optical fibers within Al matrices. Fibers can also be placed and embedded between dissimilar materials, as seen in Fig. 9.17. In the case of dissimilar materials, the presence of a stiff fiber exacerbates the plastic deformation between the stiffer and less stiff material, causing the material with a lower flow stress to deform more than the higher flow stress material. In addition, in contrast to the case of embedment between similar materials where the fiber center is typically aligned with the foil interfaces, the fiber is offset into the softer material (compare Figs. 9.16 and 9.17).

Embedded ceramic fibers are typically mechanically entrapped within metal matrices, without any chemical bonding between fiber and matrix materials. As a result of this mechanical entrapment, friction aids in the transfer of tensile loads from the matrix to the fiber, thus strengthening the part, whereas the lack of chemical bonding means that there is little resistance to shear loading at the fiber/matrix interface, thus weakening the structure for this failure mode.

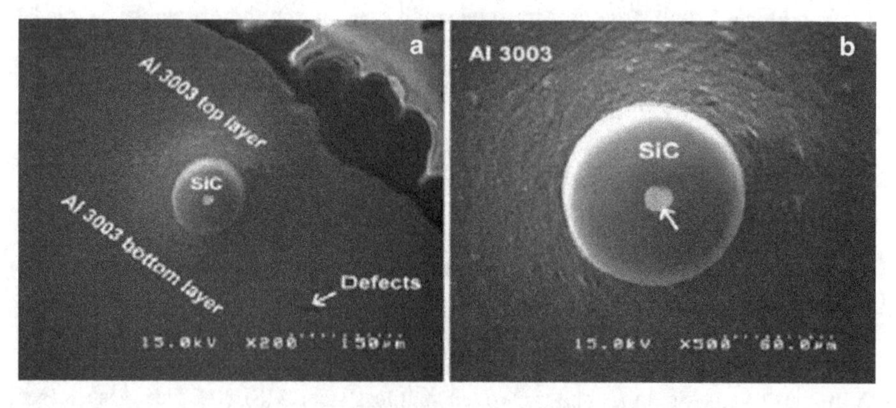

Fig. 9.16 SEM microstructures of Al 3003/SiC: (**a**) SiC fiber embedded between Al 3003 layers showing a lack of defects near the fiber and (**b**) the same SiC fiber at a higher magnification showing excellent bonding near the fiber. © Emerald Group Publishing Limited, "Use of Ultrasonic Consolidation for Fabrication of Multi-Material Structures," G.D. Janaki Ram, Chris Robinson, Yanzhe Yang and Brent Stucker, Rapid Prototyping Journal, 13 (4), pp. 226–235, 2007

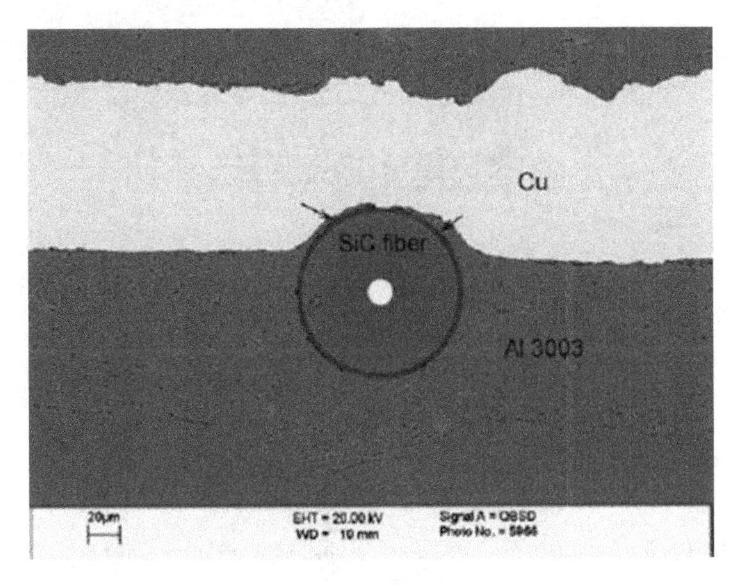

Fig. 9.17 SiC Fiber embedded between copper and aluminum using UAM. Black arrows denote regions where the softer Al extruded around the fiber during embedment, resulting in displacement of the fiber away from the interface into the Al base material

UAM is a candidate manufacturing process for fabrication of long-fiber-reinforced metal matrix composites (MMC). However, to utilize UAM to make end-use MMC parts, several technical difficulties need to be overcome, including automatic fiber feeding and alignments mechanisms, and the ability to change the fiber/foil direction between layers.

Optical fibers have been successfully embedded by many researchers worldwide. Since UAM operates at relatively low processing temperatures, many types of optical fibers can be deposited without damage, thus enabling data and energy to be optically transported through the metal structure.

9.4.5.4 Smart Structures

Smart structures are structures which can sense, transmit, control, and/or react to data, such as environmental conditions. In a smart structure, sensors, actuators, processors, thermal management devices, and more can be integrated to achieve a desired functionality (see Fig. 9.18). Fabrication of smart structures is difficult for conventional manufacturing processes, as they do not enable full three-dimensional control over geometry, composition and/or placement of components. AM processes are inherently suited to the fabrication of smart structures and UAM, in particular, offers several advantages. Primarily due to the fact that UAM is the only AM process whereby metal structures can be formed at low temperatures, UAM offers excellent processing capability for fabrication of smart structures. In addition to traditional internal self-supporting features (honeycomb structures, cooling channels, etc.), larger internal cavities can designed to enable placement of

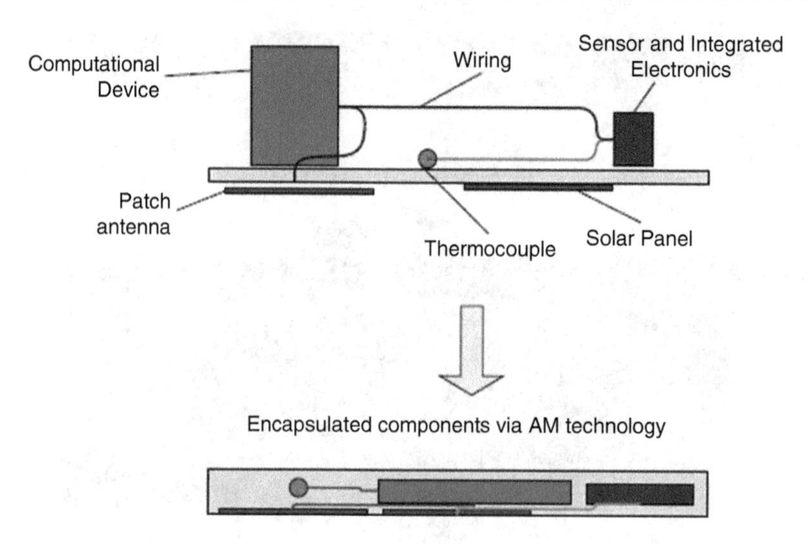

Fig. 9.18 Schematic illustrating the creation of a smart structure using UAM

electronics, actuators, heat pipes, or other features at optimum location within a structure [17]. Many types of embedded electronics, sensors, and thermal management devices have been inserted into UAM cavities. Sensors for recording temperature, acceleration, stress, strain, magnetism, and other environmental factors have been fully encapsulated and have remained functional after UAM embedment. In addition to prefabricated electronics, it is feasible to fabricate customized electronics in UAM with the integration of direct write technologies (see Chap. 11). By combining UAM with direct write, electronic features (conductors, insulators, batteries, capacitors, etc.) can be directly created within or on UAM-made structures in an automated manner.

9.5 Conclusions

As illustrated in this chapter, a broad range of sheet lamination techniques exist. From the initial LOM paper-based technology to the more recent UAM approach, sheet lamination processes have shown themselves to be robust, flexible, and valuable for many applications and materials. The basic method of trimming a sheet of material to form a cross-sectional layer is inherently fast, as trimming only occurs at the layer's outline rather than needing to melt or cure the entire cross-sectional area to form a layer. This means that sheet lamination approaches exhibit the speed benefits of a layer-wise process while still utilizing a point-wise energy source. Many variations of sheet lamination processes have been demonstrated, which have proved to be suitable for many different types of metal, ceramic, polymer, and paper materials.

Future variations of sheet lamination techniques will likely include better materials, new bonding methods, novel support material strategies, new sheet placement mechanisms, and new forming/cutting techniques. As these developments occur, sheet lamination techniques will likely move from the fringe of AM to a more central role in the future of many types of products.

9.6 Exercises

1. Discuss the benefits and drawbacks of bond-then-form versus form-then-bond approaches. In your discussion, include discussion of processes which can use secondary support material and those which do not.
2. Find four papers not mentioned in the references to this chapter which discuss the creation of tooling from laminated sheets of metal. Discuss the primary benefits and drawbacks identified in these papers to this approach to tooling. Based upon this, what do you think about the commercial viability of this approach?
3. Find three examples where SDM was used to make a complex component. What about this approach proved to be useful for these components? How might these beneficial principles be better applied to AM today?
4. What are the primary benefits and drawbacks of UAM compared to other metal AM processes? Discuss UAM and at least three other metal AM processes in your comparison.
5. Develop several different machine architectures for paper sheet lamination processes. Start with the Helisys and Mcor Technologies examples. Investigate form-then-bond and bond-then-form approaches. Include ink-jet printing capability for color part fabrication. Evaluate the pros and cons of each technology and compare with the machine architectures of commercial machines, if you can find them.

References

1. Wimpenny DI, Bryden B, Pashby IR (2003) Rapid laminated tooling. J Mater Process Technol 138:214
2. Solidimension. www.solidimension.com
3. Stratoconception. www.stratoconception.com
4. Yi S et al (2004) Study of the key technologies of LOM for functional metal parts. J Mater Process Technol 150:175
5. Himmer T, Nakagawa T, Anzai M (1999) Lamination of metal sheets. Comput Ind 39:27
6. Himmer T et al (2004) Metal laminated tooling – a quick and flexible tooling concept. In: Proceedings of the solid freeform fabrication symposium, Austin, TX, p 304
7. Obikawa T (1998) Rapid manufacturing system by sheet steel lamination. In: Proceedings of the 14th international conference computer aided production engineering, Tokyo, Japan, p 265
8. Yamasaki H (2000) Applying laminated die to manufacture automobile part in large size. Die Mould Technol 15:36

9. Weiss L, Prinz F (1998) Novel applications and implementations of shape deposition manufacturing. Naval Research Reviews, Office of Naval Research, Three/1998, Vol L
10. Blaha F, Langenecker B (1966) Plasticity test on metal crystals in an ultrasonic field. Acta Metall 7:93–100
11. Janaki Ram GD, Robinson C, Yang Y, Stucker B (2007) Use of ultrasonic consolidation for fabrication of multi-material structures. Rapid Prototyping J 13(4):226–235
12. Robinson CJ, Zhang C, Janaki Ram GD, Siggard EJ, Stucker B, Li L (2006) Maximum height to width ratio of freestanding structures built using ultrasonic consolidation. Proceedings of the 17th solid freeform fabrication symposium, August, Austin, TX
13. Weare NE, Antonevich JN, Monroe RE (1960) Fundamental studies of ultrasonic welding. Welding J 39:331s–341s
14. Flood G (1997) Ultrasonic energy welds copper to aluminum. Welding J 76:761–766
15. Gunduz I, Ando T, Shattuck E, Wong P, Doumanidis C (2005) Enhanced diffusion and phase transformations during ultrasonic welding of zinc and aluminum. Scr Mater 52:939–943
16. Joshi KC (1971) The formation of ultrasonic bonds between metals. Welding J 50:840–848
17. Robinson C, Stucker B, Coperich-Branch K, Palmer J, Strassner B, Navarrete M, Lopes A, MacDonald E, Medina F, Wicker R (2007) Fabrication of a mini-SAR antenna array using ultrasonic consolidation and direct-write. Second international conference on rapid manufacturing. Loughborough, England

Directed Energy Deposition Processes 10

10.1 Introduction

Directed energy deposition (DED) processes enable the creation of parts by melting material as it is being deposited. Although this basic approach can work for polymers, ceramics, and metal matrix composites, it is predominantly used for metal powders. Thus, this technology is often referred to as "metal deposition" technology.

DED processes direct energy into a narrow, focused region to heat a substrate, melting the substrate and simultaneously melting material that is being deposited into the substrate's melt pool. Unlike powder bed fusion techniques (see Chap. 5) DED processes are NOT used to melt a material that is pre-laid in a powder bed but are used to *melt materials as they are being deposited.*

DED processes use a focused heat source (typically a laser or electron beam) to melt the feedstock material and build up three-dimensional objects in a manner similar to the extrusion-based processes from Chap. 6. Each pass of the DED head creates a track of solidified material, and adjacent lines of material make up layers. Complex three-dimensional geometry requires either support material or a multiaxis deposition head. A schematic representation of a DED process using powder feedstock material and laser is shown in Fig. 10.1.

Commercial DED processes include using a laser or electron beam to melt powders or wires. In many ways, DED techniques can be used in an identical manner to laser cladding and plasma welding machines. For the purposes of this chapter, however, DED machines are considered those which are designed to create complex 3D shapes directly from CAD files, rather than the traditional welding and cladding technologies, which were designed for repair, joining, or to apply coatings and do not typically use 3D CAD data as an input format.

A number of organizations have developed DED machines using lasers and powder feeders. These machines have been referred to as Laser Engineered Net Shaping (LENS) [1], Directed Light Fabrication (DLF) [2], Direct Metal Deposition (DMD), 3D Laser Cladding, Laser Generation, Laser-Based Metal Deposition

I. Gibson et al., *Additive Manufacturing Technologies*,
DOI 10.1007/978-1-4939-2113-3_10

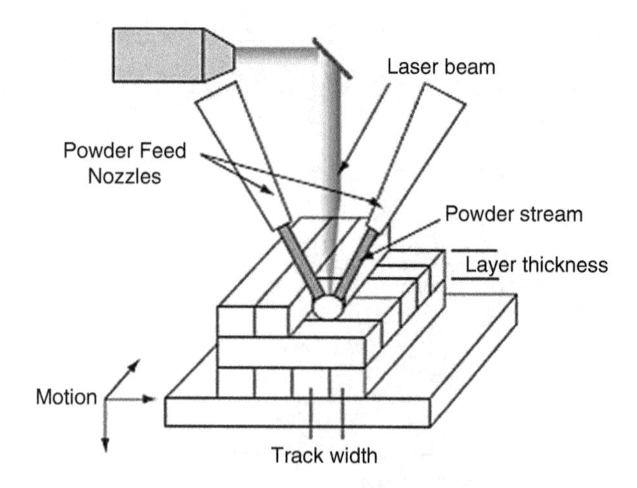

Fig. 10.1 Schematic of a typical laser powder DED process

(LBMD), Laser Freeform Fabrication (LFF), Laser Direct Casting, LaserCast [3], Laser Consolidation, LasForm, and others. Although the general approach is the same, differences between these machines commonly include changes in laser power, laser spot size, laser type, powder delivery method, inert gas delivery method, feedback control scheme, and/or the type of motion control utilized. Because these processes all involve deposition, melting and solidification of powdered material using a traveling melt pool, the resulting parts attain a high density during the build process. The microstructure of parts made from DED processes (Figs. 10.2 and 10.3) are similar to powder bed fusion processes (*see* Fig. 5.14), wherein each pass of the laser or heat source creates a track of rapidly solidified material.

As can be seen from Figs. 10.2 and 10.3, the microstructure of a DED part can be different between layers and even within layers. In the Ti/TiC deposit shown in Fig. 10.2, the larger particles present in the microstructure are unmelted carbides. The presence of fewer unmelted carbides in a particular region is due to a higher overall heat input for that region of the melt pool. By changing process parameters, it is possible to create fewer or more unmelted carbides within a layer, and by increasing laser power, for instance, a greater amount of the previously deposited layer (or substrate for the first layer) will be remelted. By comparing the thickness of the last-deposited layer with the first- or second-deposited layer (such as in Fig. 10.3a), an estimate of the proportion of a layer that is remelted during subsequent deposition can be made. Each of these issues is discussed in the following section.

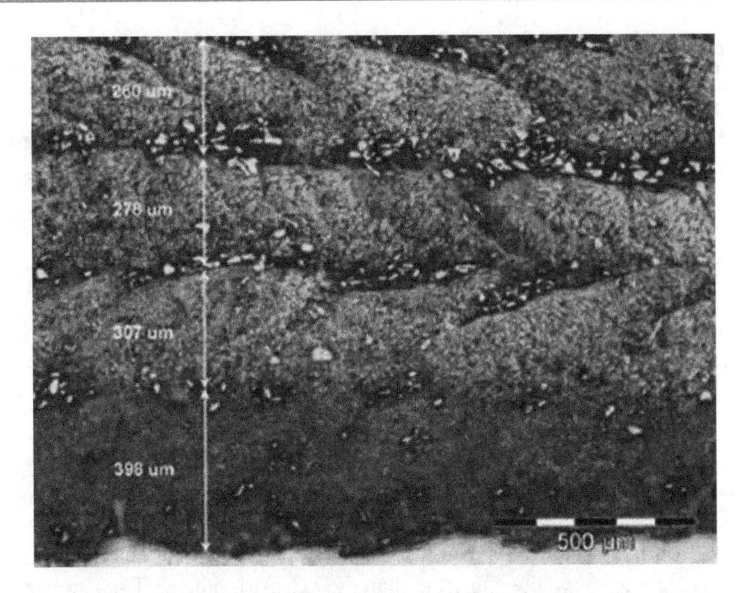

Fig. 10.2 LENS-deposited Ti/TiC metal matrix composite structure (four layers on top of a Ti substrate)

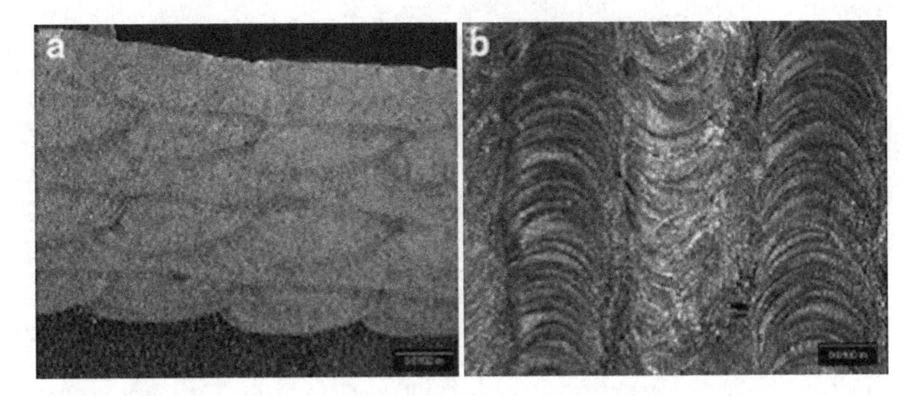

Fig. 10.3 CoCrMo deposit on CoCrMo: (**a**) side view (every other layer is deposited perpendicular to the previous layer using a 0,90,0 pattern) and (**b**) top view of deposit

10.2 General DED Process Description

As the most common type of DED system is a powder-based laser deposition system optimized for metals, we will use a typical LBMD process as the paradigm process against which other processes will be compared. In LBMD, a "deposition head" is utilized to deposit material onto the substrate. A deposition head is typically an integrated collection of laser optics, powder nozzle(s), inert gas tubing, and in some cases, sensors. The substrate can be either a flat plate on which a new part will be fabricated or an existing part onto which additional geometry will be added. Deposition is controlled by relative differential motion between the substrate and deposition head. This differential motion is accomplished by moving the deposition head, by moving the substrate, or by a combination of substrate and deposition head motion. 3-axis systems, whereby the deposition occurs in a vertical manner, are typical. However, 4- or 5-axis systems using either rotary tables or robotic arms are also available. In addition, numerous companies have started to sell LBMD deposition heads as "tools" for inclusion in multi-tool-changer CNC milling machines. By integration into a CNC milling machine, a LBMD head can enable additive plus subtractive capabilities in one apparatus. This is particularly useful for overhaul and repair (as discussed below).

The kinetic energy of powder particles being fed from a powder nozzle into the melt pool is greater than the effect of gravity on powders during flight. As a result, nonvertical deposition is just as effective as vertical deposition. Multiaxis deposition head motion is therefore possible and indeed quite useful. In particular, if the substrate is very large and/or heavy, it is easier to accurately control the motion of the deposition head than the substrate. Conversely, if the substrate is a simple flat plate, it is easier to move the substrate than the deposition head. Thus, depending on the geometries desired and whether new parts will be fabricated onto flat plates or new geometry will be added to existing parts, the optimum design of a LBMD apparatus will change.

In LBMD, the laser generates a small molten pool (typically 0.25–1 mm in diameter and 0.1–0.5 mm in depth) on the substrate as powder is injected into the pool. The powder is melted as it enters the pool and solidifies as the laser beam moves away. Under some conditions, the powder can be melted during flight and arrive at the substrate in a molten state; however, this is atypical and the normal procedure is to use process parameters that melt the substrate and powder as they enter the molten pool.

The typical small molten pool and relatively rapid traverse speed combine to produce very high cooling rates (typically 10^3–10^5 °C/s) and large thermal gradients. Depending upon the material or alloy being deposited, these high cooling rates can produce unique solidification grain structures and/or nonequilibrium grain structures which are not possible using traditional processing. At lower cooling rates, such as when using higher beam powers or lower traverse speeds—which is typical when using electron beams rather than lasers for DED—the grain features grow and look more like cast grain structures.

The passing of the beam creates a thin track of solidified metal deposited on and welded to the layer below. A layer is generated by a number of consecutive overlapping tracks. The amount of track overlap is typically 25 % of the track width (which results in re-melting of previously deposited material) and typical layer thicknesses employed are 0.25–0.5 mm. After each layer is formed, the deposition head moves away from the substrate by one layer thickness.

10.3 Material Delivery

DED processes can utilize both powder and wire feedstock material. Each has limitations and drawbacks with respect to each other.

10.3.1 Powder Feeding

Powder is the most versatile feedstock, and most metal and ceramic materials are readily available in powder form. However, not all powder is captured in the melt pool (e.g., less than 100 % powder capture efficiency), so excess powder is utilized. Care must be taken to ensure excess powder is recaptured in a clean state if recycling is desired.

Excess powder feeding is not necessarily a negative attribute, as it makes DED processes geometrically flexible and forgiving. This is due to the fact that excess powder flow enables the melt pool size to dynamically change. As described below, DED processes using powder feeding can enable overlapping scan lines to be used without the swelling or overfeeding problems inherent in material extrusion processes (discussed in Sect. 6.3).

In DED, the energy density of the beam must be above a critical amount to form a melt pool on the substrate. When a laser is focused to a small spot size, there is a region above and below the focal plane where the laser energy density is high enough to form a melt pool. This region is labeled in Fig. 10.4. If the substrate surface is either too far above or too far below the focal plane, no melt pool will form. Similarly, the melt pool will not grow to a height that moves the surface of the melt pool outside this region.

Within this critical beam energy density region, the height and volume of the deposit melt pool is dependent upon melt pool location with respect to the focal plane, scan rate, laser power, powder flow rate, and surface morphology. Thus, for a given set of parameters, the deposit height approaches the layer thickness offset value only after a number of layers of deposition. This is evident, for instance, in Fig. 10.2, where a constant layer thickness of 200 μm was used as the deposition head z-offset for each layer. The substrate was initially located within the buried spot region, but not far enough within it to achieve the desired thickness for the layers shown (i.e., the laser power, scan rate, and powder flow settings caused the deposit to be thicker than the layer thickness specified). Thus, deposit thickness approached the layer thickness z-offset as the spot became effectively more "buried" during each subsequent layer addition. In Fig. 10.2, however, too few layers were deposited to reach the steady-state layer thickness value.

If the laser and scanning parameters settings used are inherently incapable of producing a deposit thickness at least as thick as the layer thickness z-offset value, subsequent layers will become thinner and thinner. Eventually, no deposit will occur when scanning for the next layer starts outside the critical energy density region (i.e., when the substrate starts out below the exposed spot region, there is insufficient energy density to form a melt pool on the substrate).

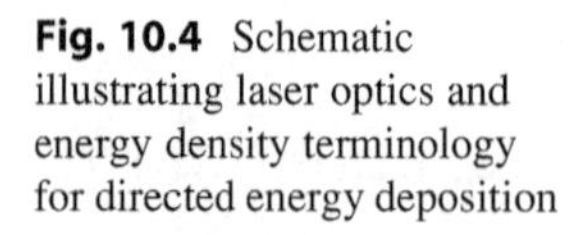

Fig. 10.4 Schematic illustrating laser optics and energy density terminology for directed energy deposition

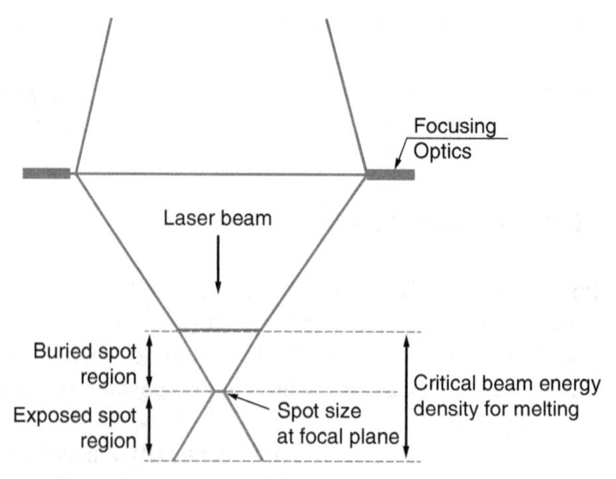

In practice, when the first layer is formed on a substrate, the laser focal plane is typically buried below the surface of the substrate approximately 1 mm. In this way, a portion of the substrate material is melted and becomes a part of the melt pool. The first layer, in this case, will be made up of a mixture of melted substrate combined with material from the powder feeders, and the amount of material added to the surface for the first layer is dependent upon process parameters and focal plane location with respect to the substrate surface. If little mixing of the substrate and deposited material is desired, then the focal plane should be placed at or above the substrate surface to minimize melting of the substrate—resulting in a melt pool made up almost entirely of the powdered material. This may be desirable, for instance, when depositing a first layer of "material A" on top of "material B" that might form "intermetallic AB" if mixed in a molten state. In order to suppress intermetallic formation, a sharp transition from A to B is typically required.

In summary, the first few layers may be thicker or thinner than the layer thickness set by the operator, depending upon the focal plane location with respect to the substrate surface and the process parameters chosen. As a result, the layer thickness converges to the steady-state layer thickness setting after several layers or, if improper parameters are utilized, the laser "walks away" from the substrate and deposition stops after a few layers.

The dynamic thickness benefits of powder feeding also help overcome the corrugated surface topology associated with DED. This corrugated topology can be seen in Fig. 10.3b and is a remnant of the set of parallel, deposited tracks (beads) of material which make up a layer. As in extrusion-based AM processes, in DED a subsequent layer is typically deposited in a different orientation than the previous layer. Common scan patterns from layer to layer are typically multiples of 30, 45, and 90° (e.g., 0, 90, 0, 90...; 0, 90, 180, 270, 360...; 0, 45, 90...315, 360...; and 0, 30, 60...330, 360...). Layer orientations can also be randomized between layers at preset multiples. The main benefits of changing orientation from layer to layer are the elimination of preferential grain growth (which otherwise makes the properties anisotropic) and minimization of residual stresses.

Changing orientation between layers can be accomplished easily when using powders, as the presence of excess powder flow provides for dynamic leveling of the deposit thickness and melt pool at each region of the deposited layer. This means that powdered material feedstock allows the melt pool size to dynamically change to fill the bottoms of the corrugated texture without growing too thick at the top of each corrugation. This is not as easy for wire feeding.

Powder is typically fed by first fluidizing a container of powder material (by bubbling up a gas through the powder and/or applying ultrasonic vibration) and then using a pressure drop to transfer the fluidized powder from the container to the laser head through tubing. Powder is focused at the substrate/laser interaction zone using either coaxial feeding, 4-nozzle feeding, or single nozzle feeding. In the case of coaxial feeding, the powder is introduced as a toroid surrounding the laser beam, which is focused to a small spot size using shielding gas flow, as illustrated in Fig. 10.5a. The two main benefits of coaxial feeding are that it enables a higher capture efficiency of powder, and the focusing shielding gas can protect the melt pool from oxidation when depositing in the presence of air. Single nozzle feeding involves a single nozzle pointed at the interaction zone between the laser and substrate. The main benefits of single nozzle feeding are the apparatus simplicity (and thus lower cost), a better powder capture efficiency than 4-nozzle feeding, and the ability to deposit material into tight locations (such as when adding material to the inside of a channel or tube). The main drawback of single nozzle feeding is that the melt pool geometry is direction specific (i.e., the melt pool is different when feeding towards the nozzle, versus away from the nozzle or at right angles to the nozzle). 4-nozzle feeding involves 4 separate nozzle heads equally spaced at 90° increments around the laser beam, focused to intersect at the melt pool. The main benefit of a 4-nozzle feeding system is that the flow characteristics of 4-nozzle feeding gives more consistency in build height for complex and arbitrary 3D geometries that involve combinations of thick and thin regions.

10.3.2 Wire Feeding

In the case of wire feeding, the volume of the deposit is always the volume of the wire that has been fed, and there is 100 % feedstock capture efficiency (minus a little "splatter" from the melt pool). Wires are most effective for simple geometries, "blocky" geometries without many thin/thick transitions, or for coating of surfaces. When complex, large, and/or fully dense parts are desired, geometry-related process parameters (such as hatch width, layer thickness, wire diameter, and wire feed rate) must be carefully controlled to achieve a proper deposit size and shape. Just as in extrusion-based processes, large deposits with geometric complexity must have porosity designed into them to remain geometrically accurate.

For certain geometries, it is not possible to control the geometry-related process parameters accurately enough to achieve both high accuracy and low porosity with

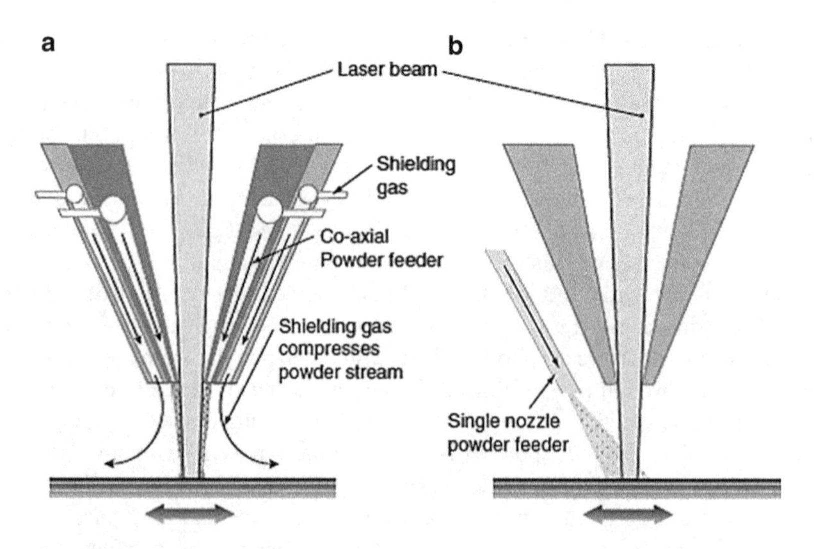

Fig. 10.5 Illustration of powder nozzle configurations: (**a**) coaxial nozzle feeding and (**b**) single-nozzle feeding

a wire feeder unless periodic subtractive processing (such as CNC machining) is done to reset the geometry to a known state. For most applications of DED low porosity is more important than geometric accuracy. Thus, wire-based DED scan processes are designed to be pore free, at the sacrifice of dimensional accuracy. Thus, the selection of a wire feeding system versus a powder feeding system is best done after determining what type of deposit geometries are required, whether dimensional accuracy is critical, and whether a subtractive milling system will be integrated with the additive deposition head.

10.4 DED Systems

10.4.1 Laser Based Metal Deposition Processes

One of the first commercialized DED processes, LENS, was developed by Sandia National Laboratories, USA, and commercialized by Optomec, USA. Optomec's "LENS 750" machine was launched in 1997. Subsequently, the company launched its "LENS 850" and "LENS 850-R" with larger build volume (460 × 460 × 1,070 mm) and dual laser head capability. Optomec's machines originally used an Nd-YAG laser, but more recent machines, such as their MR-7 research system, utilize fiber lasers.

LENS machines process materials in an enclosed inert gas chamber (*see* Fig. 10.6). An oxygen removal, gas recirculation system is used to keep the oxygen concentration in the gas (typically argon) near or below 10 ppm oxygen. This is several orders of magnitude cleaner than the inert gas systems used in powder bed

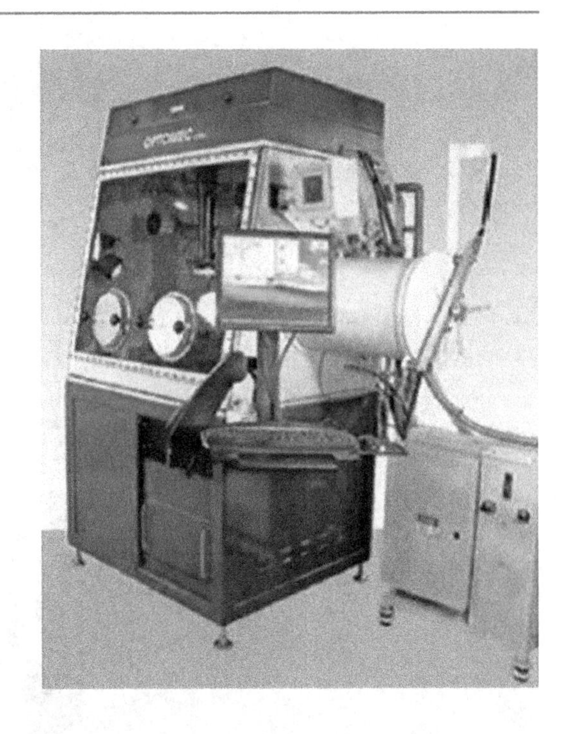

Fig. 10.6 Optomec LENS® 750 system (courtesy Optomec)

fusion machines. The inert gas chamber, laser type, and 4-nozzle feeder design utilized by Optomec make their LENS machines some of the most flexible platforms for DED, as many materials can be effectively processed with this combination of laser type and atmospheric conditions. Most LENS machines are 3-axis and do not use closed-loop feedback control, however 5-axis "laser wrist" systems can enable deposition from any orientation, and systems for monitoring build height and melt pool area can be used to dynamically change process parameters to maintain constant deposit characteristics.

POM Group, USA, is another company building LBMD machines, although they were recently acquired by a company called DM3D Technology. Their DED machines with 5-axis, coaxial powder feed capability build parts using a shielding gas approach. A key feature of DM3D Technology machines has always been the integrated closed-loop control system (*see* Fig. 10.7). The feedback control system adjusts process variables such as powder flow rate, deposition velocity, and laser power to maintain deposit conditions. Closed loop control of DED systems has been shown to be effective at maintaining build quality. Thus, not only do DM3D Technology and Optomec machines offer this option, but so do competing LBMD machine manufacturers as well as companies building electron beam DED machines that utilize wire feeders. DM3D Technology machines have historically utilized CO_2 lasers, which have the benefit of being an economical, high-powered heat source. But the absorptivity of most materials is much less at CO_2 laser wavelengths than for Nd-YAG or fiber lasers (as discussed in Chap. 5 and shown in Fig. 5.10) and thus almost all new DED machines now utilize fiber, diode,

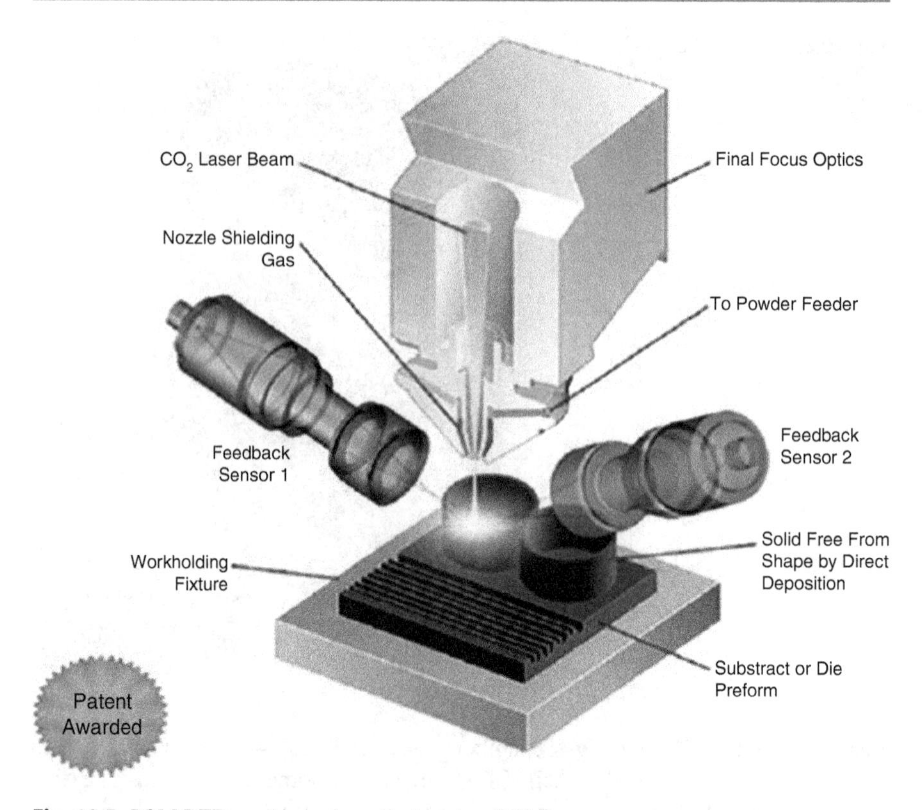

Fig. 10.7 POM DED machine schematic (courtesy POM)

or Nd-YAG lasers. For machines where CO_2 lasers are still used, in order to compensate for their lower absorptivity a larger amount of laser energy is applied, resulting in a larger heat-affected zone and overall heat input.

Another company which was involved early in the development of DED machines was AeroMet Inc., USA—until the division was closed in 2005. The AeroMet machine was specifically developed for producing large aerospace "rib-on-plate" components using prealloyed titanium powders and an 18 kW CO_2 laser (*see* Fig. 10.8). Although they were able to demonstrate the effectiveness of building rib-on-plate structures cost-effectively, the division was not sustainable financially and was closed. The characteristics of using such a high-powered laser are that large deposits can be made quite quickly, but at the cost of geometric precision and a much larger heat-affected zone. Companies which today are interested in the high-deposition-rate characteristics of the AeroMet machine typically choose to use electron beam energy sources with wire feed, and thus there are no commercially available LBMD machines similar to AeroMet's machines.

The benefits behind adding features to simple shapes to form aerospace and other structures with an otherwise poor "buy-to-fly" ratio is compelling. The term buy-to-fly refers to the amount of wrought material that is purchased as a block that

Fig. 10.8 AeroMet System (courtesy MTS Systems Corp.)

is required to form a complex part. In many cases, 80 % or more of the material is machined away to provide a stiff, lightweight frame for aerospace structures. By building ribs onto flat plates using DED, the amount of waste material can be reduced significantly. This has both significant cost and environmental benefits. This is also true for other geometries where small features protrude from a large object, thus requiring a significant waste of material when machined from a block. This benefit is illustrated in the electronics housing deposited using LENS on the hemispherical plate shown in Fig. 10.9.

Another example of LBMD is the laser consolidation process from Accufusion, Canada. The key features of this process are the small spot-size laser, accurate motion control, and single-nozzle powder feeding. This enables the creation of small parts with much better accuracy and surface finish than other DED processes, but with the drawback of a significantly lower deposition rate.

Controlled Metal Buildup (CMB) is a hybrid metal deposition process developed by the Fraunhofer Institute for Production Technology, Germany. It illustrates an integrated additive and subtractive manufacturing approach that a number of research organizations are experimenting with around the globe. In CMB, a diode laser beam is used and the build material is introduced in the form of a wire. After depositing a layer, it is shaped to the corresponding slice contour by a high-speed milling cutter. The use of milling after each deposited layer eliminates the geometric drawbacks of a wire feeder and enables highly accurate parts to be built. The process has been applied primarily to weld repairs and modifications to tools and dies. Subsequent to CMB's success, several of the DED machine manufacturers now offer their LBMD deposition heads for integration into subtractive machine

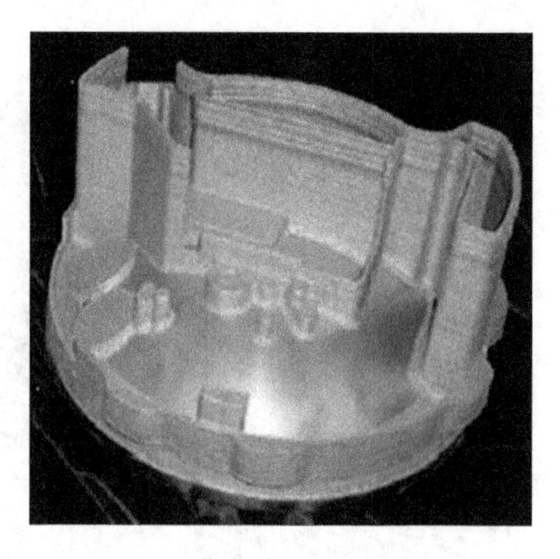

Fig. 10.9 Electronics Housing in 316SS (Courtesy Optomec and Sandia National Laboratories)

tools, and a new company, Hybrid Manufacturing Technologies, was formed specifically to design, build, and integrate LBMD heads into existing CNC machine tool platforms.

10.4.2 Electron Beam Based Metal Deposition Processes

Electron Beam Freeform Fabrication (EBF^3) was developed by NASA Langley, USA, as a way to fabricate and/or repair aerospace structures both terrestrially and in future space-based systems. Using an electron beam as a thermal source and a wire feeder, EBF^3 is capable of rapid deposition under high current flows, or more accurate depositions using slower deposition rates. The primary considerations which led to the development of EBF^3 for space-based applications include: electron beams are much more efficient at converting electrical energy into a beam than most lasers, which conserves scarce electrical resources; electron beams work effectively in a vacuum but not in the presence of inert gases and thus are well suited for the space environment; and powders are inherently difficult to contain safely in low-gravity environments and thus wire feeding is preferred.

Sciaky, USA, has developed a number of electron beam based DED machines which utilize wire feedstock. These Sciaky machines are built inside very large vacuum chambers and enable depositions within a build volume exceeding 6 m in their largest dimension. These machines are excellent at building large, bulky deposits very quickly, enabling large rib-on-plate structures and other deposits to be produced (typically for aerospace applications), eliminating the long lead-times needed for the forged components they replace.

10.4.3 Other DED Processes

Several research groups have investigated the use of welding and/or plasma-based technologies as a heat source for DED. One such group at Southern Methodist University, USA, has utilized gas metal arc welding combined with 4-½ axis milling to produce three-dimensional structures. Similar work has also been demonstrated by the Korea Institute of Science and Technology, which demonstrated combined CO_2 arc welding and 5-axis milling for part production. These approaches are viable and useful as lower-cost alternatives to laser and electron beam approaches, however the typically larger heat-affected zone and other process control issues have kept these approaches from widespread commercialization.

A number of other DED machine architectures and materials have been explored. In addition to multiple investigators who have demonstrated the processing of ceramics using a standard LENS process, other researchers have investigated almost any type of powder which can theoretically be melted using a thermal energy sources. For instance, plastic powder or even table sugar can be blown into a melt pool produced by a hot air gun to produce plastic or sugar parts. Although to date only metal-focused systems have been commercialized, it is likely only a matter of time before different material systems and machine architectures for DED become commercially viable.

10.5 Process Parameters

Most AM machines come pre-programmed with optimized process parameters for materials sold by the machine vendors, but DED machines are sold as flexible platforms; and thus DED users must identify the correct process parameters for their application and material. Optimum process parameters are material dependent and application/geometry dependent. Important process parameters include track scan spacing, powder feed rate, beam traverse speed, beam power, and beam spot size. Powder feed rate, beam power, and traverse speed are all interrelated; for instance, an increase in feed rate has a similar effect to lowering the beam power. Likewise, increasing beam power or powder feed rate and decreasing traverse speed all increase deposit thickness. From an energy standpoint, as the scan speed is increased, the input beam energy decreases because of the shorter dwell time, resulting in a smaller melt pool on the substrate and more rapid cooling.

Scan patterns also play an important role in part quality. As mentioned previously, it may be desirable to change the scan orientation from layer to layer to minimize residual stress buildup. Track width hatch spacing must be set so that adjacent beads overlap, and layer thickness settings must be less than the melt pool depth to produce a fully dense product. Sophisticated accessory equipment for melt pool imaging and real-time deposit height measurement for accurately monitoring the melt pool and deposit characteristics are worthwhile additions for repeatability, as it is possible to use melt pool size, shape, and temperature as feedback control inputs to maintain desired pool characteristics. To control deposit thickness, travel speed can be dynamically changed based upon sensor feedback. Similarly to control solidification rate, and thus microstructure and properties, the melt pool size can be monitored and then controlled by dynamically changing laser power.

10.6 Typical Materials and Microstructure

DED processes aim to produce fully dense functional parts. Any powder material or powder mixture which is stable in a molten pool can be used for construction of parts. In general, metals with high reflectivities and thermal conductivities are difficult to process, such as gold and some alloys of aluminum and copper. Most other metals are quite straightforward to process, unless there is improper atmospheric preparation and bonding is inhibited by oxide formation. Generally, metallic materials that exhibit reasonably good weldability are easy to process.

Ceramics are more difficult to process, as few can be heated to form a molten pool. Even in the event that a ceramic material can be melted, cracking often occurs during cooling due to thermal shock. Thus, most ceramics that are processed using DED are processed as part of a ceramic or metal matrix composite.

For powder feedstock, the powder size typically ranges from approximately 20–150 μm. It is within this range that powder particles can be most easily fluidized and delivered using a flowing gas. Blended elemental powders can be used to produce an infinite number of alloy combinations, or prealloyed powders can be used. Elemental powders can be delivered in precise amounts to the melt zone using separate feeders to generate various alloys and/or composite materials in-situ. When using elemental powders for generation of an alloy in-situ, the enthalpy of mixing plays an important role in determining the homogeneity of the deposited alloy. A negative enthalpy of mixing (heat release) promotes homogeneous mixing of constituent elements and, therefore, such alloy systems are quite suitable for processing using elemental powders.

The fruitfulness of creating multimaterial or gradient material combinations to investigate material properties quickly is illustrated in Figs. 10.10 and 10.11. Figure 10.10 illustrates a tensile bar made with a smooth 1D transition between Ti-6-4 and Ti-22-23, where Fig. 10.11 illustrates the yield strength of various combinations of these alloys. Using optical methods, localized stress and strain fields can be calculated during a tensile test and correlated back to the alloy combination for that location. Using this methodology, the properties of a wide range of alloy combinations can be investigated in a single experiment. Creating larger samples with 2D transitions of alloys (alloy transitions both longitudinally and transversely to the test axis using 3 or 4 powder feeders) can enable even greater numbers of alloy combinations to be investigated simultaneously.

DED processes can involve extremely high solidification cooling rates, from 10^3 to as high as 10^5 °C/s. (This is also true for metal PBF processes, and thus the following discussion is also relevant to parts made using metal PBF.) High cooling rates can lead to several microstructural advantages, including: (a) suppression of

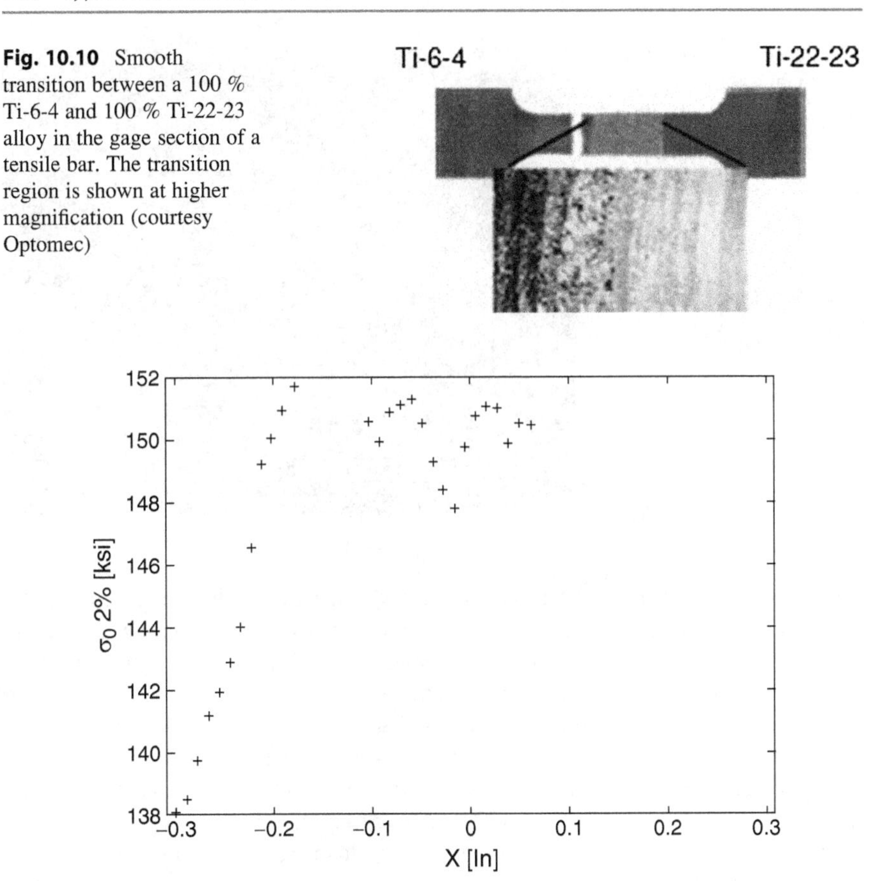

Fig. 10.10 Smooth transition between a 100 % Ti-6-4 and 100 % Ti-22-23 alloy in the gage section of a tensile bar. The transition region is shown at higher magnification (courtesy Optomec)

Fig. 10.11 Yield strength at various locations along the tensile bar from Fig. 10.10 representing the mechanical properties for different combinations of Ti-6-4 and Ti-22-23 (courtesy Optomec)

diffusion controlled solid-state phase transformations; (b) formation of supersaturated solutions and nonequilibrium phases; (c) formation of extremely fine microstructures with dramatically reduced elemental segregation; and (d) formation of very fine secondary phase particles (inclusions, carbides, etc.). Parts produced using DED experience a complex thermal history in a manner very similar to multi-pass weld deposits. Changes in cooling rate during part construction can occur due to heat buildup, especially in thin-wall sections. Also, energy introduced during deposition of subsequent layers can reheat previously deposited material, changing the microstructure of previously deposited layers. The thermal history, including peak temperatures, time at peak temperature, and cooling rates, can be different at each point in a part, leading to phase transformations and a variety of microstructures within a single component.

As shown in Figs. 10.2, 10.3, and 10.10, parts made using DED typically exhibit a layered microstructure with an extremely fine solidification substructure. The interface region generally shows no visible porosity and a thin heat-affected zone

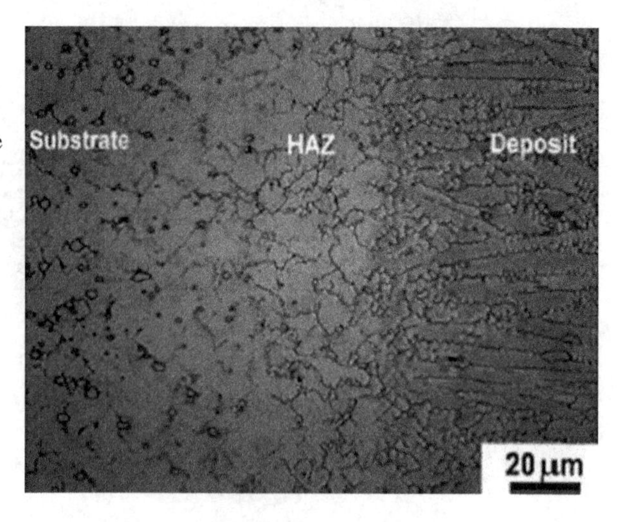

Fig. 10.12 CoCrMo LENS deposit on a wrought CoCrMo substrate of the same composition (deposit occurred from the right of the picture)

(HAZ), as can be seen, for example, on the microstructure at the interface region of a LENS deposited medical-grade CoCrMo alloy onto a CoCrMo wrought substrate of the same composition (Fig. 10.12). Some materials exhibit pronounced columnar grain structures aligned in the laser scan direction, while some materials exhibit fine equiaxed structures. The deposited material generally shows no visible porosity, although gas evolution during melting due to excess moisture in the powder or from entrapped gases in gas-atomized powders can cause pores in the deposit. Pores can also result if excess energy is utilized, resulting in material vaporization and "keyholing." Parts generally show excellent layer-to-layer bonding, although lack-of-fusion defects can form at layer interfaces when the process parameters are not properly optimized and insufficient energy density is utilized.

Residual stresses are generated as a result of solidification, which can lead to cracking during or after part construction. For example, LENS deposited TiC ceramic structures are prone to cracking as a result of residual stresses (Fig. 10.13). Residual stresses pose a particularly significant problem when dealing with metallurgically incompatible dissimilar material combinations.

Formation of brittle intermetallic phases formed at the interface of dissimilar materials in combination with residual stresses can also lead to cracking. This can be overcome by suppressing the formation of intermetallics using appropriate processing parameters or by the use of a suitable interlayer. For instance, in several research projects, it has been demonstrated that it is possible to suppress the formation of brittle intermetallics when depositing Ti on CoCrMo by placing the focal plane above the CoCrMo substrate during deposition of the first layer, and depositing a thin coating of Ti using a low laser power and rapid scan rate. Subsequent layers are likewise deposited using relatively thin deposits at high scan rates and low laser power to avoid reheating of the Ti/CoCrMo interface. Once a sufficient Ti deposit is accumulated, normal process parameters for higher deposition rate can be utilized. However, if excess heat is introduced either during

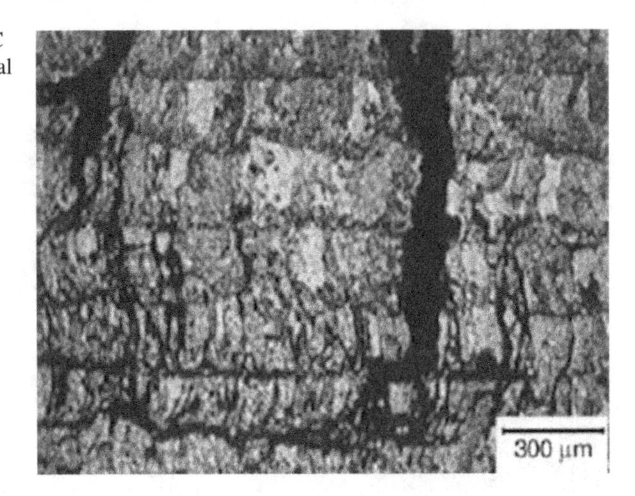

Fig. 10.13 Cracks in a TiC LENS deposit due to residual stresses [4] (SCRIPTA MATERIALIA by Weiping Liu, and J. N. DuPont. Copyright 2003 by Elsevier Science & Technology Journals. Reproduced with permission of Elsevier Science & Technology Journals in the format Textbook via Copyright Clearance Center)

the deposition of subsequent layers or in subsequent heat treatment, equilibrium intermetallics will form and cracking and delamination occurs. In other work, CoCrMo has been successfully deposited on a porous Ta substrate when employing Zr as an interlayer material, a combination that is otherwise prone to cracking and delamination.

It is common for laser deposited parts to exhibit superior yield and tensile strengths because of their fine grain structure. Ductility of DED parts, however, is generally considered to be inferior to wrought or cast equivalents. Layer orientation can have a great influence on % elongation, with the worst being the *z* direction. However, in many alloys ductility can be recovered and anisotropy minimized by heat treatment—without significant loss of strength in most cases.

10.7 Processing–Structure–Properties Relationships

Parts produced in DED processes exhibit high cooling rate cast microstructures. Processing conditions influence the solidification microstructure in ways that can be predicted in part by rapid solidification theory. For a specific material, solidification microstructure depends on the local solidification conditions, specifically the solidification rate and temperature gradient at the solid/liquid interface. By calculating the solidification rate and thermal gradient, the microstructure can be predicted based upon calibrated "solidification maps" from the literature.

To better understand solidification microstructures in DED processes, Beuth and Klingbeil [5] have developed procedures for calculating thermal gradients, G, and solidification rates, R, analytically and numerically. These calculated G and R values can then be plotted on solidification maps to determine the types of microstructures which can be achieved with different DED equipment, process parameters, and material combinations. Solutions for both thin walls [5] and bulky deposits [6] have been described. For brevity's sake, the latter work by Bontha

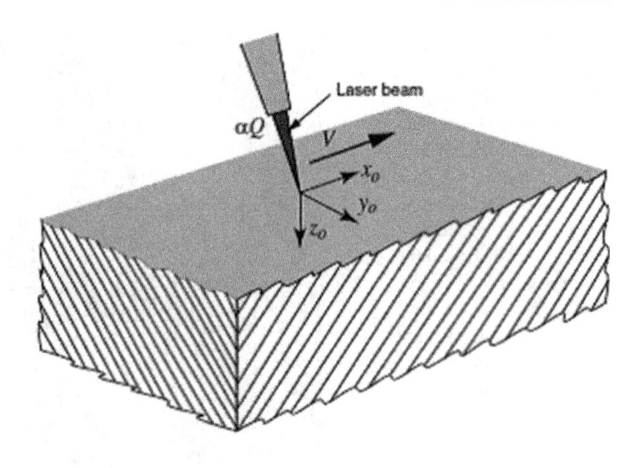

Fig. 10.14 3D Rosenthal geometry considered

et al. [6] based upon the 3D Rosenthal solution for a moving point heat source on an infinite substrate will be introduced here (Fig. 10.14). This 3D Rosenthal solution also has been applied to PBF techniques in an identical fashion.

In this simplified model, material deposition is ignored. The model considers only heat conduction within the melt pool and substrate due to a traveling heat source moving at velocity, V. The fraction of impinging energy absorbed is αQ, which is a simplification of the physically complex temperature-dependent absorption of the beam by regions of the melt pool and solid, absorption of energy by powder in flight, and other factors. Thus a single parameter, α, represents the fraction of impinging beam energy power absorbed.

It is assumed the beam moves only in the x direction, and thus the beam's relative coordinates (x_0, y_0, z_0) from Fig. 10.14 are related to the fixed coordinates (x, y, z) at any time t as $(x_0, y_0, z_0) = (x - Vt, y, z)$.

With the above conditions, the Rosenthal solution for temperature T at time t for any location in an infinite half-space can be expressed in dimensionless form as:

$$\overline{T} = \frac{e^{-\left(\overline{x}_0 + \sqrt{\overline{x}_0^2 + \overline{y}_0^2 + \overline{z}_0^2}\right)}}{2\sqrt{\overline{x}_0^2 + \overline{y}_0^2 + \overline{z}_0^2}} \tag{10.1}$$

where

$$\begin{aligned} &\overline{T} = \frac{T - T_0}{(\alpha Q/\pi k)(\rho c V/2k)}, \\ &\overline{x}_0 = \frac{x_0}{(2k/\rho c V)}, \overline{y}_0 = \frac{y_0}{(2k/\rho c V)} \text{ and } \overline{z}_0 = \frac{z_0}{(2k/\rho c V)}. \end{aligned} \tag{10.2}$$

In these equations, T_0 is the initial temperature, and ρ, c, and k are density, specific heat, and thermal conductivity of the substrate, respectively. In this simplified model, the thermophysical properties are assumed to be temperature independent, and are often selected at the melting temperature, since cooling rate and thermal gradient at the solid/liquid interface is of greatest interest.

The parameters of interest are solidification cooling rate and thermal gradient. The dimensionless expression for cooling rate becomes:

$$\frac{\partial \overline{T}}{\partial \overline{t}} = \frac{1}{2}\frac{e^{-\left((\overline{x}-\overline{t})+\sqrt{(\overline{x}-\overline{t})^2+\overline{y}_0^2+\overline{z}_0^2}\right)}}{\sqrt{(\overline{x}-\overline{t})^2+\overline{y}_0^2+\overline{z}_0^2}} \times \left\{1+\frac{(\overline{x}-\overline{t})}{\left(\sqrt{(\overline{x}-\overline{t})^2+\overline{y}_0^2+\overline{z}_0^2}\right)}+\frac{(\overline{x}-\overline{t})}{\left((\overline{x}-\overline{t})^2+\overline{y}_0^2+\overline{z}_0^2\right.}\right\}. \tag{10.3}$$

where the dimensionless x coordinate is related to the dimensionless x_0 by $\overline{x} = \overline{x}_0 + \overline{t}$ where $\overline{t} = \left(t/\left(2k/\rho c V^2\right)\right)$ and the dimensionless cooling rate is related to the actual cooling rate by:

$$\frac{\partial \overline{T}}{\partial \overline{t}} = \left(\frac{2k}{\rho c V}\right)^2 \left(\frac{\pi k}{\alpha Q V}\right)\frac{\partial T}{\partial t}. \tag{10.4}$$

The dimensionless thermal gradient is obtained by differentiating (10.1) with respect to the dimensionless spatial coordinates, giving

$$|\nabla \overline{T}| = \sqrt{\left(\frac{\partial \overline{T}}{\partial \overline{x}_0}\right)^2 + \left(\frac{\partial \overline{T}}{\partial \overline{y}_0}\right)^2 + \left(\frac{\partial \overline{T}}{\partial \overline{z}_0}\right)}, \tag{10.5}$$

where

$$\frac{\partial \overline{T}}{\partial \overline{x}_0} = -\frac{1}{2}\frac{e^{-\left(\overline{x}_0+\sqrt{\overline{x}_0^2+\overline{y}_0^2+\overline{z}_0^2}\right)}}{\sqrt{\overline{x}_0^2+\overline{y}_0^2+\overline{z}_0^2}} \times \left\{1+\frac{\overline{x}_0}{\left(\sqrt{\overline{x}_0^2+\overline{y}_0^2+\overline{z}_0^2}\right)}+\frac{\overline{x}_0}{\left(\overline{x}_0^2+\overline{y}_0^2+\overline{z}_0^2\right.}\right\}, \tag{10.6}$$

$$\frac{\partial \overline{T}}{\partial \overline{y}_0} = -\frac{1}{2}\frac{\overline{y}_0 e^{-\left(\overline{x}_0+\sqrt{\overline{x}_0^2+\overline{y}_0^2+\overline{z}_0^2}\right)}}{\left(\overline{x}_0^2+\overline{y}_0^2+\overline{z}_0^2\right)}\left\{1+\frac{1}{\left(\sqrt{\overline{x}_0^2+\overline{y}_0^2+\overline{z}_0^2}\right)}\right\}, \tag{10.7}$$

and

$$\frac{\partial \overline{T}}{\partial \overline{z}_0} = -\frac{1}{2}\frac{\overline{z}_0 e^{-\left(\overline{x}_0+\sqrt{\overline{x}_0^2+\overline{y}_0^2+\overline{z}_0^2}\right)}}{\left(\overline{x}_0^2+\overline{y}_0^2+\overline{z}_0^2\right)}\left\{1+\frac{1}{\left(\sqrt{\overline{x}_0^2+\overline{y}_0^2+\overline{z}_0^2}\right)}\right\}. \tag{10.8}$$

As defined above, the relationship between the dimensionless thermal gradient $|\overline{\nabla T}|$ and the actual thermal gradient $|\nabla T|$ is given by

$$|\overline{\nabla T}| = \left(\frac{2k}{\rho c V}\right)^2\left(\frac{\pi k}{\alpha Q}\right)|\nabla T|. \tag{10.9}$$

Using this formulation, temperature, cooling rates, and thermal gradients can be solved for any location (x, y, z) and time (t).

For microstructure prediction purposes, solidification characteristics are of interest; and thus we need to know the cooling rate and thermal gradients at the boundary of the melt pool. The roots of (10.1) can be solved numerically for temperature T equal to melting temperature T_m to find the dimensions of the melt pool. Similarly to (10.2) for normalized temperature, normalized melting temperature can be represented by:

$$\overline{T}_m = \frac{T_m - T_0}{(\alpha Q/\pi k)(\rho c V/2k)}. \tag{10.10}$$

Given cooling rate $\frac{\partial T}{\partial t}$ from (10.4) and thermal gradient, G, defined as $G = |\nabla T|$, we can define the solidification velocity, R, as

$$R = \frac{1}{G}\frac{\partial T}{\partial t}. \tag{10.11}$$

We can now solve these sets of equations for specific process parameters (i.e., laser power, velocity, and material properties) for a machine/material combination of interest. After this derivation, Bontha et al. [6] used this analytical model to demonstrate the difference between solidification microstructures which can be achieved using a small scale DED process with a lower-powered laser beam, such as utilized in a LENS machine, compared to a high-powered laser beam system, such as practiced by AeroMet for Ti–6Al–4 V. Assumptions included the thermophysical properties of Ti–6Al–4 V at $T_m = 1{,}654$ °C, a room temperature initial substrate temperature $T_0 = 25$ °C, fraction of energy absorbed $\alpha = 35$, laser power from 350 to 850 W, and beam velocity ranging from 2.12 to 10.6 mm/s. For the high-powered beam system, a laser power range from 5 to 30 kW was selected. A set of graphs representing microstructures with low-powered systems is shown in Fig. 10.15. Microstructures from high-powered systems are shown in Fig. 10.16 for comparison.

As can be seen from Fig. 10.15, lower-powered DED systems cannot create mixed or equiaxed Ti–6Al–4 V microstructures, as the lower overall heat input

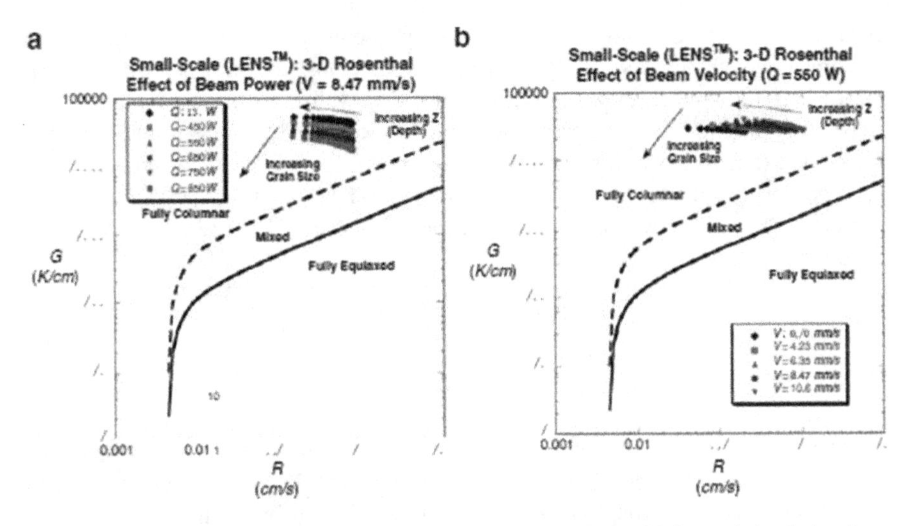

Fig. 10.15 Process maps showing microstructures predicted by the 3D Rosenthal solution for a lower-powered (LENS-like) directed energy deposition system for Ti–6Al–4V (MATERIALS SCIENCE & ENGINEERING. A. STRUCTURAL MATERIALS: PROPERTIES, MICRO-STRUCTURE AND PROCESSING by Srikanth Bontha, Nathan W. Klingbeil, Pamela A. Kobryn and Hamish L. Fraser. Copyright 2009 by Elsevier Science & Technology Journals. Reproduced with permission of Elsevier Science & Technology Journals in the format Textbook via Copyright Clearance Center)

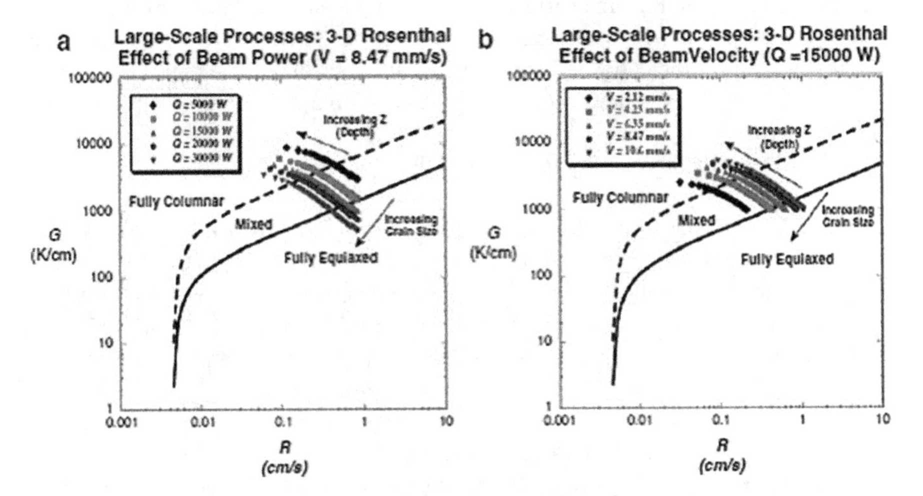

Fig. 10.16 Process maps showing microstructures predicted by the 3D Rosenthal solution for a higher-powered (AeroMet-like) directed energy deposition system for Ti–6Al–4V (MATERIALS SCIENCE & ENGINEERING. A. STRUCTURAL MATERIALS: PROPERTIES, MICRO-STRUCTURE AND PROCESSING by Srikanth Bontha, Nathan W. Klingbeil, Pamela A. Kobryn and Hamish L. Fraser. Copyright 2009 by Elsevier Science & Technology Journals. Reproduced with permission of Elsevier Science & Technology Journals in the format Textbook via Copyright Clearance Center)

means that there are very large thermal gradients. For higher-powered DED systems (relevant to AeroMet-like processes and most electron beam DED process), it is possible to create dendritic, mixed or fully equiaxed microstructures depending upon the process parameter combinations used. As a result, without the need for extensive experimentation, process maps such as these, when combined with appropriate modeling, can be used to predict the type of DED system (specifically the scan rates and laser power) needed to achieve a desired microstructure type for a particular alloy.

10.8 DED Benefits and Drawbacks

DED processes are capable of producing fully dense parts with highly controllable microstructural features. These processes can produce functionally graded components with composition variations in the *X*, *Y*, and *Z* directions.

The main limitations of DED processes are poor resolution and surface finish. An accuracy better than 0.25 mm and a surface roughness of less than 25 μm (arithmetic average) are difficult with most DED processes. Slower build speed is another limitation. Build times can be very long for these processes, with typical deposition rates as low as 25–40 g/h. To achieve better accuracies, small beam sizes and deposition rates are required. Conversely, to achieve rapid deposition rates, degradation of resolution and surface finish result. Changes in laser power and scan rate to achieve better accuracies or deposition rates may also affect the microstructures of the deposited components, and thus finding an optimum deposition condition necessitates tradeoffs between build speed, accuracy, and microstructure.

Examples of the unique capabilities of DED include:

- DED offers the capability for unparalleled control of microstructure. The ability to change material composition and solidification rate by simply changing powder feeder mixtures and process parameters gives designers and researchers tremendous freedom. This design freedom is further explored in Chap. 17.
- DED is capable of producing directionally solidified and single crystal structures.
- DED can be utilized for effectively repairing and refurbishing defective and service damaged high-technology components such as turbine blades.
- DED processes are capable of producing in-situ generated composite and heterogeneous material parts. For example, Banerjee et al. [7] have successfully produced Ti–6Al–4V/TiB composite parts using the LENS process employing a blend of pure prealloyed Ti–6Al–4 V and elemental B powders (98 wt% Ti–6Al–4V + 2 wt% B). The deposited material exhibited a homogeneous refined dispersion of nanoscale TiB precipitates within the Ti–6Al–4V α/β matrix.
- DED can be used to deposit thin layers of dense, corrosion resistant, and wear resistant metals on components to improve their performance and lifetime. One

example includes deposition of dense Ti/TiC coatings as bearing surfaces on Ti biomedical implants, as illustrated in Fig. 10.2.

When contrasted with other AM processes, DED processes cannot produce as complex of structures as powder bed fusion processes. This is due to the need for more dense support structures (or multiaxis deposition) for complex geometries and the fact that the larger melt pools in DED result in a reduced ability to produce small-scale features, greater surface roughness, and less accuracy.

Post-processing of parts made using DED typically involves removal of support structures or the substrate, if the substrate is not intended to be a part of the final component. Finish machining operations because of relatively poor part accuracy and surface finish are commonly needed. Stress relief heat treatment may be required to relieve residual stresses. In addition, depending upon the material, heat treatment may be necessary to produce the desired microstructure(s). For instance, parts constructed in age-hardenable materials will require either a direct aging treatment or solution treatment followed by an aging treatment to achieve precipitation of strengthening phases.

DED processes are uniquely suited among AM process for repair and feature addition. As this AM process is formulated around deposition, there is no need to deposit on a featureless plate or substrate. Instead, DED is often most successful when used to add value to other components by repairing features, adding new features to an existing component and/or coating a component with material which is optimized for the service conditions of that component in a particular location.

As a result of the combined strengths of DED processes, practitioners of DED primarily fall into one of several categories. First, DED has been highly utilized by research organizations interested in the development of new material alloys and the application of new or advanced materials to new industries. Second, DED has found great success in facilities that focus on repair, overhaul, and modernization of metallic structures. Third, DED is useful for adding features and/or material to existing structures to improve their performance characteristics. In this third category, DED can be used to improve the life of injection molding or die casting dies by depositing wear-resistant alloys in high-wear locations; it is being actively researched by multiple biomedical companies for improving the characteristics of biomedical implants; and it is used to extend the wear characteristics of everything from drive shafts to motorcycle engine components. Fourth, DED is increasingly used to produce near net structures in place of wrought billets, particularly for applications where conventional manufacturing results in a large buy-to-fly ratio.

10.9 Exercises

1. Discuss three characteristics where DED is similar to extrusion-based processes and three characteristics where DED is different than extrusion-based processes.

2. Read reference [4] related to thin-wall structures made using DED. What are the main differences between modeling thin wall and bulky structures? What ramifications does this have for processing?
3. Why is solidification rate considered the key characteristic to control in DED processing?
4. From the literature, determine how solidification rate is monitored. From this information, describe an effective, simple closed-loop control methodology for solidification rate.
5. Why are DED processes particularly suitable for repair?

References

1. Keicher DM, Miller WD (1998) Metal powder report 53:26
2. Lewis GK et al (1994) Directed light fabrication. In: Proceedings of the ICALEO'94. Laser Institute of America, Orlando, p 17
3. House MA, et al (1996) Rapid laser forming of titanium near shape articles: LaserCast. In: Proceedings of the solid freeform fabrication symposium, Austin, TX, p 239
4. Liu W, DuPont JN (2003) Fabrication of functionally graded TiC/Ti composites by Laser Engineered Net Shaping. Scr Mater 48:1337
5. Beuth J, Klingbeil N (2001) The role of process variables in laser based direct metal solid freeform fabrication. J Met 9:36–39
6. Bontha S, Klingbeil NW, Kobryn PA, Fraser HL (2009) Effects of process variables and size-scale on solidification microstructure in beam-based fabrication of bulky 3D structures. Mater Sci Eng A 513–514:311–318
7. Banerjee R et al (2003) Direct laser deposition of in situ Ti–6Al–/4V–/TiB composites. Mater Sci Eng A A358:343

Direct Write Technologies

11

11.1 Direct Write Technologies

The term "Direct Write" (DW) in its broadest sense can mean any technology which can create two- or three-dimensional functional structures directly onto flat or conformal surfaces in complex shapes, without any tooling or masks [1]. Although directed energy deposition, material jetting, material extrusion, and other AM processes fit this definition; for the purposes of distinguishing between the technologies discussed in this chapter and the technologies discussed elsewhere in this book, we will limit our definition of DW to those technologies which are designed to build freeform structures or electronics with feature resolution in one or more dimensions below 50 μm. This "small-scale" interpretation is how the term direct write is typically understood in the additive manufacturing community. Thus, for the purposes of this chapter, DW technologies are those processes which create meso-, micro-, and nanoscale structures using a freeform deposition tool.

Although freeform surface modification using lasers and other treatments in some cases can be referred to as direct write [2] we will only discuss those technologies which add material to a surface. A more complete treatment of direct write technologies can be found in books dedicated to this topic [3].

11.2 Background

Although the initial use of some DW technologies predate the advent of AM, the development of DW technologies was dramatically accelerated in the 1990s by funding from the US Defense Advanced Research Projects Agency (DARPA) and its Mesoscopic Integrated Conformal Electronics (MICE) program. DARPA recognized the potential for creating novel components and devices if material deposition technologies could be further developed to enable manufacture of complex electronic circuitry and mesoscale devices onto or within flexible or complex three-dimensional objects. Many different DW technologies were

I. Gibson et al., *Additive Manufacturing Technologies*,
DOI 10.1007/978-1-4939-2113-3_11

developed or improved following funding from DARPA, including Matrix-Assisted Pulsed Laser Evaporation (MAPLE), nScrypt 3De, Maskless Mesoscale Materials Deposition (M^3D, now known as Aerosol Jet), and Direct Write Thermal Spraying. As a result, most people have come to consider DW technologies as those devices which are designed to write or print passive or active electronic components (conductors, insulators, batteries, capacitors, antennas, etc.) directly from a computer file without any tooling or masks. However, DW devices have found broad applicability outside the direct production of circuitry and are now used to fabricate structures with tailored thermal, electrical, chemical, and biological responses, among other applications.

DW processes can be subdivided into five categories, including ink-based, laser transfer, thermal spray, beam deposition, liquid-phase, and beam tracing processes. Most of these use a 3D programmable dispensing or deposition head to accurately apply small amounts of material automatically to form circuitry or other useful devices on planar or complex geometries. The following sections of this chapter describe these basic approaches to DW processing and commercial examples, where appropriate.

11.3 Ink-Based DW

The most varied, least expensive, and most simple approaches to DW involve the use of liquid inks. These inks are deposited on a surface and contain the basic materials which become the desired structure. A significant number of ink types are available, including, among others:

- Colloidal inks
- Nanoparticle-filled inks
- Fugitive organic inks
- Polyelectrolyte inks
- Sol–gel inks

After deposition, these inks solidify due to evaporation, gelation, solvent-driven reactions, or thermal energy to leave a deposit of the desired properties. A large number of research organizations, corporations, and universities worldwide are involved in the development of new and improved DW inks.

DW inks are typically either extruded as a continuous filament through a nozzle (see Chap. 6) or deposited as droplets using a printing head (see Chap. 7). Important rheological properties of DW inks include their ability to (1) flow through the deposition apparatus, (2) retain shape after deposition, and (3) either span voids/gaps or fill voids/gaps, as the case may be. To build three-dimensional DW structures it is highly desirable for the deposited inks to be able to form a predictable and stable 3D deposition shape, and to bridge small gaps. For 2D electronic structures built onto a surface, it is highly desirable for the deposited inks to maintain a constant and controllable cross section, as this will determine the

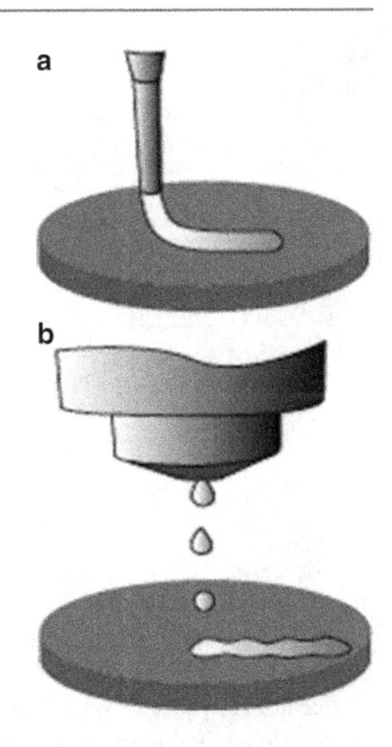

Fig. 11.1 Schematic illustration of direct ink writing techniques: (**a**) continuous filament writing and (**b**) droplet jetting [4] (courtesy nScrypt)

material properties (e.g., conductivity, capacitance, etc.). In general, this means that viscoelastic materials which flow freely under shear through a nozzle but become rigid and set up quickly after that shear stress is released are preferred for DW inks.

DW inks must be transformed after deposition to achieve the desired properties. This transformation may be due to the natural environment surrounding the deposit (such as during evaporation or gelation) but in many cases external heating using a thermal source or some other post-processing step is required.

Figure 11.1 illustrates the two most common methodologies for DW ink dispensing. Continuous dispensing, as in (a), has the merits of a continuous cross-sectional area and a wider range of ink rheologies. Droplet dispensing, as in (b), can be parallelized and done in a very rapid fashion; however, the deposit cross sections are discontinuous, as the building blocks are basically overlapping hemispherical droplet splats, and the rheological properties must be within a tighter range (as discussed in Sect. 7.4). Nozzle dispensing and quill processes both create continuous deposits from DW inks. Printing and aerosol jet processes both create droplets from DW inks. These four approaches are discussed in more detail below.

11.3.1 Nozzle Dispensing Processes

Nozzle DW processes are technologies which use a pump or syringe mechanism to push DW inks through an orifice for deposition onto a substrate. A three-axis motion control system is typically used with these nozzle systems to enable deposition onto complex surfaces or to build-up scaffolds or other 3D geometry,

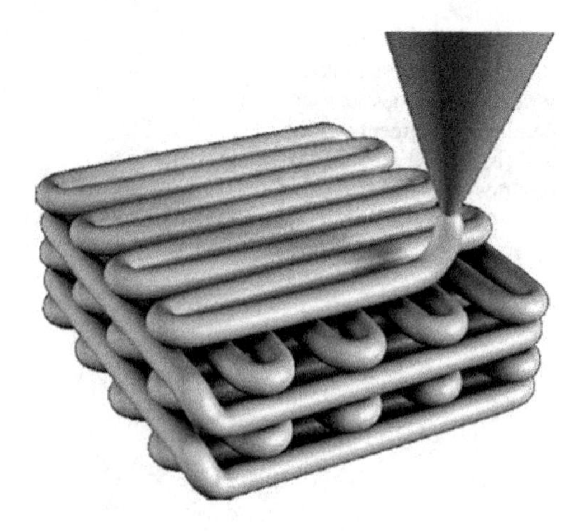

Fig. 11.2 A schematic drawing showing the deposition of a scaffold using a nozzle process [4] (courtesy nScrypt)

as illustrated in Fig. 11.2. Some nozzle devices are packaged with a scanning system that first determines the topology of the substrate on which the deposit is to be made, and then deposits material conformally over that substrate surface based on the scan data.

For nozzle processes, the main differentiating factors between devices are the: (1) nozzle design, (2) motion control system, and (3) pump design. Nozzle design determines the size and shape of the deposit, directly influences the smallest feature size, and has a large effect on the types of inks which can be used (i.e., the viscosity of the ink and the size and type of fillers which can be used in the inks). The motion controller determines the dimensional accuracy and repeatability of the deposit, the maximum size of the deposit which can be made, and the speed at which deposition can occur. The pump design determines the volumetric control and repeatability of dispensing, how accurately the deposits can be started and stopped, and the speed at which deposition can occur. The difference between these three factors for different manufacturers and designs determines the price and performance of a nozzle-based DW process.

Micropen and nScrypt are two companies with well-developed extrusion nozzles and deposition systems for DW. Micropen stopped selling machines in 2008 and currently sells DW services and solutions. nScrypt markets and sells nozzles, pumps, and integrated scanning, dispensing and motion control systems for DW. A wide range of nozzle designs are available, and feedback systems help ensure that the stand-off distance between the nozzle and the substrate remains substantially constant to enable repeatable and accurate deposition of traces across conformal surfaces.

One characteristic of nScrypt systems is their Smart Pump™ design, which has 20 pL control of deposition volume and has an aspirating function, causing the material to be pulled back into the print nozzle at the end of a deposition path. This aspiration function enables precise starts and stops. In addition, a conical nozzle design enables a large range of viscous materials to be dispensed. The pump and nozzle design, when combined, enable viscosities which are processable over six orders of magnitude, from 1 to 1,000,000 cp (the equivalent of processing materials ranging from water to peanut butter). This means that virtually any electronic ink or paste can be utilized. Materials ranging from electronic inks to quick setting concrete have been deposited using nScrypt systems.

Simple DW nozzle devices can be built using off-the-shelf syringes, pumps, and three-axis motion controllers for a few thousand dollars, such as by using an inexpensive material extrusion systems such as the Fab@Home system developed at Cornell University [6]. These enable one to experiment with nozzle-based DW processes for a relatively low-capital investment. Fully integrated devices with multiple nozzles capable of higher dimensional accuracy, dispensing repeatability, and wider range of material viscosities can cost significantly more than $250,000.

Nozzle DW processes have successfully been used to fabricate devices such as integrated RC filters, multilayer voltage transformers, resistor networks, porous chemical sensors, biological scaffolds, and other components. Three aspects of nozzle-based processes make them interesting candidates for DW practitioners: (1) these processes can deposit fine line traces on nonplanar substrates, (2) they work with the largest variety of inks of any DW technology and (3) they can be built-up from interchangeable low-cost components, and integrated easily onto various types of multiaxis motion control systems. The main drawback of nozzle-based systems is that the inks must typically be thermally post-processed to achieve the robust properties desired for most end-use applications. Thus, a thermal or laser post-deposition-processing system is highly beneficial. Although the types of materials which have been deposited successfully using nozzle-based processes are too numerous to list, examples include [5]:

- Electronic Materials—metal powders (silver, copper, gold, etc.) or ceramic powders (alumina, silica, etc.) suspended in a liquid precursor that after deposition and thermal post-processing form resistors, conductors, antennas, dielectrics, etc.
- Thermoset Materials—adhesives, epoxies, etc. for encapsulation, insulation, adhesion, etc.
- Solders—lead-free, leaded, etc. as electrical connections.
- Biological Materials—synthetic polymers and natural polymers, including living cells.
- Nanomaterials—nanoparticles suspended in gels, slurries, etc.

11.3.2 Quill-Type Processes

DW inks can be deposited using a quill-type device, much like a quill pen can be used to deposit writing ink on a piece of paper. These processes work by dipping the pen into a container of ink. The ink adheres to the surface of the pen and then, when the pen is put near the substrate, the ink is transferred from the pen to the substrate. By controlling the pen motion, an accurate pattern can be produced. The primary DW method for doing this is the dip-pen nanolithography (DPN) technique

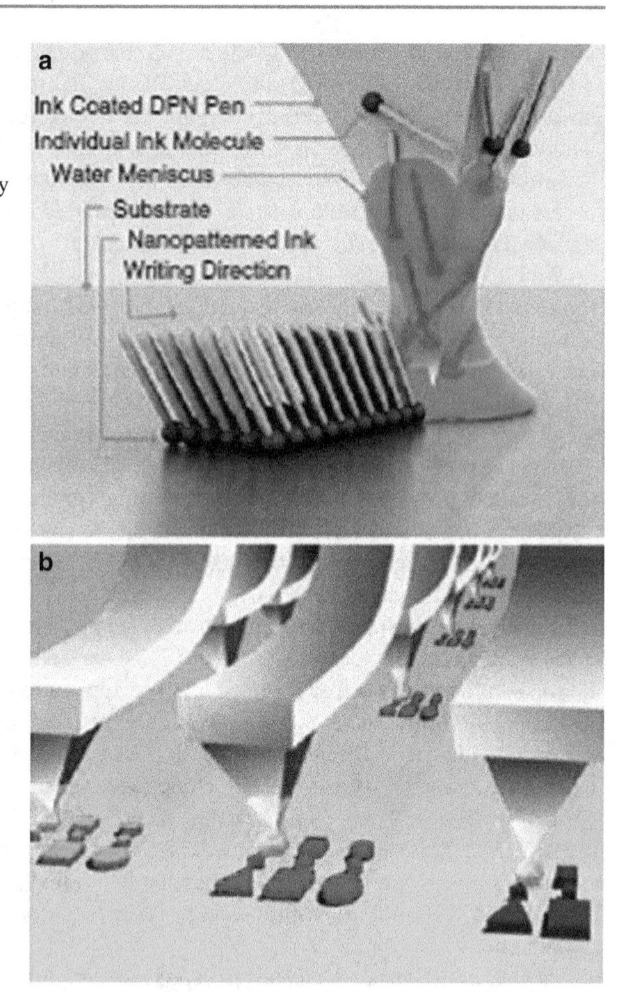

Fig. 11.3 (**a**) A schematic showing how an AFM tip is used to write a pattern on a substrate. (**b**) An illustration of a 2D array of print heads (55,000 per cm^2) [7] (courtesy NanoInk)

developed by a number of universities and sold by Nanoink, Inc. This process works by dipping an atomic force microscope (AFM) tip into an inkwell of specially formulated DW ink. The ink adheres to the AFM tip, and then is used to write a pattern onto a substrate, as illustrated in Fig. 11.3a. Nanoink-ceased operations in 2013, but a number of organizations continue to use DPN for nanoscale DW research and development.

DPN is capable of producing 14 nm line widths with 5 nm spatial resolution. It is typically used to produce features on flat surfaces (although uneven topography at the nm scale is unavoidable even on the so-called flat surfaces). Various 1D and 2D arrays of pen tips are available, with some 1D 8-pen designs capable of individual tip actuation (either "on" or "off" with respect to each other by lifting individual AFM tips using a thermal bimorph approach) so that not all print heads produce the deposition pattern being traced by the motion controller or so that unused pens can be used for AFM scanning. The scalability of DPN was demonstrated using the

PrintArray™, which had 55,000 AFM quills in a square centimeter, enabling 55,000 identical patterns to be made at one time. This array, however, did not enable individual tip actuation.

One use of DPN is the placement of DNA molecules in specific patterns. DNA is inherently viscous, so the pens used for these materials must be stiffer than for most nano inks. Also, unique inkwell arrays have been developed to enable multiple tips to be charged with the same ink, or different inks for different tips. When combining a multimaterial inkwell with an actuated pen array, multimaterial nanoscale features can be produced. Based on the physics of adhesion to AFM tips at very small length scales, most inks developed for other DW processes cannot be used with DPN.

11.3.3 Inkjet Printing Processes

Hundreds, if not thousands, of organizations around the world practice the deposition of DW precursor inks using inkjet printing [3, 8]. This is primarily done to form complex electronic circuitry on flat surfaces, as deposition onto a conformal substrate is difficult. The inkjet printing approach to DW fabrication is comparable to the direct printing class of additive manufacturing technologies discussed in Chap. 7. In the case of DW, the print heads and motion control systems are optimized for printing high-accuracy electronic traces from DW inks onto relatively flat substrates in one or just a few layers rather than the build-up of three-dimensional objects from low-melting-point polymers or photopolymers.

The primary benefits of inkjet approaches to direct write are their speed and low cost. Parallel sets of inkjet print heads can be used to very rapidly deposit DW inks. By setting up arrays of print heads, very large areas can be printed rapidly. In addition, there are numerous suppliers for inkjet print heads.

The primary drawbacks of inkjet approaches to direct write are the difficulty inherent with printing on conformal surfaces, the use of droplets as building blocks which can affect material continuity, more stringent requirements on ink rheology, and a limited droplet size range. Since inkjet print heads deposit material in a droplet-by-droplet manner, the fundamental building block is hemispherical (see Fig. 11.1b). In order to produce consistent conductive paths, for instance, the droplets need a repeatable degree of overlap. This overlap is relatively easy to control between droplets that are aligned with the print head motion, but for deposits that are at an angle with respect to the print head motion there will be a classic "stair-step" effect, resulting in a change in cross-sectional area at locations in the deposit. This can be overcome by using only a single droplet print head and controlling its motion so that it follows the desired traces (similar to the Solidscape approach to material jetting). It can also be overcome using a material removal system (such as a laser) to trim the deposits after their deposition to a highly accurate, repeatable cross section, giving consistent conductivity, resistivity, or other properties throughout the deposit. However, these solutions to stair stepping mean that one cannot take advantage of the parallel nature of inkjet printing, or a more complicated apparatus is needed.

Most inkjet print heads work best with inks of low viscosity at or near room temperature. However, the rheological properties (primarily viscosity) which are needed to print a DW ink can often only be achieved when printing is done at elevated temperatures. The modeling introduced in Chap. 7 is useful for determining the material types which can be considered for inkjet DW.

11.3.4 Aerosol DW

Aerosol DW processes make deposits from inks or ink-like materials suspended as an aerosol mist. The commercialized version of this approach is the Aerosol Jet process developed by Optomec. The Aerosol Jet process begins with atomization of a liquid molecular precursor or a colloidal suspension of materials, which can include metal, dielectric, ferrite, resistor, or biological materials. The aerosol stream is delivered to a deposition head using a carrier gas. The stream is surrounded by a coaxial sheath air flow, and exits the chamber through an orifice directed at the substrate. This coaxial flow focuses the aerosol stream onto the substrate and allows for deposition of features with dimensions as small as 5 μm. Typically either laser chemical decomposition or thermal treatment is used to process the deposit to the desired state.

The Aerosol Jet process can be controlled to be gentle enough to deposit living cells. A schematic illustration of the Aerosol Jet process is shown in Fig. 11.4.

The Aerosol Jet process was initially conceived as a process which made use of the physics of laser guidance. When photons of light interact with free-floating or suspended small particles there is a slight amount of "force" applied to these particles, and these particles move in the direction of photon motion. When applied to aerosol DW, a laser is transmitted through the mist into a hollow fiber optic. The laser propels tiny droplets from the mist into and through the hollow fiber, depositing the droplets onto a substrate where the fiber ends [9]. Laser guidance, however, entrains and moves droplets slowly and inefficiently. To overcome this drawback,

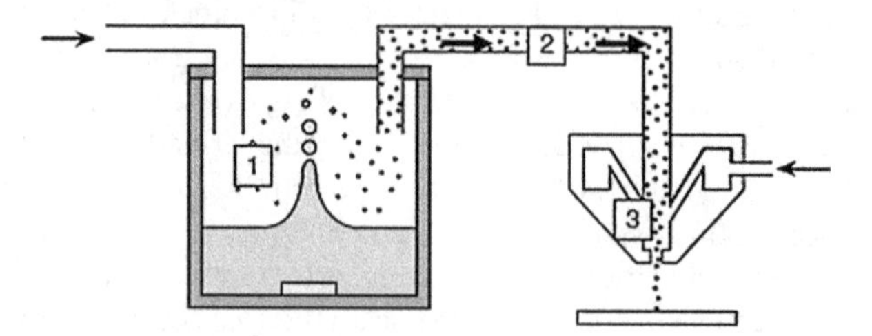

Fig. 11.4 Aerosol Jet System. (1) Liquid material is placed into an atomizer, creating a dense aerosol of tiny droplets 1–5 μm in size. (2) The aerosol is carried by a gas flow to the deposition head (with optional in-flight laser processing). (3) Within the deposition head, the aerosol is focused by a second gas flow and the resulting high-velocity stream is jetted onto the substrate creating features as small as 10 μm in size (Courtesy of Optomec)

Table 11.1 Key benefits and drawbacks of ink-based approaches to DW

	Nozzle	Quill	Inkjet printing	Aerosol
Manufacturer	nScrypt	Nanoink	Various	Optomec
Key benefits	Greatest range of viscosities, simplicity, capable of 3D lattice structures	Nanoscale structures, massive parallelization is possible	Speed due to parallelization of print heads, numerous manufacturers	Widest range of working distances and line widths, coaxial laser treatment
Key drawbacks	Knowledge of surface topography needed to maintain constant stand-off distance	Only relevant at very small length scales, requires precise motion controllers and custom inks	Need flat plates or low-curvature substrates, limited ink viscosity ranges	Complex apparatus. Requires inks which can be aerosolized

further iterations with the technology involved pressurizing the atomizer and using a pressure drop and flow of gas through the tube between the atomizer and the deposition head as the primary means of droplet propulsion. Lasers can still be used, however, to provide in-flight energy to the droplets, or to modify them thermally or chemically. The ability to laser-process the aerosol droplets in-flight and/or on the substrate enables the deposition of a wider variety of materials, as both untreated and coaxially laser-treated materials can be considered.

One benefit of a collimated aerosol spraying process is its high stand-off distance and large working distance. The nozzle can be between 1 and 5 mm from the substrate with little variation in deposit shape and size within that range. This means that repeatable deposits are possible on substrates which have steps or other geometrical features on their surface. A major application for Aerosol Jet is creating interconnects for solar panels; which makes use of its unique ability to deposit conductive traces on a substrate with widely varying stand-off distances.

The Aerosol Jet process is also more flexible than inkjet printing processes, as it can process a wide range of material viscosities (0.7–2,500 cPs), it has variable line widths from 5 to 5,000 μm, and layer thicknesses between 0.025 and 10 μm. The main drawback of the Aerosol Jet process is its complexity compared to other ink-based processes. The Aerosol Jet process has been parallelized to include large numbers of printheads in an array, so the process can be made quite fast and flexible, in spite of its complexity.

Table 11.1 summarizes the key benefits and drawbacks of ink-based approaches to DW.

11.4 Laser Transfer DW

When a focused high-energy laser beam is absorbed by a material, that material may be heated, melted, ablated, or some combination thereof. In the case of ablation, there is direct evaporation (or transformation to plasma) of material.

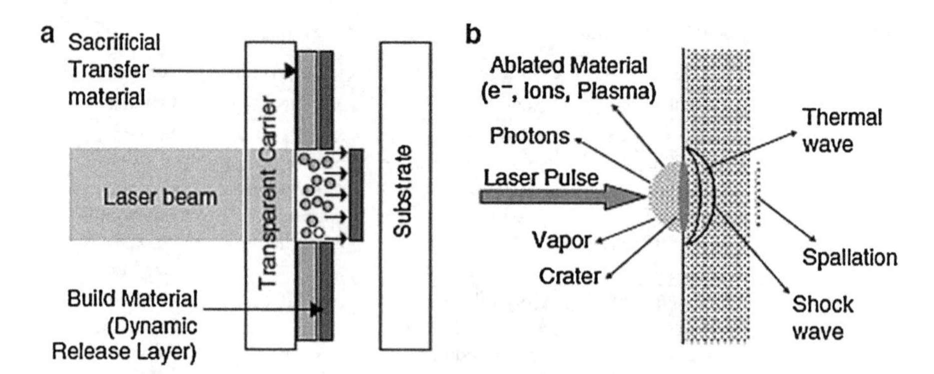

Fig. 11.5 (**a**) Mechanism for laser transfer using a sacrificial transfer material (based on [10]). (**b**) Mechanism for laser transfer using thermal shock and spallation (based on [11]) (courtesy Douglas B. Chrisey)

During ablation, a gas or plasma is formed, which expands rapidly as further laser energy is added. This rapid expansion can create a shock wave within a material or it can propel a material. By focusing the expansion of the material during ablation (utilizing shock waves produced by laser ablation) or taking advantage of rapid thermal expansion inherent with laser heating, materials can be accurately transferred in a very repeatable and accurate manner from one location to another. Laser transfer DW makes use of these phenomena by transferring material from a foil, tape, or plate onto a substrate. By carefully controlling the energy and location of the impinging laser, complex patterns of transferred material can be formed on a substrate.

Two different mechanisms for laser transfer are illustrated in Fig. 11.5. Figure 11.5a illustrates a laser transfer process where a transparent carrier (a foil or plate donor substrate which is transparent to the laser wavelength) is coated with a sacrificial transfer material and the dynamic release layer (the build material). The impinging laser energy ablates the transfer material (forming a plasma or gas), which propels the build material towards the substrate. The material impacts the substrate and adheres, forming a coating on the substrate. When using a pulsed laser, a precise amount of material can be deposited per pulse.

Figure 11.5b shows a slightly different mechanism for material transfer. In this case the laser pulse ablates a portion of the surface of a foil. This ablation and absorption of thermal energy creates thermal waves and shock waves in the material. These waves are transmitted through the material and cause a portion of the material on the opposing side to fracture from the surface in a brittle manner (known as spallation). The fractured material is propelled towards the substrate, forming a deposit coating on the substrate (not shown).

The Matrix-Assisted Pulsed Laser Evaporation Direct Write (MAPLE-DW) process was developed to make use of these laser transfer phenomena [12]. A schematic of the MAPLE-DW process is shown in Fig. 11.6. In this process, a laser transparent quartz disc or polymer tape is coated on one side with a film (a few

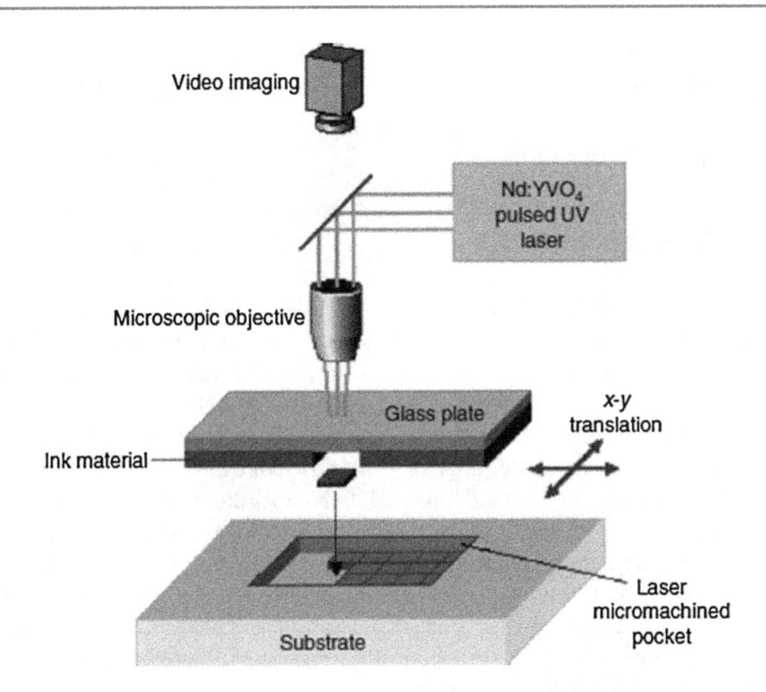

Fig. 11.6 Matrix-Assisted Pulsed Laser Evaporation Direct Write (MAPLE-DW) process [13] (Courtesy PennWell Corp., Laser Focus World)

microns thick), which consists of a powdered material that is to be deposited and a polymer binder. The coated disc or tape is placed in close proximity and parallel to the substrate. A laser is focused onto the coated film. When a laser pulse strikes the coating, the polymer is evaporated and the powdered material is deposited on the substrate, firmly adhering to it. By appropriate control of the positions of both the ribbon and the substrate, complex patterns can be deposited. By changing the type of ribbon, multimaterial structures can be produced.

Laser transfer processes have been used to create deposits of a wide variety of materials, including metals, ceramics, polymers, and even living tissues. The main drawbacks of a laser transfer process are the need to form a tape with the appropriate transfer and/or deposit materials, and the fact that the unused portions of the tape are typically wasted.

The benefits of the laser transfer process are that it produces a highly repeatable deposit (the deposit is quantized based on the laser pulse energy), it can be as accurate as the laser scanners used to manipulate the laser beam, and the deposited materials may not need any further post-processing. In addition, the laser can be used to either simply propel the material onto the substrate without thermally affecting the substrate or it can be used to laser treat the deposit (including heating, cutting, etc.) to modify the properties or geometry of the deposit during or after deposition. In the case of a rigid tape (such as when using a glass plate) the plate is typically mechanically suspended above the substrate. When a flexible polymer tape is used, it can be laid directly onto the substrate before laser processing and then peeled from the substrate after laser processing, leaving behind the desired pattern.

11.5 Thermal Spray DW

Thermal spray is a process that accelerates material to high velocities and deposits them on a substrate, as shown in Fig. 11.7. Material is introduced into a combustion or plasma flame (plume) in powder or wire form. The plume melts and imparts thermal and kinetic energy to the material, creating high-velocity droplets. By controlling the plume characteristics and material state (e.g., molten or softened) it is possible to deposit a wide range of metals, ceramics, polymers, or composites thereof. Particles can be deposited in a solid or semisolid state, which enables the creation of useful deposits at or near room temperature. Thermal spray techniques for DW have been commercialized by MesoScribe Technologies, Inc. [14].

A deposit is built-up by successive impingement of droplets, which yield flattened, solidified platelets, referred to as splats. The deposit microstructure, and thus its properties, strongly depends on the processing parameters utilized. Key characteristics of thermal spray DW include: (1) a high volumetric deposition rate, (2) material flexibility, (3) useful material properties in the as-deposited state (without thermal treatment or curing), and (4) moderate thermal input during processing, allowing for deposition on a variety of substrates.

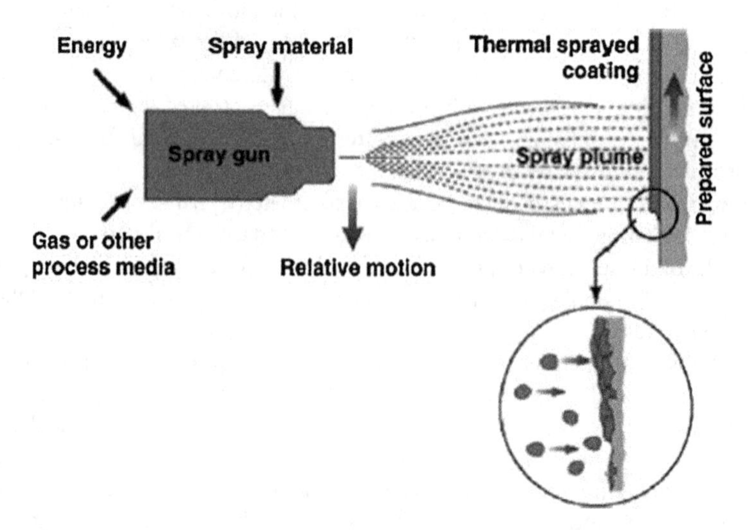

Fig. 11.7 General apparatus for thermal spray [15] (Courtesy of and (C) Copyright Sulzer Metco. All rights reserved)

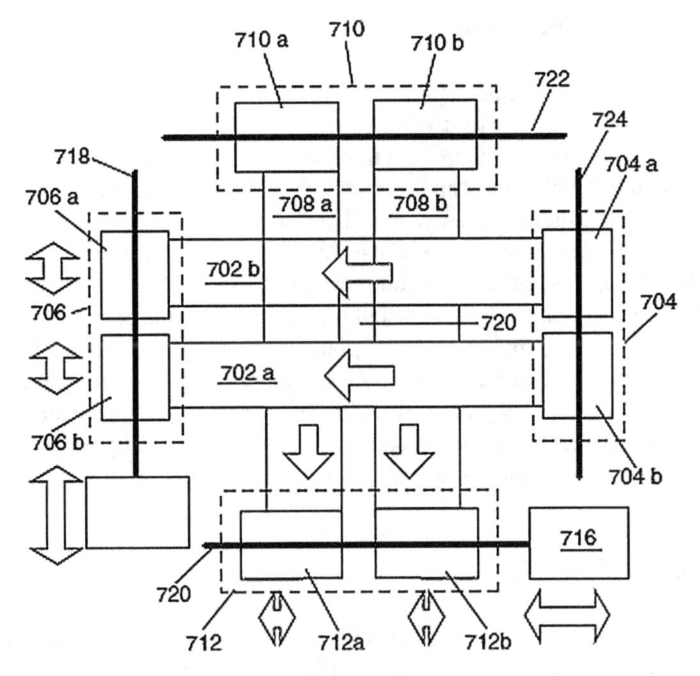

Fig. 11.8 Schematic aperture apparatus for direct write thermal spray (US patent 6576861). Foils 702a, b and 708a, b are in constant motion and are adjusted to allow different amounts of spray to reach the substrate through hole 720 in the center

DW thermal spray differs from traditional thermal spray in that the size and shape of the deposit is controlled by a unique aperture system. A schematic aperture system from an issued patent is shown in Fig. 11.8. This aperture is made up of adjustable, moving metal foils (702a and 702b moving horizontally and 708a and 708b moving vertically) which constrain the plume to desired dimensions (region 720). The distance between the moving foils determines the amount of spray which reaches the substrate. The foils are in constant motion to avoid overheating and build-up of the material being sprayed. The used foils become a waste product of the system.

Because the temperature of the substrate is kept low and no post-treatment is typically required, DW thermal spray is well-suited to produce multilayer devices formed from different materials. It is possible to create insulating layers, conductive/electronic layers, and further insulating layers stacked one on top of the other (including vias for signal transmission between layers) by changing between various metal, ceramic, and polymer materials. DW thermal spray has been used to successfully fabricate thermocouples, strain gages, antennas, and other devices for harsh environments directly from precursor metal and ceramic powders. In addition, DW thermal spray, combined with ultrafast laser micromachining, has been shown to be capable of fabricating thermopiles for power generation [16].

11.6 Beam Deposition DW

Several direct write procedures have been developed based upon vapor deposition technologies using, primarily, thermal decomposition of precursor gases. Vapor deposition technologies produce solid material by condensation, chemical reaction, or conversion of material from a vapor state. In the case of chemical vapor deposition (CVD), thermal energy is utilized to convert a reactant gas to a solid at a substrate. In the regions where a heat source has raised the temperature above a certain threshold, solid material is formed from the surrounding gaseous precursor reactants. The chemical composition and properties of the deposit are related to the thermal history during material deposition. By moving a localized heat source across a substrate (such as by scanning a laser) a complex geometry can be formed. A large number of research groups over almost 20 years have investigated the use of vapor deposition technologies for additive manufacturing purposes [17]. A few examples of these technologies are described below.

11.6.1 Laser CVD

Laser Chemical Vapor Deposition (LCVD) is a DW process which uses heat from a laser to selectively transform gaseous reactants into solid materials. In some systems, multiple gases can be fed into a small reactant chamber at different times to form multimaterial structures, or mixtures of gases with varying concentrations can be used to form gradient structures. Sometimes flowing jets of gas are used to create a localized gaseous atmosphere, rather than filling a chamber with the gaseous precursor materials.

The resolution of an LCVD deposit is related to the laser beam diameter, energy density, and wavelength (which directly impact the size of the heated zone on the substrate) as well as substrate thermal properties. Depending on the gases present at the heated reactive zone, many different metals and ceramics can be deposited, including composites. LCVD has been used to deposit carbon fibers and multilayered carbon structures in addition to numerous types of metal and ceramic structures.

A LCVD system developed at the Georgia Institute of Technology is displayed in Fig. 11.9. This design constrains the reactant gas (which is often highly corrosive, and/or biologically harmful) to a small chamber that is separated from the motion controllers and other mechanisms. This small, separated reaction chamber has multiple benefits, including an ability to quickly change between reagent gas materials for multimaterial deposition, and better protection of the hardware from corrosion. By monitoring the thermal and dimensional characteristics of the deposit, process parameters can be controlled to create deposits of desired geometry and material properties.

LCVD systems are capable of depositing many types of materials, including carbon, silicon carbide, boron, boron nitride, and molybdenum onto various substrates including graphite, grafoil, zirconia, alumina, tungsten, and silicon

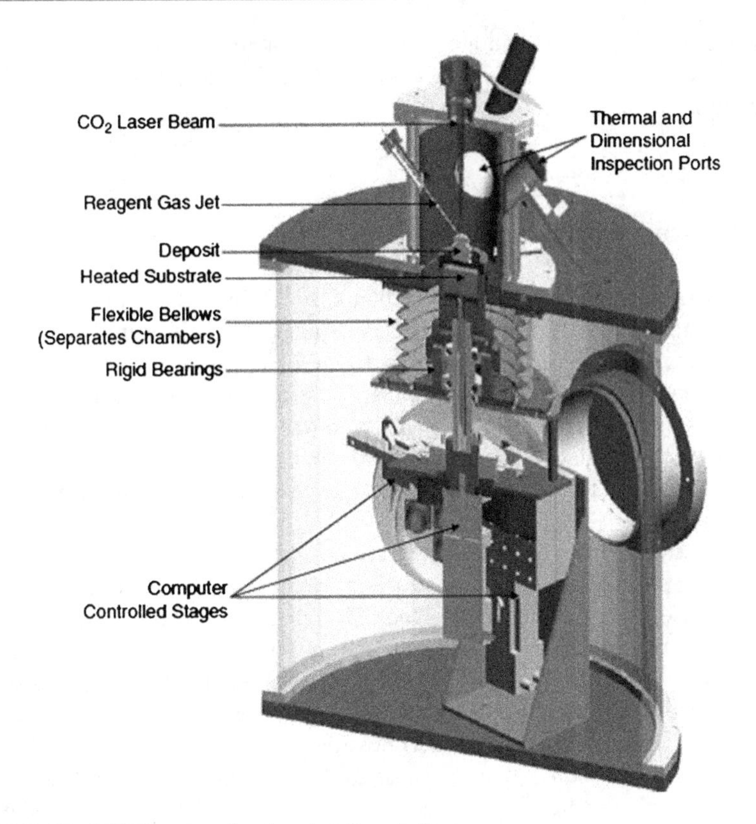

Fig. 11.9 The LCVD system developed at Georgia Tech

[18]. Direct write patterns as well as fibers have been successfully deposited. A wide variety of materials and deposit geometries make LCVD a viable technology for further direct write developments. LCVD is most comparable to microthermal spray, in that deposits of metals and ceramics are directly formed without post-treatment, but without the "splat" geometry inherent in thermal spray. The benefits of LCVD are the unique materials and geometries it can deposit. However, LCVD has a very low deposition rate and a relatively high system complexity and cost compared to most DW approaches (particularly ink-based technologies). High-temperature deposition can be another disadvantage of the process. In addition, the need to deposit on surfaces that are contained within a controlled-atmosphere chamber limits its ability to make deposits on larger preexisting structures.

LCVD can be combined with layer-wise deposition of powders (similar to the binder jetting techniques in Chap. 8) to more rapidly fabricate structures than when using LCVD alone. In this case the solid material created from the vapor phase is used to bind the powdered material together in regions where the laser has heated the powder bed. This process is known as Selective Area Laser Deposition Vapor Infiltration (SALDVI). In SALDVI, the build-rates are much higher than when the entire structure is fabricated from LCVD-deposited materials only; but the resultant structures may be porous and are composite in nature. The build-rate difference between LCVD and SALDVI is analogous to the difference between binder jetting and material jetting.

11.6.2 Focused Ion Beam CVD

A focused ion beam (FIB) is a beam of ionized gallium atoms created when a gallium metal source is placed in contact with a tungsten needle and heated. Liquid gallium wets the needle, and the imposition of a strong electric field causes ionization and emission of gallium atoms. These ions are accelerated and focused in a small stream (with a spot size as low as a few nanometers) using electrostatic lenses. A FIB is similar in conceptualization to an electron beam source, and thus FIB is often combined with electron beams, such as in a dual-beam FIB-scanning electron microscope system.

FIB processing, when done by itself, can be destructive, as high-energy gallium ions striking a substrate will cause sputtering and removal of atoms. This enables FIB to be used as a nanomachining tool. However, due to sputtering effects and implantation of gallium atoms, surfaces near the machining zone will be changed by deposition of the removed material and ion implantation. This sputtering and ion implantation, if properly controlled, can also be a benefit for certain applications.

Direct write deposition using FIB is possible in a manner similar to LCVD. By scanning the FIB source over a substrate in the presence of CVD gaseous precursors, solid materials are deposited onto the substrate (and/or implanted within the surface of the substrate) [19, 20]. These deposits can be submicron in size and feature resolution. FIB CVD for DW has been used to produce combinations of metallic and dielectric materials to create three-dimensional structures and circuitry. In addition, FIB CVD is being used in the integrated circuits (IC) industry to repair faulty circuitry. Both the machining and deposition features of FIB are used for IC repairs. In the case of short-circuits, excess material can be removed using a FIB. In the case of improperly formed electrical contacts, FIB CVD can be used to draw conductive traces to connect electrical circuitry.

11.6.3 Electron Beam CVD

Electron beams can be used to induce CVD in a manner similar to FIB CVD and LCVD. Electron beam CVD is slower than laser or FIB CVD; however, FIB CVD and electron beam CVD both have a better resolution than LCVD [21].

11.7 Liquid-Phase Direct Deposition

Similarly to the vapor techniques described above, thermal or electrical energy can be used to convert liquid-phase materials into solid materials. These thermochemical and electrochemical techniques can be applied in a localized manner to create prescribed patterns of solid material.

Drexel University illustrated the use of thermochemical means for DW traces using ThermoChemical Liquid Deposition (TCLD). In TCLD, liquid reactants are sprayed through a nozzle onto a hot substrate. The reactants thermally decompose or react with one another on the hot surface to form a solid deposit on the substrate. By controlling the motion of the nozzle and the spraying parameters, a 3D shape of deposited material can be formed. This is conceptually similar to the ink-based DW approaches discussed above, but requires a high-temperature substrate during deposition.

A second Electrochemical Liquid Deposition (ECLD) approach was also tested at Drexel. In ECLD, a conductive substrate is submerged in a plating bath and connected to a DC power source as the cathode, as in Fig. 11.10. A pin made up of the material to be deposited is used as the anode. By submerging the pin in the bath near the substrate and applying an appropriate voltage and current, electrochemical decomposition and ion transfer results in a deposit of the pin material onto the substrate. By moving the pin, a prescribed geometry can be traced. As electrochemical plating is a slow process, the volumetric rate of deposition for ELCD can be increased by putting a thin layer of metal powder in the plating bath on the surface of the substrate (similar conceptually to SALDVI described above). In this case, the

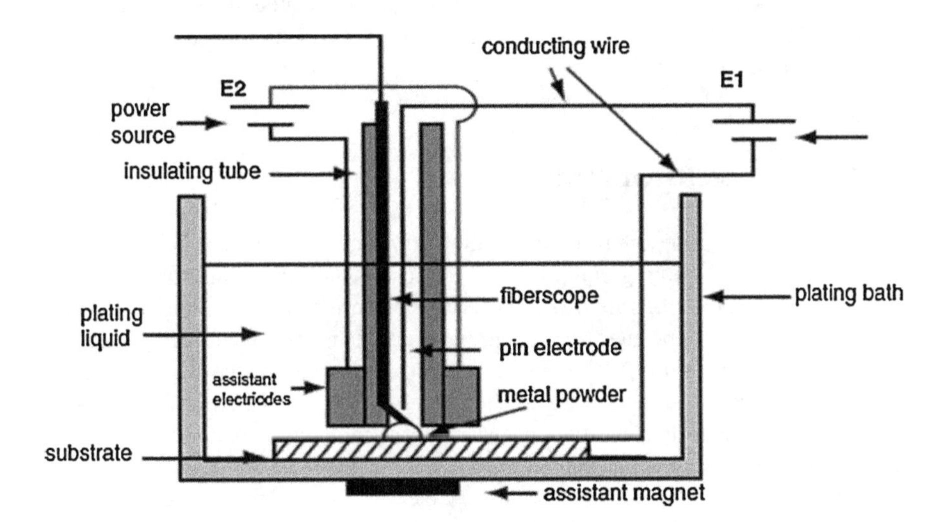

Fig. 11.10 Schematic of an electrochemical liquid deposition system [22] (MATERIALS & DESIGN by Zongyan He, Jack G. Zhou and Ampere A. Tseng. Copyright 2000 by Elsevier Science & Technology Journals. Reproduced with permission of Elsevier Science & Technology Journals in the format Textbook via Copyright Clearance Center.)

deposited material acts as a binder for the powdered materials, and the volumetric rate of deposition is significantly increased [22].

Thermochemical and electrochemical techniques can be used to produce complex-geometry solids at small length scales from any metal compatible with thermochemical or electrochemical deposition, respectively. These processes are also compatible with some ceramics. However, these approaches are not available commercially and may have few benefits over the other DW techniques described above. Drawbacks of TCLD-based approaches are the need for a heated substrate and the use of chemical precursors which may be toxic or corrosive. Drawbacks of ECLD-based approaches include the slow deposition rate of electrochemical processes and the fact that, when used as a binder for powders, the resultant product is porous and requires further processing (such as sintering or infiltration) to achieve desirable properties.

11.8 Beam Tracing Approaches to Additive/Subtractive DW

By combining layer-wise additive approaches with freeform beam (electron, FIB, or laser) subtractive approaches, it is possible to create DW features. Many coating techniques exist to add a thin layer of material to a substrate. These include physical vapor deposition, electrochemical or thermochemical deposition, CVD, and other thin film techniques used in the fabrication of integrated circuits. Once these layers are added across the surface of a substrate, a beam can be used to trim each layer into the prescribed cross-sectional geometry. These micro- or nanodiameter beams are used to selectively cure or remove materials deposited in a layer-by-layer fashion. This approach is conceptually similar to the bond-then-form sheet lamination techniques discussed in Chap. 9.

11.8.1 Electron Beam Tracing

Electron beams can be used to either cure or remove materials for DW. Standard spin-on deposition coating equipment can be used to produce thin films between 30 and 80 nm. These films are then exposed to a prescribed pattern using an electron beam. Following exposure, the uncured material is removed using standard IC-fabrication techniques. This methodology can produce line-edge definition down to 3.3 nm. A converse approach can also be used, whereby the exposed material is removed and the unexposed material remains behind. In the case of curing, low-energy electrons can be utilized (and are often considered more desirable, to reduce the occurrence of secondary electron scattering). These techniques fit well within existing IC-fabrication methodologies and enable maskless IC fabrication.

Another variant of electron beam tracing is to produce a thin layer of the desired material using physical vapor deposition or a similar approach and then to use high-powered electron beams to remove portions of the coating to form the desired pattern.

Electron beams are not particularly efficient for either curing or removing layers of material, however. Thus, electron beam tracing techniques for DW are quite slow.

11.8.2 Focused Ion Beam Tracing

As discussed above, a FIB can be utilized to machine materials in a prescribed pattern. By combining the steps of layer-wise deposition with FIB machining, a multilayer structure can be formed. If the deposited material is changed layer-by-layer, then a multimaterial or gradient structure can be formed.

11.8.3 Laser Beam Tracing

Short-wavelength lasers can be utilized to either cure layers of deposited materials or ablatively remove materials to form micro and nanoscale DW features. To overcome the diffraction limit of traditional focusing optics, a number of nanopatterning techniques have been developed to create features that are smaller than half the optical wavelength of the laser [23]. These techniques include multiphoton absorption, near-field effects, and Bessel beam optical traps. Although these techniques can be used to cure features at the nano scale, inherent problems with alignment and positioning at these length scales make it difficult to perform subwavelength nanopatterning in practice.

These laser approaches are conceptually identical to the electron beam and FIB-based additive plus subtractive approaches just mentioned. Some of the benefits of lasers for beam tracing DW are that they can process materials much more rapidly than electron beams, and they can do so without introducing FIB gallium ions. The main drawback of lasers is their relatively large spot size compared to electron and FIBs.

11.9 Hybrid Technologies

As in most additive manufacturing techniques, there is an inherent trade-off between material deposition speed and accuracy for most DW processes. This will remain true until techniques are developed for line wise or layer wise deposition (such as is done with mask projection stereolithography using a DLP system, as described in Chap. 4). Thus, to achieve a good combination of deposition speed and accuracy, hybrid technologies are often necessary. Some examples of hybrid technologies have already been mentioned. These include the additive/subtractive beam tracing methods described above and the use of a laser in the Aerosol Jet aerosol system.

The primary form of hybrid technology used in DW is to form deposits quickly and inexpensively using an ink-based technique and then trim those deposits using a short-wavelength laser. This results in a good combination of build speed, accuracy, and overall cost for a wide variety of materials. In addition, the laser used to trim the ink-based deposits has the added benefit of being an energy source for curing the deposited inks, when used in a lower power or more diffuse manner. If DW is integrated with an AM process that includes a laser, such as stereolithography, the laser can be used to modify the DW traces [24].

One of the fastest hybrid DW approaches is the "roll-to-roll" approach. As the name suggests, roll-to-roll printing is analogous to high-speed 2D printing. In roll-to-roll DW, a paper or plastic sheet of material is used as a substrate material and moved through printing rolls that deposit patterns of DW ink that are subsequently thermally processed, such as with a flash lamp. Inkjet printing and other DW techniques can be added into the roll-to-roll facility to add ink patterns in a more flexible manner than the repeated patterns printed by a printing roll. When DW techniques and flexible laser systems are integrated into a roll-to-roll facility, this gives the combination of the speed of line-wise AM via the rolls plus the flexibility of point-wise AM via another DW technique.

11.10 Applications of Direct Write Technologies

The applications of DW processes are growing rapidly [25]. There is a growing variety of materials which are available, including semiconductors, dielectric polymers, conductive metals, resistive materials, piezoelectrics, battery elements, capacitors, biological materials, and others. These can be deposited onto various substrate materials including plastic, metal, ceramic, glass, and even cloth. The combination of these types of materials and substrates means that the applications for DW are extremely broad.

The most often cited applications for DW techniques are related to the fabrication of sensors. DW approaches have been used to fabricate thermocouples, thermistors, magnetic flux sensors, strain gages, capacitive gages, crack detection sensors, accelerometers, pressure gages, and more [3, 14, 16, 26].

A second area of substantial interest is in antenna fabrication. Since DW, like other AM techniques, enables fabrication of complex geometries directly from CAD data, antenna designs of arbitrary complexity can be made on the surface of freeform objects; including, for instance, fractal antennas on conformal surfaces. Figure 11.11 illustrates the fabrication of a fractal antenna on the abdomen of a worker honeybee using MAPLE-DW.

Another area of interest for DW is as a freeform tool to connect combinations of electronic components on freeform surfaces. One area where this is particularly useful is in harsh environments, as shown in Fig. 11.12. In this example, direct write thermal spray is used as a method for producing and connecting a series of electronic components that monitor and feed back information about the state of a turbine blade. A thermocouple, labeled High-Temperature Sensor in the figure, is

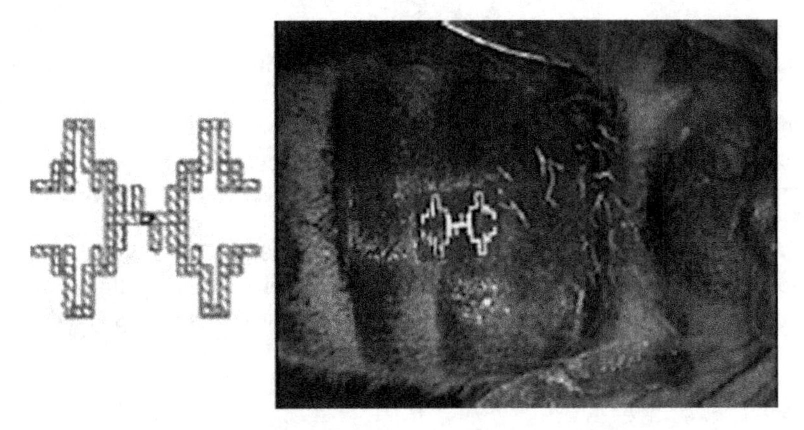

Fig. 11.11 35-GHz fractal antenna design (*left*) and MAPLE-DW printed antenna on the abdomen of a dead drone honeybee (*right*). (Courtesy Douglas B. Chrisey)

Fig. 11.12 A direct write sensor and associated wiring on a turbine blade structure. Signal conditioning electronics are positioned on a more shielded spot (Courtesy MesoScribe Technologies, Inc. and Arkansas Power Electronics Int.)

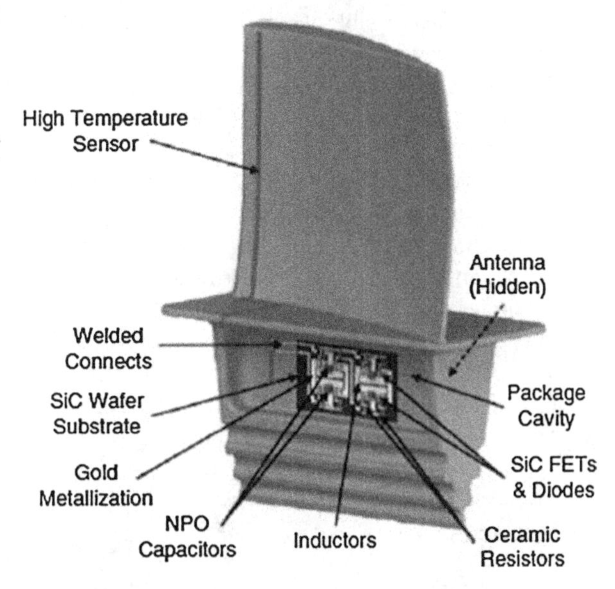

deposited on the hot region of the blade, whereas the supporting electronics are deposited on the cold regions of the blade. DW-produced conductors are used to transmit signals between regions and components.

Although most DW processes can produce thermal and strain sensors, there is still opportunity for improved conductors, insulators, antennas, batteries, capacitors, resistors, and other electronic circuitry. In addition, in every case where a conductive path is not possible between a power source and a deposited sensor, some form of local power generation is necessary. Several researchers have demonstrated the ability to create systems which use electromagnetic or thermopile power generation schemes using DW [16]. If designs for energy harvesting devices can be made robustly using DW, then the remote monitoring and sensing of components and parts is possible. For instance, the ability to create a thermal sensor with integrated power harvesting and antenna directly onto an internally rotating component (such as a bearing) within a transmission could provide feedback to help optimize performance of systems from power plants to motor vehicles to jet engines. In addition, this type of remote sensing could notify the operator of thermal spikes before catastrophic system failure, thus saving time and money.

DW techniques are rapidly growing. Ongoing investments in DW R&D indicate these technologies will continue to expand their application potential to become common methods for creating nano-, micro-, and mesoscale features and devices.

11.10.1 Exercises

1. From an internet search, identify two DW inks for conductive traces, one ink for resistors and one for dielectric traces, that are commonly used in nozzle-based systems. Make a table which lists their room-temperature properties and their primary benefits and drawbacks.
2. For the inks identified in problem 1, estimate the printing number (7.7). List all of your assumptions. Can any of the inks from problem 1 be used in an inkjet printing system?
3. Would you argue that DW techniques are a subset of AM technologies (like PBF or DED) or are they more an application of AM technologies? Why?
4. Two techniques for accelerating DW were discussed in this chapter where DW deposition was used to bind powders to form an object. What other DW techniques might be accelerated by the use of a similar approach? How would you go about doing this? What type of machine architecture would you propose?
5. Research thermocouple types that can withstand 1,000 °C. Based on the materials that are needed, which DW techniques could be used to make these thermocouples and which could not?

References

1. Mortara L, Hughes J, Ramsundar PS, Livesey F, Probert DR (2009) Proposed classification scheme for direct writing technologies. Rapid Prototyping J 15(4):299–309
2. Abraham MH, Helvajian H (2004) Laser direct write of SiO/sub 2/MEMS and nano-scale devices. Proceedings of SPIE, vol 5662, Fifth international symposium on laser precision microfabrication, October 2004, pp. 543–550
3. Pique A, Chrisey DB (2001) Direct write technologies for rapid prototyping applications. In: Sensors, electronics and integrated power sources. Academic, Boston
4. Li B, Dutta Roy T, Clark PA, Church KH (2007) A robust true direct-print technology for tissue engineering. Proceedings of the 2007 international manufacturing science and engineering conference MSEC2007, October 15–17, 2007, ASME, Atlanta, paper # MSEC2007-31074
5. www.nscryptinc.com
6. www.fabathome.org
7. www.nanoink.net

8. Szczech JB et al (2000) Manufacture of microelectronic circuitry by drop-on-demand dispensing of nanoparticle liquid suspensions. In: Proceedings of the materials research society symposium, vol 624, p. 23
9. Essien M, Renn MJ (2002) Development of mesoscale processes for direct write fabrication of electronic components. In: Keicher D et al (eds) Proceedings of the conference on metal powder deposition for rapid prototyping, p. 209
10. http://materials.web.psi.ch/Research/Thin_Films/Methods/LIFT.htm
11. Young D, Chrisey DB (2000) Issues for tissue engineering by direct-write technologies. http://www.fractal.org/Fractal-Research-and-Products/Biomanufacturing.pdf
12. Fitz-Gerald JM et al (2000) Matrix assisted pulsed laser evaporation direct write (MAPLE DW): a new method to rapidly prototype active and passive electronic circuit elements. In: Proceedings of the materials research society symposium, vol 624, p. 143
13. http://www.laserfocusworld.com/display_article/204194/12/none/none/OptWr/Direct-write-laser-processing-creates-tiny-electrochemical-system
14. Sampath S et al (2000) Thermal spray techniques for fabrication of meso-electronics and sensors. In: Proceedings of the Materials research society symposium, vol 624, p. 181
15. www.sulzermetco.com
16. Chen Q, Longtin JP, Tankiewicz S, Sampath S, Gambino RJ (2004) Ultrafast laser micromachining and patterning of thermal spray multilayers for thermopile fabrication. J Micromech Microeng 14:506–513
17. Kadekar V, Fang W, Liou F (2004) Deposition technologies for micromanufacturing: a review. J Manuf Sci Eng 126(4):787–795
18. Duty C, Jean D, Lackey WJ (2001) Laser chemical vapor deposition: materials, modeling, and process control. Int Mater Rev 46(6):271–287
19. Hoffmann P et al (2000) Focused ion beam induced deposition of gold and rhodium. In: Proceedings of the materials research society symposium, vol 624, p. 171
20. Longo DM, Hull R (2000) Direct focused ion beam writing of printheads for pattern transfer utilizing microcontact printing. In Proceedings of the materials research society symposium, vol 624, p. 157
21. Bhushan B (2007) Springer handbook of nanotechnology. Springer, New York, p 179
22. He Z, Zhou JG, Tseng A (2000) Feasibility study of chemical liquid deposition based solid freeform fabrication. J Materials Design 21:83–92
23. McLeod E, Arnold CB (2008) Laser direct write near-field nanopatterning using optically trapped microspheres. Lasers and Electro-Optics, 2008 and 2008 conference on quantum electronics and laser science. CLEO/QELS 2008. 4–9 May 2008
24. Palmer JA et al (2005) Stereolithography: a basis for integrated meso manufacturing. In: Proceedings of the 16th solid freeform fabrication symposium, Austin
25. Church KH et al (2000) Commercial applications and review for direct write technologies. In: Proceedings of the materials research society symposium, p. 624
26. Lewis JA (2006) Direct ink writing of 3D functional materials. Adv Funct Mater 16:2193–2204

The Impact of Low-Cost AM Systems 12

Abstract

Media attention over additive manufacturing is at an all-time high. Much of this is to do with the vast increase in the availability of the technology due to massive reductions in the technology costs. By making it possible for individuals to afford them for their own personal use, the true potential has been, to some extent, uncovered. This chapter will discuss some of the issues surrounding the low-cost technologies, including machine developments due to patent expiry, the rise of the Maker movement and some of the new business models that have resulted.

12.1 Introduction

When the first additive manufacturing machines came on to the market for the purposes of rapid prototyping, they were, not surprisingly, very expensive. The fact that they were aimed at early adopters, based around complex and new technologies, like lasers, and only produced in small volumes meant that purchase of such machines left you with little change from a quarter of a million US dollars or even more. Furthermore, the perceived value of these machines to these early adopters was also very high. Even at these prices the Return on Investment (ROI) was often only a matter of months or attributable to a small selection of high-value projects. An automotive manufacturer could, for example, achieve the ROI just by proving the AM technology ensured a new vehicle was launched on or ahead of time. Such perceived value did little to bring the prices of these machines down.

As the technology became more popular, the market became more competitive. However, demand for these new machines was also high, particularly from the traditional market drivers of automotive, aerospace, and medicine as mentioned in previous chapters. Vendors did find themselves in competition with each other, but there were many different customers. Furthermore, different machines were exhibiting different strengths and weaknesses that the vendors exploited to develop

I. Gibson et al., *Additive Manufacturing Technologies*,
DOI 10.1007/978-1-4939-2113-3_12

differing markets. For most of these markets, the more successful vendors also ensured they had excellent intellectual property (IP) protection. All of this served to maintain the technology at a high cost.

Ultimately it was the issues surrounding IP that led to the current situation. Many of the original technologies were protected by patents that prevented other companies from copying them. The vendor companies were also very aggressive in defending their IP as well as buying up related IP and their associated companies where they existed. This competitive market did have the effect of slowly eroding the vendor companies' profit margins, but huge changes have recently taken place as a result of these patents lapsing. We will go on in this chapter to discuss how this has impacted the low-cost AM technology marketplace. The key patents will be discussed as well as the more prominent players in this field. This activity has fuelled a huge amount of interest by the media, which will be discussed in terms of how it has in turn impacted the industry. Much of this interest can be associated with what is becoming known as disruptive innovation. AM certainly fits with a number of other technologies to form the basis for disruptive business models, which we will discuss. AM also is a huge enabling technology that has assisted many home users to solve many of their own technical problems at home. Sharing these experiences and even profiting from them has spearheaded what is being commonly called the Maker Movement, which we shall also examine. We will go on to consider how this branch of AM may develop in the future.

12.2 Intellectual Property

As mentioned in the introduction and in other parts of this book, the key patents with the most protection originated in the USA. While there was activity in other countries, 3D Systems, Stratasys, DTM, and Helisys were the principal vendor names for much of the world in the early days. Other companies were also present, like EOS, Sony, Sanders, and Objet, but they either came along at a later date or were in close IP conflict with these American vendors.

European and US patent law for technology are somewhat similar to each other in that they refer to a 20-year term from the initial filing date of the patent in most cases. Charles Hull filed his first Stereolithography patent in 1984 [1], which therefore expired in 2004. Scott Crump patented the Fused Deposition Modeling process in 1989 [2], which means that patent expired in 2009. The major difference between these technologies, in this context, is that the melt extrusion process is much easier to replicate at a low cost compared with the laser-cured photopolymer systems. It is evident therefore that while the door to widespread development was opened in 2004, it was not opened wide until 2009 and beyond. This is evident from the figures quoted by Terry Wohlers from 2008 to 2013, shown in Fig. 12.1, concerning the number of low-cost AM machines purchased over that period [3].

Of course this is not the complete story. While the original patents were filed in the 1980s, additional patents have been constantly filed ever since. If one notices that there have been many copies of the original FDM patent, all the resulting

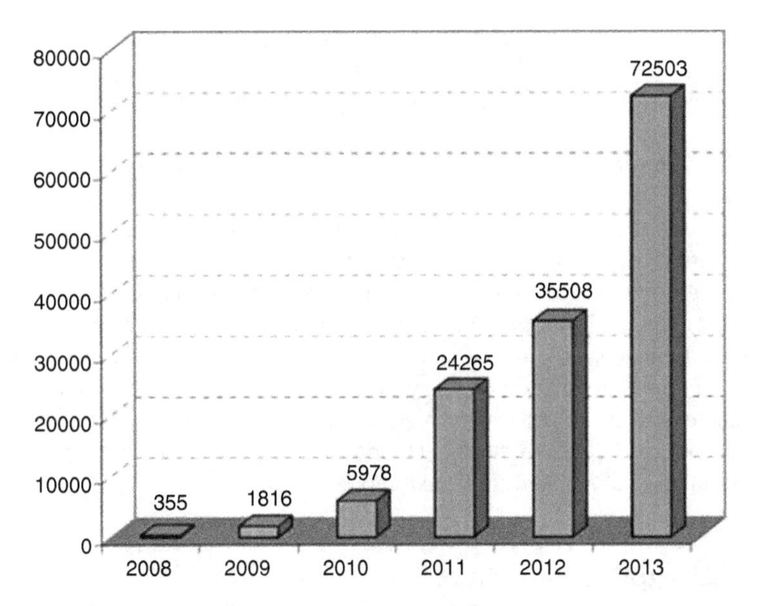

Fig. 12.1 Wohlers' data showing how the numbers of low-cost AM machines has increased from virtually nothing in 2008

machines have thus far not implemented an environmental control apparatus. This is because the patent for this development was not filed by Crump until a few years later.

There is an excellent review of the recently expiring and shortly expiring patents carried out by Hornick and Roland [4]. In it they make a number of interesting observations that lead to the following discussion:

- There is a huge number of patents involved and it is a very difficult process to navigate through them. Many patents make multiple claims across numerous platforms. They often do not adhere to a single process.
- Earlier patents discuss broad-based approaches that are in fact quite easy to defend and/or find ways around. It is quite easy to distinguish one process from another since they do not have the benefits that are gained through experience of actual application of the technology. Later patents obviously result from discoveries resulting from use of these base technologies, which discuss subtle features like soluble supports or fill patterns.
- While 3D Systems patents are among the earliest to expire, there are obvious technical complexities in them that would make them difficult to replicate without significant industrial backing. An example relates to the formulation of resins to speed up the curing and build process or another to facilitate even spreading of resins.

– Some key droplet deposition and powder sintering by laser patents expired in late 2014, early 2015. Expiration of these patents may help to open up the high-end, high-quality AM market by bringing the machine costs down.

At the time of writing this chapter it is merely speculation, but it will be interesting to see how some of the large printing or equipment manufacturing companies view these as opportunities. One can anticipate that a large printer company may possess the necessary knowledge, resources, and infrastructure to produce high-quality AM equipment in volume at very reasonable costs. It is worth noting that although low-cost AM started with material extrusion technology, since they were the easiest to produce and some of the first patents to expire, low-cost versions of photopolymerization (Formlabs), polymer laser sintering (Norge), and even metal machines such as metal laser sintering (Matterfab) and lower-cost directed energy deposition heads (Hybrid Manufacturing Technologies or the LENS print engine) are starting to proliferate.

12.3 Disruptive Innovation

12.3.1 Disruptive Business Opportunities

Disruptive innovation and disruptive technology are terms that were originally defined by Christensen to describe activities or technology that create new markets [5]. A very obvious example of this is how the Internet made it possible to create online businesses which could not have existed before. However, the effects may be much more subtle and it is possible to create a disruptive business merely by using existing technology in a different way. Often this process can be achieved by early adopters of technology or by those who have skills that are either difficult to learn or not commonly used in a disruptive way. Here we can say that although there are many musicians, there is only one David Bowie or Lady Gaga, who made use of their artistic talents to generate additional business opportunities.

There is no doubt that AM is a disruptive technology, which can be combined with other technologies to generate new businesses. Improvements and more widespread use of CAD technology has made it possible for individuals to design products with minimal cost and training. Google Sketchup and Tinkercad are online design tools that are basically free to use. While they are not as powerful and versatile as paid CAD software, these accessible and simple to use tools have opened up a new market for home designers, who would then like to find outlets for these designs. While eventually this may mean all of these people will have machines in their homes, we are not quite there yet. Some people may have a low-cost AM machine at home, but even these may not meet the functional requirements and so will look to outside services to build and supply their designs. This has led to the establishment of companies who provide online services where designers can not only have their models made but also find an outlet where their designs can also be sold. Shapeways and i.materialise (see Fig. 12.2) both operate in

Fig. 12.2 Web site image for i.materialise, showing how designers can post their ideas on the web and have them built and made available for others to buy

this space, using techniques developed for social media platforms, where viewers can "like" other people's designs as well as discuss, share, and promote them. For those wishing to share designs but not concerned with the commercial aspects, there are also portals like the Thingiverse web site. Issues surrounding copyright of designs are regularly discussed around these web sites. Replicas or models inspired by merchandising for TV and movie shows for example can be regularly found on these sites and some sharing sites will have more or less control than others.

12.3.2 Media Attention

Since disruptive innovation is always going to attract media attention, it is worth considering some of the more prominent stories that have attracted interest and therefore structured how the general public view AM.

The first example that springs to many people's minds would be the use of AM to create firearms. A great deal of attention was directed towards AM when it was realized that it was possible to create firearms using the technology, the most well-known of these being the Liberator, single shot pistol [6]. Here, it was discovered that this gun could be largely manufactured using an AM machine and that the plans for its manufacture were posted online. A primary reason for this attracting so much attention was the obvious contentious nature of the topic, backed up by the fact that this design can be easily shared on the Internet. It was particularly interesting to note that the attention was focused on the negative impact of AM rather than that of the Internet. There are a number of issues that are worth discussing here:

- While the gun was indeed built, certain items like the firing pin and obviously the ammunition would need to be added to the design.
- The original gun was created using a relatively high-end AM machine by a skilled operator. Sharing of the design online does not include the build parameters and a study by the New South Wales police in Australia revealed that the user is at great risk if the gun is not built correctly [7].
- Improvised firearms have been possible for many years and can be constructed by anyone with a small amount of technical knowledge and access to simple manufacturing equipment [8].

Admittedly, AM can be used like any technology, for good or for bad. However, we can see here how the media can latch on to one example and confuse the public image. These examples did bring the technology into a much wider public domain than previously and therefore allowed specialists in the field the opportunity to more properly explain the true impact of AM.

In stark contrast to the use of AM for destructive purposes, there have been numerous articles that describe how AM can be used to create replacement body parts [9]. As described in the chapter on Applications in this book, there is a huge potential for AM to contribute in this direction. The problem in the media coverage however is the time frame attached to this. Some applications have been implemented where AM has made significant improvements in medical and healthcare. However, this cannot be easily generalized into a conclusion that all medical problems can be solved this way. We must expect significant developments in the technology before we can make that conclusion and the AM machines of the future will look nothing like the machines we have today. Furthermore, we need parallel efforts in the biomedical sciences as well because they are far from being ready to plug in directly to AM devices. Specialists in these fields need to understand that, while it is reasonable to speculate that AM for body parts is on the horizon, they must be wary of that message being misconstrued.

Another sector that has attracted the media attention is the "cool gadget" area. One early example that again elevated general public awareness was The Economist magazine front-page article titled "Print me a Stradivarius" (see Fig. 12.3). This was probably the first mass-market article that highlighted how AM could be used for truly functional applications. While other media sites have included similar articles somewhere in their portfolios, the fact that this was on the front-page of an international magazine certainly had an impact. It is also worthwhile taking note that the BBC have published regular articles about AM over the years, averaging around 1 every 2 months or so in a wide variety of areas.

In relation to this chapter, the reason for including these articles is because of the increasing number of personal users of AM machines. All of these articles have one thing in common: they imply that you will eventually be able to create solutions for yourself.

Fig. 12.3 The Economist magazine front cover

12.4 The Maker Movement

As alluded to earlier in this chapter, the number of users who have their own AM machines has increased dramatically and they are driving much of the innovation that we are seeing. The technology is often aimed originally at providing solutions for problems that the user has experienced around his or her home or workplace. Social media and other outlets have allowed these users to demonstrate these solutions and thus inspire other users to do the same. While social media does help in dissemination, physical demonstration is usually a much more effective way of presenting your designs. To this effect, Maker Faires® have almost literally taken the world by storm, as can be seen in Fig. 12.4.

Maker Faires® are events where "Makers" congregate to display, demonstrate, and trade in items that they have designed and built themselves. While there is a lot of emphasis on technology, there is equal emphasis on design, environment, engagement, and fun in these fairs. Additive manufacturing is certainly a component in this, but so are Arduino and Raspberry Pi microcontrollers, laser cutting, conventional machine and hand tools, and home crafting skills like carving, sewing, and knitting. Many Makers merge technology with more conventional craft and design to come up with often personalized, often quirky systems that they would like others to see. The low-cost AM technologies have found a huge following

Fig 12.4 Maker Faire web page

within this Maker movement that is fuelling a large amount of innovation and even spinning off into new commercial ventures.

Perhaps not surprisingly this Maker movement originated in the USA, originally promoted by Make magazine [10] for the first event in California in 2006. The USA has a long and distinguished culture of innovation, perhaps spawning from the original frontier mentality of having to make do with whatever is around you, and this can also be seen in terms of how innovation is an accepted part of everyday life there. However, the concept of the fair is now a worldwide phenomenon with dozens of events held yearly attracting thousands of exhibitors and hundreds of thousands of attendees. The Maker Faire® has become the spiritual home for the Makerbot and other similar low-cost AM technologies, including the RepRap designs [11], and for design sharing portals like Thingiverse.

A huge number of commercial entry-level machines can thank the RepRap project for their origins. It would be almost impossible to identify every company, there are so many with varying levels of success. Designs have evolved considerably from the original RepRap machines, but the basic principle remains very much the same, exploiting the hot-melt extrusion process that was facilitated by the FDM patent expiry.

The Maker movement, along with the FabLab [12] and Idea to Product (I2P [13]) concepts, has done much to highlight the benefits of AM and associated technologies to the general public. The FabLab and the I2P labs are walk-in facilities aimed at providing an environment that encourages people to experiment with accessible manufacturing technologies and develop their ideas. Often the costs are at least partially absorbed by local authorities or donations. These are different from what is referred to as hardware incubators, where the costs are sourced from an investor network. All of these recognize in some way that there is a need to foster the creative processes.

12.5 The Future of Low-Cost AM

Low-cost AM has done much to bring the technology into the public domain. While it may be that it has taken some time to get to this stage with many of the current issues having been dealt with by early adopters, there has never been a better time to get involved in AM. A lot of confusion still surrounds what AM can and cannot do but even some of the negative press coverage has helped to promote the technology in some manner. While the majority of AM machines have exploited the Stratasys FDM, melt extrusion process, recent patent expirations of other AM technologies have started to open the doors to some interesting new low-cost technologies in the near future.

12.6 Exercises

1. What low-cost technologies are available in your area? Are there any local vendors and are their machines different in any way to the standard Makerbot and RepRap variants?
2. Examine articles about AM in the press. Is the information presented accurate? Would you write the article in a different way? Does inaccurate reporting have any impact on how the general public views AM technology?
3. What other technologies are represented at Maker Faires?
4. Consider the copyright infringement issues that surrounded the development of YouTube. Could similar things happen with respect to model sharing sites in the future? How can this be regulated?

References

1. Apparatus for production of three-dimensional objects by stereolithography google.com.au/patents/US4575330
2. Apparatus and method for creating three-dimensional objects google.com/patents/US5121329
3. Wohlers T (2014) Wohlers report 2014: additive manufacturing and 3D printing state of the industry, annual worldwide progress report. Wohlers Associates, Inc., Fort Collins, CO
4. Hornick, Roland. Many 3D printing patents are expiring soon: here's a round up & overview of them. 3dprintingindustry.com
5. Christensen CM. The innovator's dilemma: when new technologies cause great firms to fail. Boston: Harvard Business School; 1997. ISBN 978-0-87584-585-2.
6. en.wikipedia.org/wiki/Liberator_(gun)
7. theguardian.com/uk-news/2013/oct/25/3d-printed-guns-risk-user
8. en.wikipedia.org/wiki/Improvised_firearm
9. telegraph.co.uk/technology/news/10629531/The-next-step-3D-printing-the-human-body.html
10. Make Magazine. makezine.com
11. RepRap. http://reprap.org/
12. Gershenfeld N (2007) Fab: the coming revolution on your desktop. ReadHowYouWant Limited. p. 360. ISBN 978-1-4596-1057-6
13. I2P Lab, facebook.com/Idea2ProductLab

Guidelines for Process Selection 13

Abstract
AM processes, like all materials processing, are constrained by material properties, speed, cost, and accuracy. The performance capabilities of materials and machines lag behind conventional manufacturing technology (e.g., injection molding machinery), although the lag is decreasing. Speed and cost, in terms of time to market, are where AM technology contributes, particularly for complex or customized geometries.

13.1 Introduction

The initial purpose of rapid prototyping technology was to create parts as a means of visual and tactile communication. Since those early days of rapid prototyping, the applications of additive manufacturing processes have expanded considerably. According to Wohlers and Associates [1], parts from AM machines are used for a number of purposes, including:

- Visual aids
- Presentation models
- Functional models
- Fit and assembly
- Patterns for prototype tooling
- Patterns for metal castings
- Tooling components
- Dircct digital/rapid manufacturing

AM processes, like all materials processing, are constrained by material properties, speed, cost, and accuracy. The performance capabilities of materials and machines lag behind conventional manufacturing technology (e.g., injection molding machinery), although the lag is decreasing. Speed and cost, in terms of

I. Gibson et al., *Additive Manufacturing Technologies*,
DOI 10.1007/978-1-4939-2113-3_13

time to market, are where AM technology contributes, particularly for complex or customized geometries.

With the growth of AM, there is going to be increasing demand for software that supports making decisions regarding which machines to use and their capabilities and limitations for a specific part design. In particular, software systems can help in the decision-making process for capital investment of new technology, providing accurate estimates of cost and time for quoting purposes, and assistance in process planning.

This chapter deals with three typical problems involving AM that may benefit from decision support:

1. *Quotation support*. Given a part, which machine and material should I use to build?
2. *Capital investment support*. Given a design and industrial profile, what is the best machine that I can buy to fulfill my requirements?
3. *Process planning support*. Given a part and a machine, how do I set it up to work in the most efficient manner alongside my other operations and existing tasks?

Examples of systems designed to fulfill the first two problems are described in detail. The third problem is much more difficult and is discussed briefly.

13.2 Selection Methods for a Part

13.2.1 Decision Theory

Decision theory has a rich history, evolving in the 1940s and 1950s from the field of economics [2]. Although there are many approaches taken in the decision theory field, the focus in this chapter will be only on the utility theory approach. Broadly speaking, there are three elements of any decision [3]:

- Options—the items from which the decision maker is selecting
- Expectations—of possible outcomes for each option
- Preferences—how the decision maker values each outcome

Assume that the set of decision options is denoted as $A = \{A_1, A_2, \ldots, A_n\}$. In engineering applications, one can think of outcomes as the performance of the options as measured by a set of evaluation criteria. More specifically, in AM selection, an outcome might consist of the time, cost, and surface finish of a part built using a certain AM process, while the AM process itself is the option. Expectations of outcomes are modeled as functions of the options, $X = g(A)$, and may be modeled with associated uncertainties.

Preferences model the importance assigned to outcomes by the decision maker. For example, a designer may prefer low cost and short turn-around times for a concept model, while being willing to accept poor surface finish. In many ad hoc decision support methods, preferences are modeled as weights or importances, which are represented as scalars. Typically, weights are specified so that they sum to 1 (normalized). For simple problems, the decision maker may just choose weights, while for more complex decisions, more sophisticated methods are used, such as pair-wise comparison [4]. In utility theory, preferences are modeled as utility functions on the expectations. Expectations are then modeled as expected utility. The best alternative is the one with the greatest expected utility.

F. Mistree, J.K. Allen, and their coworkers have been developing the Decision Support Problem (DSP) Technique over the last 20 years. The advantages of DSPs, compared with other decision formulations, are that they provide a means for mathematically modeling design decisions involving multiple objectives and supporting human judgment in designing systems [5, 6]. The formulation and solution of DSPs facilitate several types of decisions, including:

Selection—the indication of preference, based on multiple attributes, for one among several alternatives, [7].

Compromise—the improvement of an alternative through modification [4, 6].

Coupled and hierarchical—decisions that are linked together, such as selection–selection, compromise–compromise, or selection–compromise [6].

The selection problems being addressed in this chapter will be divided into two related selection subproblems. First, it is necessary to generate feasible alternatives, which, in this case, include materials and processes. Second, given those feasible alternatives, a quantification process is applied that results in a rank-ordered list of alternatives. The first subproblem is referred to as "Determining Feasibility," while the second is simply called "Selection." Additional feasibility determination and selection methods will be discussed in this section as well.

13.2.2 Approaches to Determining Feasibility

The problem of identifying suitable materials and AM machines with which to fabricate a part is surprisingly complex. As noted previously, there are many possible applications for an AM part. For each application, one should consider the suitability of available materials, fabrication cost and time, surface finish and accuracy requirements, part size, feature sizes, mechanical properties, resistance to chemicals, and other application-specific considerations. To complicate matters, the number and capability of commercial materials and machines continues to increase. So, in order to solve AM machine and process chain selection problems, one must navigate the wide variety of materials and machines, comparing one's needs to their capabilities, while ensuring that the most up-to-date information is available.

To date, most approaches to determining feasibility have taken a knowledge-based approach in order to deal with the qualitative information related to AM process capability. One of the better developed approaches was presented by Deglin and Bernard [8]. They presented a knowledge-based system for the generation, selection, and process planning of production AM processes. The problem as they defined it was: "To propose, from a detailed functional specification, different alternatives of rapid manufacturing processes, which can be ordered and optimized when considering a combination of different specification criteria (cost, quality, delay, aspect, material, etc.)." Their approach utilized two reasoning methods, case-based and the bottom-up generation of processes; the strengths of each compensated for the other's weaknesses. Their system was developed on the KADVISER platform and utilized a relational database system with extensive material, machine, and application information.

A group at the National University of Singapore (NUS) developed an AM decision system that was integrated with a database system [9, 10]. Their selection system was capable of identifying feasible material/machine combinations, estimating manufacturing cost and time, and determining optimal part orientations. From the feasible material/machines, the user can then select the most suitable combination. Their approach to determining feasible materials and processes is broadly similar to the work of Deglin and Bernard. The NUS group utilized five databases, each organized in a hierarchical, object-oriented manner: three general databases (materials, machines, and applications) and two part-specific databases (geometric information and model specifications).

Several web-based AM selection systems are available. One was developed at the Helsinki University of Technology (see http://ltk.hut.fi/RP-Selector/). Through a series of questions, the selector acquires information about the part accuracy, layer thickness, geometric features, material, and application requirements. The user chooses one of 4–5 options for each question. Additionally, the user specifies preferences for each requirement using a 5-element scale from insignificant to average to important. When all 10–12 questions are answered, the user receives a set of recommended AM machines that best satisfy their requirements.

The problem of determining process and material feasibility can be represented by the Preliminary Selection Decision Support Problem (ps-DSP) [11]. The word formulation of the ps-DSP is given in Fig. 13.1. This is a structured decision formulation and corresponds to a formal decision method based on decision theory. Qualitative comparisons among processes and materials, with respect to decision criteria, are sufficient to identify feasible alternatives and eliminate infeasible ones. After more quantitative information is known, more detailed evaluations of alternatives can be made, as described in the next subsection.

Given: a set of concepts

Identify: The principal criteria influencing selection.
The relative importance of the criteria

Capture: Experience-based knowledge about the concepts with respect to a datum and the established criteria.

Rank: The concepts in order of preference based on multiple criteria and their relative importance.

Fig. 13.1 Preliminary selection decision support problem word formulation

The key step in the ps-DSP is how to capture and apply experience-based knowledge. One chooses a datum concept against which all other concepts are compared. Qualitative comparisons are performed, where a concept is judged as better, worse, or about the same (+1, −1, 0, respectively) as the datum with respect to the principal criteria for the selection problem. Then, a weighted sum of comparisons with the datum is computed. Typically, this procedure is repeated for several additional choices of datums. In this manner, one gets a good understanding of the relative merits and deficiencies of each concept.

The ps-DSP has been applied to various engineering problems, most recently for a problem to design an AM process to fabricate metal lattice structures [12].

13.2.3 Approaches to Selection

As stated earlier, there have been a number of approaches taken to support the selection of AM processes for a part. Most aid selection, but only in a qualitative manner, as described earlier. Several methods have been developed in academia that are based on the large literature on decision theory. For an excellent introduction to this topic, see the book by Keeney and Raiffa [2]. In this section, the selection DSP is covered in some detail and selection using utility theory is summarized.

While the basic advantages of using DSPs of any type lie in providing context and structure for engineering problems, regardless of complexity, they also facilitate the recording of viewpoints associated with these decisions, for completeness and future reference, and evaluation of results through post solution sensitivity analysis. The standard Selection Decision Support Problem (s-DSP) has been applied to many engineering problems and has recently been applied to AM selection [13]. The word formulation of the standard s-DSP is given in Fig. 13.2. Note that the decision options for AM selection are feasible material-process combinations. Expectations are determined by rating the options against the attributes. Preferences are modeled using simple importance values. Rank ordering of options is determined using a weighted-sum expression of importance and attribute ratings. An extension to include utility theory has recently been accomplished, as described next.

For the Identify step, evaluation attributes are to be specified. For example, accuracy, cost, build time, tensile strength, and feature detail (how small of a feature can be created) are typical attributes. Scales denote how the attribute is to

Given: Set of AM processes/machines and materials (alternatives)

Identify: Set of evaluation attributes. Create scales and determine importances.

Rate: Each alternative relative to each attribute.

Rank: AM methods from most to least promising

Fig. 13.2 Word formulation of the selection decision support problem

be measured. For example, the cost scale is typically measured in dollars and is to be minimized. Tensile strength is measured in MPa and is to be maximized. These are examples of ratio scales, since they are measured using real numbers. Interval scales, on the other hand, are measured using integers. Complexity capability is an example attribute that could be measured using an interval scale from 1 to 10, where 10 represents the highest complexity. The decision maker should formulate interval scales carefully so that many of the integers in the scale have clear definitions. In addition to specifying scales, the decision maker should also specify minimum and maximum values for each attribute. Finally, the decision maker is to specify preferences using importance values or weights for each attribute.

For the Rate step of the s-DSP, each alternative AM process or machine should be evaluated against each attribute. From the Identify step, each attribute, a_i, has minimum and maximum values specified, $a_{i,\min}$ and $a_{i,\max}$, respectively. The decision maker specifies a rating value for attribute a_{ij} for each alternative, j, that lies between $a_{i,\min}$ and $a_{i,\max}$. The final step is to normalize the ratings so that they always take on values between 0 and 1. For cases where the attribute is to be maximized, (13.1) is used to normalize each attribute rating, where r_{ij} is the normalized rating for attribute i and alternative j. (13.2) is used to normalize attribute ratings when the attribute is to be minimized.

$$r_{ij} = \frac{a_{ij} - a_{ij,\min}}{a_{ij,\max} - a_{ij,\min}} \tag{13.1}$$

$$r_{ij} = \frac{a_{ij,\max} - a_{ij}}{a_{ij,\max} - a_{ij,\min}} \tag{13.2}$$

After all attributes are rated, the total merit for each alternative is computed using a weighted-sum formulation, as shown in (13.3). The I_i are the importances, or weights. Note that the merit value M_i is always normalized between 0 and 1.

$$M_j = \sum_{j=1} I_i r_{ij} \tag{13.3}$$

After computing the merit of each alternative, the alternatives can be rank ordered from the most favorable to the least. If two or more alternatives are close to the highest rank, additional investigation should be undertaken to understand under which conditions each alternative may be favored over the others. Additionally, the alternatives could be developed further so that more information about them is known. It is also helpful to run multiple sets of preferences (called scenarios) to understand how emphasis on certain attributes can lead to alternatives becoming favored.

Decision theory has a rich history, evolving in the 1940s and 1950s from the field of economics [2]. In order to provide a rigorous, preference-consistent alternative to the traditional merit function for considering alternatives with uncertain attribute values, the area of utility theory is often applied. This requires the satisfaction of a

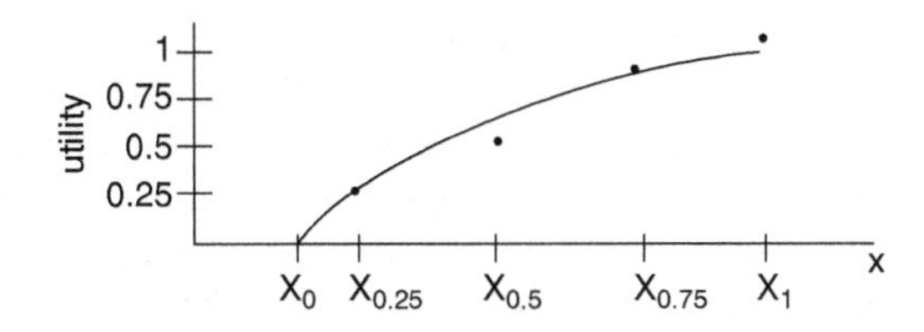

Fig. 13.3 Utility curve from five data points

set of axioms such as those proposed by von Neumann and Morgenstern [14], Luce and Raiffa [15], or Savage [16], describing the preferences of rational individuals. Once satisfied, there then exists a utility function with the desirable property of assigning numerical utilities to all possible consequences.

In utility theory, preferences are modeled as utility functions on the expectations. Mathematically, let alternative A_i result in outcome $x_i \in X$ with probability p_i, if outcomes are discrete. Otherwise, expectations on outcomes are modeled using probability density functions, $f_i = f_i(x_i)$. Utility is denoted $u(x)$. Expectations are then modeled as expected utility as shown in (13.4).

$$\begin{aligned} E[u(x)] &= \sum p_i u(x_i) \quad \text{for discrete outcomes} \\ E[u(x)] &= u(x)g(x) \text{ for continuous outcomes} \end{aligned} \tag{13.4}$$

This leads to the primary decision rule of utility theory:

Select the alternative whose outcome has the largest expected utility.

Note that expected utility is a probabilistic quantity, not a certain quantity, so there is always risk inherent in these decisions.

Utility functions are constructed by determining points that represent the decision maker's preferences then fitting a utility curve to these points. The extreme points indicate ideal and unacceptable values. These points are labeled as x_* and x_0, respectively, and are assigned utilities of 1 and 0, respectively, in Fig. 13.3. The remaining points are usually obtained by asking the decision maker a series of questions (for more information, please consult a standard reference on utility theory, e.g., [2]). Specifically, a decision maker is asked to identify his/her certainty equivalent for a few 50–50 lotteries. A lottery is a hypothetical situation in which the outcome of a decision is uncertain; it is used to assess a decision maker's preferences. A certainty equivalent is the level of an attribute for which the decision maker would be indifferent between receiving that attribute level for *certain* and receiving the results of a specified 50–50 lottery. For example, to obtain the value of $x_{0.5}$ in Fig. 13.3, the decision maker is asked to identify his/her certainty equivalent to the lottery. Generally, at least five points are identified along the decision maker's utility curve. This preference assessment procedure must be repeated for each of the attributes of interest. In the 5-point form, typical utility functions have the form

$$u(x) = c + b\mathrm{e}^{-ax}. \tag{13.5}$$

By complementing the standard selection DSP with utility theory, an axiomatic basis is provided for accurately reflecting the preferences of a designer for tradeoffs and uncertainty associated with multiple attributes. The utility selection DSP has been formulated and applied to several engineering problems, including AM selection [5, 17]. Quite a few other researchers have applied utility theory to engineering selection problems; one of the original works in this area is [18].

13.2.4 Selection Example

In this section, we present an example of a capital investment decision related to the application of metal AM processes to the production manufacture of steel caster wheels. This selection problem is very similar to a quotation problem, but includes a range of part dimensions, not single dimension values for one part. In this scenario, the caster wheel manufacturer is attempting to select an AM machine that can be used for production of its small custom orders. It is infeasible to stock all the combinations of wheels that they want to offer, thus they need to be able to produce these quickly, while also keeping the price down for the customer. The technologies under consideration are Direct Metal Deposition, Direct Metal Laser Sintering, Electron Beam Melting, Laser Engineered Net Shaping, Selective Laser Melting, and Selective Laser Sintering. A readily available stainless steel material (whatever was commercially available for the process) was used for this example. The processes will be numbered randomly (Processes 1–6) for the purposes of presentation, since this example was developed in the mid-2000s [19] and, as a result, the data are obsolete.

Before beginning the selection process, the uncertainty involved in the customization process was considered. Since these caster wheels will be customized, there is a degree of geometric uncertainty involved.

A model of a caster wheel is displayed in the Fig. 13.4a, while its main dimensions are shown in Fig. 13.4b. In this example, we have decided to only allow customization of certain features. Only standard 12 mm diameter × 100 mm length bolts will be used for the inner bore, therefore, these dimensions will be constrained. Customers will be allowed to customize all other features of the caster wheel within allowable ranges for this model wheel, as displayed in Table 13.1.

The alternative AM technologies will be evaluated based on 7 attributes that span a typical range of requirements, as shown in the following section. Scale type refers to the method used to quantify the attribute. For example, ultimate tensile strength is a ratio scale, meaning that it is represented by a real number, in this case with units of MPa. Geometric complexity is an example of an interval scale, in this case with ratings between 1 and 10, with 1 meaning the lowest complexity and 10 meaning the greatest amount of complexity.

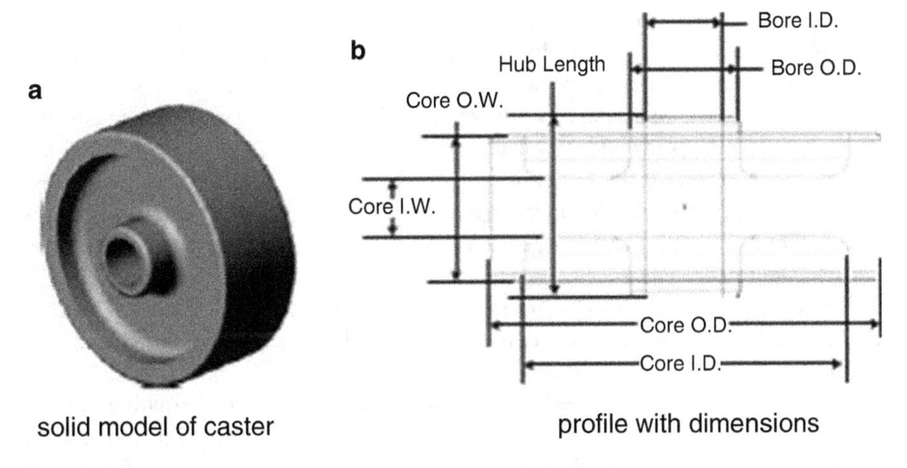

Fig. 13.4 Model of steel caster wheel

Table 13.1 Caster wheel dimensions

	Dimensions (mm)	
	Min	Max
Core outer diameter	100	150
Core inner diameter	90	140
Bore outer diameter	38	58
Bore inner diameter	32	32
Hub length	64	64
Core outer width	38	125
Core inner width	12	32

- Ultimate Tensile Strength (UTS): UTS is the maximum stress reached before a material fracture. Ratio scale [MPa].
- Rockwell Hardness C (Hard): Hardness is commonly defined as the resistance of a material to indentation. Ratio scale [HRc].
- Density (Dens.): The density refers to the final density of the part after all processing steps. This density is proportional to the amount of voids found at the surface. These voids cause a rough surface finish. Ratio scale [%].
- Detail Capability (DC): The detail capability is the smallest feature size the technology can make. Ratio scale [mm].
- Geometric complexity (GC): The geometric complexity is the ability of the technology to build complex parts. More specifically, in this case, it is used to refer to the ability to produce overhangs. Interval scale (1–10).
- Build Time (Time): The build time refers to the time required to fabricate a part, not including post-processing steps. Ratio scale [h].
- Part Cost (Cost): The part cost is the cost it takes to build one part with all costs included. These costs include manufacturing cost, material cost, machine cost, and operation cost. Ratio scale [$].

In this example, we examine two weighting scenarios (relative importance ratings). In Scenario 1, geometric complexity was most heavily weighted because of the significant overhangs present in the build orientation of the casters. Build time and part cost were also heavily weighted because of their importance to the business structure surrounding customization of caster wheels. Because of the environment of use of the caster wheels, UTS was also given a high weighting. Detail capability was weighted least because of the lack of small, detailed features in the geometry of the caster wheels. In Scenario 2, all selection attributes were equally weighted.

Table 13.2 shows the results of the evaluation of the alternatives with respect to the attributes. Weights for the two scenarios, called Relative Importances, are included under the attribute names.

On the basis of these ratings, the overall merit for each alternative can be computed. Merit values for each scenario are given in Table 13.3, along with their rankings. Note that slightly different rankings are evident from the different scenarios. This indicates the importance of accurately capturing decision maker's preferences. Process 4 is the top ranking process in both scenarios. However, the second choice could be Process 2, 3, or 6, depending upon preferences. In cases like this, it is a good idea to run additional scenarios in order to understand the trade-offs that are relevant.

This capital investment example illustrated the application of selection decision support methods. As mentioned, it is very important to explore several scenarios (sets of preferences) to understand the sensitivities of ratings and rankings to changes in preferences. Modifications to the method are straightforward to achieve target values, instead of minimizing or maximizing an attribute, and to incorporate other types of uncertainty.

13.3 Challenges of Selection

The example from the previous section illustrates some of the difficulties and limitations of straightforward application of decision methods to real decision-making situations. The complex relationships among attributes, and the variations that can arise when building a wide range of parts, make it difficult to decouple decision attributes and develop structured decision problems. Nonetheless, with a proper understanding of technologies and attributes, and how to relate them together, meaningful information can be gained. This section takes a brief look at these issues.

Different AM systems are focused on slightly different markets. For example, there are large, expensive machines that can fabricate parts using a variety of materials with relatively good accuracy and/or material properties and with the ability to fine-tune the systems to meet specific needs. In contrast, there are cheaper systems, which are designed to have minimal setup and to produce parts of acceptable quality in a predictable and reliable manner. In this latter case, parts may not have high accuracy, material strength, or flexibility of use.

Table 13.2 Rating alternatives with respect to attributes

Attributes						Geom. compl.		Build time		Part cost	
		UTS	Hardness	Density	Detail cap.	mm	max	mm	max	mm	max
Rel. Imp.	Scen 1	0.167	0.143	0.071	0.024	0.214	0.214	0.19	0.19	0.19	0.19
	Scen 2	0.143	0.143	0.143	0.143	0.143	0.143	0.143	0.143	0.143	0.143
Alternatives	Proc 1	600	21	95	0.3	7	10	2.8	59	390	2,050
	Proc 2	1,430	50	100	1.2	7	10	1.42	30	134	510
	Proc 3	1,700	53	100	0.762	4	6	0.26	5	64	310
	Proc 4	2,000	60	99.5	0.15	7	10	1.9	39	240	1,340
	Proc 5	600	15	100	0.6	7	10	1.4	24	180	890
	Proc 6	1,800	53	100	1	4	6	0.12	2	30	170
Scales	Type	Ratio	Ratio	Ratio	Ratio	Interval		Ratio		Ratio	
	Low	500	10	95	0.1	1		2		25	
	High	2,500	70	100	2	10		120		1,000	
	Pref.	High	High	High	Low	High		Low		Low	
	Units	MPa	HRc	Percent	mm	nmu		h		$	

Table 13.3 Merit values and rankings

	Scenario 1		Scenario 2	
	Merit	Rank	Merit	Rank
Proc 1	0.254	6	0.284	6
Proc 2	0.743	2	0.667	4
Proc 3	0.689	4	0.703	2
Proc 4	0.753	1	0.808	1
Proc 5	0.528	5	0.539	5
Proc 6	0.72	3	0.697	3

Different users will require different things from an AM machine. Machines vary in terms of cost, size, range of materials, accuracy of part, time of build, etc. It is not surprising to know that the more expensive machines provide the wider range of options and, therefore, it is important for someone looking to buy a new machine to be able to understand the costs vs. the benefits so that it is possible to choose the best machine to suit their needs.

Approaching a manufacturer or distributor of AM equipment is one way to get information concerning the specification of their machine. Such companies are obviously biased towards their own product and, therefore, it is going to be difficult to obtain truly objective opinions. Conventions and exhibitions are a good way to make comparisons, but it is not necessarily easy to identify the usability of machines. Contacting existing users is sometimes difficult and time consuming, but they can give very honest opinions. This approach works best if you are already equipped with background information concerning your proposed use of the technology.

When looking for advice about suitable selection methods or systems, it is useful to consider the following points. One web-based system was developed to meet these considerations [20]. An alternative approach will be presented in the next section.

- The information in the system should be unbiased wherever possible.
- The method/system should provide support and advice rather than just a quantified result.
- The method/system should provide an introduction to AM to equip the user with background knowledge as well as advice on different AM technologies.
- A range of options should be given to the user in order to adjust requirements and show how changes in requirements may affect the decision.
- The system should be linked to a comprehensive and up-to-date database of AM machines.
- After the search process has completed, the system should give guidance on where to look next for additional information.

The process of accessing the system should be as beneficial to the user as the answers it gives. However, this is not as easy a task as one might first envisage. If it were possible to decouple the attributes of the system from the user specification, then it would be a relatively simple task to select one machine against another. To illustrate that this is not always possible, consider the following scenarios:

1. In a Powder Bed Fusion (PBF) machine, warm-up and cool down are important stages during the build cycle that do not directly involve parts being fabricated. This means that large parts do not take proportionally longer times to build compared with smaller ones. Large builds are more efficient than small ones. In Vat Photopolymerization (VP) and Material Extrusion (ME) machines, there is a much stronger correlation between part size and build time. Small parts would therefore take less time on a VP or ME machine than when using PBF, if considered in isolation. Many users, however, batch process their builds and the ability to vertically stack parts in a PBF machine makes it generally possible to utilize the available space more efficiently. The warm-up and cool down overheads are less important for larger builds and the time per layer is generally quicker than most SL and ME machines. As a result of this discussion, it is not easy to see which machine would be quicker without carefully analyzing the entire process plan for using a new machine.
2. Generally, it costs less to buy a Dimension or other low-end ME machine compared to a Binder Jetting (BJ) machine. There are technical differences between these machines that make them suitable for different potential applications. However, because they are in a similar price bracket, they are often compared for similar applications. Dimension ME machines use a cartridge-based material delivery system that requires a complete replacement of the cartridge when empty. This makes material use much more expensive when compared with the 3D Systems BJ (aka "ZCorp") machine. For occasional use, it is therefore perhaps better to use a Dimension machine when all factors are equal. On the other hand, the more parts you build, the more cost-effective the BJ machine becomes.
3. Identifying a new application or market can completely change the economics of a machine. For example, in the metals area, directed energy deposition machines (e.g., LENS, DMD) tend to be slower and have worse feature detail capability than powder bed metal machines (e.g., SLM, EBM, DMLS). This has led to many more machine sales for ARCAM, Renishaw, and EOS. However, some companies identified a market for repairing molds and metal parts, which is very difficult, if not impossible, with a powder bed machine.

These examples indicate that selection results depend to a large extent on the user's knowledge of AM capabilities and applications. Selection tools that include expert systems may have an advantage over tools based on straightforward decision methods alone. Expert systems attempt to embody the expertise resulting from extensive use of AM technology into a software package that can assist the user in overcoming at least some of the learning curve quickly and in a single stage. See [20] for a more complete coverage of this idea.

13.4 Example System for Preliminary Selection

A preliminary selection tool was developed for AM, called AMSelect, that walks the user through a series of questions to identify feasible processes and machines [21]. Build times and costs are computed, but quantitative rating and rank-ordering are not performed. More specifically, the software enables designers, managers, and service bureau personnel to:

- Explore AM technologies for their application in a possible DDM project
- Identify candidate materials and processes
- Explore build times, build options, costs
- Explore manufacturing and life-cycle benefits of AM
- Select appropriate AM technologies for DDM applications
- Explore case studies, anticipate benefits
- Support Quotation and Capital Investment decisions

Figure 13.5 illustrates the logic underlying AMSelect. A database of machine types and capabilities is read, which represents the set of machines that the software will consider. The software supports a qualitative assessment of the suitability of DDM for the application, then enables the user to explore the performance of various AM machines. Build time and cost estimates are provided, which enable the user to make a selection decision.

To use AMSelect, the user first enters information about the production project, including production rate (parts per week), target part cost, how long the part is expected to be in production, and the useful life of the part. After the user enters

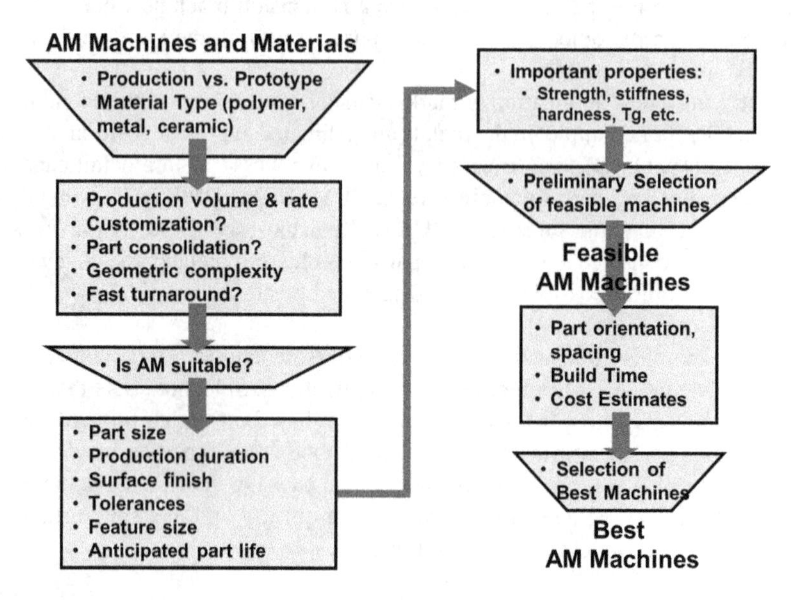

Fig. 13.5 Flowchart of AMSelect operation

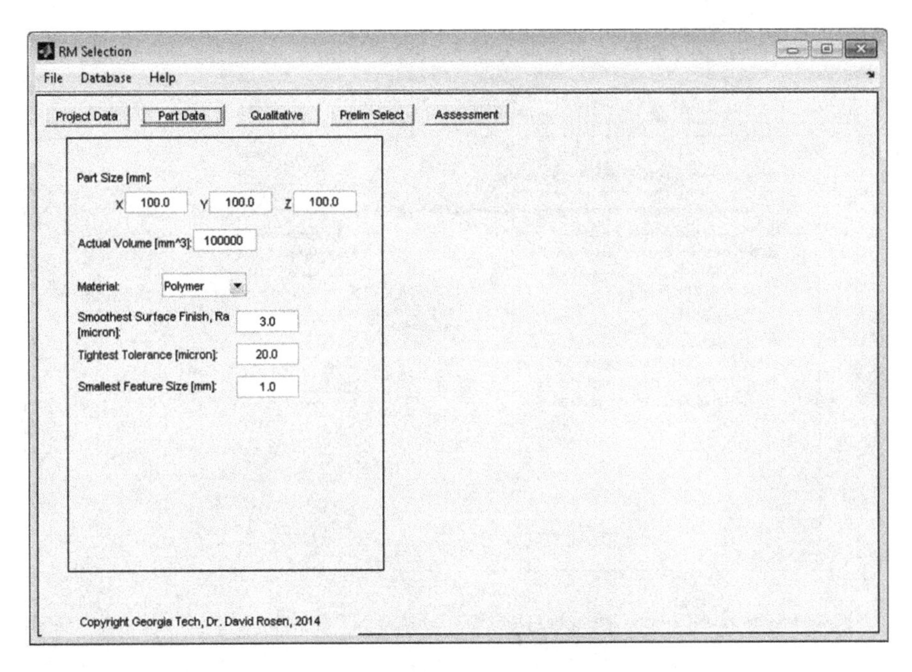

Fig. 13.6 Part Data entry screen for AMSelect

information about the part to be produced and its desired characteristics (Fig. 13.6), the user answers questions about how the application may take advantage of the unique capabilities of AM processes, as shown in Figs. 13.7 and 13.8. In this version of the software, the questions ask about part shape similarity across the production volume, part geometric complexity, the extent of part consolidation compared to a design for conventional manufacturing processes, and the part delivery time. Based on the responses, the software responds with general statements about the likelihood of AM processes being suitable for the user's application; for example, see the responses for the fictitious problem from Fig. 13.6. If the user is satisfied that his/her application is suitable for DDM, then they can proceed with a more quantitative exploration of AM machines.

The AMSelect software enables the user to explore the capabilities of various AM machines for their application. As shown in Fig. 13.9, AMSelect first segregates machines that appear to be feasible from those that are infeasible, based on material, surface finish, accuracy, and minimum feature size requirements. The user can select from both the sets of feasible and infeasible machines, which can be useful for comparison purposes.

If the user wants to see the layout of parts in a machine's build chamber, they can hit the Display button, while the machine of interest is selected. For example, the build chamber of an SLA ProX 950 machine is shown in Fig. 13.10 for parts with bounding box dimensions of 100 × 100 × 100 mm (part size from Fig. 13.6). Since the ProX 950 has platform dimensions of 1,500 × 550 mm, a total of 78 parts can fit

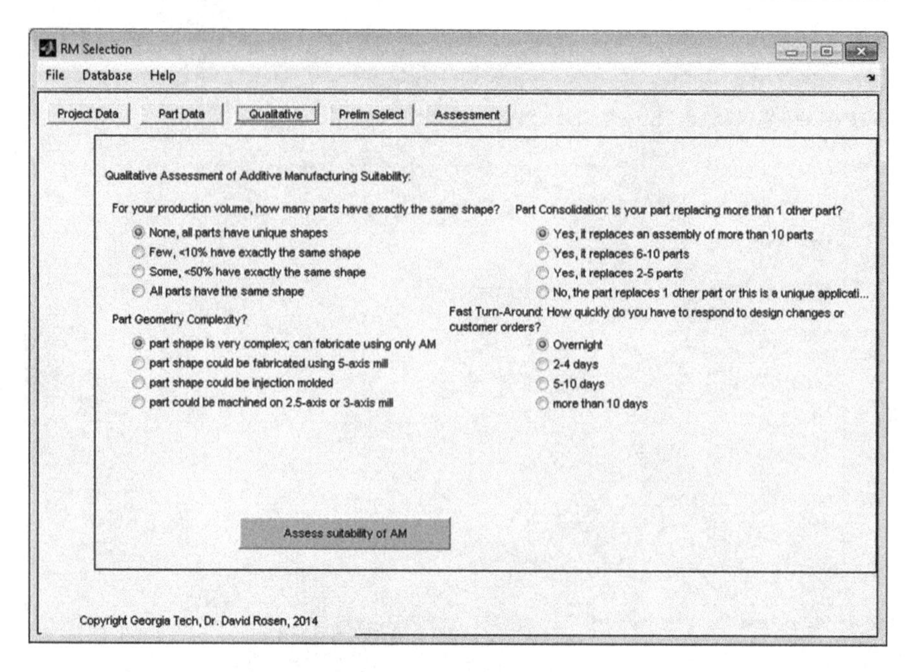

Fig. 13.7 Qualitative assessment question screen

on the platform with 10 mm spacing, as shown. The user has control over the spacing between parts. Entering negative spacing values effectively "nests" parts within one another, which may be useful if parts are shaped like drinking cups, for example. Note that AMSelect will stack parts vertically if that is a typical build mode for the technology; parts are often fabricated in stacked layers in PBF as one example. Also note that serial manufacturing of end-use products is assumed for the AMSelect software. As such, the software assumes that a large quantity of parts must be produced and fills the platform or build chamber with only one type of part (part described in the Part Data screen, Fig. 13.6). The user can also change the part orientation in an attempt to fit additional parts into a build.

In the last major step in AMSelect, the Assessment button (see Figs. 13.6, 13.7 and 13.9) can be selected to estimate the build time for the platform of parts, as well as the cost per part. These assessments can be particularly useful in comparing technologies and machines for an application. Considerable uncertainty exists regarding build speeds so ranges of build times are calculated based on typical ranges of scanning speeds, delay times, recoating speeds, etc. Part costs are broken down into machine, material, operation, and maintenance costs, similar to the cost model to be presented in Chap. 16.

As shown in Fig. 13.11, long build times do not necessarily translate to high part costs, particularly if many parts can be built at once. The SLA iPro 8000, SLS sPro 230, and the FDM Fortus 900mc have similar build times for a platform full of parts, but part costs are several times smaller for the sPro 230, compared to the SLA

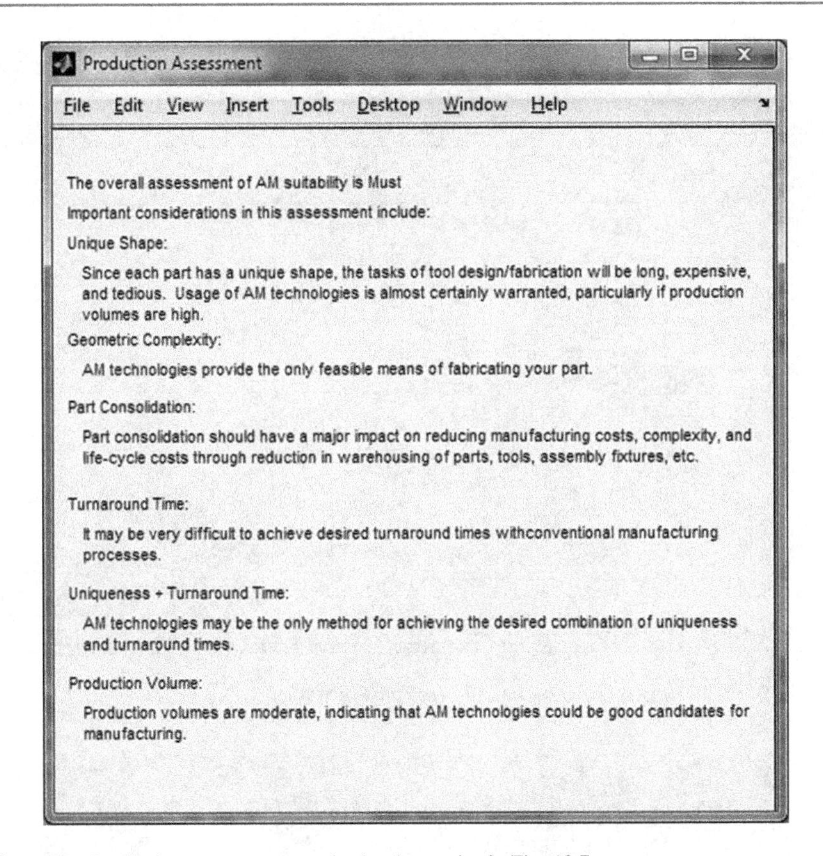

Fig. 13.8 Qualitative assessment results for the entries in Fig. 13.7

iPro 8000, since many more parts can be built in about the same amount of time. On the other hand, the SLA Viper Si2 takes almost as long for its platform, but parts are very expensive since the Viper is building in high-resolution mode (built-in assumption) and few parts can fit on its relatively small platform.

Maintenance of the database for AMSelect can be problematic, since machine capabilities may be upgraded, costs may be reduced, and new machines developed. AMSelect allows users to edit its database, either by modifying existing machines or by creating new ones. The screen that shows this capability is shown in Fig. 13.12.

Armed with these results, the user can make a selection of AM machines to explore further. The decision may be based on part cost. But, the user needs to take all relevant information into account. Recall that the SinterStation machines were not feasible for this application (due to feature size requirements, although this was not shown). This was why the SinterStation appeared in the infeasible column in Fig. 13.9. With these results, the user can determine whether or not he/she wants to relax the feature size requirement to reduce costs, or maintain requirements with a potential cost penalty. Hence, trade-off scenarios can be explored with AMSelect.

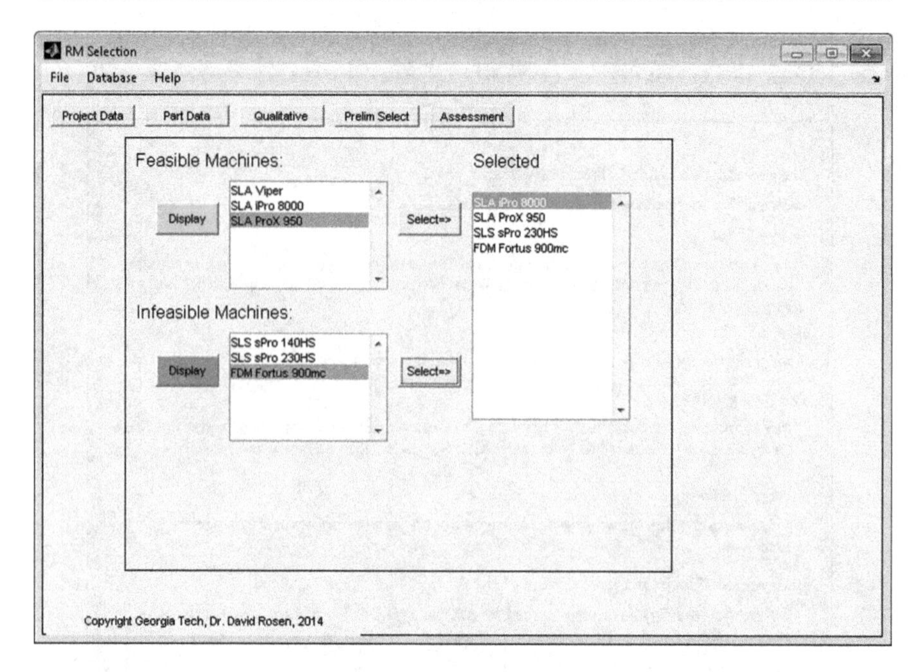

Fig. 13.9 Preliminary selection of machines to consider further

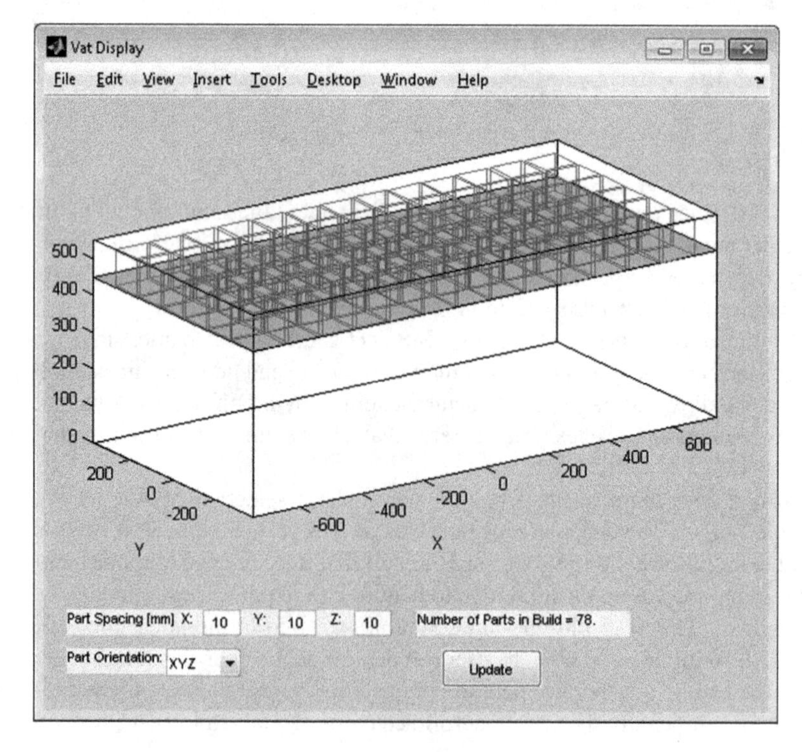

Fig. 13.10 Layout of parts on the machine platform

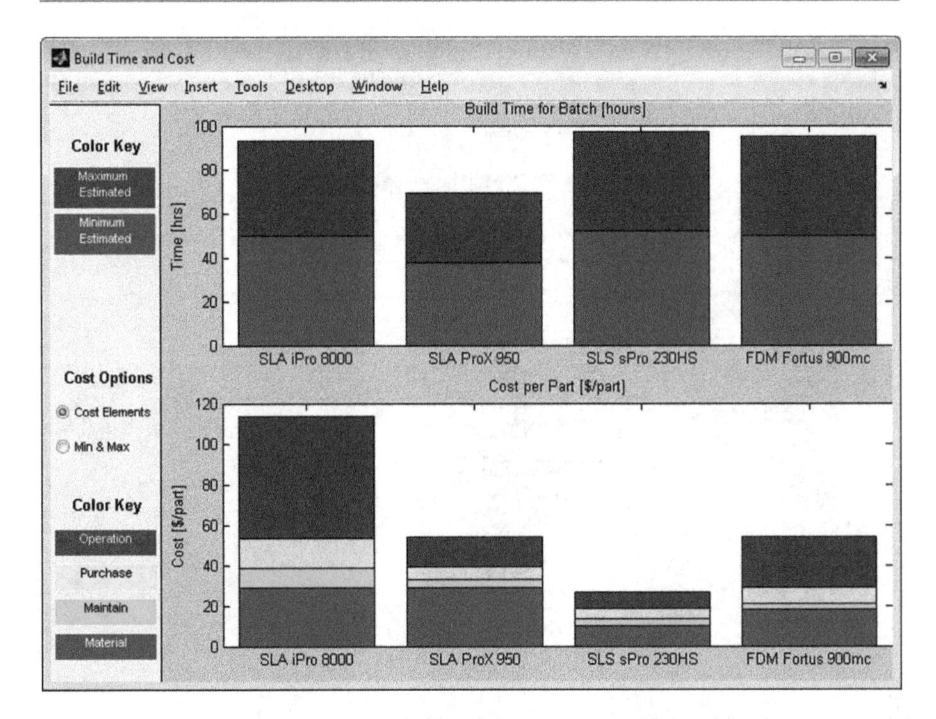

Fig. 13.11 Build time and cost results for the example

In fact, a tool like AMSelect can be used by machine vendors to explore new product development. They can "create" new machines by adding a machine to the database with the characteristics of interest. Then, they can test their new machine by quantitatively comparing it with existing machines based on build times and part costs.

13.5 Production Planning and Control

This section addresses the third type of selection decision introduced in the Introduction, namely support for process planning. It is probably most relevant to the activities of service bureaus (SBs), including internal organizations in manufacturing companies that operate one or more AM machines and processes. The SB may know which machine and material a part is to be made from, but in most circumstances, the part cannot be considered in isolation. When any new part is presented to the process planner at the SB, it is likely that he has already committed to build a number of parts. A decision support software system may be useful in keeping track and optimizing machine utilization.

Consider the process when a new part is presented to the SB for building. In general, the information presented to the process planner will include the following:

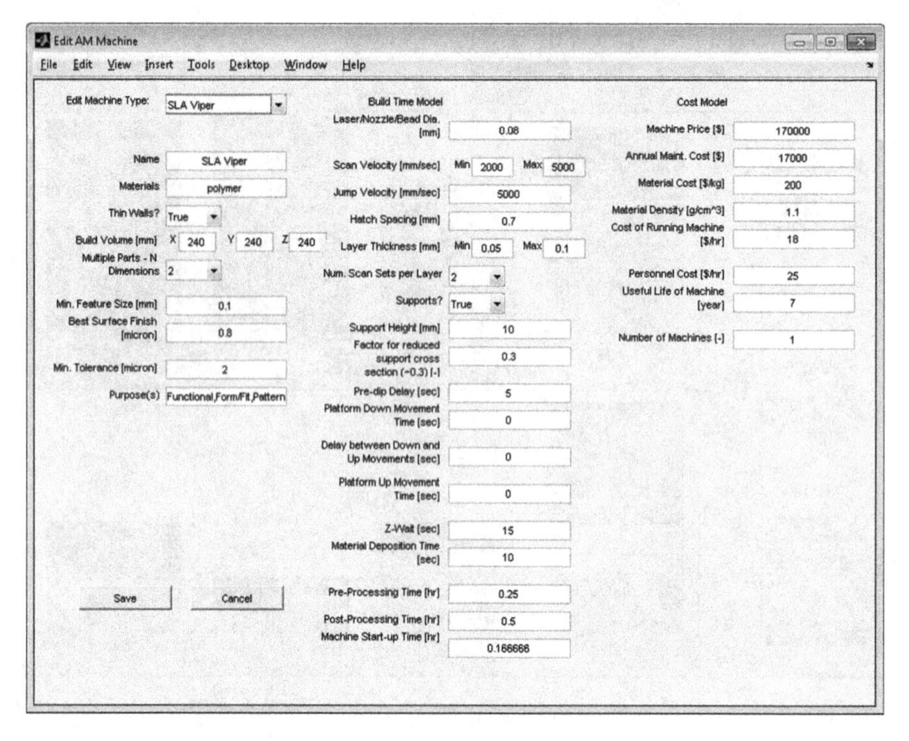

Fig. 13.12 Screen for adding or editing an AM machine definition

- Part geometry
- Number of parts
- Delivery date or schedule for batches of parts
- Processes other than AM to be carried out (pre-processing and post-processing)
- Expectations of the user (accuracy, degree of finish, etc.)

Furnished with this small amount of data, it is possible to start integrating the new job with the existing jobs and available resources. Four topics will be explored further, namely production planning, pre-processing, part build, and post-processing.

13.5.1 Production Planning

Several related decision are needed early in the process. A suitable AM process and machine must be identified from among those in the facility. This was probably done during the quoting stage before the customer selected the SB. After that is settled, AM machine availability must be considered. If the SB has more than one suitable machine, a choice must be made as to which machine to use. If the job is for a series of part batches, the SB may choose to run all batches on the same machine, or on multiple machines. If multiple machines, the SB must ensure that all selected machines can provide repeatable results, which is not always the case. Otherwise, potentially lengthy calibration builds may be needed to ensure consistent part quality from all machines used.

A job scheduling system should be used, particularly for production manufacturing applications, so that part batches can be produced to meet deadlines. If the SB has insufficient resources, it may need to invest in further capacity, necessitating a machine selection scenario. Alternatively, the SB could retain the services of other SBs if the economics of further machine investments is questionable.

13.5.2 Pre-processing

Pre-processing means software-based manipulation. This will be carried out on the file that describes the geometry of the part. Such manipulation can generally be divided into two areas, modification of the design and determination of build parameters.

Modification of the design may be required for two reasons. First, part details may need adjustment to accommodate process characteristics. For example, shaft or pin diameters may need to be reduced, to increase clearance for assembly, when building in many processes since most processes are material safe (i.e., features become oversized). Second, models may require repair if the STEP, IGES, AMF, or STL file has problems such as missing triangles, incorrectly oriented surfaces, or the like.

Determination of the build parameters is very specific to the AM process to be used. This includes selecting a part orientation, support generation, setting of build styles, layer thickness selection, and temperature setting. In general, this is either a very quick process or it takes a predictable length of time to set up. On occasion, and for some particular types of machine, this process can be very time consuming. This usually corresponds to instances when the user expectations closely meet the upper limits of the machine specification (high accuracy, build strength, early delivery date, etc.). Under such conditions, the user must devote more time and attention to parameter setting. The decision support software should make the process planner aware under which circumstances this may occur and allocate resources appropriately.

13.5.3 Part Build

For some processes, like material extrusion or LENS, it does not really matter in terms of time whether parts are built one after another (batches of 1) or parts are grouped together in batches. However, most processes will vary significantly regarding this factor. This may be due to significant preparation time before the build process takes place (such as powder bed heating in PBF), or because there is a significant delay between layers. In the latter case, it is obvious that the cumulative number of layers should be as low as possible to minimize the overall build time for many parts.

Another factor is part orientation. It is well known that because of anisotropic properties caused by most AM processes, parts will generally build more effectively in one orientation compared with another. This can cause difficulties when organizing the batch production of parts. Orientation of parts so that they fit efficiently within the work volume does not necessarily mean optimal build quality and vice versa. For those machines that need to use support structures during the build process, this represents an additional problem, both in terms of build time (allocation of time to build the support structures for different orientations) and post-processing time (removing the supports). Many researchers have discussed these dilemmas [22].

What this generally means to a process planner is compromise. Compromise is not unusual to a process planner; in fact, it is a typical characteristic, but the degree of flexibility provided by many AM machines makes this a particularly interesting problem. Just because an AM machine is being used constantly does not mean it is being used efficiently.

13.5.4 Post-processing

All AM parts require a degree of post-processing. At the low end, this may require removal of support structures or excess powder for those who merely want quick, simple verification. At the high end, the AM process may be a very insignificant time overhead in the overall process. Parts may require a large amount of skilled manual work in terms of surface preparation and coating. Alternatively, the AM part may be one stage in a complex rapid tooling process that requires numerous manual and automated stages. All this can result from the same source machine. It can even be an iterative process involving all of the above steps at different stages in the development cycle based on the same part CAD data.

13.5.5 Summary

It is clear that only process planners who have a very detailed understanding of all the roles that AM parts can play will be able to utilize the resources effectively and efficiently. Even then it may be difficult to perform this task reliably given the large number of variables involved. A software system to assist in this difficult task would be a very valuable tool.

13.6 Open Problems

Some summary statements and open problems that motivate continued research are presented here.

- Selection methods and systems are only as good as the information that is utilized to make suggestions. Maintaining up to date and accurate machine and material databases will likely be an ongoing problem. Centralized databases and standard database and benchmarking practices will help to mitigate this issue.
- Customers (people wanting parts made) have a wide range of intended applications and needs. A better representation of those needs is required in order to facilitate better selection decisions. Improved methods for capturing and modeling user preferences are also needed.
- Related to the wide range of applications is the wide range of manufacturing process chains that could be used to construct parts. Better, more complete methods of generating, evaluating, and selecting process chains are needed for cases where multiple parts or products are needed (10–100) or when complex prototypes need to be constructed. An example of the latter case is a functional prototype of a new product that consists of electronic and mechanical subsystems. Many options likely exist for fabricating individual parts or modules.
- More generally, integration of selection methods, with databases and process chain exploration methods would be very beneficial.
- Methods are needed that hide the complexity associated with the wide variety of process variables and nuances of AM technologies. This is particularly important for novice users of AM machines or even for knowledgeable users who work in production environments. Alternatively, knowledgeable users must have access to all process variables if necessary to deal with difficult builds.
- Better methods are needed that enable users to explore trade-offs (compromises) among build goals and to find machine settings that enable them to best meet their goals. These methods should work across the many different types of AM machines and materials.
- It is not uncommon for AM customers to want parts that are at the boundaries of AM machine capabilities. Tools that recognize when capability limits are reached or exceeded would be very helpful. Furthermore, these tools should provide guidance that assists users in identifying process settings that are likely to yield the best results. Providing estimates of part qualities (e.g., part detail actual sizes vs. desired sizes, actual surface finish vs. desired surface finish, etc.) would also be helpful.

13.7 Exercises

1. You have been assigned to fabricate several prototypes of a cell phone housing for assembly and functional testing purposes. Discuss the advantages and disadvantages of commercial AM processes. Identify the most likely to succeed processes.
2. Repeat Exercise 1 for a laptop housing.
3. Repeat Exercise 1 for metal copings for dental restorations (e.g., crowns and bridges). Realize that accuracy requirements are approximately 10 μm. Titanium or cobalt-chrome materials are used typically.
4. For the selection example in Section 13.2.4.
   (a) Update the information used using information sources at your disposal (web sites, etc.).
   (b) Repeat the selection process using your updated information. Develop your new versions of Tables 13.2 and 13.3.
5. Repeat Exercise 3 using the selection DSP method and the updated information that you found for Exercise 4.

References

1. Wohlers T (2013) State of the industry—2013 Worldwide progress report, Wohlers Associates, Fort Collins
2. Keeney RL, Raiffa H (1976) Decisions with multiple objectives: preferences and value tradeoffs. Wiley, New York
3. Hazelrigg G (1996) Systems engineering: an approach to information-based design. Prentice Hall, Upper Saddle River
4. Mistree F, Smith WF, Bras BA (1993) A decision-based approach to concurrent engineering. In: Paresai HR, Sullivan W (eds) Handbook of concurrent engineering. Chapman and Hall, New York, pp 127–158
5. Marston M, Allen JK, Mistree F (2000) The decision support problem technique: integrating descriptive and normative approaches. Eng Valuation Cost Anal, Special Issue on Decisions-Based Design: Status and Promise 3:107–129
6. Mistree F, Smith WF, Bras BA, Allen JK, Muster D (1990) Decision-based design: a contemporary paradigm for ship design. Trans Soc Naval Arch Marine Eng 98:565–597
7. Bascaran E, Bannerot RB, Mistree F (1989) Hierarchical selection decision support problems in conceptual design. Eng Optim 14:207–238
8. Deglin A, Bernard A (2000) A knowledge-based environment for modelling and computer-aided process planning of rapid manufacturing processes, CE'2000 conference, Lyon
9. Fuh JYH, Loh HT, Wong YS, Shi DP, Mahesh M, Chong TS (2002) A web-based database system for RP machines, processes, and materials selection, Chap 2. In: Gibson I (ed) Software solutions for RP. Professional Engineering, London
10. Xu F, Wong YS, Loh HT (1999) A knowledge-based decision support system for RP&M process selection. In: Proceedings solid freeform fabrication symposium, Austin, Aug 9–11
11. Allen JK (1996) The decision to introduce new technology: the fuzzy preliminary selection decision support problem. Eng Optim 26(1):61–77
12. Williams CB (2007) Design and development of a layer-based additive manufacturing process for the realization of metal parts of designed mesostructure, PhD dissertation, Georgia Institute of Technology

13. Herrmann A, Allen JK (1999) Selection of rapid tooling materials and processes in a distributed design environment. ASME design for manufacturing conference, paper #DETC99/DFM-8930, Las Vegas, Sept. 12–15
14. von Neumann J, Morgenstern O (1947) The theory of games and economic behavior. Princeton University, Princeton
15. Luce RD, Raiffa H (1957) Games and decisions. Wiley, New York
16. Savage LJ (1954) The foundations of statistics. Wiley, New York
17. Fernandez MG, Seepersad CC, Rosen DW, Allen JK, Mistree F (2001) Utility-based decision support for selection in engineering design. ASME design automation conference, paper #DETC2001/DAC-21106, Pittsburgh, Sept. 9–12
18. Thurston DL (1991) A formal method for subjective design evaluation with multiple attributes. Res Eng Des 3(2):105–122
19. Wilson J, Rosen DW (2005) Selection for rapid manufacturing under epistemic uncertainty. Proceedings ASME design automation conference, paper DETC2005/DFMLC-85264, Long Beach, Sept. 24–28
20. Rosen DW, Gibson I (2002) Decision support and system selection for RP, Chapter 4. In: Gibson I (ed) Software solutions for RP. Professional Engineering, London
21. Rosen DW (2005) Direct digital manufacturing: issues and tools for making key decisions. Proceedings SME rapid prototyping and manufacturing conference, Dearborn, May 9–12
22. Dutta D, Prinz FB, Rosen D, Weiss L (2001) Layered manufacturing: current status and future trends. ASME J Comput Inf Sci Eng 1(1):60–71

Post-processing 14

14.1 Introduction

Most AM processes require post-processing after part building to prepare the part for its intended form, fit and/or function. Depending upon the AM technique, the reason for post-processing varies. For purposes of simplicity, this chapter will focus on post-processing techniques which are used to enhance components or overcome AM limitations. These include:

1. Support material removal
2. Surface texture improvements
3. Accuracy improvements
4. Aesthetic improvements
5. Preparation for use as a pattern
6. Property enhancements using non-thermal techniques
7. Property enhancements using thermal techniques

The skill with which various AM practitioners perform post-processing is one of the most distinguishing characteristics between competing service providers. Companies which can efficiently and accurately post-process parts to a customer's expectations can often charge a premium for their services; whereas, companies which compete primarily on price may sacrifice post-processing quality in order to reduce costs.

14.2 Support Material Removal

The most common type of post-processing in AM is support removal. Support material can be broadly classified into two categories: (a) material which surrounds the part as a naturally occurring by-product of the build process (natural supports),

I. Gibson et al., *Additive Manufacturing Technologies*,
DOI 10.1007/978-1-4939-2113-3_14

and (b) rigid structures which are designed and built to support, restrain, or attach the part being built to a build platform (synthetic supports).

14.2.1 Natural Support Post-processing

In processes where the part being built is fully encapsulated in the build material, the part must be removed from the surrounding material prior to its use. Processes which provide natural supports are primarily powder-based and sheet-based processes. Specifically, all powder bed fusion (PBF) and binder jetting processes require removal of the part from the loose powder surrounding the part; and bond-then-form sheet metal lamination processes require removal of the encapsulating sheet material.

In polymer PBF processes, after the part is built it is typically necessary to allow the part to go through a cool-down stage. The part should remain embedded inside the powder to minimize part distortion due to nonuniform cooling. The cool-down time is dependent on the build material and the size of the part(s). Once cool-down is complete, there are several methods used to remove the part(s) from the surrounding loose powder. Typically, the entire build (made up of loose powder and fused parts) is removed from the machine as a block and transported to a "breakout" station where the parts are removed manually from the surrounding powdered material. Brushes, compressed air, and light bead blasting are commonly used to remove loosely adhered powder; whereas, wood-working tools and dental cleaning tools are commonly used to remove powders which have sintered to the surface or powder entrapped in small channels or features. Internal cavities and hollow spaces can be difficult to clean and may require significant post-processing time.

With the exception of an extended cool-down time, natural support removal techniques for binder jetting processes are identical to those used for PBF. In most cases, parts made using binder jetting are brittle out of the machine. Thus, until the parts have been strengthened by infiltration the parts must be handled with care. This is also true for PBF materials that require post-infiltration, such as some elastomeric materials, polystyrene materials for investment casting, and metal and ceramic green parts.

More recently, automated loose powder removal processes have been developed. These can be stand-alone apparatuses or integrated into the build chamber. One of the first ZCorp (now 3D Systems) binder jetting machines with this capability is illustrated in Fig. 14.1. Several metal PBF machine manufacturers have started to integrate semi-automated powder removal techniques into their machines as well. Current trends suggest that many future PBF and binder jetting machines will incorporate some form of automated powder removal after part completion.

Bond-then-form sheet lamination processes, such as Mcor's machines, also require natural support material removal prior to use. If complex geometries with overhanging features, internal cavities, channels or fine features are used, the

Fig. 14.1 Automated powder removal using vibratory and vacuum assist in a ZCorp 450 machine (Courtesy Z Corporation)

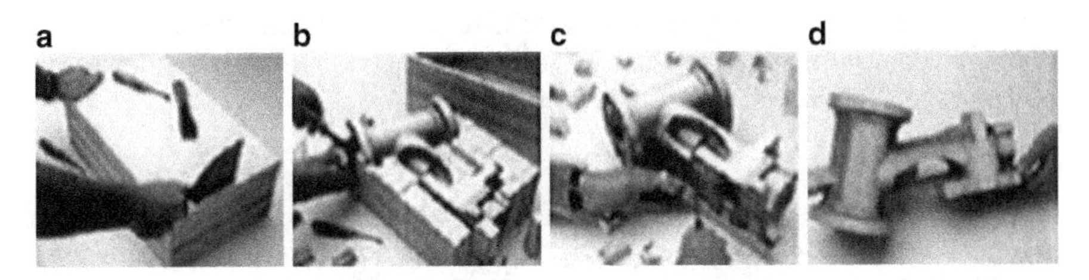

Fig. 14.2 LOM support removal process (de-cubing), showing: (**a**) the finished block of material; (**b**) removal of cubes far from the part; (**c**) removal of cubes directly adjacent to the part; (**d**) the finished product (Courtesy Worldwide Guide to Rapid Prototyping web site © Copyright Castle Island Co., All rights reserved. Photo provided by Cubic Technologies.)

support removal may be tedious and time-consuming. If enclosed cavities or channels are created, it is often necessary to delaminate the model at a specific *z*-height in order to gain access to de-cube the internal feature; and then re-glue it after removing excess support materials. An example of de-cubing operation for LOM is shown in Fig. 14.2.

14.2.2 Synthetic Support Removal

Processes which do not naturally support parts require synthetic supports for overhanging features. In some cases, such as when using PBF techniques for metals, synthetic supports are also required to resist distortion. Synthetic supports can be made from the build material or from a secondary material. The development of secondary support materials was a key step in simplifying the removal of

Fig. 14.3 Flat FDM-produced aerospace part. White build material is ABS plastic and black material is the water-soluble WaterWorks™ support material (Courtesy of Shapeways. Design by Nathan Yo Han Wheatley.)

synthetic supports as these materials are designed to be either weaker, soluble in a liquid solution, or to melt at a lower temperature than the build material.

The orientation of a part with respect to the primary build axis significantly affects support generation and removal. If a thin part is laid flat, for instance, the amount of support material consumed may significantly exceed the amount of build material (see Fig. 14.3). The orientation of supports also affects the surface finish of the part, as support removal typically leaves "witness marks" (small bumps or divots) where the supports were attached. Additionally, the use of supports in regions of small features may lead to these features being broken when the supports are removed. Thus, orientation and location of supports is a key factor for many processes to achieve desirable finished part characteristics.

14.2.2.1 Supports Made from the Build Material

All material extrusion, material jetting, and vat photopolymerization processes require supports for overhanging structures and to connect the part to the build platform. Since these processes are used primarily for polymer parts, the low strength of the supports allows them to be removed manually. These types of supports are also commonly referred to as breakaway supports. The removal of supports from downward-facing features leaves witness marks where the supports were attached. As a result, these surfaces may require subsequent sanding and polishing. Figure 14.4 shows breakaway support removal techniques for parts made using material extrusion and vat photopolymerization techniques.

PBF and DED processes for metals and ceramics also typically require support materials. An example of dental framework, oriented so that support removal does not mar the critical surfaces, is shown in Fig. 14.5. For these processes the metal supports are often too strong to be removed by hand; thus, the use of milling, band-saws, cut-off blades, wire-EDM, and other metal cutting techniques are widely employed. As discussed in Chap. 5, parts made using electron beam melting have fewer supports than those made using metal laser sintering, since EBM holds the part at elevated temperature throughout the build process and less residual stresses are induced.

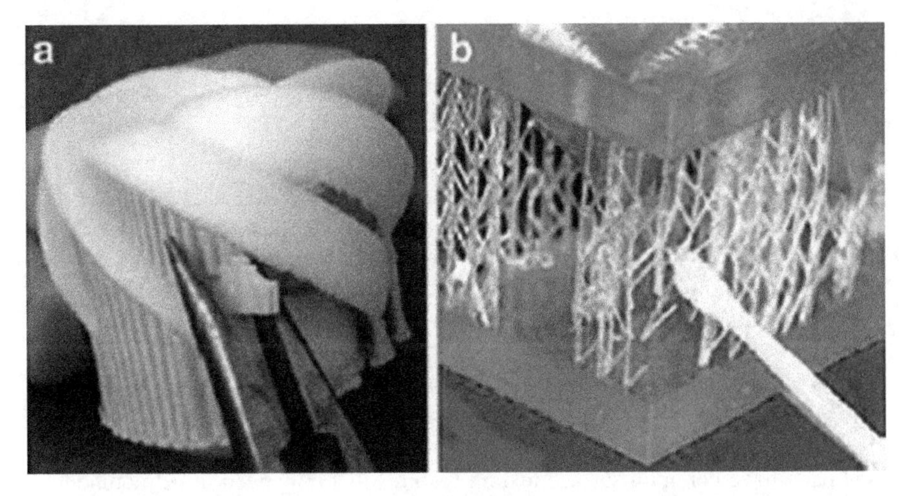

Fig. 14.4 Breakaway support removal for (**a**) an FDM part (courtesy of Jim Flowers) and (**b**) an SLA part (Courtesy Worldwide Guide to Rapid Prototyping web site. © Copyright Castle Island Co., All rights reserved. Photo provided by Cadem A.S., Turkey)

Fig. 14.5 SLM dental framework (© Emerald Group Publishing Limited) [1]

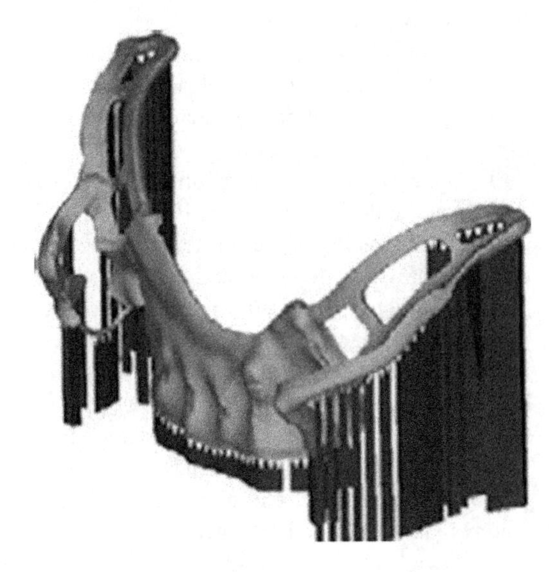

14.2.2.2 Supports Made from Secondary Materials

A number of secondary support materials have been developed over the years in order to alleviate the labor-intensive manual removal of support materials. Two of the first technologies to use secondary support materials were the Cubital layer-wise vat photopolymerization process and the Solidscape material jetting process. Their use of wax support materials enabled the block of support/build to be placed in a warm water bath; thus, melting or dissolving the wax yields the final parts.

Since that time, secondary supports have become common commercially in material extrusion (Fig. 14.3) and material jetting processes. Secondary supports have also been demonstrated for form-then-bond sheet metal lamination and DED processes in research environments.

For polymers, the most common secondary support materials are polymer materials which can be melted and/or dissolved in a water-based solvent. The water can be jetted or ultrasonically vibrated to accelerate the support removal process. For metals, the most common secondary support materials are lower-melting-temperature alloys or alloys which can be chemically dissolved in a solvent (in this case the solvent must not affect the build material).

14.3 Surface Texture Improvements

AM parts have common surface texture features that may need to be modified for aesthetic or performance reasons. Common undesirable surface texture features include: stair-steps, powder adhesion, fill patterns from material extrusion or DED systems, and witness marks from support material removal. Stair-stepping is a fundamental issue in layered manufacturing, although one can choose a thin layer thickness to minimize error at the expense of build time. Powder adhesion is a fundamental characteristic of binder jetting, PBF, and powder-based DED processes. The amount of powder adhesion can be controlled, to some degree, by changing part orientation, powder morphology, and thermal control technique (such as modifying the scan pattern).

The type of post-processing utilized for surface texture improvements is dependent upon the desired surface finish outcome. If a matte surface finish is desired, a simple bead blasting of the surface can help even the surface texture, remove sharp corners from stair-stepping, and give an overall matte appearance. If a smooth or polished finish is desired, then wet or dry sanding and hand-polishing are performed. In many cases, it is desirable to paint the surface (e.g., with cyanoacrylate, or a sealant) prior to sanding or polishing. Painting the surface has the dual benefit of sealing porosity and, by viscous forces, smoothing the stair-step effect, thus making sanding and polishing easier and more effective.

Several automated techniques have been explored for surface texture improvements. Two of the most commonly utilized include tumbling for external features and abrasive flow machining for, primarily, internal features. These processes have been shown to smooth surface features nicely, but at the cost of small feature resolution, sharp corner retention, and accuracy.

14.4 Accuracy Improvements

There is a wide range of accuracy capabilities between AM processes. Some processes are capable of submicron tolerances, whereas others have accuracies around 1 mm. Typically, the larger the build volume and the faster the build speed the worse the accuracy. This is particularly noticeable, for instance, in directed energy deposition processes where the slowest and most accurate DED processes have accuracies approaching a few microns; whereas, the larger bulk deposition machines have accuracies of several millimeters.

14.4.1 Sources of Inaccuracy

Process-dependent errors affect the accuracy of the *X–Y* plane differently from the *Z*-axis accuracy. These errors come from positioning and indexing limitations of specific machine architectures, lack of closed-loop process monitoring and control strategies, and/or from issues fundamental to the volumetric rate of material addition (such as melt pool or droplet size). In addition, for many processes, accuracy is highly dependent upon operator skill. Future accuracy improvements in AM will require fully automatic real-time control strategies to monitor and control the process, rather than the need to rely on expert operators as a feedback mechanism. Integration of additive plus subtractive processing is another method for process accuracy improvement.

Material-dependent phenomena also play a role in accuracy, including shrinkage and residual stress-induced distortion. Repeatable shrinkage and distortion can be compensated by scaling the CAD model; however, predictive capabilities at present are not accurate enough to fully understand and compensate for variations in shrinkage and residual stresses that are scan pattern or geometry dependent. Quantitative understanding of the effects of process parameters, build style, part orientation, support structures, and other factors on the magnitude of shrinkage, residual stress, and distortion is necessary to enhance these predictive capabilities. In the meantime, for parts which require a high degree of accuracy, extra material must be added to critical features, which is then removed via milling or other subtractive means to achieve the desired accuracy.

In order to meet the needs of applications where the benefits of AM are desired with the accuracy of a CNC machined component, a comprehensive strategy for achieving this accuracy can be adopted. One such strategy involves pre-processing of the STL file to compensate for inaccuracies followed by finish machining of the final part. The following sections describe steps to consider when seeking to establish a comprehensive finish machining strategy.

14.4.2 Model Pre-processing to Compensate for Inaccuracy

For many AM processes, the position of the part within the build chamber and the orientation will influence part accuracy, surface finish, and build time. Thus, translation and rotation operations are applied to the original model to optimize the part position and orientation.

Shrinkage often occurs during AM. Shrinkage also occurs during the post-process furnace operations needed for indirect processing of metal or ceramic green parts. Pre-process manipulation of the STL model will allow a scale factor to be used to compensate for the average shrinkage of the process chain. However, when compensating for average shrinkage, there will always be some features which shrink slightly more or less than the average (shrinkage variation).

In order to compensate for shrinkage variation, if the highest shrinkage value is used then ribs and similar features will always be at least as big as the desired geometry. However, channels and holes will be too large. Thus, simply using the largest shrinkage value is not an acceptable solution.

In order to make sure that there is enough material left on the surface to be machined, adding "skin" to the original model is necessary. This skin addition, such that there is material left to machine everywhere, can be referred to as making the part "steel-safe." Many studies have shown that shrinkage variations are geometry dependent, even when using the same AM or furnace post-processing parameters. Thus, compensating for shrinkage variation requires offsetting of the original model to guarantee that even the features with the largest shrinkage levels and all channels and holes are steel-safe.

There are two primary methods for adding a skin to the surface of a part. The first is to offset the surfaces and then recalculate all of the surface intersections. This methodology, though the most common, has many drawbacks for STL files made up of triangular facets. In answer to these drawbacks, an algorithm developed for offsetting all of the individual vertices of an STL file by using the normal vector information for the connected triangles, then reconstructing the triangles by using new vertex values, has been developed [2]. See Chap. 15 for more information on STL files and software systems to manipulate them.

In an STL file, each vertex is typically shared by several triangles whose unit normal vectors are different. When offsetting the vertices of a model, the new value of each vertex is determined by the unit normal values of its connected triangles.

Suppose $\overline{V}_{\text{offset}}$ is the unit vector from the original position to the new position of the vertex which is to be moved, and $N_1, N_2, \ldots N_n$ are the unit normal vectors of the triangles which share that vertex; $\overline{V}_{\text{offset}}$ can be calculated by the weighted mean of those unit normal vectors,

$$\overline{V}_{\text{offset}} = \sum_{i=1}^{n} W_i \overline{N}_i \tag{14.1}$$

where W_i are coefficients whose values are determined to satisfy the equation,

$$\overline{V}_{\text{offset}} \cdot \overline{N} = 1 \quad (i = 1, 2, \ldots n) \tag{14.2}$$

After solving for $\overline{V}_{\text{offset}}$, the new position P_{new} of the vertex is given by the equation,

$$P_{\text{new}} = P_{\text{orignal}} + \overline{V}_{\text{offset}} * d_{\text{offset}} \tag{14.3}$$

where d_{offset} is the offset dimension set by the user.

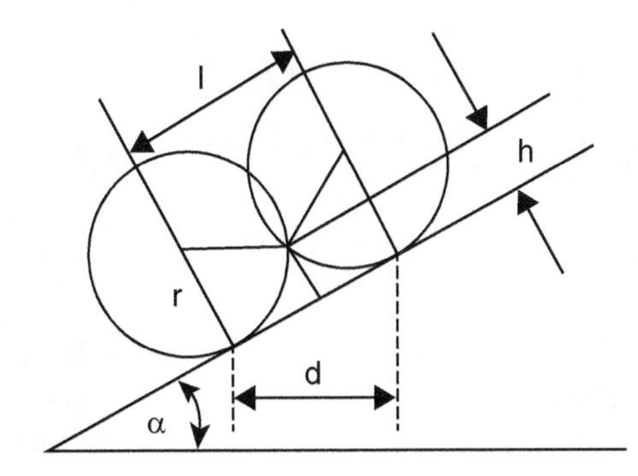

Fig. 14.6 Illustration for determining stepover distance (© Emerald Group Publishing Limited) [3]

The above procedure is repeated until the new position values for all vertices are calculated. The model is then reconstructed using the new triangle information.

Thus, to use this offset methodology, one need to only enter a d_{offset} value that is the same as the largest shrinkage variation anticipated. In practical terms, d_{offset} should be set equal to 2 or 3 times the absolute standard deviation of shrinkage measured for a particular machine/material combination.

14.4.3 Machining Strategy

Machining strategy is very important for finishing AM parts and tools. Considering both accuracy and machine efficiency, adaptive raster milling of the surface, plus hole drilling and sharp edge contour machining can fulfill the needs of most parts.

14.4.3.1 Adaptive Raster Milling

When raster machining is used for milling operations, stepover distance between adjacent toolpaths is a very important parameter that controls the machining accuracy and surface quality. It is known that higher accuracy and surface quality require a smaller stepover distance. Normally, the cusp height of material left after the model is machined is used as a measurement of the surface quality.

Figure 14.6 shows a triangle face being machined with a ball endmill. The relationship between cusp height h, cutter radius r, stepover distance d, and incline angle α is given in the following equation:

$$d = 2.0\sqrt{r^2 - (r - h)^2} \cos\alpha \tag{14.4}$$

α is determined by the triangle surface normal and stepover direction. Suppose $\overline{N}_{\text{triangle}}$ is the unit normal vector of the triangle surface, and $\overline{N}_{\text{Stepover}}$ is the unit vector along stepover direction, then

$$\cos\left(\frac{\pi}{2} - \alpha\right) = \sin \alpha = \left|\overline{N}_{\mathrm{Triangle}} \cdot \overline{N}_{\mathrm{Stepover}}\right| \tag{14.5}$$

From (14.4) and (14.5), the following equation for stepover distance is derived,

$$d = 2.0\sqrt{h(2r - h)\left(1 - \left(\overline{N}_{\mathrm{Triangle}} \cdot \overline{N}_{\mathrm{Stepover}}\right)^2\right)} \tag{14.6}$$

When machining the model, the cutter radius and milling direction are the same for all triangle surfaces. If given the user-set maximum cusp height h, d is only related to the triangle normal vector. For surfaces with different normal vectors, the stepover distance obtained will be different.

If using a constant stepover distance, in order to guarantee a particular machining tolerance, the minimum calculated d should be used for the entire part. However, using the minimum stepover distances will lead to longer programs and machining times. Therefore, an adaptive stepover distance for milling operations according to local geometry should be used to allow for both accuracy and machine efficiency. This means that stepover distances are calculated dynamically for each just-finished tool pass, using the maximum cusp height to determine the stepover distance for the next tool pass.

An example of the use of this type of algorithm for tool-path generation is shown in Fig. 14.7. As can be seen, for tool paths which pass through a region of high angle, the tools paths are more closely spaced; whereas, for toolpaths that only cross relatively flat regions, the tool paths are widely spaced.

14.4.3.2 Sharp Edge Contour Machining

Sharp edges are often the intersection curves between features and surfaces. Normally, these edges define the critical dimensions. When using raster milling, the edges parallel to the milling direction can be missed, causing large errors. As shown in Fig. 14.8, when a stepover distance d is used to machine a part with slot width W, even when the CNC machine is perfectly aligned (i.e., ignoring machine positioning errors), the slot width error will be at least,

$$W_{\mathrm{error}} = 2d - \delta_1 - \delta_2 \tag{14.6}$$

where δ_1, δ_2 represent the offset between the actual and desired edge location.

When δ_1, δ_2 become 0, $W_{\mathrm{error}} = 2d$. This means that the possible maximum error for a slot using raster milling is approximately two times the stepover distance.

For complicated edges not parallel to the milling direction, raster milling is ineffective for creating smooth edges, as the edge will have a stair-step appearance, with the step size equal to the local stepover distance, d. Thus, after raster milling, it is advantageous to run a machining pass along the sharp edges (contours) of the part [3]. In order to machine along sharp edges, all sharp edges must first be identified from the STL model. The normal vector information of each triangle is used to check the property of an edge. The angle between normal vectors of two

Fig. 14.7 Finish machining using adaptive raster milling of a copper-filled polyamide part made using polymer laser sintering. (**a**) CAD model, (**b**) tool paths, (**c**) machined part (© Emerald Group Publishing Limited, from "Raster Milling Tool-Path Generation from STL Files," Xiuzhi Qu and Brent Stucker, *Rapid Prototyping Journal*, 12 (1), pp. 4–11, 2006)

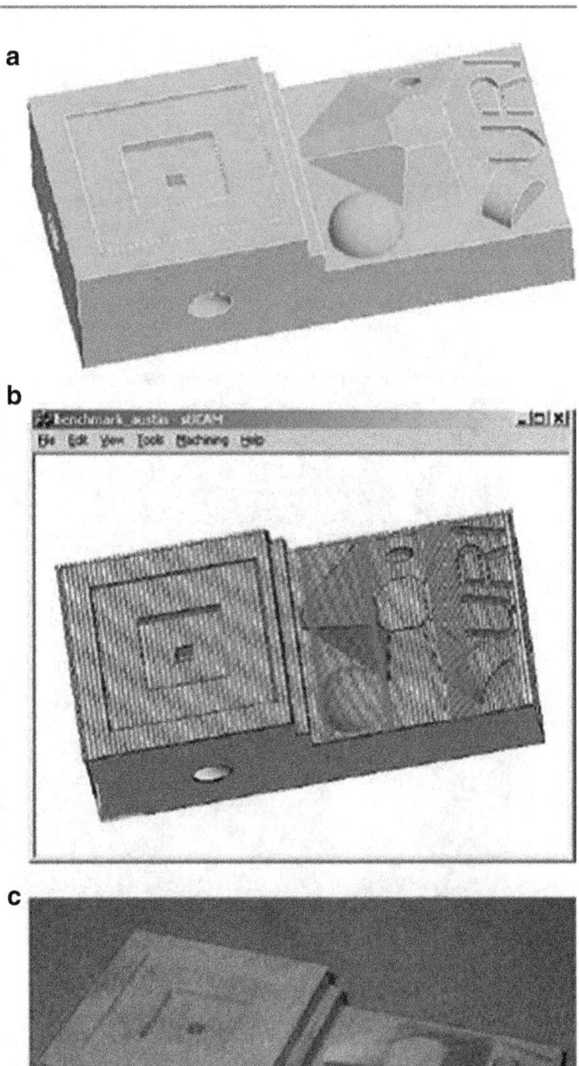

neighboring triangles is calculated. If this angle is larger than a user-specified angle, the edge shared by these two triangles will be marked as a sharp edge. All triangle edges are checked this way to generate a sharp edge list. Hidden edges and redundant tool paths are eliminated before tool paths are calculated. By offsetting the edges by the cutter radius, the x, y location of the endmill is obtained. The

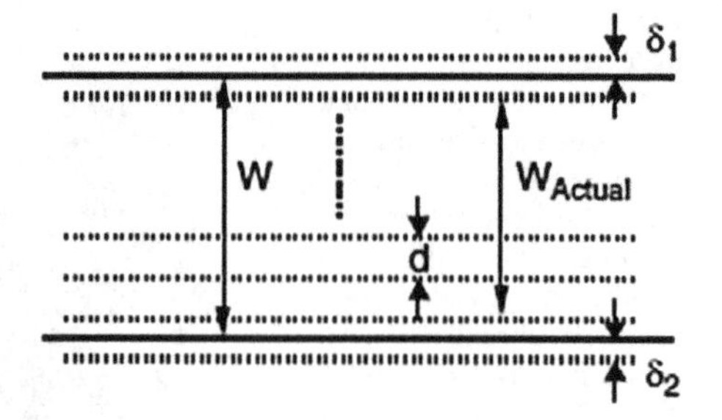

Fig. 14.8 Influence of stepover distance on dimensional accuracy (© Emerald Group Publishing Limited) [3]

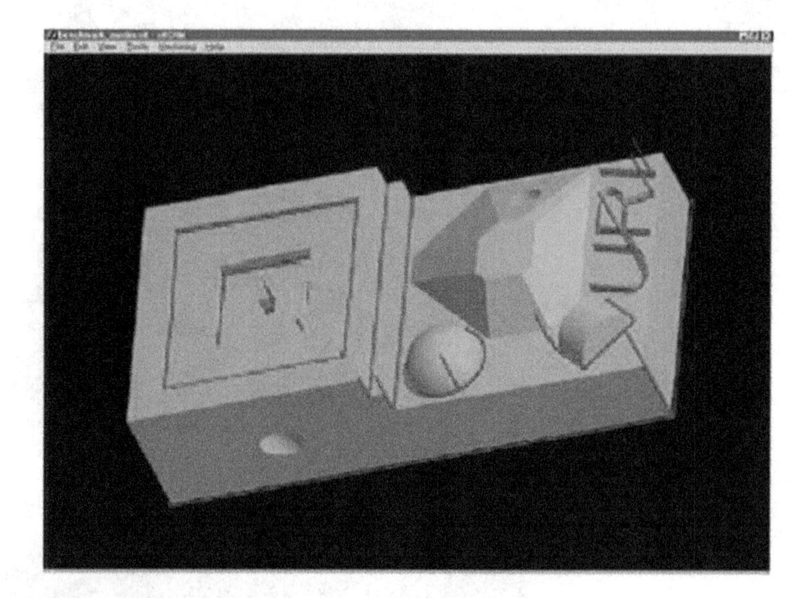

Fig. 14.9 Sharp edge contours identified for milling

z value is determined by calculating the intersection with the 3D model and finding the corresponding maximum z value. Using this approach, sharp edges can be identified and easily finish machined. Figure 14.9 shows the part from Fig. 14.7 with sharp edge contour paths highlighted.

14.4.3.3 Hole Drilling

Circular holes are common features in parts and tools. Using milling tools to create holes is inefficient and the circularity of the holes is poor. Therefore, a machining strategy of identifying and drilling holes is preferable. The most challenging aspect is to recognize holes in an STL or AMF file, as the 3D geometry is represented by a collection of unordered triangular planar facets (and thus all feature information is lost).

The intersection curve between a hole and a surface is typically a closed loop. By using this information, a hole recognition algorithm begins by identifying all closed loops made up of sharp edges from the model. These closed loops may not necessarily be the intersection curves between holes and a surface, so a series of hole-checking rules are used to remove the loops that do not correspond to drilled holes. The remaining loops and their surface normal vectors are used to determine the diameter, axis orientation, and depth for drilling. From this information, tool paths can be automatically generated [3].

Thus, by pre-processing an STL file using a shrinkage and surface offset value, and then post-processing the part using adaptive raster milling, contour machining, and hole drilling, an accurate part can be made. In many cases, however, this type of comprehensive strategy is not necessary. For instance, for a complex part where only one or two features must be made accurately, the part could be pre-processed using the average shrinkage value as a scaling factor and a skin can be added only to the critical features. These critical features could then be manually machined after AM part creation, leaving the other features as is. Thus, the finish machining strategy adopted will depend greatly upon the application and part-specific design requirements.

14.5 Aesthetic Improvements

Many times AM is used to make parts which will be displayed for aesthetic or artistic reasons or used as marketing tools. In these and similar instances, the aesthetics of the part is of critical importance for its end application.

Often the desired aesthetic improvement is solely related to surface finish. In this case, the post-processing options discussed in Sect. 14.2 can be used. In some cases, a difference in surface texture between one region and another may be desired (this is often the case in jewelry). In this case, finishing of selected surfaces only is required (such as for the cover art for this book).

In cases where the color of the AM part is not of sufficient quality, several methods can be used to improve the part aesthetics. Some types of AM parts can be effectively colored by simply dipping the part into a dye of the appropriate color. This method is particularly effective for parts created from powder beds, as the inherent porosity in these parts leads to effective absorption. If painting is required, the part may need to be sealed prior to painting. Common automotive paints are quite effective in these instances.

Another aesthetic enhancement (which also strengthens the part and improves wear resistance) is chrome plating. Figure 14.10 shows a stereolithography part before and after chrome plating. Several materials have been electroless coated to AM parts, including Ni, Cu, and other coatings. In some cases, these coatings are thick enough that, in addition to aesthetic improvements, the parts are robust enough to use as tools for injection molding or as EDM electrodes.

Fig. 14.10 Stereolithography part (**a**) before and (**b**) after chrome plating (Courtesy of Artcraft Plating)

14.6 Preparation for Use as a Pattern

Often parts made using AM are intended as patterns for investment casting, sand casting, room temperature vulcanization (RTV) molding, spray metal deposition, or other pattern replication processes. In many cases, the use of an AM pattern in a casting process is the least expensive way to use AM to produce a metal part, as many of the metal-based AM processes are still expensive to own and operate.

The accuracy and surface finish of an AM pattern will directly influence the final part accuracy and surface finish. As a result, special care must be taken to ensure the pattern has the accuracy and surface finish desired in the final part. In addition, the pattern must be scaled to compensate for any shrinkage that takes place in the pattern replication steps.

14.6.1 Investment Casting Patterns

In the case of investment casting, the AM pattern will be consumed during processing. In this instance, residue left in the mold as the pattern is melted or burned out is undesirable. Any sealants used to smooth the surface during pattern preparation should be carefully chosen so as not to inadvertently create unwanted residue.

AM parts can be printed on a casting tree or manually added to a casting tree after AM. Figure 14.11 shows rings made using a material jetting system. In the first

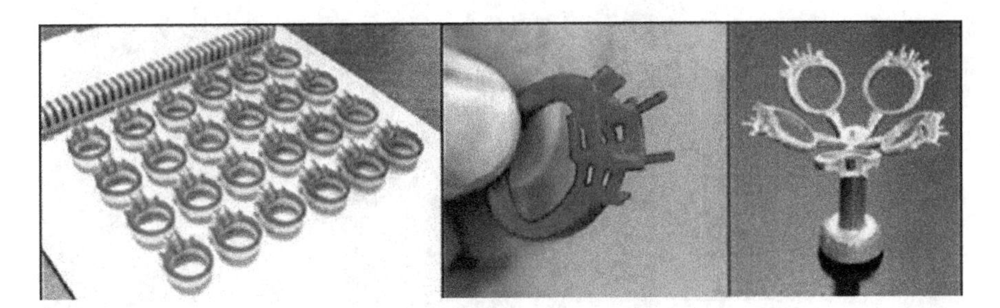

Fig. 14.11 Rings for investment casting, made using a ProJet® CPX 3D Printer (Courtesy 3D Systems)

picture, a collection of rings is shown on the build platform; each ring is supported by a secondary support material in white. In the second picture, a close-up of the ring pattern is shown. The third picture shows metal rings still attached to a casting tree. In this instance, the rings were added to the tree after AM, but before casting.

When using the stereolithography Quickcast build style, the hollow, truss-filled shell patterns must be drained of liquid prior to investment. The hole(s) used for draining must be covered to avoid investment entering the interior of the pattern. Since photopolymer materials are thermosets, they must be burned out of the investment rather than melted.

When powdered materials are used as investment casting patterns, such as polystyrene from a polymer laser sintering process or starch from a binder jetting process, the resulting part is porous and brittle. In order to seal the part and strengthen it for the investment process, the part is infiltrated with an investment casting wax prior to investment.

14.6.2 Sand Casting Patterns

Both binder jetting and PBF processes can be used to directly create sand mold cores and cavities by using a thermosetting binder to bind sand in the desired shape. One benefit of these direct approaches is that complex-geometry cores can be made that would be very difficult to fabricate using any other process, as illustrated in Fig. 14.12.

In order to prepare AM sand casting patterns for casting, loose powder is removed and the pattern is heated to complete cross-linking of the thermoset binder and to remove moisture and gaseous by-products. In some cases, additional binders are added to the pattern before heating, to increase the strength for handling. Once the pattern is thermally treated, it is assembled with its corresponding core(s) and/or cavity, and hot metal is poured into the mold. After cooling, the sand pattern is removed using tools and bead blasting.

In addition to directly producing sand casting cores and cavities, AM can be used to create parts which are used in place of the typical wooden or metal patterns around which a sand casting mold is created. In this case, the AM part is built as one

Fig. 14.12 Sand casting pattern for a cylinder head of a V6, 24-valve car engine (*left*) during loose powder removal and (*right*) pattern prepared for casting alongside a finished casting (Joint project between CADCAM Becker GmbH and VAW Südalumin GmbH, made on an EOSINT S laser-sintering machine, courtesy EOS)

or more portions of the part to be cast, split along the parting line. The split part is placed in a box, sand mixed with binder is poured around the part, and the sand is compressed (pounded) so that the binder holds the sand together. The box is then disassembled, the sand mold is removed from the box, and the pattern is removed from the mold. The mold is then reassembled with its complementary mold half and core(s) and molten metal is poured into the mold.

14.6.3 Other Pattern Replication Methods

There are many pattern replication methods which have been utilized since the late 1980s to transform the weak "rapid prototypes" of those days into parts with useful material properties. As the number of AM technologies has increased and the durability of the materials that they can produce has improved substantially, these replication processes are finding less use, as people prefer to directly produce a usable part if possible. However, even with the multiplication of AM technologies and materials, pattern replication processes are widely used among service bureaus and companies who need parts from a specific material that is not directly processable in AM.

Probably, the most common pattern replication methods are RTV molding or silicone rubber molding. In RTV molding, as shown in Fig. 14.13, the AM pattern is given visual markers (such as by using colored tape) to illustrate the parting line locations for mold disassembly; runners, risers, and gates are added; the model is suspended in a mold box; and a rubber-like material is poured around the model to encapsulate it. After cross-linking, the solid translucent rubber mold is removed from the mold box, a knife is used to cut the rubber mold into pieces according to the parting line markers, and the pattern is removed from the mold. In order to

Fig. 14.13 RTV molding process steps (Courtesy MTT Technologies Group)

complete the replication process the mold is reassembled and held together in a box or by placing rubber bands around the mold and molten material is poured into the mold and allowed to solidify. After solidification, the mold is opened, the part is removed and the process is repeated until a sufficient number of parts are made. Using this process, a single pattern can be used to make 10s or 100s of identical parts.

If the part being made in the RTV mold is a wax pattern, it can subsequently be used in an investment or plaster casting process to produce a metal part. Thus, by combining RTV molding and investment casting, one AM pattern can be replicated into a large number of metal parts for a relatively modest cost.

Metal spray processes have also been used to replicate geometry from an AM part into a metal part. In the case of metal spray, only one side of the pattern is replicated into the metal part. This is most often used for tooling or parts where one side contains all the geometric complexity and the rest of the tool or part is made up of flat edges. Using spray metal or electroless deposition processes, an AM pattern can be replicated to form an injection molding core or cavity, which can then be used to mold other parts.

14.7 Property Enhancements Using Non-thermal Techniques

Powder-based and extrusion-based processes often create porous structures. In many cases, that porosity can be infiltrated by a higher-strength material, such as cyanoacrylate (Super Glue®). Proprietary methods and materials have also been developed to increase the strength, ductility, heat deflection, flammability resistance, EMI shielding, or other properties of AM parts using infiltrants and various types of nano-composite reinforcements.

A common post-processing operation for photopolymer materials is curing. During processing, many photopolymers do not achieve complete polymerization. As a result, these parts are put into a Post-Cure Apparatus, a device that floods the part with UV and visible radiation in order to completely cure the surface and subsurface regions of the part. Additionally, the part can undergo a thermal cure in a low temperature oven, which can help completely cure the photopolymer and in some cases greatly enhance the part's mechanical properties.

14.8 Property Enhancements Using Thermal Techniques

After AM processing, many parts are thermally processed to enhance their properties. In the case of DED and PBF techniques for metals, this thermal processing is primarily heat treatment to form the desired microstructures and/or to relieve residual stresses. In these instances, traditional recipes for heat treatment developed for the specific metal alloy being employed are often used. In some cases, however, special heat treatment methods have been developed to retain the fine-grained microstructure within the AM part while still providing stress relief and ductility enhancement.

Before the advent of DED and PBF techniques capable of directly processing metals and ceramics, many techniques were developed for creating metal and ceramic green parts using AM. These were then furnace post-processed to achieve dense, usable metal and ceramic parts. Binder jetting is the only AM process which is commonly used for these purposes. The basic approach to furnace processing of green parts was illustrated in Fig. 5.7. In order to prepare a green part for furnace processing, several preparatory steps are typically done. Figure 14.14 shows the steps for preparing a metal green part made from LaserForm ST-100 for furnace infiltration.

Figure 14.15 shows an injection molding tool made from an ExOne binder jetting process after furnace debinding, sintering, and infiltration (same as Fig. 8.5). The use of cooling channels which follow the contours of the surface (conformal cooling channels) in an injection mold has been shown to significantly

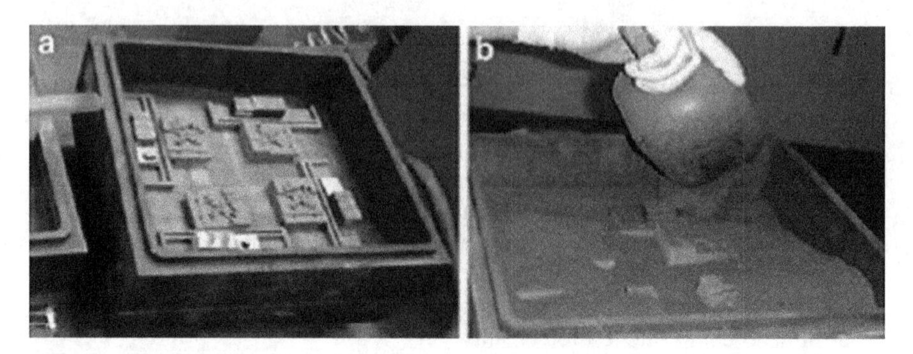

Fig. 14.14 LaserForm ST-100 green parts. (**a**) Parts are placed next to "boats" on which the bronze infiltrant is placed. The bronze infiltrates through the boat into the part. (**b**) The parts are often covered in aluminum oxide powder before placing them in a furnace to help support fragile features during debinding, sintering, and infiltration, and to help minimize thermal gradients

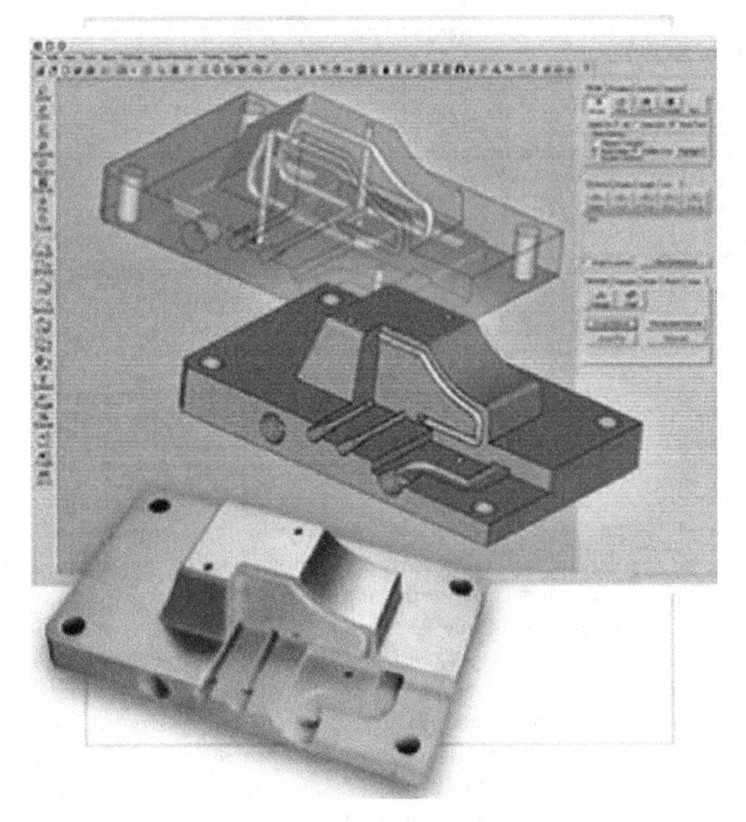

Fig. 14.15 Cross-section of a ExOne ProMetal injection molding tool showing CAD files and finished, infiltrated component with internal conformal cooling channels (Courtesy ProMetal LLC, an ExOne Company)

increase the productivity of injection mold tooling by decreasing the cooling time and part distortion. Thus, the appropriate use of conformal cooling channels enables many companies to utilize AM-produced tools to increase their productivity.

Control of shrinkage and dimensional accuracy during furnace processing is complicated by the number of process parameters that must be optimized and the multiple steps involved. Figure 14.16 illustrates the complicated nature of optimization for this type of furnace processing. The y-axis, $(F1–F2)/F1$ represents the dimensional changes during the final furnace stage of infiltration of stainless steel (RapidSteel 2.0) parts using bronze. $F1$ is the dimension of the brown part before infiltration and $F2$ is the dimension after infiltration. The data represent thousands of measurements across both internal (channel-like) and external (rib-like) features ranging from 0.3 to 3.0 in. Although many factors were studied, only two were found to be statistically significant for the infiltration step, atmospheric pressure in the furnace, and infiltrant amount. The atmospheric pressure ranged between 10 and 800 Torr. The amount of infiltrant used ranged from a low of 85 % to a high of

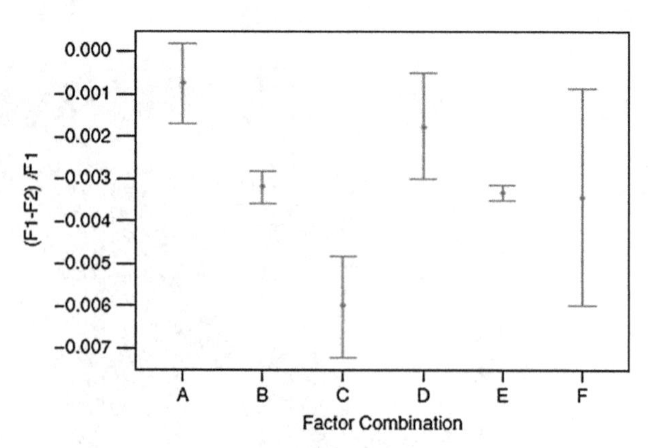

Fig. 14.16 Ninety-five percent confidence intervals for variation in shrinkage for stainless steel (RapidSteel 2.0) infiltration by bronze (Factor combinations are: (**a**) 10 Torr, 80 %; (**b**) 10 Torr, 95 %; (**c**) 10 Torr, 110 %; (**d**) 800 Torr, 80 %; (**e**) 800 Torr, 95 %; (**f**) 800 Torr, 110 %)

110 %, where the percentage amount was based upon the theoretical amount of material needed to fully fill all of the porosity in the part, based upon measurements of the weight and the volume of the part just prior to infiltration.

It can be seen from Fig. 14.16 that the factor combinations with the lowest overall shrinkage were not the factors with the lowest shrinkage variation. Factor combination A had the lowest average shrinkage, while factor combination E had the lowest shrinkage variation. As average shrinkage can be easily compensated using a scaling factor, the optimum factor combination for highest accuracy and precision would be factor combination E. If the accuracy strategy discussed in Sect. 14.3 is followed, the skin offset d_{offset} would be determined by identifying the shrinkage variation for the entire process (green part fabrication using AM, plus sintering and infiltration) using a similar approach and then setting d_{offset} equal to the maximum shrinkage variation at the desired confidence interval.

In addition to the thermal processes discussed earlier, a number of other procedures have been developed over the years to combine AM with furnace processing to produce metal or ceramic parts. One example approach utilized laser sintering to produce porous parts with gas impermeable skins. By scanning only the outside contours of a part during fabrication by SLS, a metal "can" filled with loose powder is made. These parts are then post-processed to full density using hot isostatic pressing (HIP). This in situ encapsulation results in no adverse container–powder interactions (as they are made from the same bed of powder), reduced pre-processing time, and fewer post-processing steps compared to conventional HIP of canned parts. The SLS/HIP approach was successfully used to produce complex 3D parts in Inconel 625 and Ti–6Al–4 V for aerospace applications [4].

Laser sintering has also been used to produce complex-shaped ZrB_2/Cu composite EDM electrodes. The approach involved (a) fabrication of a green part from polymer coated ZrB_2 powder using the laser sintering, (b) debinding and sintering of the ZrB_2, and (c) infiltration of the sintered, porous ZrB_2 with liquid copper. This manufacturing route was found to result in a more homogeneous structure compared to a hot pressing route.

14.9 Conclusions

Most AM-produced parts require post-processing prior to implementation in their intended use. Effective utilization of AM processes requires not only a knowledge of AM process benefits and limitations, but also of the requisite post-processing operations necessary to finalize the part for use.

When considering the intended form, fit and function for an AM-produced part, post-processing is typically required. To achieve the correct form, support material removal, surface texture improvements, and aesthetic improvements are commonly required. To achieve the correct fit, accuracy improvements, typically via milling, are commonly required. To achieve the correct function, the AM part may require preparation for use as a pattern, property enhancements using non-thermal techniques, or property enhancements using thermal techniques. Whether using automated secondary support material removal, labor-intensive de-cubing, high-temperature furnace processing, or secondary machining, choosing and properly implementing the best AM process, material and post-processing combination for the intended application is critical to success.

14.10 Exercises

1. What are the key material property considerations when selecting a secondary support material for material jetting and material extrusion? Would these considerations change when considering supporting metals deposited using a directed energy deposition process?
2. What are the primary benefits and drawbacks when offsetting triangle surfaces versus triangle vertices? (Note: You will need to find this information by finding and reading a relevant paper, as the details are not in this chapter.) Which approach would be better for freeform surfaces, such as the hood of a car or the profile of a face?
3. Assuming that the total shrinkage in an AM process is represented by Fig. 14.16, what shrinkage value and what surface offset value would you choose for pre-processing a model for each of the Factors A–F?
4. Why is contour milling beneficial for parts if adaptive raster milling ensures that all cusp heights are within acceptable values?
5. In AM processes often a larger shrinkage value is found in the X–Y plane than in the Z direction before post-processing. Why might this be the case?

References

1. Bibb R, Eggbeer D, Williams R (2006) Rapid manufacture of removable partial denture frameworks. Rapid Prototyping J 12(2):95–99
2. Qu X, Stucker B (2003) A 3D surface offset method for STL-format models. Rapid Prototyping J 9(3):133–141
3. Stucker B, Qu X (2003) A finish machining strategy for rapid manufactured parts and tools. Rapid Prototyping J 9(4):194–200
4. Das S et al (1999) Processing of titanium net shapes by SLS/HIP. Mater Des 20:115

Software Issues for Additive Manufacturing 15

Abstract

This chapter deals with the software that is commonly used for additive manufacturing technology. In particular we will discuss the STL file format that is commonly used by many of the machines to describe the model input data. These files are manipulated in a number of machine-specific ways to create slice data and for support generation and the basic principles are covered here including some discussion on common errors and other software that can assist with STL files. Finally, we consider some of the limitations of the STL format and how it may be replaced by something more suitable in the future like the newly developed Additive Manufacturing File format.

15.1 Introduction

It is clear that additive manufacturing would not exist without computers and would not have developed so far if it were not for the development of 3D solid modeling CAD. The quality, reliability, and ease of use of 3D CAD have meant that virtually any geometry can be modeled, and it has enhanced our ability to design. Some of the most impressive models made using AM are those that demonstrate the capacity to fabricate complex forms in a single stage without the need to assemble or to use secondary tooling. As mentioned in Chap. 1, the WYSIWYB (What You See Is What You Build) capability allows users to consider the design with fewer concerns over how it can be built.

Virtually every commercial solid modeling CAD system has the ability to output to an AM machine. This is because, for most cases, the only information that an AM machine requires from the CAD system is the external geometric form. There is no requirement for the machine to know how the part was modeled, any of the features or any functional elements. So long as the external geometry can be defined, the part can be built.

I. Gibson et al., *Additive Manufacturing Technologies*,
DOI 10.1007/978-1-4939-2113-3_15

This chapter will describe the fundamentals for creating output files for AM. It will discuss the most common technique, which is to create the STL file format, explaining how it works and typical problems associated with it. Some of the numerous software tools for use with AM will be described and possible effects of new concepts in AM on the development of associated software tools in the future will be discussed. Finally, there will be a discussion concerning the Additive Manufacturing File (AMF) format, which has been developed to address the needs of future AM technology where more than just the geometric information may be required.

15.2 Preparation of CAD Models: The STL File

The STL file is derived from the word STereoLithography, which was the first commercial AM process, produced by the US company 3D Systems in the late 1980s [1], although some have suggested that STL should stand for Stereolithography Tessellation Language. STL files are generated from 3D CAD data within the CAD system. The output is a boundary representation that is approximated by a mesh of triangles.

15.2.1 STL File Format, Binary/ASCII

STL files can be output as either binary or ASCII (text) format. The ASCII format is less common (due to the larger file sizes) but easier to understand and is generally used for illustration and teaching. Most AM systems run on PCs using Windows. The STL file is normally labeled with a ".STL" extension that is case insensitive, although some AM systems may require a different or more specific file definition. These files only show approximations of the surface or solid entities and so any information concerning the color, material, build layers, or history is ignored during the conversion process. Furthermore, any points, lines, or curves used during the construction of the surface or solid, and not explicitly used in that solid or surface, will also be ignored.

An STL file consists of lists of triangular facets. Each triangular facet is uniquely identified by a unit normal vector and three vertices or corners. The unit normal vector is a line that is perpendicular to the triangle and has a length equal to 1.0. This unit length could be in mm or inches and is stored using three numbers, corresponding to its vector coordinates. The STL file itself holds no dimensions, so the AM machine operator must know whether the dimensions are mm, inches, or some other unit. Since each vertex also has three numbers, there are a total of 12 numbers to describe each triangle. The following file shows a simple ASCII STL file that describes a right-angled, triangular pyramid structure, as shown in Fig. 15.1.

Note that the file begins with an object name delimited as a solid. Triangles can be in any order, each delimited as a facet. The facet line also includes the normal

```
solid triangular_pyramid
     facet normal 0.0 -1.0 0.0
       outer loop
         vertex 0.0 0.0 0.0
         vertex 1.0 0.0 0.0
         vertex 0.0 0.0 1.0
       endloop
     endfacet
     facet normal 0.0 0.0 -1.0
       outer loop
         vertex 0.0 0.0 0.0
         vertex 0.0 1.0 0.0
         vertex 1.0 0.0 0.0
       endloop
     endfacet
     facet normal 0.0 0.0 -1.0
       outer loop
         vertex 0.0 0.0 0.0
         vertex 0.0 0.0 1.0
         vertex 0.0 1.0 0.0
       endloop
     endfacet
     facet normal 0.577 0.577 0.577
       outer loop
         vertex 1.0 0.0 0.0
         vertex 0.0 1.0 0.0
         vertex 0.0 0.0 1.0
       endloop
     endfacet
   endsolid
```

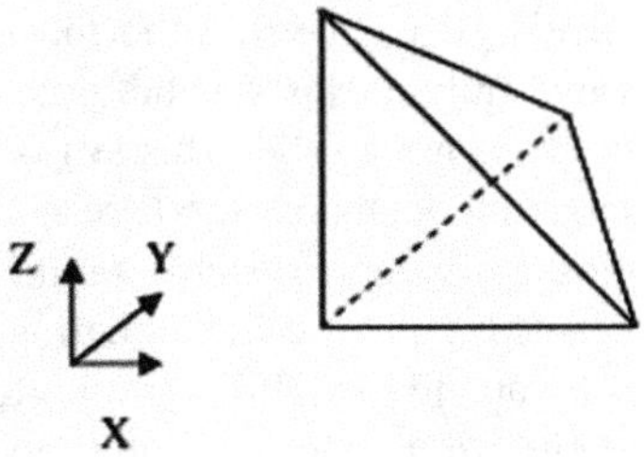

Fig. 15.1 A *right-angled triangular pyramid* as described by the sample STL file. Note that the *bottom left-hand* corner coincides with the origin and that every vertex coming out of the origin is of unit length

vector for that triangle. Note that this normal is calculated from any convenient location on the triangle and may be from one of the vertices or from the center of the triangle. It is defined that the normal is perpendicular to the triangle and is of unit length. In most systems, the normal is used to define the outside of the surface of the solid, essentially pointing to the outside. The group of three vertices defining the triangle is delimited by the terms "outer loop" and "endloop." The outside of the triangle is best defined using a right-hand rule approach. As we look at a triangle from the outside, the vertices should be listed in a counterclockwise order. Using the right hand with the thumb pointing upwards, the other fingers curl in the direction of the order of the vertices, the starting vertex being arbitrary. This approach is becoming more popular since it avoids having to do any calculations with an additional number (i.e., the facet normal) and therefore STL files may not even require the normal to prevent ambiguity.

A binary STL file can be described in the following way:

- An 80-byte ASCII header that can be used to describe the part
- A 4 byte unsigned long integer that indicates the number of facets in the object
- A list of facet records, each 50 bytes long

The facet record will be presented in the following way:

- 3 floating values of 4 bytes each to describe the normal vector
- 3 floating values of 4 bytes each to describe the first vertex
- 3 floating values of 4 bytes each to describe the second vertex
- 3 floating values of 4 bytes each to describe the third vertex
- One unsigned integer of 2 bytes, that should be zero, used for checking

15.2.2 Creating STL Files from a CAD System

Nearly all geometric solid modeling CAD systems can generate STL files from a valid, fully enclosed solid model. Most CAD systems can quickly tell the user if a model is not a solid. This test is particularly necessary for systems that use surface modeling techniques, where it can be possible to create an object that is not fully closed off. Such systems would be used for graphics applications where there is a need for powerful manipulation of surface detail (like with Autodesk AliasStudio software [2] or Rhino from Robert McNeel & Associates [3]) rather than for engineering detailing. Solid modeling systems, like SolidWorks, may use surface modeling as part of the construction process, but the final result is always a solid that would not require such a test.

Most CAD systems use a “Save as” or “Export” function to convert the native format into an STL file. There is typically some control over the size of the triangles to be used in the model. Since STL uses planar surfaces to approximate curved surfaces, then obviously the larger the triangles, the looser that approximation becomes. Most CAD systems do not directly limit the size of the triangles since it is also obvious that the smaller the triangle, the larger the resulting file for a given object. An effective approach would be to minimize the offset between the triangle and the surface that it is supposed to represent. A perfect cube with perfectly sharp edges and points can be represented by 12 triangles, all with an offset of 0 between the STL file and the original CAD model. However, few designs would be that convenient and it is important to ensure a good balance between surface approximation and excessively large file. Figure 15.2 shows the effect of changing the triangle offset parameter on an STL file. The exact value of the required offset would largely depend on the resolution or accuracy of the AM process to be used. If the offset is smaller than the basic resolution of the process, then making it smaller will have no effect on the precision of the resulting model. Since many AM processes operate around the 0.1 mm layer resolution, then a triangle offset of 0.05 mm or slightly lower will be acceptable for fabrication of most parts.

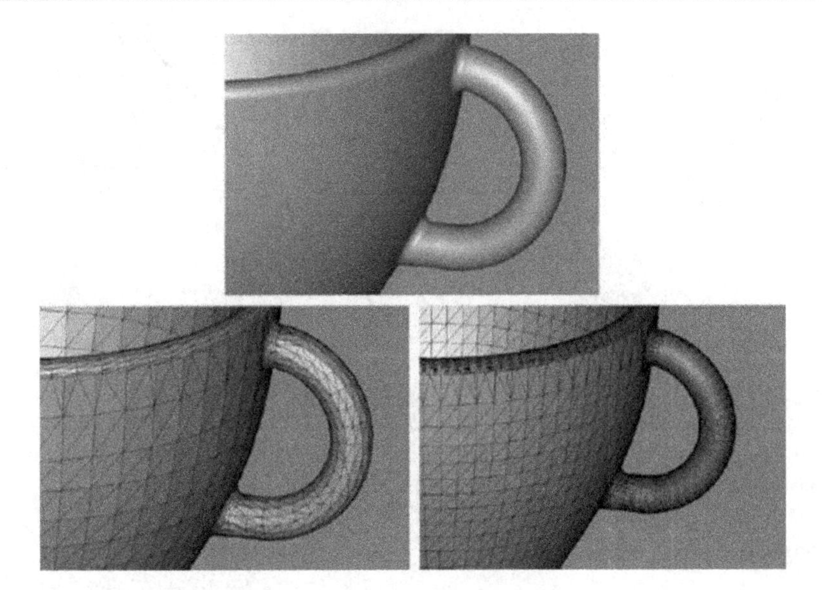

Fig. 15.2 An original CAD model converted into an STL file using different offset height (cusp) values, showing how the model accuracy will change according to the triangle offset

15.2.3 Calculation of Each Slice Profile

Virtually every AM system will be able to read both binary and ASCII STL files. Since most AM processes work by adding layers of material of a prescribed thickness, starting at the bottom of the part and working upwards, the part file description must therefore be processed to extract the profile of each layer. Each layer can be considered a plane in a nominal XY Cartesian frame. Incremental movement for each layer can then be along the orthogonal *Z* axis.

The *XY* plane, positioned along the *Z* axis, can be considered as a cutting plane. Any triangle intersecting this plane can be considered to contribute to the slice profile. An algorithm like the one in Flowchart 15.1 can be used to extract all the profile segments for a given STL file.

The resultant of this algorithm is a set of intersecting lines that are ordered according to the set of intersecting planes. A program that is written according to this algorithm would have a number of additional components, including a way of defining the start and end of each file and each plane. Furthermore, there would be no order to each line segment, which would be defined in terms of the *XY* components and indexed by the plane that corresponds to each *Z* value. Also, the assumption is that the STL file has an arbitrary set of triangles that are randomly distributed. It may be possible to pre-process each file so that searches can be carried out in a more efficient manner. One way to optimize the search would be to order the triangles according to the minimum *Z* value. A simple check for intersection of a triangle with a plane would be to check the *Z* value for each vertex. If the *Z* value of any vertex in the triangle is less than or equal to the *Z* value of the plane,

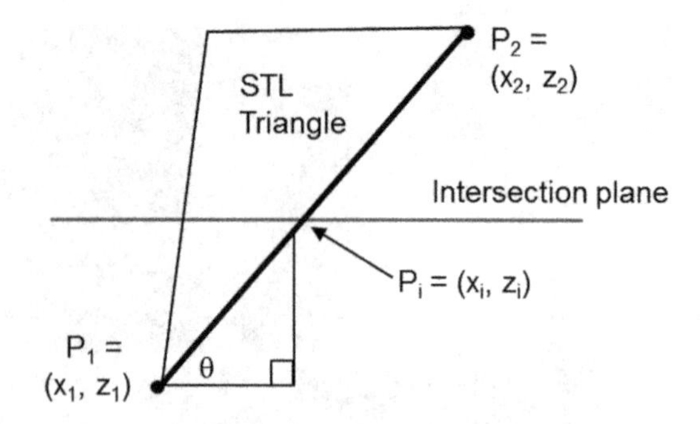

Fig. 15.3 A vertex taken from an STL triangle projected onto the $y=0$ plane. Since the height z_i is known, we can derive the intersection point x_i. A similar case can be done for y_i in the $x=0$ plane

then that triangle may intersect the plane. Using the above test, once it has been established that a triangle does not intersect with the cutting plane, then every other triangle is known to be above that triangle and therefore does not require checking. A similar check could be done with the maximum Z value of a triangle.

There are a number of discrete scenarios describing the intersection of each triangle with the cutting plane:

1. All the vertices of a triangle lie above or below the intersecting plane. This triangle will not contribute to the profile on this plane.
2. A single vertex directly lies on the plane. In this case, there is one intersecting point, which can be ignored but the same vertex will be included in other triangles satisfying another condition below.
3. Two vertices lie on the plane. Here one of the edges of the corresponding triangle lies on that plane and that edge contributes fully to the profile.
4. Three vertices lie on the plane. In this case, the whole triangle contributes wholly to the profile, unless there are one or more adjacent triangles also lying on the plane, in which case the included edges can be ignored.
5. One vertex lies above or below the intersecting plane and the other two vertices lie on the opposite side of the plane. In this case, an intersecting vector must be calculated from the edges of the triangle.

Most triangles will conform to Scenario 1 or 5. Scenarios 2–4 may be considered special cases and require special treatment. Assuming that we have performed appropriate checks and that a triangle corresponds to scenario 5, then we must take action and generate a corresponding intersecting profile vector. In this case, there will be two vectors defined by the triangle vertices and these vectors will intersect with the cutting plane. The line connecting these two intersection points will form part of the outline for that plane.

The problem to be solved is a classical line intersection with a plane problem. In this case, the line is defined using Cartesian coordinates in (x, y, z). The plane is defined in (x, y) for a specific constant height, z. In a general case, we can therefore project the line and plane onto the $x=0$ and $y=0$ planes. For the $y=0$ plane, we can obtain something similar to Fig. 15.3. Points P_1 and P_2 correspond to two points of the intersecting triangle. P_p is the projected point onto the $y = 0$ plane to form a unique right-angled triangle. The angle θ can be calculated from

$$\tan\theta = \frac{(z_2 - z_1)}{(x_2 - x_1)} \tag{15.1}$$

since we know the z height of the plane, we can use the following equation:

$$\tan\theta = \frac{(z_i - z_1)}{(x_i - x_1)} \tag{15.2}$$

and solve for x_i

A point y_i can also be found after projecting the same line on to the $x = 0$ plane to fully define the intersecting point P_i. A second intersecting point can be determined using another line of the triangle that intersects the plane. These two points will make up a line on the plane that forms part of the outline of the model. It is possible to determine directionality of this line by correct use of the right-hand rule, thus turning this line segment into a vector. This may be useful for determining whether a completed curve forms part of an enclosing outline or corresponds to a hole.

Once all intersecting lines have been determined according to Flowchart 15.1, then these lines must be joined together to form complete curves. This would be done using an algorithm based on that described in Flowchart 15.2. In this case, each line segment is tested to determine which segment is closest. A "closest point" algorithm is necessary since calculations may not exactly locate points together, even though the same line would normally be used to determine the start location of one segment and the end of another. Note that this algorithm should really have further nesting to test whether a curve has been completed. If a curve is complete, then any remaining line segments would correspond to additional curves. These additional curves could form a nest of curves lying inside or outside others, or they could be separate. The two algorithms mentioned here focus on the intersection of triangular facets with the cutting plane. A further development of these algorithms could be to use the normal vectors of each triangle. In this way, it would be possible to establish the external direction of a curve. This would be helpful in determining nested curves. The outermost curve will be pointing outside the part. If a curve set is pointing inwards on itself then it is clear there must be a further curve enveloping this one (see Fig. 15.3). Use of the normal vectors may also be helpful in organizing curve sets that are in very close proximity to each other.

Once this stage has been completed, there will be a file containing an ordered set of vectors that will trace complete outlines corresponding to the intersecting plane. How these outlines are used depends somewhat on which AM technology is to be used. Many machines can use the vectors generated in Flowchart 15.2 to control a plotting process to draw the outlines of each layer. However, most machines would

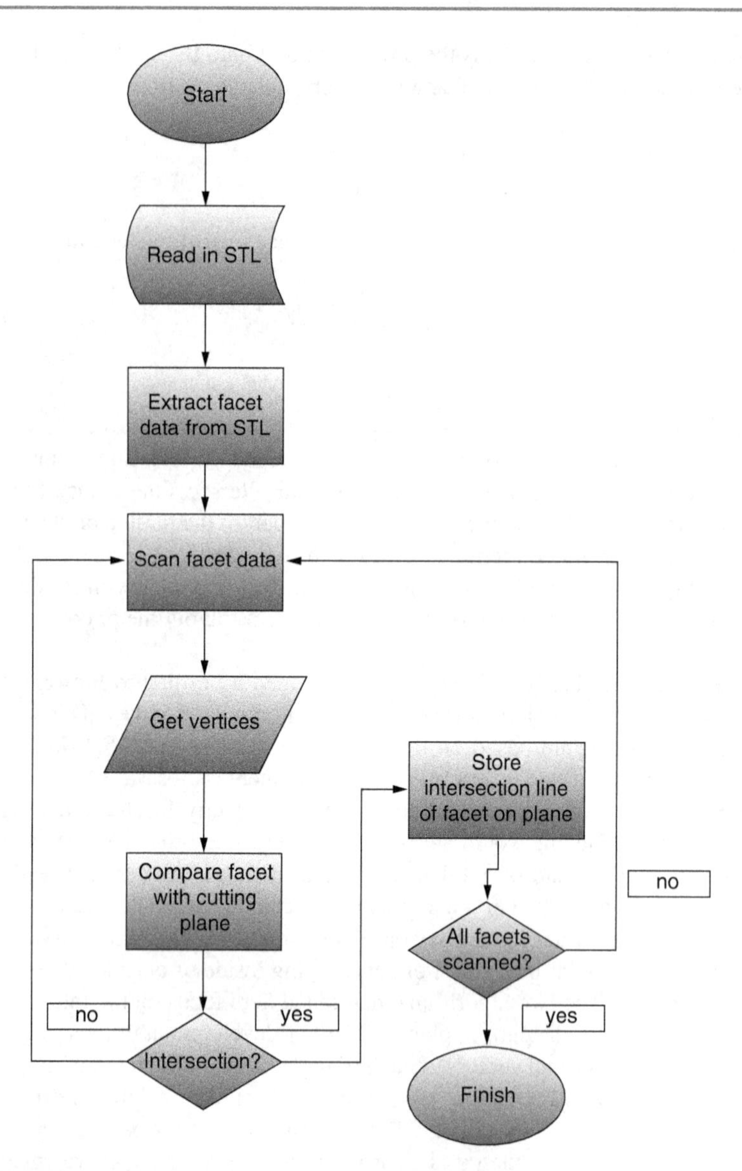

Flowchart 15.1 Algorithm for testing triangles and generating line intersections. The result will be an unordered matrix of intersecting lines (adapted from [4])

also need to fill in these outlines to make a solid. Flowchart 15.3 uses an inside/outside algorithm to determine when to switch on a filling mechanism to draw scanning lines perpendicular to one of the planar axes. The assumption is that the part is fully enclosed inside the build envelope and therefore the default fill is switched off.

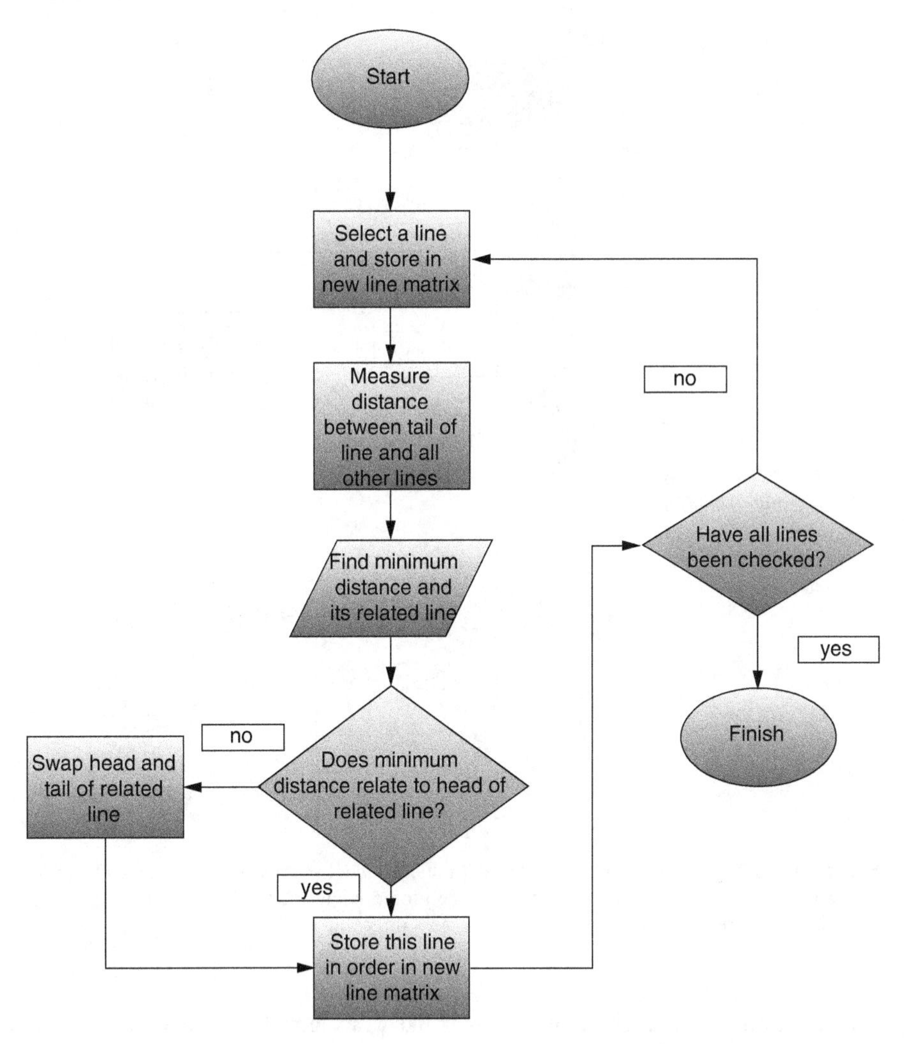

Flowchart 15.2 Algorithm for ordering the line intersections into complete outlines. This assumes there is only a single contour in each plane (adapted from [4])

15.2.4 Technology-Specific Elements

Flowcharts 15.1, 15.2, and 15.3 are basic algorithms that are generic in nature. These algorithms need to be refined to prevent errors and to tailor them to suit a particular process. Other refinements may be employed to speed the slicing process up by eliminating redundancy, for example.

As mentioned in previous chapters, many AM systems require parts built using support structures. Supports are normally a loose-woven lattice pattern of material placed below the region to be supported. Such a lattice pattern could be a simple

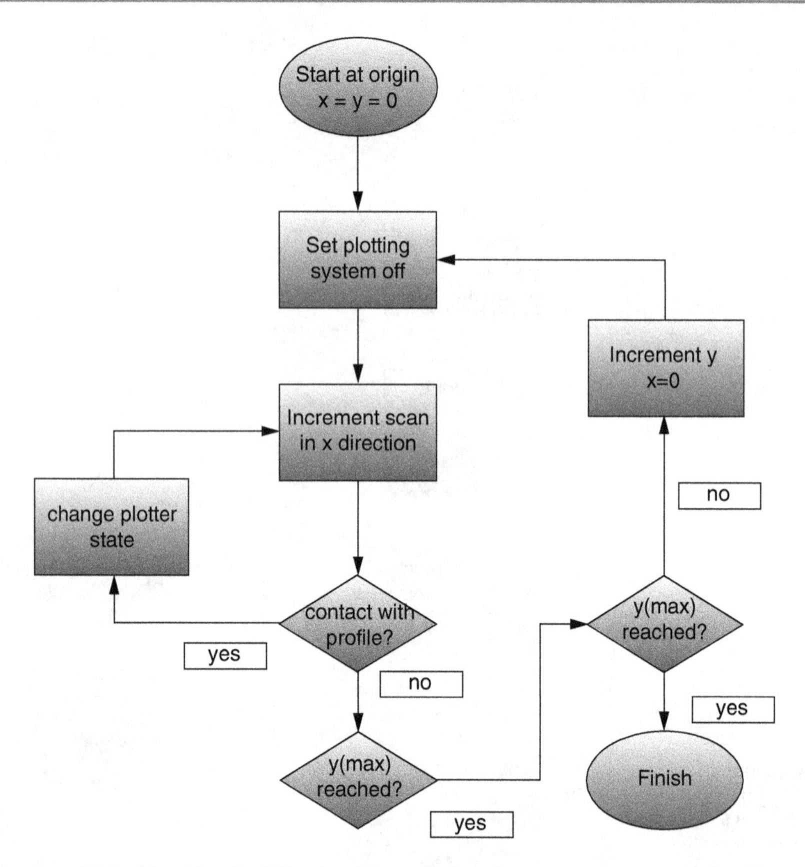

Flowchart 15.3 Algorithm for filling in a 2D profile based on vectors generated using Flowchart 15.2 and a raster scanning approach. Assume the profile fits inside the build volume, the raster scans in the *X* direction, and lines increment in the *Y* direction

square pattern or something more complex like a hexagonal or even a fractal mesh. Furthermore, the lattice could be connected to the part with a tapered region that may be more convenient to remove when compared with thicker connecting edges.

Determination of the regions to be supported can be made by analyzing the angle of the triangle normals. Those normals that are pointing downwards at some previously defined minimum angle would require supports. Those triangles that are sloping above that angle would not require supports. Supports are extended until they intersect either with the base platform or another upward facing surface of the part. Supports connecting with the upward facing surface may also have a taper that enables easy removal. The technique that would normally be used would be to extend supports from the entire build platform and eliminate any supports that do not intersect with the part at the minimum angle or less (see Fig. 15.4).

The support structures could be generated directly as STL models and can be incorporated into the slicing algorithms already mentioned. However, they are more

Fig. 15.4 Supports generated for a part build

likely to be directly generated by a proprietary algorithm within the slicing process. Some other processing requirements that would be dependent for different AM technologies include:

- *Raster scanning*: While many technologies would use a simple raster scan for each layer, there are alternatives. Some systems use a switchable raster scan, scanning in the X direction of an XY plane for one layer and then moving to the Y direction for alternate layers. As discussed in Chap. 5, some systems subdivide the fill area into smaller square regions and use switchable raster scans between squares.
- *Patterned vector scanning*: Material extrusion technology requires a fill pattern to be generated within an enclosed boundary. This is done using vectors generated using a patterning strategy. For a particular layer, a pattern would be determined by choosing a specific angle for the vectors to travel. The fill is then a zigzag pattern along the direction defined by this angle. Once a zigzag has reached an end there may be a need for further zigzag fills to complete a layer (see Fig. 15.5 for an example of zigzag scan pattern).
- *Hatching patterns*: Sheet lamination processes like the, now discontinued, Helisys LOM and the Solid Centre machine from Kira require material surrounding the part to be hatched with a pattern that allows it to be de-cubed once the part has been completed (see Fig. 15.6).

15.3 Problems with STL Files

Although the STL format is quite simple, there can still be errors in files resulting from CAD conversion. The following are typical problems that can occur in bad STL files:

Unit changing: This is not strictly a result of a bad STL file. Since US machines still commonly use imperial measurements and most of the rest of the world uses metric, some files can appear scaled because there is no explicit mention of the units

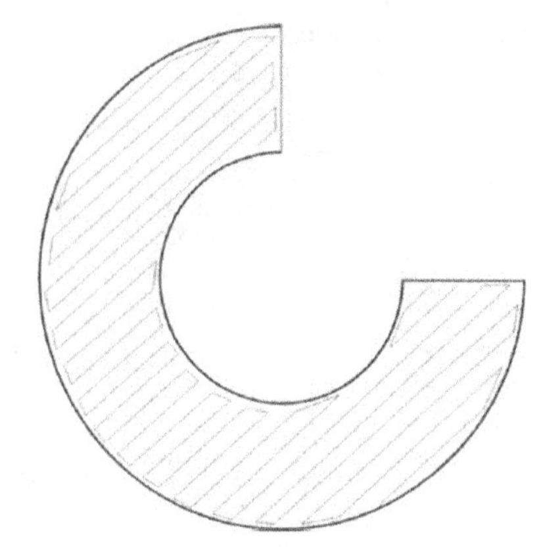

Fig. 15.5 A scan pattern using vector scanning in material extrusion. Note the outline drawn first followed by a small number of zigzag patterns to fill in the space

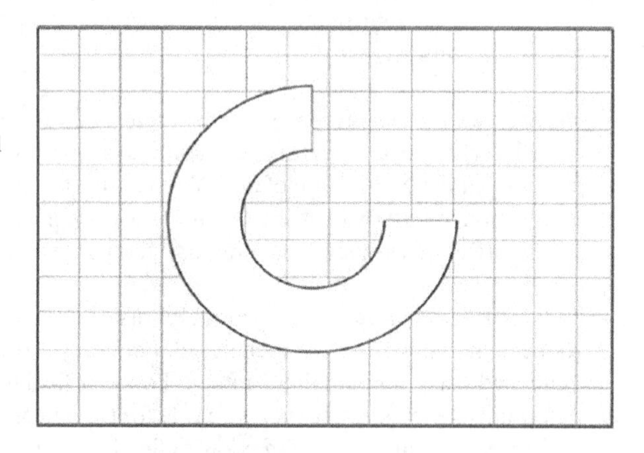

Fig. 15.6 Hatching pattern for LOM-based (sheet lamination) processes. Note the outside hatch pattern that will result in cubes which will be separated from the solid part during post-processing

used in the STL format. If the person building the model is unaware of the purpose of the part then he may build it approximately 25 times too large or too small in one direction. Furthermore, units must correspond to the location of the origin within the machine to be used. This normally means that the physical origin of the machine lies in the bottom left-hand corner and so all triangle coordinates within an STL file must be positive. However, this may not be the case for a particular part made in the CAD system and so some adjustment offset of the STL file may be required.

Vertex to vertex rule: Each triangle must share two of its vertices with each of the triangles adjacent to it. This means that a vertex cannot intersect the side of another, like that shown in Fig. 15.7. This is not something that is explicitly stated in the STL file description and therefore STL file generation may not adhere to this rule. However, a number of checks can be made on the file to determine whether this rule has been violated. For example, the number of faces of a proper solid defined using STL must be an even number. Furthermore, the number of edges must be divisible by three and follow the equation:

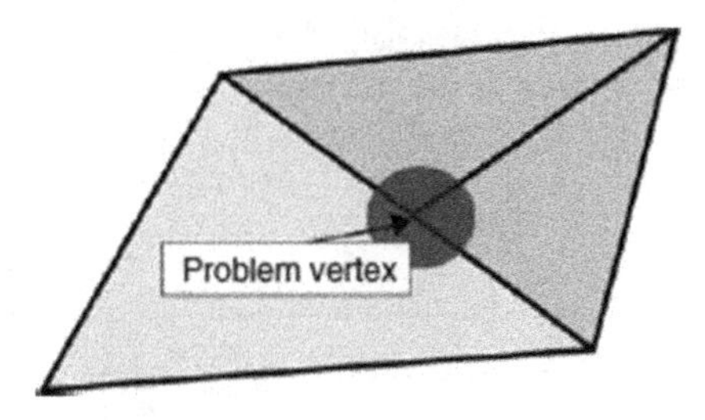

Fig. 15.7 A case that violates the vertex-to-vertex rule

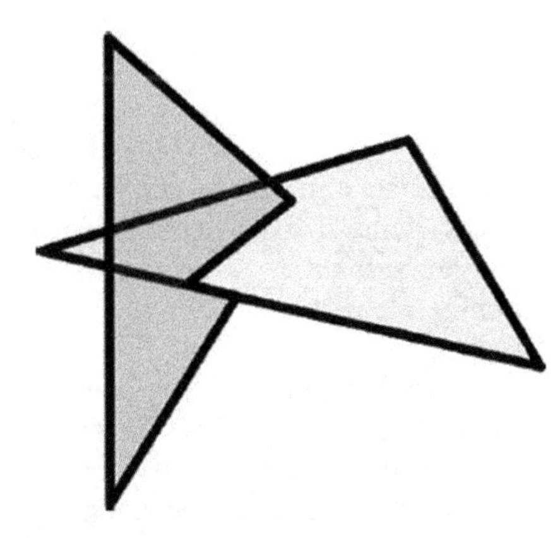

Fig. 15.8 Two *triangles* intersecting each other in 3D space

$$\frac{\text{No. of faces}}{\text{No. of edges}} = \frac{3}{2} \tag{15.3}$$

Leaking STL files: As mentioned earlier, STL files should describe fully enclosed surfaces that represent the solids generated within the originating CAD system. In other words, STL data files should construct one or more manifold entities according to Euler's Rule for solids:

$$\text{No. of faces} - \text{No. of edges} + \text{No. of vertices} = 2 \times \text{No. of bodies} \tag{15.4}$$

If this rule does not hold then the STL file is said to be leaking and the file slices will not represent the actual model. There may be too few or too many vectors for a particular slice. Slicing software may add in extra vectors to close the outline or it may just ignore the extra vectors. Small defects can possibly be ignored in this way. Large leaks may result in unacceptable final models.

Leaks can be generated by facets crossing each other in 3D space as shown in Fig. 15.8. This can result from poorly generated CAD models, particularly those that do not use Boolean operations when generating solids.

A CAD model may also be generated using a method which stitches together surface patches. If the triangulated edges of two surface patches do not match up with each other, then holes, like in Fig. 15.9, may occur.

Degenerated facets: These facets normally result from numerical truncation. A triangle may be so small that all three points virtually coincide with each other.

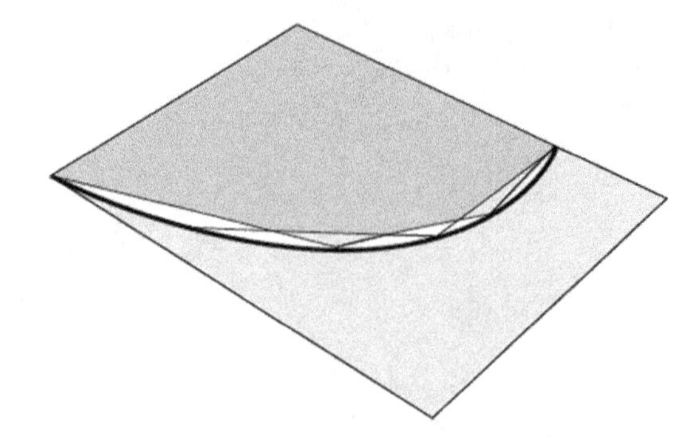

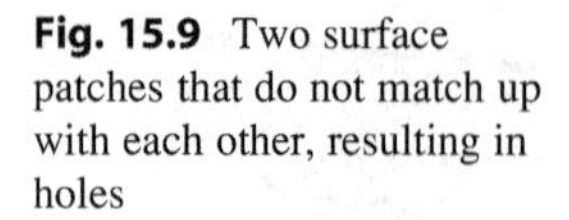

Fig. 15.9 Two surface patches that do not match up with each other, resulting in holes

After truncation, these points lay on top of each other causing a triangle with no area. This can also occur when a truncated triangle returns no height and all three vertices of the triangle lie on a single straight line. While the resulting slicing algorithm will not cause incorrect slices, there may be some difficulties with any checking algorithms and so such triangles should really be removed from the STL file.

It is worth mentioning that, while a few errors may creep into some STL files, most professional 3D CAD systems today produce high-quality and error-free results. In the past, problems more commonly occurred from surface modeling systems, which are now becoming scarcer, even in fields outside of engineering CAD-like computer graphics and 3D gaming software. Also, in earlier systems, STL generation was not properly checked and faults were not detected within the CAD system. Nowadays, potential problems are better understood and there are well-known algorithms for detecting and correcting such problems. However, the recent surge in home-use 3D Printers has resulted in a large selection of software routines that are freely available but not thoroughly tested and could suffer from the problems described in this chapter.

15.4 STL File Manipulation

Once a part has been converted into STL there are only a few operations that can be performed. This is because the triangle-based definition does not permit radical changes to the data. Associations between individual triangles are through the shared points and vertices only. A point or vertex can be moved, which will affect the connected triangles, but creating a regional affect on larger groups of points would be more difficult. Consider the modeling of a simple geometry, like the cut cylinder in Fig. 15.10a. Making a minor change in one of the measurements may result in a very radical change in distribution of the triangles. While it is possible to simplify the model by reducing the number of triangles, it is quite easy to see that defining boundaries in most models cannot be easily done. The addition of a fillet in Fig. 15.10b shows an even more radical change in the STL file. Furthermore, if one

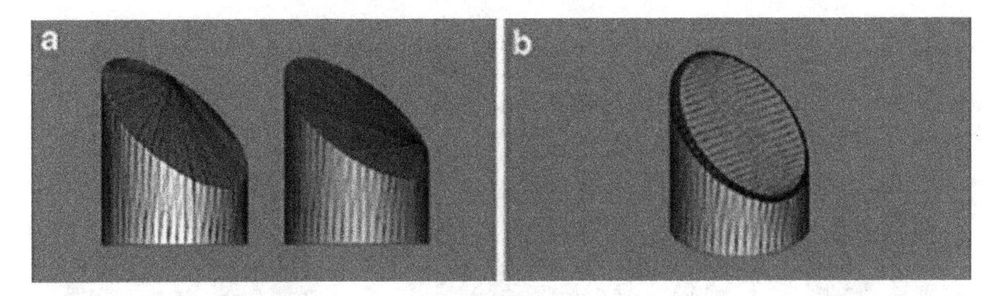

Fig. 15.10 STL files of a cut cylinder. Note that although the two models in (**a**) are very similar, the location of the triangles is very different. Addition of a simple filet in (**b**) shows even greater change in the STL file

were to attempt to move the oval that represents the cut surface, the triangles representing the filet would no longer show a constant-radius curve.

Building models using AM is often done by people working in departments or companies that are separate from the original designers. It may be that whoever is building the model may not have direct access to the original CAD data. There may therefore be a need to modify the STL data before the part is to be built. The following sections discuss STL tools that are commonly used.

15.4.1 Viewers

There are a number of STL viewers available, often as a free download. An example is STLview from Marcam [5] (see Fig. 15.11). Like many other systems, this software allows limited access to the STL file, making it possible to view the triangles, apply shading, show sections, etc. By purchasing the full software version, other tools are possible, for example, allowing the user to measure the part at various locations, annotate the part, display slice information, and detect potential problems with the data. Often the free tools allow passive viewing of the STL data, while the more advanced tools permit modification of the data, either by rewriting the STL or supplying additional information with the STL data (like measurement information, for example). Often these viewers are connected to part building services and provided by the company as an incentive to use these services and to help reduce errors in data transfer, either from incorrect STL conversion or from wrong interpretation of the design intent.

15.4.2 STL Manipulation on the AM Machine

STL data for a part consist of a set of points defined in space, based on an arbitrarily selected point of origin. This origin point may not be appropriate to the machine the part is to be built on. Furthermore, even if the part is correctly defined within the machine space, the user may wish to move the part to some other location or to

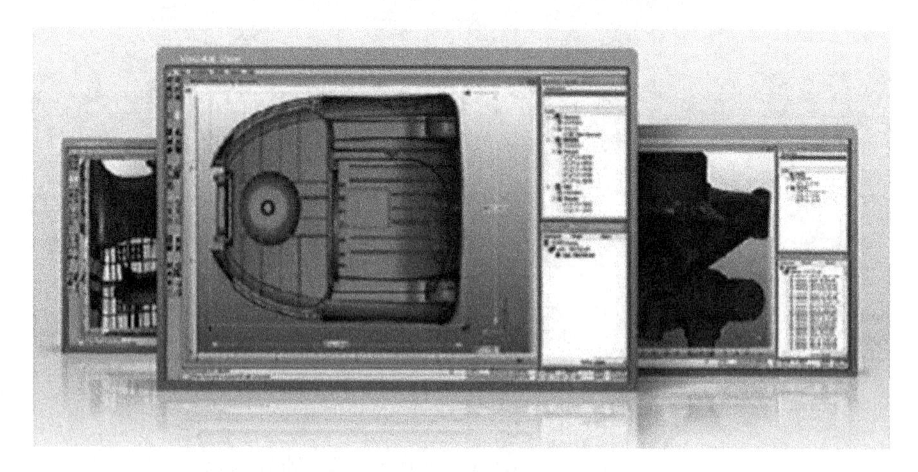

Fig. 15.11 The VisCAM viewer from Marcam that can be used to inspect STL models

make a duplicate to be built beside the original part. Other tasks, like scaling, changing orientation, and merging with other STL files are all things that are routinely done using the STL manipulation tools on the AM machine.

Creation of the support structures is also something that would normally be expected to be done on the AM machine. This would normally be done automatically and would be an operation applied to downward-facing triangles. Supports would be extended to the base of the AM machine or to any upward facing triangle placed directly below. Triangles that are only just veering away from the vertical (e.g., less than 10°) may be ignored for some AM technologies. Note, for example, the supports generated around the cup handle in Fig. 15.4.

With some AM operating systems there is little or no control over placement of supports or manipulation of the model STL data. Considering Fig. 15.4 again, it may be possible to build the handle feature without so many supports, or even with no supports at all. A small amount of sagging around the handle may be evident, but the user may prefer this to having to clean up the model to remove the support material. If this kind of control is required by the user, it may be necessary to purchase additional third party software, like the MAGICS and 3-matic systems from Materialise [6].

Such third party software may also be used to undertake additional roles. MAGICS, for example, has a number of modules useful to many AM technologies. Other STL file manipulators may have similar modules:

- Checking the integrity of STL files based on the problems described above.
- Incorporating support structures including tapered features on the supports that may make them easier to remove.
- Optimizing the use of AM machines, like ensuring the machine is efficiently filled with parts, the amount of support structures is minimized, etc.

- Adding in features like serial numbers and identifying marks onto the parts to ensure correct identification, easy assembly, etc.
- Remeshing STL files that may have been created using Reverse Engineering software or other non-CAD-based systems. Such files may be excessively large and can often be reduced in size without compromising the part accuracy.
- Segmenting large models or combining multiple STL files into a single model data set.
- Performing Boolean tasks like subtracting model data from a tool insert blank model to create a mold.

15.5 Beyond the STL File

The STL definition was created by 3D Systems right at the start of the development history of AM technology and has served the industry well. However, there are other ways in which files can be defined for creation of the slice. Furthermore, the fact that the STL file only represents the surface geometry may cause problems for parts that require some heterogeneous content. This section will discuss some of the issues surrounding this area.

15.5.1 Direct Slicing of the CAD Model

Since generation of STL files can be tedious and error-prone, there may be some benefit from using inbuilt CAD tools to directly generate slice data for the AM machines. It is a trivial task for most 3D solid modeling CAD systems to calculate the intersection of a plane with a model, thus extracting a slice. This slice data would ordinarily need to be processed to suit the drive system of the AM technology, but this can be handled in most CAD systems with the use of macros. Support structures can be generated using standard geometry specifications and projected onto the part from a virtual representation of the AM machine build platform.

Although this approach has never been a popular method for creating slice data, it has been investigated as a research topic [7] and even developed to suit a commercialized variant of the Stereolithography process by a German company called Fockle & Schwarz. The major barrier to using this approach is that every CAD system must include a suite of different algorithms for direct slicing for a variety of machines or technologies. This would be a cumbersome approach that may require periodic updates of the technology as new machines become available. There may be some benefit in the future in creating an integrated design and manufacturing solution, especially for niche applications or for low-cost solutions. However, at present it is more sensible to separate the development of the design tools from that of the AM technology by using the STL format.

15.5.2 Color Models

Currently, several AM technologies are available on the market that can produce parts with color variations, including full color output, including the color binder jetting technology from 3D Systems, material jetting machines from 3D Systems and Stratasys, and sheet lamination systems from Mcor Technologies. Colored parts have proven to be very popular, and it is likely that other color AM machines will make their way to the market. The conventional STL file contains no information pertaining to the color of the part or any features thereon. Coloring of STL files is possible and there are in fact color STL file definitions available [8], but you would be limited by the fact that a single triangle can only be one specific color. It is therefore much better to use the VRML painting options that allow you to assign bitmap images to individual facets [8]. In such a way, it is possible to take advantage of the color possibilities that the AM machine can give you.

15.5.3 Multiple Materials

Carrying on from the previous section, color is one of the simplest examples of multiple material products that AM is capable of producing. As has been mentioned in other chapters, parts can be made using AM from composite materials, with varying levels of porosity or indeed with regions containing discretely different materials. For many of these new AM technologies, STL is starting to become an impediment. Since the STL definition is for surface data only, the assumption therefore is that the solid material between these surfaces is homogeneous. As we can see from the above this may not be the case. While there has been significant thought applied to the problem of representations for heterogeneous solid modeling [9], there is still much to be considered before we can arrive at a standard to supersede STL for future AM technology, as discussed in Sect. 15.7.

15.5.4 Use of STL for Machining

STL is used for applications beyond just converting CAD to additive manufacturing input. Reverse Engineering packages can also be used to convert point cloud data directly into STL files without the need for CAD. Such technology connected directly to AM could conceivably form the basis for a 3D Fax machine. Another technology that can easily make use of STL files is subtractive manufacturing.

Subtractive manufacturing systems can readily make use of the surface data represented in STL to determine the boundaries for machining. With some additional knowledge concerning the dimensions of the starting block of material, tool, machining center, etc., it is possible to calculate machining strategies for creating a 3D surface model. As mentioned in earlier chapters, the likelihood is that it may not be possible to fully machine complex geometries due to undercutting features, internal features, etc. but there is no reason why STL files cannot be used to create

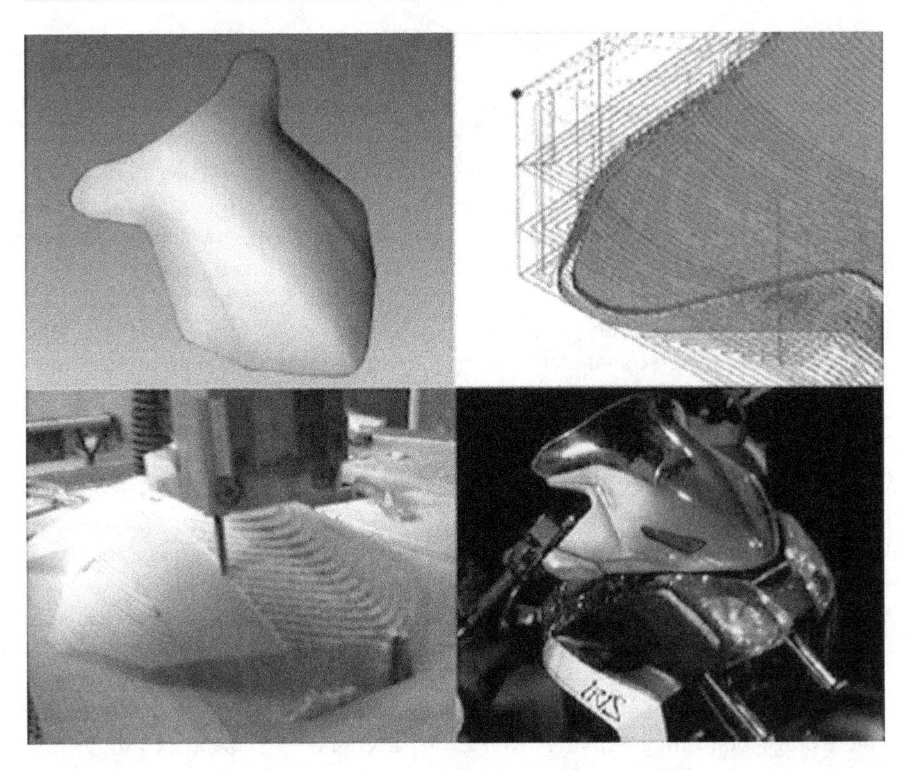

Fig. 15.12 DeskProto software being used to derive machine tool paths from STL file data to form a mold for creating the windscreen of a motorcycle

Computer Aided Manufacturing (CAM) profiles for machining centers. Delft Spline [10] has been using STL files to create CAM profiles for a number of years now. Figure 15.12 shows the progression of a model through to a tool to manufacture a final product using their DeskProto software. Another technology that uses a hybrid of subtractive and additive processes to fabricate parts is the SRP (Subtractive Rapid Prototyping) technique developed by Roland [11] for their desktop milling machines. The creation of finish machining tool paths for AM parts from STL files is also discussed in Chap. 14.

15.6 Additional Software to Assist AM

As well as directly controlling the manufacturing process, other software systems may be helpful in running an effective and efficient AM-based facility.

15.6.1 Survey of Software Functions

Such software can include one or more of the following functions:

Simulation: Many operating systems can perform a simulation of the machines operations in a build process, showing how the layers will be formed step by step in accelerated time. This can allow the user to detect obvious errors in the slice files and determine whether critical features can be built. This may be particularly important for processes like material extrusion where the hatch patterns can have a critical effect on thin wall features, for example. Some work has been carried out to simulate AM systems to get a better impression of the final result, including rendered images to give an understanding of the surface roughness for a given layer thickness, for example [12].

Build-time estimation: AM is a highly automated process and the latest machines are very reliable and can operate unattended for long periods of time. For effective process planning it is very important to know when a build is going to be completed. Knowing this will help in determining when operators will be required to change over jobs. Good estimates will also help to balance builds; adding or subtracting a part from the job batch may ensure that the machine cycle will complete within a day-shift, for example, making it possible to keep machines running unattended at night. Also, if you are running multiple machines, it would be helpful to stagger the builds throughout the shift to optimize the manual work required. Early build-time estimation software was extremely unreliable, performing rolling calculations of the average build time per layer. Since the layer time is dependent on the part geometry, such estimates could be very imprecise and vary wildly, especially at the beginning of a build. Later software versions saw the benefit in having more precise build-time estimations. A simplified build-time model is discussed in Chap. 16.

Machine setup: While every AM machine has an operating system that makes it possible to set up a build, such systems can be very basic, particularly in terms of manipulation of the STL files. Determining build parameters based on a specific material is normally very comprehensive however.

Monitoring: This is a relatively new feature for most AM systems. Even though nearly every AM machine will be connected either directly or indirectly to the Internet, this has traditionally been for uploading of model files for building. Export of information from the machine to the Internet or within an Intranet has not been common except in the larger, more expensive machines. The simplest monitoring systems would provide basic information concerning the status of the build and how much longer before it is complete. However, more complex systems may tell you about how much material is remaining, the current status parameters like temperatures, laser powers, etc. and whether there is any need for manual intervention through an alerting system. Some monitoring systems may also provide video feedback of the build.

Planning: Having a simulation of the AM process running on a separate computer may be helpful to those working in process planning. Process planners may be able to determine what a build could look like, thus allowing the possibility of planning for new jobs, variability analysis, or quoting.

Once again such software may be available from the AM system vendor or from a third party vendor. The advantage of third party software is that it is more likely to be modified to suit the exact requirements of the user.

15.6.2 AM Process Simulations Using Finite Element Analysis

Finite element analysis (FEA) techniques are increasingly popular tools to predict how the outcome of various manufacturing processes change with changing process parameters, geometry and/or material. Commercial software packages such as SYSWELD, COMSOL, MoldFlow, ANSYS, and DEFORM are used to predict the outcomes of welding, forming, molding, casting, and other processes.

AM technologies are particularly difficult to simulate using predictive finite element tools. For instance, the multiscale nature of metal powder bed fusion approaches such as metal laser sintering and electron beam melting are incredibly time consuming to accurately simulate using physics-based FEA. AM processes are inherently multiscale in nature, and fine-scale finite element meshes that are 10 μm or smaller in size are required to accurately capture the solidification physics around the melt pool, while the overall part size can be 10,000 times larger than the element size. In three dimensions, this means that if we apply a uniform 10 μm mesh size, we would need 10^8 elements in the first layer and more than 10^{12} elements in total to capture the physics for a single part that fills much of a powder bed. Since rapid movement of a point heat source is used to create parts, capturing the physics requires a time step of 10 ms or less during laser/electron beam melting, which for a complete build would require more than 10^{10} total time steps. To solve a problem with this, number of elements for this many time steps on a relatively high-speed supercomputer would take billions of years. Thus, to date all AM simulation tools are limited to predictions of only a small fraction of a part, or very small, simplified geometries.

Using existing FEA tools, several researchers are looking for ways to make assumptions whereby they can cut and paste solutions from simplified geometries to form a solution for large, complex geometries. This approach has the benefit of faster solution time; however, for large, complex geometries these types of predictions fail to accurately capture the effects of changing scan patterns, complex accumulation of residual stresses, and localized thermal characteristics. In addition, minor changes to input conditions can make the simplified solutions invalid. Thus a simulation infrastructure which can quickly build up a "new" answer for any arbitrary geometry, input condition, and scan pattern is the ultimate goal for a predictive AM simulation tool.

Recently, researchers have begun using dynamic, multiscale moving meshes to accelerate FEA analysis for AM [13]. These types of multiscale simulations are many orders of magnitude faster than standard FEA simulations. However multiscale simulations alone are still too slow to enable a complete part simulation, even on the world's fastest supercomputers. Thus a new computational approach for AM is needed, which can extend FEA beyond its historical capabilities. 3DSIM, a new software start-up, is seeking to do that for the AM industry. 3DSIM software tools include: (1) a new approach to formulating and solving multiscale moving meshes, (2) a novel finite element-based Eigensolver which predicts thermal evolution and residual stresses very quickly in regions of low thermal gradients, (3) an insignificant number truncation Cholesky module which eliminates the "multiplication by zero" calculations that occur when solving sparse finite element matrices, and (4) an Eigenmodal approach to identifying periodicity in AM computations to enable feed-forward "insertion" of solutions into regions where periodicity is present and the solution is already known from a prior time step. These approaches are reported to reduce the solution time for large-scale AM problems by orders of magnitude. Taken together, these tools should make the full part problem solvable in less than a day on a desktop GPU-based supercomputer when the algorithms are fully implemented in a combined software infrastructure. If realized, the ability to predict how process parameter and material changes affect part accuracy, distortion, residual stress, microstructure, and properties would be a significant advancement for the AM industry.

15.7 The Additive Manufacturing File Format

It has already been discussed that, while effective, there are numerous difficulties surrounding the STL format. As AM technologies move forward to include multiple materials, lattice structures, and textured surfaces, it is likely that an alternative format will be required. The ASTM Committee F42 on Additive Manufacturing Technologies released the ASTM 2915-12 AMF Standard Specification for AMF Format 1.1 in May 2011 [14]. This file format is still very much under development, but has already been implemented in some commercial and beta-stage software. Considerably more complex than the STL format, AMF aims to embrace a whole host of new part descriptions that have hindered the development of current AM technologies. These include the following features:

Curved triangles: In STL, the surface normal lies on the same plane as the triangle vertices that it is connected to. However, in AMF, the start location of the normal vector does not have to lie on the same plane. If so, the corresponding triangle must be curved. The definition of curvature is such that all triangle edges meeting at that vertex are curved so that they are perpendicular to that normal and in the plane defined by the normal and the original straight edge (i.e., if the original triangle had straight edges rather than curved ones). By specifying the triangles in this way, many fewer triangles need be used for a typical CAD model. This addresses problems associated with large STL files resulting from complex geometry models for high-resolution systems. The curved triangle approach is still an approximation since the degree of curvature cannot be too high. Overall accuracy is however significantly improved in terms of cusp height deviation.

Color: Color can be assigned in a nested way so that the main body of the part can be colored according to a function within the original design. Red, Green, and Blue coloration can be applied along with a transparency value to vertices, triangles, volumes, objects, or materials. Note that many AM processes, like vat photopolymerization, can make clear parts, so the transparency value can be an effective parameter. Color values may work along with other materials-based parameters to provide a versatile way of controlling an AM process.

Texture: The above color assignment cannot deal directly with image data assigned to objects. This can however be achieved using the texture operator. Texture is assigned first geometrically, by scaling it to the feature so that individual pixels apply to the object in a uniform way. These pixels will have intensity, which are then assigned color. It should be noted that this is an image texturing process similar to computer graphics rather than a physical texture, like ridges or dimples.

Material: Different volumes can be assigned to be made using different materials. Currently the Connex machines from Stratasys and some other extrusion-based systems have the capacity to build multiple material parts. At the moment designing parts requires a tedious redefinition process within the machine's operating system. By having a material definition within AMF, it is possible to carry this all the way through from the design stage.

Material variants: AMF operators can be used to modify the basic structure of the part to be fabricated. For example, many medical and aerospace applications may require a lattice or porous structure. An operator can be applied so that a specified volume can be constructed using an internal lattice structure or porous material. Furthermore, some AM technologies would be capable of making parts from materials that gradually blend with others. A periodic operator can be applied to a surface that will turn it into a physical texture, rather than the color mapping mentioned earlier. It is even possible to apply a random operator to provide unusual effects to the AM part.

It should be noted that a part designed and coded using the AMF will almost certainly look differently when built using different machines. This will be particularly so for parts that are coded according to different materials, colors, and textures. Each machine will have the capacity to accept and interpret the AMF design according to functionality. For example, if a part is defined with a fine texture, a lower resolution process will not be able to apply it so well. Opaque materials will not be able to make much use of the transparency function. Some machines will not be able to create parts with multiple materials, and so on. It should also be noted that machines should all be able to accept the geometry definition and make something of the AMF defined part. AMF is therefore backwards compatible so that it can recognize a simple STL file, but with the capacity to specify any conceivable design in the future.

15.8 Exercises

1. How would you adjust Flowchart 15.2 to include multiple contours?
2. Under what circumstances might you want to merge more than one STL file together?

3. Write out an ASCII STL file for a perfect cube, aligned with the Cartesian coordinate frame, starting at (0, 0, 0) and all dimensions positive. Model the same cube in a CAD system. Does it make the same STL file? What happens when you make slight changes to the CAD design?
4. Why might it be possible that a part could inadvertently be built 25 times too small or too large in any one direction?
5. Is it okay to ignore the vertex of a triangle that lies directly on an intersecting cutting plane?
6. Prove to yourself with some simple examples that the number of faces divided by the number of edges is 2/3.
7. Using the Wikipedia description of AMF [15] consider how you would code an airplane wing model that has honeycomb internal lattice for lightweight and some colored decorations on the outer skin. How might the part look using a color ZPrinter compared to using an SL machine. What design considerations would you still have to make to ensure the part is properly made on these machines?

References

1. 3D Systems Inc (1989) Stereolithography interface specification. October 1989
2. Autodesk CAD software. usa.autodesk.com
3. Rhinoceros CAD software. http://www.rhino3d.com/
4. Vatani M, Rahimi AR et al (2009) An enhanced slicing algorithm using nearest distance analysis for layer manufacturing. Proceedings of World Academy of Science, Engineering and Technology, vol 37, pp 721–726, Jan 2009. ISSN: 2070-3740
5. Marcam, VisCAM software. http://www.marcam.de
6. Materialise, AM software systems. http://www.materialise.com/materialise/view/en/92074-Magics.html
7. Jamieson R, Hacker H (1995) Direct slicing of CAD models for rapid prototyping. Rapid Prototyping J 1(2):4–12, Pub. by Emerald, ISSN 1355-2546
8. Ling WM, Gibson I (2002) Specification of VRML in color rapid prototyping. Int J CAD/CAM 1(1):1–9
9. Siu YK, Tan ST (2002) Representation and CAD modeling of heterogeneous objects. Rapid Prototyping J 8(2):70–75, ISSN 1355-2546
10. Delft Spline. http://www.spline.nl/dp/deskproto.html
11. Roland, desktop milling and subtractive RP. http://www.rolanddga.com/ASD/
12. Choi SH, Samavedam S (2001) Visualisation of rapid prototyping. Rapid Prototyping J 7 (2):99–114
13. Zeng K, Pal D, Patil N, Stucker B (2013) A new dynamic mesh method applied to the simulation of selective laser melting. Twenty fourth Annual international solid freeform fabrication symposium—an additive manufacturing conference, Austin, TX, August 2013
14. AMF, F2915-12 File Format, v1.1. http://www.astm.org/Standards/F2915.htm
15. Wikipedia AMF description. http://en.wikipedia.org/wiki/Additive_Manufacturing_File_Format

Direct Digital Manufacturing 16

Abstract

Direct digital manufacturing (DDM) is a term that describes the usage of additive manufacturing technologies for production or manufacturing of end-use components. Although it may seem that DDM is a natural extension of rapid prototyping, in practice this is not usually the case. Many additional considerations and requirements come into play for production manufacturing that are not important for prototyping. In this chapter, we explore these considerations through an examination of several DDM examples, distinctions between prototyping and production, and advantages of additive manufacturing for custom and low-volume production.

Many times, DDM applications have taken advantage of the geometric complexity capabilities of AM technologies to produce parts with customized geometries. In these instances, DDM is not a replacement for mass production applications, as customized geometry cannot be mass produced using traditional manufacturing technologies. In addition, since the economics of AM technologies do not enable economically competitive high-volume production for most geometries and applications, DDM is often most economical for low-volume production applications. Two major individual-specific medical applications of DDM will be discussed, from Align Technology and Siemens/Phonak, as well as several other applications that make use of the unique design freedom afforded by AM techniques. This will be followed by a discussion of the unique characteristics of AM technologies that lead to DDM.

16.1 Align Technology

Align Technology, in Santa Clara, California, is in the business of providing orthodontic treatment devices (www.aligntech.com). Their Invisalign treatments are essentially clear braces, called aligners, that are worn on the teeth (see Fig. 16.1). Every 1–2 weeks, the orthodontic patient receives a new set of aligners

I. Gibson et al., *Additive Manufacturing Technologies*,
DOI 10.1007/978-1-4939-2113-3_16

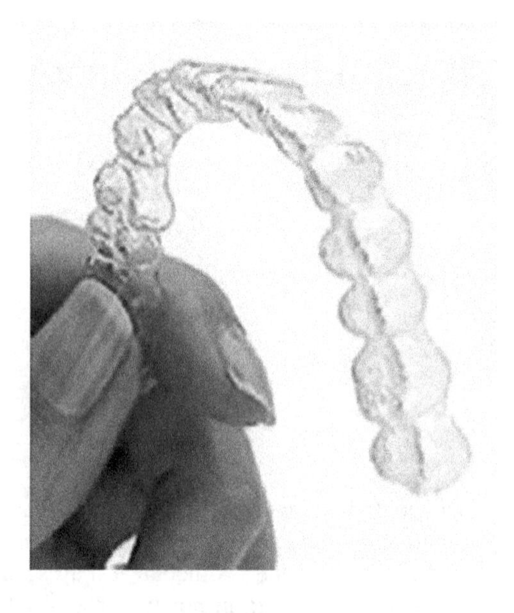

Fig. 16.1 Aligner from Align Technology (Courtesy Align Technology)

that are intended to continue moving their teeth. That is, every 1–2 weeks, new aligners that have slightly different shapes are fabricated and shipped to the patient's orthodontist for fitting. Over the total treatment time (several months to a year typically), the aligners cause the patient's teeth to move from their initial position to the position desired by the orthodontist. If both the upper and lower teeth must be adjusted for 6 months, then 26 different aligners are needed for one patient, assuming that aligners are shipped every 2 weeks.

The need for many different geometries in a short period of time requires a mass customization approach to aligner production. Align's manufacturing process has been extensively engineered. First, the orthodontist takes an impression of the patient's mouth with a typical dental clay. The impression is shipped to Align Technology where it is scanned using a laser digitizer. The resulting point cloud is converted into a tessellation (set of triangles) that describes the geometry of the mouth. This tessellation is separated into gums and teeth, then each tooth is separated into its own set of triangles. Since the data for each tooth can be manipulated separately, an Align Technology technician can perform treatment operations as prescribed by the patient's orthodontist. Each tooth can be positioned into its desired final position. Then, the motion of each tooth can be divided into a series of treatments (represented by different aligners). For example, if 13 different upper aligners are needed over 6 months, the total motion of a tooth can be divided into 13 increments. After manipulating the geometric information into specific treatments, aligner molds are built in one of Align's SLA-7000 stereolithography (SL) machines. The aligners themselves are fabricated by thermal forming of a sheet of clear plastic over SL molds in the shape of the patient's teeth.

The aligner development process is geographically distributed, as well as highly engineered. Obviously, the patient and orthodontist are separated from Align

Technology headquarters in California. Their data processing for the aligners is performed in Costa Rica, translating customer-specific, doctor-prescribed tooth movements into a set of aligner models. Each completed dataset is transferred electronically to Align's manufacturing facility in Juarez Mexico, where the dataset is added into a build on one of their SL machines. After building the mold using VP from the dataset, the molds are thermal formed. After thermal forming, they are shipped back to Align and, from there, shipped to the orthodontist or the patient.

Between its founding in 1997 and March, 2009, over 44 million aligners have been created (www.aligntech.com). At present, Align's SL machines are able to operate 24 h per day, producing approximately 100 aligner molds in one SLA-7000 build, with a total production capacity of 45,000–50,000 unique aligners per day (~17 million per year) [1]. As each aligner is unique, they are truly "customized." And by any measure, 45,000 components per day is mass production and not prototyping. Thus, Align Technology represents an excellent example of "mass customization" using direct digital manufacturing (DDM), albeit in a tooling application.

To achieve mass customization, Align needed to overcome the time-consuming pre- and post-processing steps in SL usage. A customized version of 3D Systems Lightyear control software was developed, called MakeTray; to automate most of the build preparation. Aligner mold models are laid out, supports are generated, process variables are set, and the models are sliced automatically. Typical post-processing steps, including rinsing and post-curing can take hours. Instead, Align developed several of its own post-processing technologies. They developed a rinsing station that utilizes only warm water, instead of hazardous solvents. After rinsing, conveyors transport the platforms to the special UV post-cure station that Align developed. UV lamps provide intense energy that can post-cure an entire platform in 2 min, instead of the 30–60 min that are typical in a Post-Cure Apparatus unit. Platforms traverse the entire post-processing line in 20 min. Support structures are removed manually at present, although this step is targeted for automation. The Align Technology example illustrates some of the growing pains experienced when trying to apply technologies developed for prototyping to production applications.

16.2 Siemens and Phonak

Siemens Hearing Instruments, Inc. (www.siemens-hearing.com) and Phonak Hearing Systems are competitors in the hearing aid business. In the early 2000s, they teamed up to investigate the feasibility of using polymer powder bed fusion (PBF) technology in the production of shells for hearing aids [2]. A typical hearing aid is shown in Fig. 16.2. The production of hearing aid shells (housings that fit into the ear) required many manual steps. Each hearing aid must be shaped to fit into an individual's ear. Fitting problems cause up to 1 out of every 4 hearing aids to be returned to the manufacturer, a rate that would be devastating in most other industries.

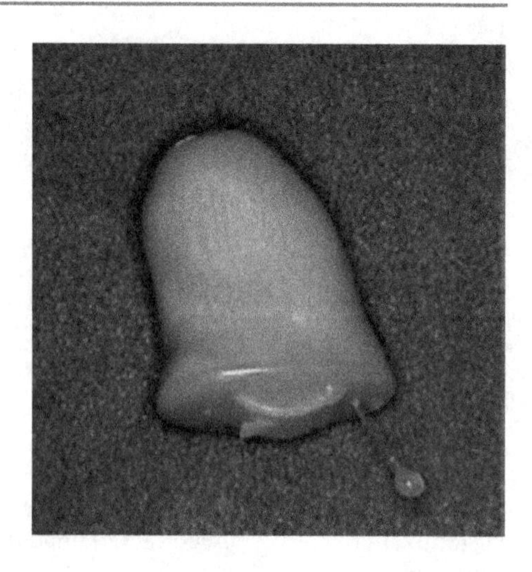

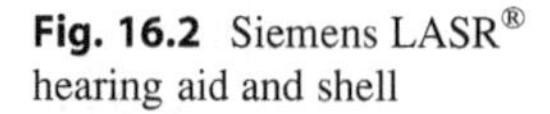

Fig. 16.2 Siemens LASR® hearing aid and shell

Traditionally, an impression is taken of a patient's ear, which is then used as a pattern to make a mold for the hearing aid shell. An acrylic material is then injected into the mold to form the shell. Electronics, controls, and a cover plate are added to complete the hearing aid. To ensure proper operation and comfort, hearing aids must fit snugly, but not too tightly, into the ear and must remain in place when the patient talks and chews (which change the geometry of the ear).

To significantly reduce return rates and improve customer satisfaction, Siemens and Phonak sought to redesign their hearing aid production processes. Since AM technologies require a solid CAD model of the design to be produced, the companies had to introduce solid modeling CAD systems into the production process. Impressions are still taken from patients' ears, but are scanned by a laser scanner, rather than used directly as a pattern. The point cloud is converted into a 3D CAD model, which is manipulated to fine-tune the shell design so that a good fit is achieved. This CAD shell model is then exported as an STL file for processing by an AM machine. A scanned point cloud is shown superimposed on a hearing aid model in Fig. 16.3.

In the mid-2000s, Siemens developed a process to produce shells using vat photopolymerization (VP) technology to complement their PBF fabrication capability. VP has two main advantages over PBF. First, VP has better feature detail, which makes it possible to fabricate small features on shells that aid assembly to other hearing aid components. Second, acrylate VP materials are similar to the materials originally used in the hearing aid industry (heat setting acrylates), which are biocompatible. As mentioned, Siemens originally adopted PBF fabrication; PBF has strengths in that the nylon polyamide materials typically used in PBF are biocompatible and the surface finish of PBF parts aided hearing aid retention in the ear, since the finish had a powder-bed texture.

In the late 2000s, Siemens Hearing Instruments produced about 250,000 hearing aids annually. In 2007, they claimed that about half of the in-the-ear hearing aids that they produced in the USA were fabricated using AM technologies. Since the

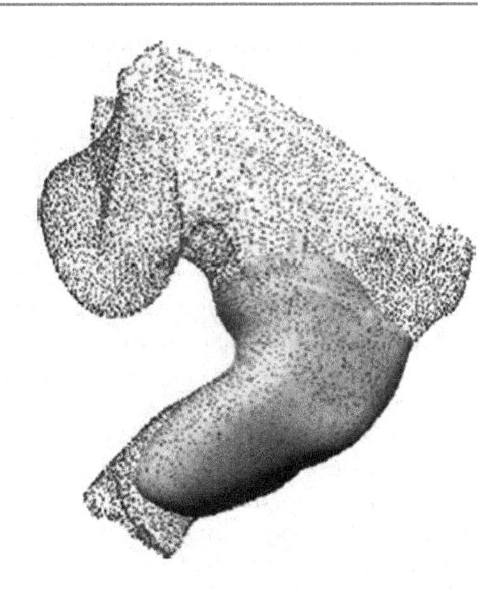

Fig. 16.3 Hearing aid within scanned point cloud

introduction of AM-fabricated hearing aid shells, most hearing aid manufacturers in the Western world have adopted AM in order to compete with Siemens and Phonak. Recent surveys estimate that about 90 % of all custom in-the-ear hearing aid shells are fabricated using AM, which totaled approximately two million AM fabricated shells in 2012 [1]. Since the adoption of AM, hearing aid return rate has fallen dramatically with improved design and manufacturing processes.

A number of technology advances have targeted this market. 3D Systems developed a variant of its SLA Viper Si2 machine to manufacture shells, called the SLA Viper HA. The machine contains two small vats, one with a red-tinted resin and the other with a blue-tinted resin. The idea is to fabricate both the left and right hearing aid shells for a patient in one build, where each shell is a different color, enabling the patient to easily distinguish them. Of course, the resins can be swapped with flesh-colored resin in both vats, if desired by a patient. EnvisionTEC has also focused on this market, developing several VP machines that are designed for hearing aid shells. Additionally, companies such as Rapid Shape, in Germany, and Carima, in South Korea, have entered the market with mask projection VP machines.

The hearing aid shell production is a great example of how companies can take advantage of the shape complexity capability of RP technologies to economically achieve mass customization. With improvements in scanning technology, it is likely that patients' ears can be scanned directly, eliminating the need for impressions [3]. If desktop AM systems can be developed, it may even be possible to fabricate custom hearing aids in the audiologist's office, rather than having to ship impressions or datasets to a central location!

16.3 Custom Footwear and Other DDM Examples

A British company called Prior 2 Lever (P2L) claims to be manufacturing the world's first custom soccer shoes for professional athletes. Polymer PBF is used to fabricate the outsoles, including cleats, for individual customers [4]. The one-piece leather uppers are also custom tailored. A model called the Assassin retailed in 2008 for £6,000 per pair; a photo is shown in Fig. 16.4. Research on PBF outsoles for custom shoes started in the early 2000s at Loughborough University; Freedom of Creation and others contributed to the development of this work. Early testing demonstrated a significant reduction in peak pressures during walking and running with personalized outsoles [5]. Custom sprinting shoes and triathlete shoes are also being developed.

In 2013, Nike developed a line of football cleats called the Vapor Laser Talon. The cleat plate was specially designed to improve player performance, particularly for speed positions. It has an intricate, lattice design and was fabricating using PBF in a proprietary polymer material.

The examples presented so far all relate to body-fitting, customized parts. However, many other opportunities exist, even in the medical arena. Many companies worldwide are investigating the use of PBF technologies for the creation of orthopedic implants. For instance, Adler Ortho Group of Italy is using Arcam's EBM system to produce stock sizes of acetabular cups for hip implants made from Ti–6Al–4V. The use of AM techniques enables a more compact design and a better transition between the solid bearing surface and the porous bone-ingrowth portion of the implant. Although a porous coating of titanium beads or hydroxyapatite on an implant's surface work well, they do not provide the optimum conditions for osseointegration. The hierarchical structure capabilities of AM enable the creation of a more optimal bone-ingrowth structure for osseointegration. As of early 2014, more than 40,000 cups have been implanted and more than 90,000 implants have been produced in series production by companies such as Adler Ortho and Lima Corporate SpA. More generally, approximately 20 different AM medical implant products have received FDA clearance for implantation in patients [1].

Fig. 16.4 Assassin model soccer shoe. Courtesy prior 2 lever

Low-volume production is often economical via AM since hard tooling does not need to be developed. This has led to the rapid adoption of DDM for low-volume components across many industries. However, the most exciting aspects of DDM are the opportunities to completely rethink how components can be shaped in order to best fulfill their functions, as discussed in Chap. 13. Integrated designs can be produced that combine several parts, eliminate assembly operations, improve performance by designing parts to utilize material efficiently, eliminate shape compromises driven by manufacturing limitations, and completely enable new styles of products to be produced. This can be true for housewares, every-day items, and even customized luxury items, as illustrated in Chap. 17 with respect to houseware and fashion products. Each of these areas will be explored briefly in this section. Many of the examples were taken, or cited, in recent Wohlers Reports [1, 6].

Stratasys developed a new class of material extrusion (ME) machines in 2007, the Fortus X00mc series. They introduced the Fortus 900mc in December and reported that 32 parts on the machine were fabricated on their ME machines. This is a novel example of how AM producers are using their own technologies in low-volume production, and the savings that can be achieved by not having to invest in tooling. Since introducing the 900mc, Stratasys has marketed several new models, presumably using in-house fabricated parts on these models as well. This is also true for other major AM manufacturers, including EOS and 3D Systems.

The aerospace industry has been the source of quite a few successful examples of DDM. The F-18 fighter jet example from Chap. 17, where Boeing and Northrop Grumman manufacture many nonstructural components using polymer PBF, is one such case. In addition, SAAB Avitronics has used polymer PBF to manufacture antenna RF boxes for an unmanned aircraft. Advantages of this approach over conventional manufacturing processes include a more compact design, 45 % reduction in mass, and integral features. Paramount Industries, a division of 3D Systems, produced PBF parts for a helicopter, including ventilation parts and electrical enclosures, and structures for unmanned aerial vehicles. The parts were manufactured on their EOSINT P 700 machine from EOS using the PA 2210 FR material (flame retardant). Additionally, thousands of parts are flying on the space shuttle, space station, and various military aircraft.

Many of the most promising commercial aircraft applications are delayed until better flame retardant materials were certified for commercial use. Now, with flame retardant PBF materials, Boeing has developed PBF components on commercial 737, 747 777, and 787 programs. In addition, large numbers of PBF components are present on several military derivative aircrafts, such as the Airborne Early Warning and Control (AEWC), C-40, AWACS, and P-8 aircraft. All told, tens of thousands of parts are flying on at least eight different military and eight different civilian models of Boeing aircraft. As another example, Northrop Grumman has identified more than 1,400 parts on a single fighter aircraft platform that could be better made using PBF than traditional methods if a suitable material with higher-temperature properties were available.

In the automotive industry, examples of DDM are emerging. Formula 16.1 teams have been using AM technologies extensively on their racecars for several years. Applications include electrical housings, camera mounts, and other aerodynamic parts. In the late 2000s, the Renault Formula 1 team used over 900 parts on racecars each racing season. Indy and NASCAR teams also make extensive use of AM parts on their cars.

Local Motors conducted a crowd-sourced car design project, with the requirement that the majority of the car would be fabricated by AM. Specifically, they intend to use a hybrid additive/subtractive machine under development at the Oak Ridge National Laboratories. The machine has a large diameter material extrusion head and subtractive machining capabilities. As of May 2014, voting on the various car modules (body, internal structure, etc.) was completed. Fabrication of body and structural components using the ORNL machine is planned for the International Machine Technology Show in September 2014.

Several automotive manufacturers use AM parts on concept cars and for other purposes. Hyundai used PBF to fabricate flooring components for their QarmaQ concept car in 2007, with assistance from Freedom of Creation. Bentley uses PBF to produce some specialty parts that are subsequently covered in leather or wood. Others use AM to fabricate replacement parts for antique cars, including Jay Leno's famous garage (www.jaylenosgarage.com). BMW uses ME extensively in production, as fixtures and tooling for automotive assembly.

In consumer-oriented industries, many specialty applications are beginning to emerge. Many service bureaus do DDM runs for customized or other specialty components. An interesting class of applications is emerging to bridge the virtual and physical worlds. The World of Warcraft is probably the largest online video game. Players can design their own characters for use in the virtual world, often adding elaborate clothing, accessories, and weapons. A company called FigurePrints (www.figureprints.com) produces 100 mm (4 in.) tall models of such characters; one example was shown in Fig. 3.3. They use binder jetting (BJ) machines from 3D Systems (formerly ZCorp BJ machines), with color printing capabilities, and sell characters for around $100 USD. Similarly, they also market fabrication services for constructions in Minecraft, a "world building" game that is very popular with children. Jujups offers custom Christmas ornaments, printed with a person's photograph. Again, color printers from 3D Systems are used for production. In many of these applications, AM can utilize the input data only after it has been converted to a usable form, as the original data was created to serve a visual purpose and not necessarily as a representation of a true 3D object. However, software producers are beginning to consider AM as an output of their games from the outset, as is evident by the Spore video game, which enables users to create characters that are fully defined in three dimensions and thus can be converted into data usable by AM technologies in a straightforward manner. Within a year after the release of the game, it was announced that players could have a 3D printout of their character, using color BJ machines, for less than $50 USD.

The gaming industry alone accounts for hundreds of millions of unique 3D virtual creations that consumers may want to have made into physical objects.

Just as the development of computer graphics has often been driven by the gaming industry, it is appearing equally likely that the further development of color DDM technologies may also be driven by the market opportunities which are enabled by the gaming industry. Quite a few online games allow players to create or customize their own characters. Tools such as Maya are used to create/modify geometry, colors, textures, etc. which are then uploaded to the game site (e.g., Dota 2 from Valve). These same files can be packaged and sent to AM service providers, such as Shapeways and Sculpteo, for printing color models of the custom characters. Make Magazine ran a very informative article on this topic in 2013 [7].

In addition to the previous lines, many other DDM applications are emerging; including in the medical and dental industries, which will be discussed further in Chap. 15. Other examples are covered in Chaps. 2, 11 and 17.

16.4 DDM Drivers

It is useful to generalize from these examples and explore how the unique capabilities of AM technologies may lead to new DDM applications. The factors that enable DDM applications include:

- Unique Shapes: parts with customized shapes.
- Complex Shapes: improved performance.
- Lot Size of One: economical to fabricate customized parts.
- Fast Turnaround: save time and costs; increase customer satisfaction.
- Digital Manufacturing: precisely duplicate CAD model.
- Digital Record: have reusable dataset.
- Electronic "Spare Parts": fabricate spare parts on demand, rather than holding inventory.
- No Hard Tooling: no need to design, fabricate, and inventory tools; economical low-volume production.

As indicated in the Align Technology and hearing aid examples, the capability to create customized, *unique geometries* is an important factor for DDM. Many AM processes are effective at fabricating platforms full of parts, essentially performing mass customization of parts. For example, 100 aligner molds fit on one SLA-7000 platform. Each has a unique geometry. Approximately 25–30 hearing aid shells can fit in the high-resolution region of an SLA Viper Si2 machine. Upwards of 4,000 hearing aid shells can be built in one PBF powder bed in one build. The medical device industry is a leading—and growing—industry where DDM and rapid tooling applications are needed due to the capability of fabricating patient-specific geometries.

The capability of building parts with *complex geometries* is another benefit of DDM. Features can be built into hearing aid shells that could not have been molded in, due to constraints in removing the shells from their molds. In many cases, it is possible to combine several parts into one DDM part due to AM's complexity capabilities. This can lead to tremendous cost savings in assembly tooling and assembly operations that would be required if multiple parts were fabricated using conventional manufacturing processes. Complexity capabilities also enable new design paradigms, as discussed with respect to acetabular cups and as seen in Chap. 17. These new design concepts will be increasingly realized in the near future.

Related to the unique geometry capability of AM, economical *lot sizes of one* are another important DDM capability. Since no tooling is required in DDM, there is no need to amortize investments over many production parts. DDM also avoids the extensive process planning that can be required for machining, so time and costs are often significantly reduced. These factors and others help make small lot sizes economical for DDM.

Fast turnaround is another important benefit of DDM. Again, little time must be spent in process planning, tooling can be avoided, and AM machines build many parts at once. All these properties lead to time savings when DDM is used. It is common for hearing aid manufacturers to deliver new hearings aids in less than 1 week from the time a patient visits an audiologist. Align Technology must deliver new aligners to patients every 1–2 weeks. Rapid response to customer needs is a hallmark of AM technologies and DDM takes advantage of this capability.

The capability of *digital manufacturing*, or precisely fabricating a mathematical model, has important applications in several areas. The medical device industry takes advantage of this; hearing aid shells must fit the patient's ear canal well, the shape of which is described mathematically. This is also important in artwork and high-end housewares, where small shape changes dictated by manufacturing limitations (e.g., draft angles for injection molding) may be unwelcome. More generally, the concept of digital manufacturing enables digital archiving of the design and manufacturing information associated with the part. This information can be transferred electronically anywhere in the world for part production, which can have important implications for global enterprises.

A *digital record* is similar in many ways to the digital manufacturing capability just discussed. The emphasis here is on the capability to archive the design information associated with a part. Consider a medical device that is unique to a patient (e.g., hearing aid, foot orthotic). The part design can be a part of the patient's digital medical records, which streamlines record keeping, sharing of records, and fabricating replacement parts.

Another way of explaining digital records and manufacturing, for engineered parts, is by using the phrase "electronic spare parts." The air handling ducts installed on F-18 fighters as part of an avionics upgrade program may be flying for another 20 years. During that time, if replacement ducts are needed, Boeing must manufacture the spare parts. If the duct components were molded or stamped, the molds or stamps must be retrieved from a warehouse to fabricate some spares. By having digital records and no tooling, it is much easier to fabricate the spare parts using AM processes; plus the fabrication can occur wherever it is most convenient. This flexibility in selecting fabrication facilities and locations is impossible if hard tooling must be used.

As mentioned several times, the advantages are numerous and significant to not requiring *tooling* for part fabrication. Note that in cases such as Align Technology, tooling is required, but the tooling itself is fabricated when and where needed, not requiring tooling inventories. The elimination of tooling makes DDM economically competitive across many applications for small lot size production.

16.5 Manufacturing Versus Prototyping

Production manufacturing environments and practices are much more rigorous than prototyping environments and practices. Certification of equipment, materials, and personnel, quality control, and logistics are all critical in a production environment. Even small considerations like part packaging can be much different than in a prototyping environment. Table 16.1 compares and contrasts prototyping and production practices for several primary considerations [8].

Certification is critical in a production environment. Customers must have a dependable source of manufactured parts with guaranteed properties. The DDM company must carefully maintain their equipment, periodically calibrate the equipment, and ensure it is always running within specifications. Processes must be engineered and not left to the informal care of a small number of skilled technicians. Experimentation on production parts is not acceptable. Meticulous records must be kept for quality assurance and traceability concerns. Personnel must be fully trained, cross-trained to ensure some redundancy, and certified to deliver quality parts.

Most, if not all, DDM companies are ISO 9000 compliant or certified. ISO 9000 is an international standard for quality systems and practices. Most customers will require such ISO 9000 practices so that they can depend on their suppliers. Many books have been written on the ISO standards so, rather than go into extensive detail here, readers should utilize these books to learn more about this topic [9].

As mentioned, personnel should be trained, certified, and periodically retrained and/or recertified. Cross-training personnel on various processes and equipment helps mitigate risks of personnel being unavailable at critical times. If multiple shifts are run, these issues become more important, since the quality must be consistent across all shifts.

Vertical integration is important, since many customers will want their suppliers to be "one stop shops" for their needs. DDM companies may rely on their own suppliers, so the supplier network may be tiered. It is up to the DDM company, however, to identify their suppliers for specialty operations, such as bonding, coating, and assembly, and ensure that their suppliers are certified.

The bottom line for a company wanting to break into the DDM industry is that they must become a production manufacturing organization, with rigorous practices. Having an informal, prototyping environment, even if they can produce high-quality prototypes, is not sufficient for success in the current DDM industry. Standard production business practices must be adopted.

Table 16.1 Contrast between rapid prototyping and direct digital manufacturing[a]

Key characteristic	RP company	DDM company
Certification		
Equipment	From equipment manufacturer	Production machines and calibration equipment
Personnel	No formal testing, certification, or training typical	On-going need for certification
Practices	Trial-and-error, no formal documentation of practices	Formal testing for each critical step, periodic recertification
Quality	Basic procedures; some inspection	ISO 9000 compliance. Extensive, thorough quality system needed
Manufacturing		
System	Basic system; controls and documentation not essential	Developed system; controls and documentation required
Planning	Basic. Requires only modest part assessment	Formal planning to ensure customer requirements are met. Developed process chains, no experimentation
Scheduling and delivery	Informally managed; critical jobs can be expedited; usually only one delivery date	Sophisticated scheduling, just-in-time delivery
Personnel	Informal training, on-the-job training; certification not necessary; redundancy not essential	Formal training for certification and periodic recertification. Redundant personnel needed for risk mitigation
Vertical integration	Helpful	From customer's perspective, should be a one-stop-shop. Qualified suppliers must be lined up ahead of time to enable integration

[a]Much of this section was adapted from Brian Hasting's presentation at the 2007 SME RAPID Conference [8]

Several industry standards, e.g., ISO 9000, have been available to assist companies in adopting quality and certification best practices. Until recently, few AM-specific standards were available, which meant that:

- Material data reported by various companies are not comparable.
- Technology users employ different process parameters to operate their equipment according to their own preferences.
- There is little repeatability of results between suppliers or service bureaus.
- There are few specifications which can be referenced by end users to help them ensure that a product is built as-desired.

In 2008, an international standards-development initiative was organized by the Society of Manufacturing Engineering; ASTM International was selected to be the organization that the AM community would work with to develop standards. The first meeting of the ASTM F42 committee on standards for Additive Manufacturing Technologies was held in May, 2009. At present, several standards have been adopted and many more are under development. Standard terminology, the AMF, reporting data for test specimens, and some specifications for PBF fabricated parts have been developed. Additional standards are being developed related to powder characterization, material qualification and traceability, parts fabricated by material extrusion and other processes, and design guidelines. With the adoption of an overall framework for AM-related standards, the F42 committee has a plan to develop an array of standards that will address all of the issues raised in the bullet items above.

16.6 Cost Estimation

From a cost perspective, AM can appear to be much more expensive for part manufacture than conventional, mass production processes. A single part out of a large VP or PBF machine can cost upwards of $5,000, if the part fills much of the material chamber. However, if parts are smaller, the time and cost of a build can be divided among all the parts built at one time. For small parts, such as the hearing aid shell, costs can be only several dollars or less. In this section, we will develop a simple cost model that applies to production manufacturing. A major component of costs is the time required to fabricate a set of parts; as such, a detailed build time model will be presented.

16.6.1 Cost Model

Broadly speaking, costs fall into four main categories: machine purchase, machine operation, material, and labor costs. In equation form, this high level cost model can be expressed, on a per-build basis, as:

$$\text{Cost} = P + O + M + L \tag{16.1}$$

or, on a per-part basis, as

$$\text{Cost} = p + o + m + l = 1/N \times (P + O + M + L) \tag{16.2}$$

where, P = machine purchase cost allocated to the build, O = machine operation cost, M = material cost, L = labor cost, N = number of parts in the build, and the lower-case letters are the per-part costs corresponding to the per-build costs expressed using capital letters. An important assumption made in this analysis is that all parts in one build are the same kind of part, with roughly the same shape and size. This simplifies the allocation of times and costs to the parts in a build.

Machine purchase and operations costs are based on the build time of the part. We can assume a useful life of the machine, denoted Y years, and apportion the purchase price equally to all years. Note that this is a much different approach than would be taken in a cash-flow model, where the actual payments on the machine would be used (assuming it was financed or leased). A typical up-time percentage needs to be assumed also. For our purposes, we will assume a 95 % up-time (the machine builds parts 95 % of the time during a year). Then, purchase price for one build can be calculated as:

$$P = \frac{\text{Purchase price} \times T_\text{b}}{0.95 \times 24 \times 365 \times Y} \tag{16.3}$$

where T_b is the time for the build in hours and 24×365 represents the number of hours in a year. Operation cost is simply the build time multiplied by the cost rate of the machine, which can be a complicated function of machine maintenance, utility costs, cost of factory floor space, and company overhead, where the operation cost rate is denoted by C_o.

$$O = T_\text{b} \times C_\text{o} \tag{16.4}$$

Material cost is conceptually simple to determine. It is the volume, v, of the part multiplied by the cost of the material per unit mass, C_m, and the mass density, ρ, as given in (16.5). For AM technologies that use powders, however, material cost can be considerably more difficult to determine. The recyclability of material that is used, the volume fraction of the build that is made up of parts versus loose powder (in the case of powder bed techniques) and/or the powder capture efficiency of the process (in the case of directed energy deposition techniques) will result in the need to multiply the volume, v, of the part by a factor ranging from a low of 1.0 to a number as high as 7.0 to accurately capture the true cost of material consumed. Thus, for powder processes where the build material is not 100 % recyclable, material cost has a complex dependency on the various factors mentioned here. The term k_r will be introduced for the purpose of modeling the additional material consumption that considers these factors. In addition, for processes that require support materials (such as ME and VP), the volume and cost of the supports needed to create each part must also be taken into account. The factor k_s takes this into account for such processes; typical values would range from 1.1 to 1.5 to include the extra material volume needed for supports. As a result, the model described in (16.5) will be used for material cost.

$$M = k_\text{s} \times k_\text{r} \times N \times v \times C_\text{m} \times \rho \tag{16.5}$$

Labor cost is the labor rate, C_l in \$/h, multiplied by the time, T_l, required for workers to set up the build, remove fabricated parts, clean the parts, clean the machine, and get the machine ready for the next build.

$$L = T_\text{l} \times C_\text{l} \tag{16.6}$$

16.6.2 Build Time Model

The major variable in this cost model is the build time of the parts. Build time (T_b) is a function of part size, part shape, number of parts in the build, and the machine's build speed. Viewed slightly differently, build time is the sum of scan or deposition time (T_s), transition time between layers (T_t), and delay time (T_e):

$$T_b = T_s + T_t + T_e \tag{16.7}$$

For this analysis, we will assume that we are given the part size in terms of its volume, v, and its bounding box, aligned with the coordinate axes: bb_x, bb_y, bb_z. Layer transition time is the easiest to deal with. The processes that build in material beds or vats have to recoat or deposit more material between layers; other processes do not need to recoat and have a T_r of 0. Recoat times for building support structures can be different than times for recoating when building parts, as indicated by (16.8).

$$T_t = L_s \times T_{ts} + L_p \times T_{tp} \tag{16.8}$$

where L_s is the number of layers of support structure, T_{ts} is the time to recoat a layer of support structures, L_p is the number of layers for building parts ($L_p = bb_z/LT$), T_{tp} is the recoat time for a part layer, and LT is the layer thickness.

Scan/deposition time is a function of the total cross-sectional area for each layer, the scan or fill strategy utilized, and the number of layers. Cross-sectional area depends upon the part volume and the number of parts. Scan/deposition time also depends upon whether the machine has to scan vectors to build the part in a point-wise fashion, as in VP, PBF, ME, or the part deposits material in a wide, line-wise swath, as in material jetting processes, or as a complete layer, as in layer-based (e.g., mask projection) vat photopolymerization processes. The equations are similar; we will present the build time model for scanning and leave the wide swath deposition and layer-based scanning processes for the exercises.

Now, we need to consider the part layout in the build chamber. Assuming a build platform, we have a 2D layout of parts on the platform. Parts are assumed to be of similar sizes and are laid out in a rectangular grid according to their bounding box sizes. Additionally, X and Y gaps are specified so that the parts do not touch. In the event that the parts can nest inside one another, gaps with negative values can be given. A 2D platform layout is shown in Fig. 16.5 showing the bounding boxes of 18 long, flat parts with gaps of 10 mm in the X direction and 20 mm in the Y direction. The number of parts on the platform can be computed as:

$$N = \left(\frac{PL_x + g_x - 20}{bb_x + g_x}\right)\left(\frac{PL_y + g_y - 20}{bb_y + g_y}\right) \tag{16.9}$$

where PL_x, PL_y are the platform sizes in X and Y, g_x, g_y are the X and Y gaps, and the -20 mm terms prevent parts from being built at the edges of the platform (10 mm

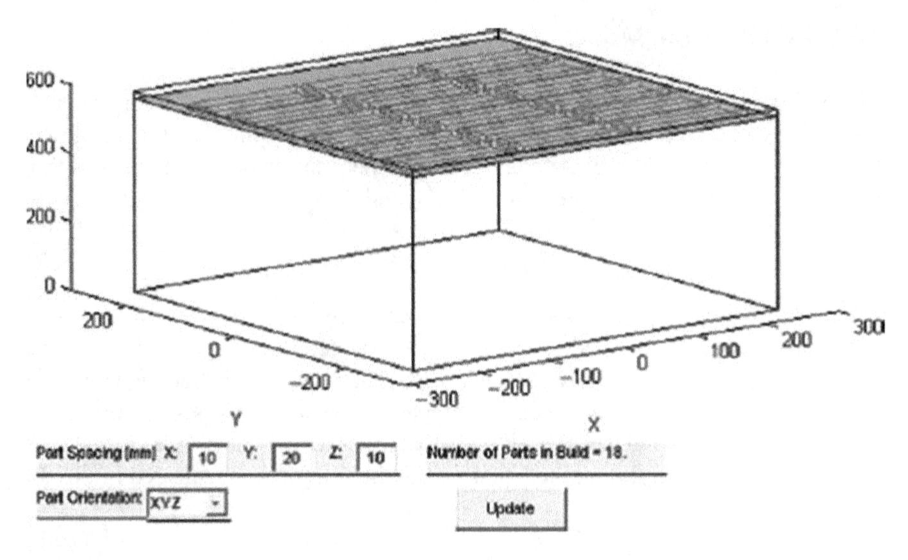

Fig. 16.5 SLA-7000 vat with 18 parts laid out on the platform

buffer area along each platform edge). This analysis can be extended to 3D build chambers for processes which enable stacking in the z direction.

The time to scan one part depends on the part cross-sectional area, the laser or deposition head diameter d, the distance between scans h, and the average scan speed ss_{avg}. Cross-sectional area, A_{avg}, is approximated by using an area correction factor γ [10], which corrects the area based on the ratio of the actual part volume to the bounding box volume, v_{bb}, $\gamma = v/v_{bb}$. The following correction has been shown to give reasonable results in many cases.

$$A_{fn} = \gamma \cdot e^{\alpha(1-\gamma)} \tag{16.10}$$

$$A_{avg} = bb_x \times bb_y \times A_{fn} \tag{16.11}$$

where α is typically taken as 1.5.

For scanning processes, it is necessary to determine the total scan length per layer. This can be accomplished by simply dividing the cross-sectional area by the diameter of the laser beam or deposited filament. Alternatively, the scan length can be determined by dividing the cross-sectional area by the hatch spacing (distance between scans). We will use the latter approach, where the hatch spacing, hr, is given as a percentage of the laser beam diameter. For support structures, we will assume that the amount of support is a constant percentage, supfac, of the cross-sectional area (assumed as about 30 %). If a process does not require supports, then the constant percentage can be taken as 0. The final consideration is the number of times a layer is scanned to fabricate a layer, denoted n_{st}. For example, in stereolithography, both X and Y scans are performed for each layer, while in material extrusion, only one scan is performed to deposit material. Scan length for one part and its support structure is determined using (16.12):

$$\mathrm{sl} = A_{\mathrm{avg}}\left(\frac{n_{\mathrm{st}}L_{\mathrm{p}}}{\mathrm{hr}\cdot d} + \mathrm{supfac}\frac{L_{\mathrm{s}}}{d}\right) \tag{16.12}$$

The final step in determining scan/deposition time is to determine scan speed. This is a function of how fast the laser or deposition head moves when depositing material, ss_{s}, as well as when moving (jumping) between scans, ss_{j}. In some cases, jump speed is much higher than typical scan speeds. To complicate this matter, many machines have a wide range of scan speeds that depend on several part building details. For example, new VP machines have scan speeds that range from 100 to 25,000 mm/s. For our purposes, we will assume a typical scan speed that is half of the maximum speed. The average scan/deposition speed will be corrected using the area correction factor determined earlier [10] as

$$\mathrm{ss}_{\mathrm{avg}} = \mathrm{ss}_{\mathrm{s}} \times \gamma + \mathrm{ss}_{\mathrm{j}}(1-\gamma) \tag{16.13}$$

With the intermediate terms determined, we can compute the scan/deposition time for all parts in the build as:

$$T_{\mathrm{s}} = \frac{N \times \mathrm{sl}}{3{,}600\ \mathrm{ss}_{\mathrm{avg}}} \tag{16.14}$$

where the 3,600 in the denominator converts from seconds to hours.

The final term in the build time expression (16.7) is the delay time, T_{e}. Many processes have delays built into their operations, such as platform move time, pre-recoat delay (T_{predelay}), post-recoat delay ($T_{\mathrm{postdelay}}$), nozzle cleaning, sensor recalibration, temperature set point delays (waiting for the layer to heat or cool to within a specified range), and more. These delays are often user specified and depend upon build details for a particular process. For example, in VP, if parts have many fine features, longer pre-recoat delays may be used to allow the resin to cure further, to strengthen the part, before subjecting fragile features to recoating stresses. Additionally, some processes require a start-up time, for example, to heat the build chamber or warm up a laser. This start-up time will be denoted T_{start}. For our purposes, delays will be given by (16.15), but it is important to realize that each process and machine may have additional or different delay terms.

$$T_{\mathrm{e}} = L_{\mathrm{p}}\left(T_{\mathrm{predelay}} + T_{\mathrm{predelay}}\right) + T_{\mathrm{start}} \tag{16.15}$$

With the cost and build time models presented, we now turn to the application of these models to VP.

Table 16.2 SLA Viper Pro parameters

	Small vat		Largest vat
PL_x (mm)	650		650
PL_y (mm)	350		750
PL_z (mm)	300		550
Purchase price ($ × 1,000)		700	
C_o ($/h)		30	
C_l ($/h)		20	
Y (yrs)		7	
	Border vectors		Hatch vectors
d (mm)	0.13		0.76
ss_s (mm/s)	3,500		25,000
ss_j (mm/s)		$2 \times V_{scan}$	
hr (hatch) (mm)		0.5	
LT (mm)		0.05–0.15	
n_{st}		2	
z_{supp} (mm)		0.10	
supfac		0.3	
$T_{predelay}$ (s)		15	
$T_{postdelay}$ (s)		10	
T_{start} (h)		0.5	
C_m ($/kg)		200	
ρ (g/cm^3)		1.1	

16.6.3 Laser Scanning Vat Photopolymerization Example

The build time and cost models presented in Sect. 16.6.2 will be applied to the case of hearing aid shell manufacturing using an iPro 8000 SLA Center stereolithography machine from 3D Systems with the smallest vat. The machine parameters are given in Table 16.2. Part information will be assumed to be as follows: bounding box = 15 × 12 × 20 mm, $v = 1{,}000$ mm^3. An average cross-sectional area of 45 mm^2 will be assumed, instead of using Eqs. (16.10) and (16.11). Layer thickness for the part is 0.05 mm. Support structures are assumed to be 10 mm tall, built with 0.1 mm layer thickness. Since the shell's walls are small, most of the scans will be border vectors; thus, an average laser beam diameter of 0.21 mm is assumed. Gaps of 4 mm will be used between shells.

With these values assumed and given, the build time will be computed first, followed by the cost per shell. We start with the total number of parts on one platform

$$N = \left(\frac{650 + 4 - 20}{15 + 4}\right)\left(\frac{350 + 4 - 20}{12 + 4}\right) = 1,393$$

The numbers of layers of part and support structure are $L_p = 400$ and $L_s = 100$. The scan length and scan speed average can be computed as: $s_l = 349{,}290$ mm,

$ss_{avg} = 6{,}230$ mm/s (linearly interpolated based on $d = 0.21$ mm). With these quantities, the scan time is:

$$T_s = \frac{1.393 \times 349{,}290}{3{,}600 \times 9986.9} = 13.53\,\mathrm{h} \tag{16.14}$$

Recoat (layer transition) time is

$$T_t = 6/3{,}600 \times (400 + 100) = 0.83333\ \mathrm{h}$$

Delay times total

$$T_e = 400/3{,}600 \times (15 + 10) + 0.5 = 3.278\,\mathrm{h}$$

Adding up the scan, recoat, and delay times gives a total build time of

$$T_b = 25.8\,\mathrm{h}$$

Part costs can be investigated now. Machine purchase price allocated to the build is \$212. Operating cost for 25.8 h is \$774. Material and labor costs for the build are \$245 and \$10, respectively. The total cost for the build is computed to be \$1,241. With 1,393 shells in the build, each shell costs about \$0.89, which is pretty low considering that the hearing aid will retail for \$400 to \$1,500. However these costs do not include support removal and finishing costs, nor the life-cycle costs discussed below.

16.7 Life-Cycle Costing

In addition to part costs, it is important to consider the costs incurred over the lifetime of the part, from both the customer's and the supplier's perspectives. For any manufactured part (not necessarily using AM processes), life-cycle costs associated with the part can be broken down into six main categories: equipment cost, material cost, operation cost, tooling cost, service cost, and retirement cost. As in Sect. 16.6, equipment cost includes the costs to purchase the machine(s) used to manufacture the part. Material and operation costs are related to the actual manufacturing process and are one-time costs associated only with one particular part. For most conventional manufacturing processes, tooling is required for part fabrication. This may include an injection mold, stamping dies, or machining fixtures. The final two costs, service and retirement, are costs that accrue over the lifetime of the part.

This section will focus on tooling, service, and retirement costs, since they have not been addressed yet. Service costs typically include costs associated with repairing or replacing a part, which can include costs related to taking the product out of service, disassembling the product to gain access to the part, repairing or replacing the part, re-assembling the product, and possibly testing the product.

Design-for-service guidelines indicate that parts needing frequent service should be easy to access and easy to repair or replace. Service-related costs are also associated with warranty costs, which can be significant for consumer products.

Let's consider the interactions between service and tooling costs. Typically, tooling is considered for part manufacture. However, tooling is also needed to fabricate replacement parts. If a certain injection molded part starts to fail in aircraft after being in service for 25 years, it is likely that no replacement parts are available "off the shelf." As a result, new parts must be molded. This requires tooling to be located or fabricated anew, refurbished to ensure it is production-worthy, installed, and tested. Assuming the tooling is available, the company would have had to store it in a warehouse for all of those years, which necessitates the construction and maintenance of a warehouse of old tools that may never be used.

In contrast, if the parts were originally manufactured using AM, no physical tooling need be stored, located, refurbished, etc. It will be necessary to maintain an electronic model of the part, which can be a challenge since forms of media become outdated; however, maintenance of a computer file is much easier and less expensive than a large, heavy tool. This aspect of life-cycle costs heavily favors AM processes.

Retirement costs are associated with taking a product out of service, dismantling it, and disposing of it. Large product dismantling facilities exist in many parts of the USA and the world that take products apart, separate parts into different material streams, and separate materials for distribution to recyclers, incinerators, and landfills. The first challenge for such facilities is collecting the discarded products. A good example of product collection is a community run electronic waste collection event, where people can discard old electronic products at a central location, such as a school or mall parking lot. Product take-back legislation in Europe offers a different approach for the same objective. For automobiles, an infrastructure already exists to facilitate disposal and recycling of old cars. For most other industries, little organized product take-back infrastructure exists in the USA, with the exceptions of paper and plastic food containers. In contrast to consumer products, recycling and disposal infrastructure exists for industrial equipment and wastes, particularly for metals, glass, and some plastics.

How recyclable are materials used in AM? Metals are very recyclable regardless of the method used to process it into a part. Thus, stainless steel, titanium alloys, and other metal parts fabricated in PBF and directed energy deposition processes can be recycled. For plastics, the situation is more complicated. The nylon blends used in PBF can be recycled, in principle. However, nylon is not as easily recycled as other common thermoplastics, such as the ABS or polycarbonate materials from ME systems. Thermoset polymers, such as photopolymers in VP and jetting processes, cannot be recycled. These materials can only be used as fillers, landfilled, or incinerated.

In general, the issue of life-cycle costing has simple aspects to it, but is also very complicated. It is clear that the elimination of hard tooling for part manufacture is a significant benefit of AM technologies, both at the time of part manufacture and over the part's lifetime since spare parts can be manufactured when needed. On the other hand, issues of material recycling and disposal become more complicated, reflecting the various industry and consumer practices across society.

16.8 Future of DDM

There is no question that we will see increasing utilization of AM technologies in production manufacturing. In the near-term, it is likely that new applications will continue to take advantage of the shape complexity capabilities for economical low production volume manufacturing. Longer time frames will see emergence of applications that take advantage of functional complexity capabilities (e.g., mechanisms, embedded components) and material complexities.

To date, tens of thousands of parts have been manufactured for the aerospace industry. Many of these parts are flying on military aircraft, space shuttles, the International Space Station, and many satellites. Several small AM service companies have been created to serve the aerospace market. Other service bureaus revamped their operations to compete in this market. The machine vendors have reconceptualized some of their machine designs to better serve manufacturing markets. An example of this is the development of the 3D Systems SinterStation Pro, and the similar public announcements by EOS that all future models of their machines will be designed with production manufacturing in mind. Flame-resistant nylon materials have been developed to enable parts manufacturing for commercial aircraft, as well as higher-temperature and higher-recyclability materials.

Other markets will emerge:

- One needs only consider the array of devices and products that are customized for our bodies to see more opportunities that are similar to aligners and hearing aids. From eye glasses and other lenses to dentures and other dental restorations, to joint replacements, the need for complex, customized geometries, hierarchical structures and complex material compositions is widespread in medical and health related areas.
- New design interfaces for non-experts have enabled individuals to design and purchase their own personal communication/computing device (e.g., cell phones) housings or covers in a manner similar to their current ability to have a physical representation of their virtual gaming characters produced. File sharing sites such as thingiverse.com and storefronts such as shapeways.com are very popular and many expansions and generalizations are to be expected.
- Structural components will have embedded sensors that detect fatigue and material degradation, warning of possible failures before they occur.
- The opportunities are bounded only by the imagination of those using AM technologies.

In summary, the capability to process material in an additive manner will drastically change some industries and produce new devices that could not be manufactured using conventional technologies. This will have a lasting and profound impact upon the way that products are manufactured and distributed, and thus on society as a whole. A further discussion of how DDM will likely affect business models, distributed manufacturing and entrepreneurship is contained in the final chapter of this book.

16.9 Exercises

1. Estimate the build time and cost for a platform of 100 aligner mold parts in an iPro 8000 SLA Center (see Chap. 4). Assume that the bounding box for each part is 11 × 12 × 8 cm and the mold volume is 75,000 mm^3. Assume a scanning speed of 5,000 mm/s and a jump speed of 20,000 mm/s. All remaining quantities are given in Sect. 16.6. What is the estimated cost per mold (two parts)?
2. A vat of hearing aid shells is to be built in an SLS Pro 140 machine (build platform size: 550 × 550 × 460 mm). How many hearing aid shells can fit in this build platform? Determine the estimated build time and cost for this build platform full of shells. Assume laser scan and jump speed of 5,000 mm/s and 20,000 mm/s, respectively. Assume the laser spot size is 0.2 mm, layer thicknesses are 0.1 mm, and only 1 scanning pass per layer is needed ($n_{st} = 1$). Assume 4 mm gaps in X, Y, and Z directions. Recall that no support structures are needed. Assume that the SLS machine needs 2 h to warm up and 2 h to cool down after the build. Assume that $T_{predelay}$ is 15 s and $T_{postdelay}$ is 2 s.
3. Develop a build time model for a jetting machine, such as the Eden models from Stratasys or the ProJets from 3D Systems. Note that this is a line-type process, in contrast to the point-wise vector scanning process used in VP or PBF. Consider that the jetting head can print material during each traversal of the build area and n_{st} may be 2 or 3 (e.g., two or three passes of the head are required to fully cover the total build area). Assume that $T_{predelay}$ and $T_{postdelay}$ are 2 s.
4. Estimate the build time and cost for a platform of hearing aid shells in an Eden 500 V machine (see Chap. 7). What is the estimated cost per shell? You will need to visit the Stratasys web site and possibly contact Stratasys personnel in order to acquire all necessary information for computing times and costs.
5. Develop a build time model for an ME machine from Stratasys, such as the Fortus 900mc. Note that this is a point-wise vector process without overlapping scans. Scan speeds can be up to 1,000 mm/s. Assume that a warm-up time of 0.5 h is needed to heat the build chamber. Assume that $T_{predelay}$ and $T_{postdelay}$ are 1 s.
6. Estimate the build time and cost for a platform of hearing aid shells in a Fortus 900mc material extrusion machine. What is the estimated cost per shell? You will need to visit the Stratasys web site and possibly contact Stratasys personnel in order to acquire all necessary information for computing times and costs.
7. Modify the model for purchase cost to incorporate net present value considerations. Rework the hearing aid shell example in Sect. 16.6.2 to use net present value. What is the estimated cost of a shell?

8. Bentley Motors has a production volume of 10,000 cars per year, over its four main models. Production volume per model per year ranges from about 200 to 4,500. Since each car may sell for $120,000 to over $500,000, each car is highly customized. Write a one-page essay on the DDM implications of such a business. The engines for these cars are shared with another car manufacturer; as such, do not focus your essay on the engines. Rather, focus on the chassis, interiors, and other parts of the car that customers will see and interact with.

References

1. Wohlers T (2014) Wohlers report 2014: additive manufacturing and 3D printing state of the industry, annual worldwide progress report. Wohlers, Fort Collins
2. Masters M (2002) Direct manufacturing of custom-made hearing instruments, SME rapid prototyping conference and exhibition, Cincinnati, OH
3. Masters M, Velde T, McBagonluri F (2006) Rapid manufacturing in the hearing industry, chap. 2. In: Hopkinson N, Hague RJM, Dickens PM (eds) Rapid manufacturing: an industrial revolution for the digital age. Wiley, Chichester
4. Wohlers T (2008) Wohlers report: state of the industry, annual worldwide progress report. Wohlers, Fort Collins
5. Hopkinson N (2005) Two projects using SLS of nylon 12, 3D Systems North American Stereolithography User Group Conference, Tucson, AZ, April 3–7
6. Wohlers T (2013) Wohlers report 2013: additive manufacturing and 3D printing state of the industry, annual worldwide progress report. Wohlers Associates, Inc., Fort Collins
7. Make Magazine (2013) http://makezine.com/2013/06/27/how-to-3d-print-a-video-game-figurine/
8. Hastings B (2007) The transition from rapid prototyping to direct manufacturing, SME RAPID Conference, Detroit, May 20–22
9. Johnson PL (1993) ISO 9000: meeting the new international standards. McGraw-Hill, New York
10. Pham DT, Wang X (2000) Prediction and reduction of build times for the selective laser sintering process. Proc Inst Mech Eng 214(B):425–430

Design for Additive Manufacturing 17

Abstract

Design for manufacture and assembly (DFM) has typically meant that designers should tailor their designs to eliminate manufacturing difficulties and minimize manufacturing, assembly, and logistics costs. However, the capabilities of additive manufacturing technologies provide an opportunity to rethink DFM to take advantage of the unique capabilities of these technologies. As mentioned in Chap. 16, several companies are now using AM technologies for production manufacturing. For example, Siemens, Phonak, Widex, and the other hearing aid manufacturers use selective laser sintering and stereolithography machines to produce hearing aid shells; Align Technology uses stereolithography to fabricate molds for producing clear dental braces ("aligners"); and Boeing and its suppliers use polymer powder bed fusion (PBF) to produce ducts and similar parts for F-17 fighter jets. For hearing aids and dental aligners, AM machines enable manufacturing of tens to hundreds of thousands of parts, where each part is uniquely customized based upon person-specific geometric data. In the case of aircraft components, AM technology enables low-volume manufacturing, easy integration of design changes and, at least as importantly, piece part reductions to greatly simplify product assembly.

The unique capabilities of AM include: *shape complexity*, in that it is possible to build virtually any shape; *hierarchical complexity*, in that hierarchical multiscale structures can be designed and fabricated from the microstructure through geometric mesostructure (sizes in the millimeter range) to the part-scale macrostructure; *material complexity*, in that material can be processed one point, or one layer, at a time; and *functional complexity*, in that fully functional assemblies and mechanisms can be fabricated directly using AM processes. These unique capabilities enable new opportunities for customization, very significant improvements in product performance, multifunctionality, and lower overall manufacturing costs. These capabilities will be expanded upon in Sects. 17.3 and 17.4.

I. Gibson et al., *Additive Manufacturing Technologies*,
DOI 10.1007/978-1-4939-2113-3_17

17.1 Motivation

Design for manufacture and assembly (DFM[1]) has typically meant that designers should tailor their designs to eliminate manufacturing difficulties and minimize manufacturing, assembly, and logistics costs. However, the capabilities of additive manufacturing technologies provide an opportunity to rethink DFM to take advantage of the unique capabilities of these technologies. As covered in Chap. 16, several companies are now using AM technologies for production manufacturing. For example, Siemens, Phonak, Widex, and the other hearing aid manufacturers use selective laser sintering and stereolithography machines to produce hearing aid shells; Align Technology uses stereolithography to fabricate molds for producing clear dental braces ("aligners"); and Boeing and its suppliers use selective laser sintering to produce ducts and similar parts for F-18 fighter jets. For hearing aids and dental aligners, AM machines enable manufacturing of tens to hundreds of thousands of parts; where each part is uniquely customized based upon person-specific geometric data. In the case of aircraft components, AM technology enables low-volume manufacturing, easy integration of design changes and, at least as importantly, piece part reductions to greatly simplify product assembly.

The unique capabilities of AM technologies enable new opportunities for customization, very significant improvements in product performance, multifunctionality, and lower overall manufacturing costs. These unique capabilities include: *shape complexity*, in that it is possible to build virtually any shape; *hierarchical complexity*, in that hierarchical multiscale structures can be designed and fabricated from the microstructure through geometric mesostructure (sizes in the millimeter range) to the part-scale macrostructure; *material complexity*, in that material can be processed one point, or one layer, at a time; and *functional complexity*, in that fully functional assemblies and mechanisms can be fabricated directly using AM processes. These capabilities will be expanded upon in Sect. 17.3.

In this chapter, we begin with a brief look at DFM to draw contrasts with Design for Additive Manufacturing (DFAM). A considerable part of the chapter is devoted to the unique capabilities of AM technologies and a variety of illustrations of these capabilities. We cover the emerging area of engineered cellular materials and relate it to AM's unique capabilities. Perhaps the most exciting aspect of AM is the design freedom that is enabled; we illustrate this with several examples from the area of industrial design (housewares, consumer products) that exhibit unique approaches to product design, resulting in geometries that can be fabricated only using AM processes. The limitations of current computer-aided design (CAD) tools are discussed, and thoughts on capabilities and technologies needed for DFAM are

[1] Design for manufacturing is typically abbreviated DFM, whereas design for manufacture and assembly is typically abbreviated as DFMA. To avoid confusion with the abbreviation for design for additive manufacturing (DFAM), we have utilized the shorter abbreviation DFM to encompass both design for manufacture and design for assembly.

presented. The chapter concludes with a discussion of design synthesis approaches to optimize designs.

17.2 Design for Manufacturing and Assembly

Design for manufacturing and assembly can be defined as the practice of designing products to reduce, and hopefully minimize, manufacturing and assembly difficulties and costs. This makes perfectly good sense, as why would one want to increase costs? However, DFM requires extensive knowledge of manufacturing and assembly processes, supplier capabilities, material behavior, etc. DFM, although simple conceptually, can be difficult and time consuming to apply. To achieve the objectives of DFM, researchers and companies have developed a large number of methods, tools, and practices. Our purpose in this chapter is not to cover the wide spectrum of DFM advances; rather, it is to convey a sense of the variety of DFM approaches so that we can compare and contrast DFAM with DFM.

Broadly speaking, DFM efforts can be classified into three categories:

- Industry practices, including reorganization of product development using integrated product teams, concurrent engineering, and the like
- Collections of DFM rules and practices
- University research in DFM methods, tools, and environments

During the 1980s and 1990s, much of the product development industry underwent significant changes in structuring product development organizations [1]. Companies such as Boeing, Pratt & Whitney, and Ford reorganized product development into teams of designers, engineers, manufacturing personnel, and possibly other groups; these teams could have hundreds or even thousands of people. The idea was to ensure good communication among the team so that design decisions could be made with adequate information about manufacturing processes, factory floor capabilities, and customer requirements. Concurrently, manufacturing engineers could understand decision rationale and start process planning and tooling development to prepare for the in-progress designs. A significant driver of this restructuring was to identify conflicts early in the product development process and reduce the need for redesign or, even worse, retooling of manufacturing processes after production starts.

The second category of DFM work, that of DFM rules and practices, is best exemplified by the Handbook for Product Design for Manufacture [2]. The 1986 edition of this handbook was over 950 pages long, with detailed descriptions of engineering materials, manufacturing processes, and rules-of-thumb. Extensive examples of good and bad practices are offered for product design for many of these manufacturing processes, such as molding, stamping, casting, forging, machining, and assembly.

University research during the 1980s and 1990s started with the development of tools and metrics for part manufacture and assembly. The Boothroyd and Dewhurst toolkit is probably the most well-known example [3]. The main concept was to develop simple tools for designers to evaluate the manufacturability of their designs. For example, injection molding DFM tools were developed that asked designers to identify how many undercuts were in a part, how much geometric detail is in a part, how many tight tolerances were needed, and similar information. From this information, the tool provided assessments of manufacturability difficulties, costs estimates, and provided some suggestions about part redesign. Similar tools and metrics were developed for many manufacturing and assembly processes, based in part on the Handbook mentioned above, and similar collections of information. Some of these tools and methods were manual, while others were automated; some were integrated into CAD systems and performed automated recognition of difficulties. For instance, Boothroy Dewhurst, Inc. markets a set of software tools that help designers conceive and modify their design to achieve lower-cost parts, taking into account the specific manufacturing process being utilized. In addition, they sell software tools which help designers improve the design of assembled components through identification of the key functional requirements of an assembled component; leading the designer through a process of design modifications with the aim of minimizing the number of parts and assembly operations used to create that assembled component. The work in this area is extensive; see for example, the ASME Journal of Mechanical Design and the ASME International Design Engineering Technical Conferences proceedings since the mid-1980s (see e.g., [4]).

The extensive efforts on DFM over many years are an indication of the difficulty and pervasiveness of the issues surrounding DFM. In effect, DFM is about the designer understanding the constraints imposed by manufacturing processes, then designing products to minimize constraint violation. Some of these difficulties are lessened when parts are manufactured by AM technologies, but some are not. Integrated product development teams that practice concurrent engineering make sense, regardless of intended manufacturing processes. Rules, methods, and tools that assist designers in making good decisions about product manufacturability have a significant role to play. However, the nature of the rules, methods, and tools should change to assist designers in understanding the design freedom allowed by AM and, potentially, aiding the designer in exploring the resulting open design spaces, while ensuring that manufacturing constraints (yes, AM technologies do have constraints) are not violated.

To illustrate the differences between DFM and DFAM, this section will conclude with two examples. The first involves typical injection molding considerations, that of undercuts and feature detail. Consider the camera spool part shown in Fig. 17.1 [5]. The various ribs and pockets are features that contribute to the time and cost of machining the mold in which the spools will be molded. Such feature detail is not relevant to AM processes since ribs can be added easily during processing in an AM machine. A similar result relates to undercuts. This spool design has at least one undercut, since it cannot be oriented in a mold consisting of only two mold pieces (core and cavity), while enabling the mold halves to be separated and the part removed. Most probably, the spool will be oriented so that

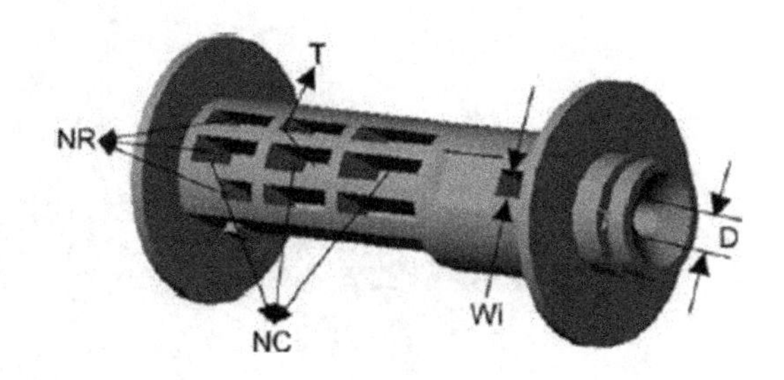

Fig. 17.1 Camera spool example

the mold closure direction is parallel to the walls of the ribs. In this manner, the core and cavity mold halves form most of the spool features, including the ribs (or pockets), the flanges near the ends, and the groove seen at the right end. In this orientation, the hole in the right end cannot be formed using the core and/or cavity. A third moving mold section, called a side action, is needed to form the hole. In AM processes, it is not necessary to be so concerned about the relative position and orientation of features, since, again, AM machines can fabricate features regardless of their position in the part.

In design for assembly, two main considerations are often offered to reduce assembly time, cost, and difficulties: minimize the number of parts and eliminate fasteners. Both considerations translate directly to fewer assembly operations, the primary driver for assembly costs [3]. To minimize parts and fasteners, integrated part designs typically become much more complex and costly to manufacture. Design for manufacture and design for assembly will often be repeated, iteratively, until an optimal solution is found; one where the increasing manufacturing costs for more complex components are no longer compensated by the assembly cost savings.

The designs in Fig. 17.2 show two very different approaches to designing ducts for aircraft [6, 7]. This example represents a design concept for conveying cooling air to electronic units in military aircraft, but could apply to many different applications. The first design is a typical approach using parts fabricated by conventional manufacturing processes (stamping, sheet metal forming, assembly using screws, etc.). In contrast, the approach on the right illustrates the benefits of taking design for assembly guidelines to their extreme: the best way to reduce assembly difficulties and costs is to eliminate assembly operations altogether! The resulting design replaces 16 parts and fasteners with 1 part that exhibits integrated flow vanes and other performance enhancing features. However, this integrated design cannot be fabricated using conventional manufacturing techniques and is only manufacturable using AM.

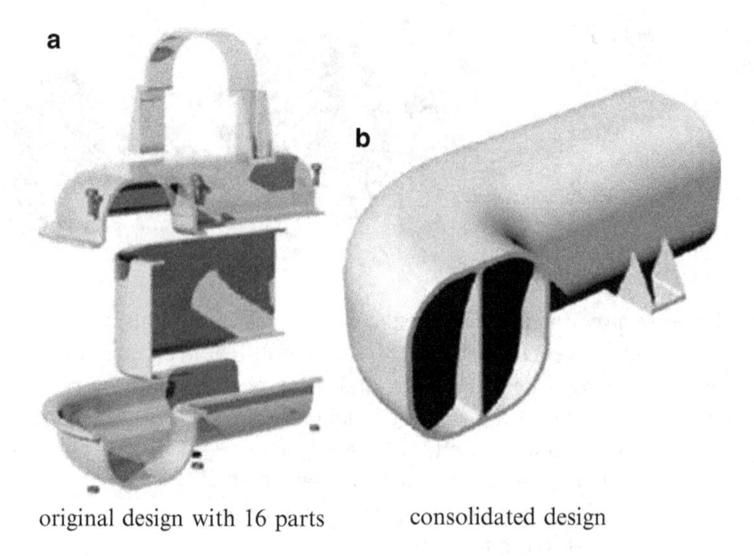

Fig. 17.2 Aircraft duct example

17.3 AM Unique Capabilities

The layer-based additive nature of AM leads to unique capabilities in comparison with most other manufacturing processes. After explaining these uniquenesses, several examples and classes of applications will be presented in the next section. The unique capabilities mentioned at the beginning of the chapter were:

- Shape complexity: it is possible to build virtually any shape
- Hierarchical complexity: features can be designed with shape complexity across multiple size scales
- Functional complexity: functional devices (not just individual piece-parts) can be produced in one build
- Material complexity: material can be processed one point, or one layer, at a time as a single material or as a combination of materials

To date, primarily shape complexity has been used to enable production of end-use parts, but applications taking advantage of the other capabilities, particularly material complexity, are being developed.

17.3.1 Shape Complexity

In AM, the capability to fabricate a layer is unrelated to the layer's shape. For example, the lasers in vat photopolymerization (VP) and powder bed fusion (PBF)

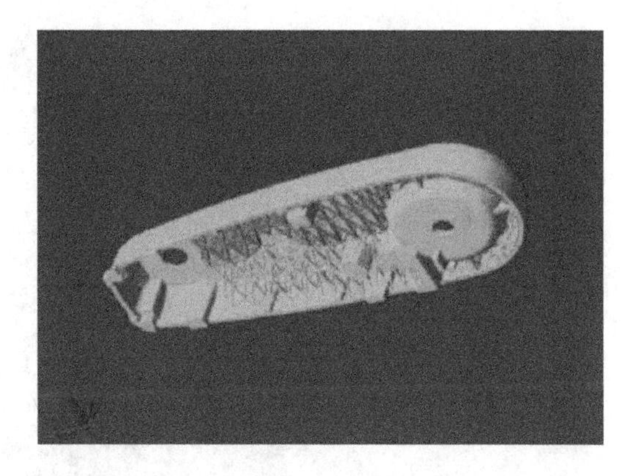

Fig. 17.3 Robot link stiffened with lattice structure

processes can reach any point in a part's cross section and process material there. As such, part complexity is virtually unlimited. This is in stark contrast to the limitations imposed by machining or injection molding, two common processes. In machining, tool accessibility is a key limitation that governs part complexity. In injection molding, the need to separate mold pieces and eject parts greatly limits part complexity.

A related capability is to enable custom-designed geometries. In production using AM, it does not matter if one part has a different shape than the previously produced part. Furthermore, no hard tooling or fixtures are necessary, which implies that lot sizes of one can be economically feasible. This is tremendously powerful for medical applications, for example, since everyone's body shape is different. Also, consider the design of a high-speed robot arm. High stiffness and low weight are desired typically. With AM, the capability is enabled to put material where it can be utilized best. The link from a commercial Adept robot (Cobra 600) shown in Fig. 17.3 has been stiffened with a custom-designed lattice structure that conforms to the link's shape. Preliminary calculations show that weight reductions of 25 % are achieved readily with this lattice structure and that much greater improvements are possible. More generally, AM processes free designers from being limited to shapes that can be fabricated using conventional manufacturing processes.

Another factor enabling lot sizes of one, and shape complexity, is the capability for automated process planning. Straightforward geometric operations can be performed on AMF or STL files (or CAD models) to decompose the part model into operations that an AM machine can perform. Although CNC has improved greatly, many more manual steps are typically utilized in process planning and generating machine code for CNC than for AM.

17.3.2 Hierarchical Complexity

Similar to shape complexity, AM enables the design of hierarchical complexity across several orders of magnitude in length scale. This includes nano/

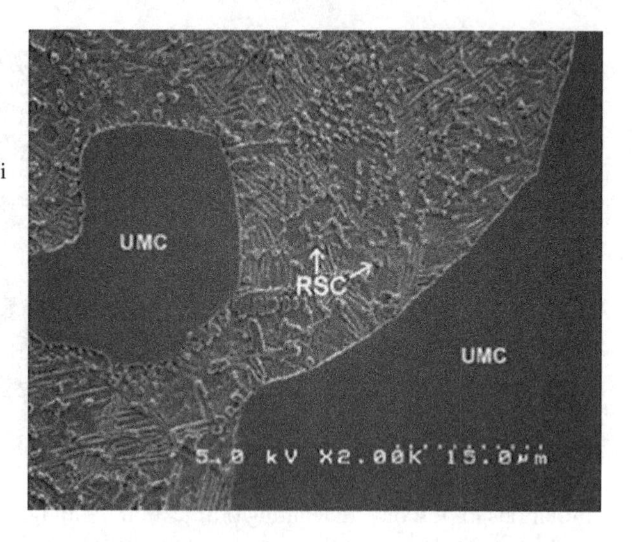

Fig. 17.4 Sixty percent CP-Ti, 40 % TiC composite made using LENS. The ratio of un-melted carbides (UMCs) to resolidified carbides (RSCs) within the Ti matrix is controlled by varying LENS process parameters

microstructures, mesostructures, and part-scale macrostructures. We will start with material microstructures.

One set of processes, which has been studied extensively with respect to hierarchical complexity, are the directed energy deposition (DED) processes. In LENS, for instance, the nano/microstructure can be tailored in a particular location by controlling the size and cooling rate of the melt pool. As a result, the size and distribution of precipitates (nanoscale) and secondary particles (microscale), for example, can be changed by locally modifying the laser power and scan rate. Figure 17.4 illustrates the types of microstructural features which can be formed when using LENS to process mixtures of TiC in Ti to form a composite structure. There are several features of the microstructure which can be controlled. In cases of lower laser energy densities, there is a greater proportion of unmelted carbide (UMC) particles within the microstructure. At higher energy densities more of the TiC particles melt and precipitate as resolidified carbides (RSC). In addition, as the RSC have a different stoichiometry (TiC transforms to $TiC_{0.65}$); for a given initial mixture of TiC and Ti, the more RSC that is present in the final microstructure, the less Ti matrix material is present. The resulting microstructures can thus have very different material and mechanical properties. If sufficient RSCs are precipitated to consume the Ti matrix material, then the structure becomes very brittle. In contrast, when most of the TiC is present as UMCs, the structure is more ductile but is less resistant to abrasive wear.

In addition to the nano/microstructure illustration above, DED technologies have been shown to be capable of producing equiaxed, columnar, directionally solidified, and single-crystal grain structures. These various types of nano/microstructures can be achieved by careful control of the process parameters for a particular material, and can vary from point to point within a structure. In many cases, for laser or electron beam PBF processes for metals, these variations are also

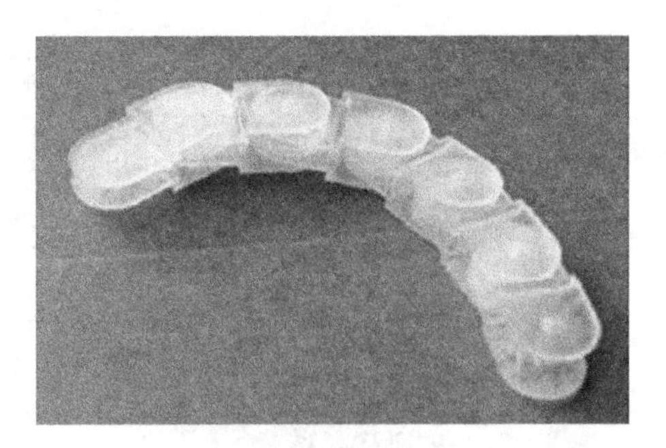

Fig. 17.5 Pulley-driven snake-like robot

achievable. Similarly, by varying either the materials present (when using a multimaterial AM system) or the processing of the materials, this type of nano/microstructure control is also possible in ME, material jetting, VP, and sheet lamination AM technologies as well. These related possibilities are further explored below with respect to material complexity.

The ability to change the mesostructure of a part is typically associated with the application of cellular structures, such as honeycombs, foams, or lattices, to fill certain regions of a geometry. This is often done to increase a part's strength to weight or stiffness to weight ratio. These structures are discussed in more detail in Sect. 17.5.2.

When considered together, the ability to simultaneously control a part's nano/microstructure, mesostructure, and macrostructure simply by changing process parameters and CAD data is a capability of AM which is unparalleled using conventional manufacturing.

17.3.3 Functional Complexity

When building parts in an additive manner, one always has access to the inside of the part. Two capabilities are enabled by this. First, by carefully controlling the fabrication of each layer, it is possible to fabricate operational mechanisms in some AM processes. By ensuring that clearances between links are adequate, revolute or translational joints can be created. Second, components have been inserted into parts being built in VP, ME, PBF, sheet lamination, and other AM machines, enabling in situ assembly.

A wide variety of kinematic joints has been fabricated directly in VP, ME, and PBF technologies, including vertical and horizontal prismatic, revolute, cylindrical, spherical, and Hooke joints. Figure 17.5 shows one example of a pulley-driven, snake-like robot with many revolute joints that was built as assembled in the SLA-3500 machine at Georgia Tech.

Similar studies have been performed using material extrusion (ME) and PBF processes. The research group at Rutgers University led by Dr. Mavroidis [8] demonstrated that the same joint geometries could be fabricated by both ME and PBF machines and similar clearances were needed in both machine types. In PBF,

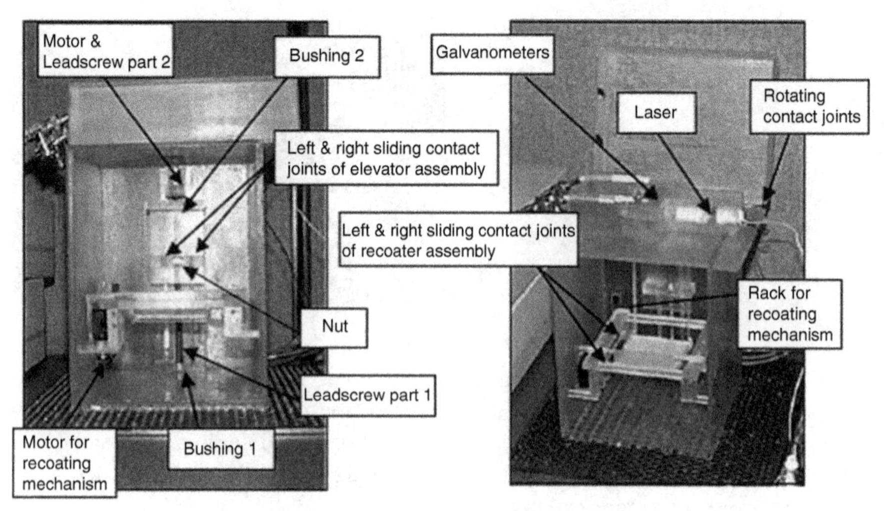

Fig. 17.6 SLA-250 model built in an SLA-250 machine with 11 embedded components

loose powder must be removed from the joint locations to enable relative joint motion. In ME, the usage of soluble support material ensures that joints can be movable after post-processing in a suitable solvent.

In the construction of functional prototypes, it is often advantageous to embed components into parts while building them in AM machines. This avoids post-fabrication assembly and can greatly reduce the number of separate parts that have to be fabricated and assembled.

For example, it is possible to fabricate VP devices with a wide range of embedded components, including small metal parts (bolts, nuts, bushing), electric motors, gears, silicon wafers, printed circuit boards, and strip sensors. Furthermore, VP resins tend to adhere well to embedded components, reducing the need for fasteners. Shown in Fig. 17.6 is a model of an SLA-250 machine that was built in the SLA-250 at Georgia Tech [9]. This 150 × 150 × 260 mm model was built at 1:¼ scale, with seven inserted components, four sliding contact joints, and one rotating contact joint. The recoating blade slides back-and-forth across the vat region, driven by an electric motor and gear train. Similarly, the elevator and platform translate vertically, driven by a second electric motor and leadscrew. The laser pointer and galvanometers worked to draw patterns on the platform, but these three components were assembled after the build, rather than subjecting them to being dipped into the resin vat. Build time was approximately 75 h, including time to pause the build and insert components.

Other researchers have also demonstrated the capability of building functional devices, including the Mavroidis group and Dr. Cutkosky's group at Stanford University [10]. Device complexity is greatly facilitated when the capability to

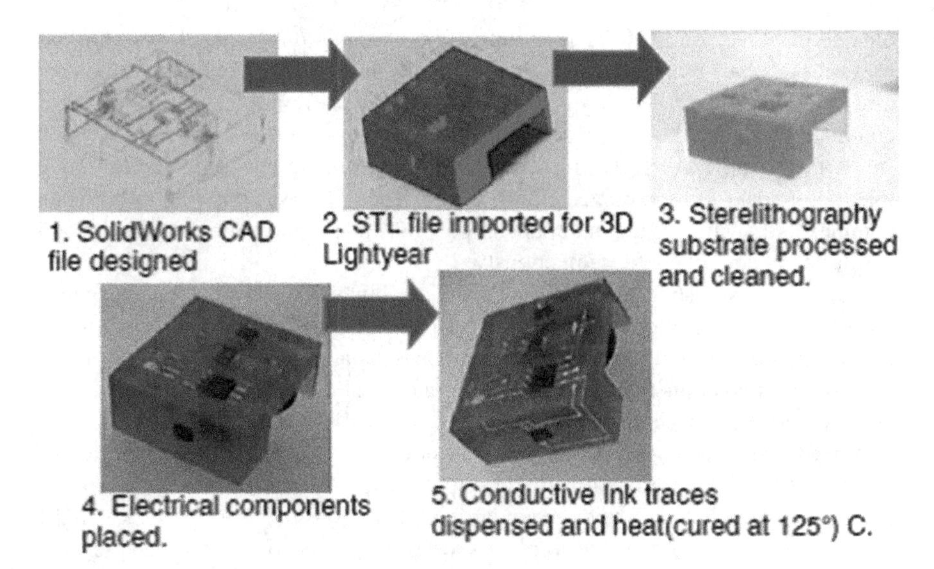

Fig. 17.7 Fabrication of a magnetic flux sensor using VP and DW (courtesy of W.M. Keck Center for 3D Innovation at The University of Texas at El Paso)

fabricate kinematic joints is coupled with embedded inserts since functional mechanisms can be fabricated entirely within the VP vat, greatly simplifying the prototyping process.

Functional complexity can also be achieved by unique combination of AM technologies to produce, for instance, 3D integrated electronics. Researchers in the W.M. Keck Center for 3D Innovation at the University of Texas at El Paso have demonstrated the ability to produce a number of working devices by novel combinations of VP or ME and DW. Figure 17.7 illustrates the process plan for fabrication of a magnetic flux sensor using VP and a nozzle-based DW process. Researchers have demonstrated similar capabilities with ME, sheet lamination, PBF, and other technologies as well.

17.3.4 Material Complexity

Since material is processed point to point in many of the AM technologies, the opportunity is available to process the material differently at different points, as illustrated above, causing different material properties in different regions of the part. In addition, many AM technologies enable changing material composition gradually or abruptly during the build process. New applications will emerge to take advantage of these characteristics.

The concept of functionally graded materials, or heterogeneous materials, has received considerable attention [11], but manufacturing useful parts from these materials often has been problematic. Consider a turbine blade for a jet engine. The outside of the blade must be resistant to high temperatures and be very stiff to prevent the blade from elongating significantly during operation. The blade root must be ductile and have high fatigue life. Blade interiors must have high heat conductivity so that blades can be cooled. This is an example of a part with complex shape that requires different material properties in different regions. No single material is ideal for this range of properties. Hence, if it was possible to fabricate complex parts with varying material composition and properties, turbine blades and similar parts could benefit tremendously.

DED processes, such as LENS and DMD machines, have demonstrated capability for fabricating graded material compositions. Ongoing work in this direction is promising. Graded and multimaterial compositions are used in the repair of damaged or worn components using DED machines, and the design and fabrication of new components is being explored around the world. One such application for improved components that is receiving considerable attention is the fabrication of higher-performance orthopedic implants. In this case, certain regions of the implant require excellent bone adhesion, whereas in other regions the bearing surfaces must be optimized to minimize the implant's wear properties. Thus, by changing the composition of the material from the bone in-growth region to the bearing surface, the overall performance of the implant can be improved.

As described in Chap. 7, Objet Geometries Ltd. (now Stratasys) introduced in 2007 the first commercial AM machine, their Connex500™ system, capable of ink-jet deposition of several polymer materials in one build. Their technology, called PolyJet Matrix™, is an evolution of their printing technology. Recall that Objet uses large arrays of printing nozzles (up to 3,000) to quickly print parts using photopolymer materials. More recently, both Stratasys and 3D Systems have introduced full color printing technology using ink-jet printing of photopolymers that exhibit a much wider range of mechanical properties than Objet's original materials.

For many years, ME machines have been shipped with multiple nozzles for multimaterial deposition. Although one or more nozzles is typically utilized for support materials and the other for build materials, many researchers and industrial practitioners have utilized different feedstock materials in two banks of nozzles to create multimaterial constructs. As can easily be imagined, it would be quite easy, conceptually, to add more nozzles, and thus easily increase the number of materials which can be deposited in a single build. In fact, this concept has been utilized by a number of researchers in their own custom-built extrusion-based machines, primarily by those investigating extrusion-based processes for biomedical materials research.

A significant issue hindering the adoption of AM's material complexity is the lack of design and CAD tools that enable representation and reasoning with multiple materials. This will be explored more completely in Sect. 17.6.

17.4 Core DFAM Concepts and Objectives

Given these unique capabilities of AM, we can articulate some core DFAM concepts and objectives. In contrast to DFM, we believe the objective of DFAM should be to:

Maximize product performance through the synthesis of shapes, sizes, hierarchical structures, and material compositions, subject to the capabilities of AM technologies.

To realize this objective, designers should keep in mind several guidelines when designing products:

- AM enables the usage of complex geometry in achieving design goals without incurring time or cost penalties compared with simple geometry.
- As a corollary to the first guideline, it is often possible to consolidate parts, integrating features into more complex parts and avoiding assembly issues.
- AM enables the usage of customized geometry and parts by direct production from 3D data.
- With the emergence of commercial multimaterial AM machines, designers should explore multifunctional part designs that combine geometric and material complexity capabilities.
- AM allows designers to ignore all of the constraints imposed by conventional manufacturing processes (although AM-specific constraints might be imposed).

17.4.1 Complex Geometry

As was discussed earlier, AM processes are capable of fabricating parts with complex geometry. The layer-by-layer fabrication approach means that the shapes of part cross sections can be arbitrarily complex, up to the resolution of the process. For example, VP and PBF processes can fabricate features almost as thin as their laser spot sizes. In material jetting processes, features in the layer can be the size of several printed droplets. In the Z direction (build direction), the discussion of feature complexity becomes more complicated. In principle, features can be as thin as a layer thickness; however, in practice, features typically are several layers thick. Stresses during the build, such as produced by recoating in VP, can limit Z resolution. Also, overcure or "bonus Z" effects occur in laser-based processes and tend to create regions that are thicker than a single layer. The need to remove the support structures necessary for some AM processes may also limit geometric complexity and/or feature size. Each AM process has its individual characteristics and will take some time to learn. But in general the geometric complexity of AM processes far exceeds that of conventional manufacturing processes.

17.4.2 Integrated Assemblies

The capability for complex geometry enables other practices. As was demonstrated at the end of Sect. 17.2, several parts can be replaced with a single, more complex part in many cases. Even when two or more components must be able to move with respect to one another, such as in a ball-and-socket joint, AM can build these components fully assembled. These capabilities enable the integration of features from multiple parts, possibly yielding better performance. Additionally, a reduction in the number of assembly operations can have a tremendous impact on production costs and difficulties for products.

As is evident from conventional DFM practices, design changes to facilitate or eliminate assembly operations can lead to much larger reductions in production costs than changes to facilitate part manufacture [3]. This is true, at least in part, due to the elimination of any assembly tooling that may have been required. Although conventional DFM guidelines for part manufacturing are not relevant to AM, the design-for-assembly guidelines remain relevant and perhaps even more important. Other advantages exist for the consolidation of parts. For example, a reduction in part count reduces product complexity from management and production perspectives. Fewer parts need to be tracked, sourced, inspected, etc. The need for spare or replacement parts decreases. Furthermore, the need to warehouse tooling to fabricate the parts can be eliminated. In summary, part consolidation can lead to significant savings across the entire enterprise.

17.4.3 Customized Geometry

Consistent with the capability of complex geometry, AM processes can fabricate custom geometries. This has been demonstrated by a series of examples throughout this book related to direct digital manufacturing and biomedical applications. A good example is that of hearing aid shells (Sect. 16.2). Each shell must be customized for an individual's particular ear canal geometry. In VP or PBF machines, hundreds or thousands of shells, each of a different geometry, can be built at the same time in a single machine. Mass customization, instead of mass production, can be realized quite readily. The lack of generic software tools for mass customization, rather than limitations of the hardware, is the key limitation when considering AM for mass customization.

17.4.4 Multifunctional Designs

Multifunctionality is simply the achievement of multiple functions, or purposes, with a single part. This is commonly achieved when performing part consolidation, but the capability of material complexity enables much more ambitious explorations of design possibilities. For example, if a part needs to be stiff in one location, but flexible in another, several AM processes could be used to fabricate such a design simply by varying material composition. Another example is a heat exchanger, that also serves a structural purpose, which could be fabricated by grading steel and copper alloys. By combining geometric and material complexity, very high performance devices can be fabricated. In many cases, designers will need to develop new design concepts and then explore them, since many domains will lack examples of previously successful designs.

17.4.5 Elimination of Conventional DFM Constraints

Since the 1980s, engineering design has changed considerably due to the impact of DFM, concurrent engineering, and integrated product-process teaming practices. A significant amount of time and funds were dedicated to learning about the capabilities and constraints imposed by other parts of the organization. As should be clear from this chapter, AM processes have the potential to reduce the burden on organizations to have integrated product development teams that spend large amounts of time resolving constraints and conflicts. With AM, designers have to learn far fewer manufacturing constraints. The embrace of DFM has resulted in a design culture where the design space is limited from the earliest conceptual design stage to those designs that are manufacturable using conventional techniques. With AM, these design constraints are no longer valid, and the designer can have much greater design freedom.

As such, the challenge in DFAM is not so much the understanding of the effects of manufacturing constraints. Rather it is the difficulty in exploring new design spaces, in innovating new product structures, and in thinking about products in unconventional ways. These do not have to be difficulties, since they are really opportunities. However, the engineering community must be open to the possibilities and learn to exercise their collective creativity.

17.5 Exploring Design Freedoms

With the unique capabilities of AM identified, we can illustrate how to utilize those capabilities through a set of examples. In one approach, companies have achieved significant part consolidation, combining several parts into a single part. In a second approach, researchers have demonstrated how hierarchical structures can result from structuring the material in parts using mesoscale or microscale features to produce the so-called cellular materials. In the third approach, industrial designers have explored new design concepts for some everyday products, such as plates, chairs, and clothing.

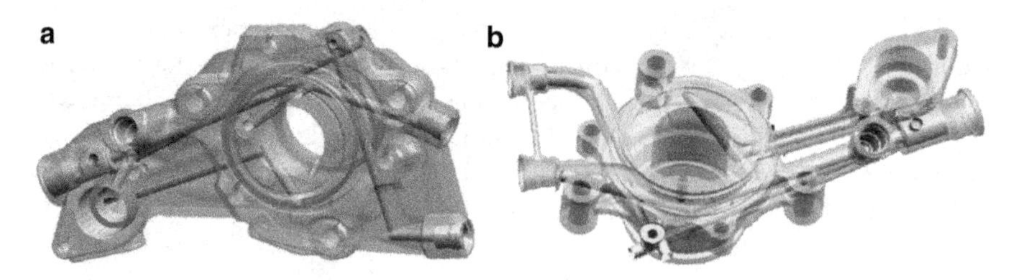

Fig. 17.8 Diesel front plate example. (**a**) Original design. (**b**) Redesigned for additive manufacturing

17.5.1 Part Consolidation and Redesign

The characteristics of geometric complexity and suitability for low-volume production combine to yield substantial benefits in many cases for consolidating parts into a smaller number of more complex parts that are then fabricated using an AM process. This has several significant advantages over designs with multiple parts. First, dedicated tooling for multiple parts is not required. Potential assembly difficulties are avoided. Assembly tooling, such as fixtures, is not needed. Fasteners can often be eliminated. Finally, it is often possible to design the consolidated parts to perform better than the assemblies.

A well-known example that illustrates these advantages was shown in Fig. 17.2, that of a prototypical duct for military aircraft [6, 7]. The design shows a typical traditional design with many formed and rotomolded plastic parts, some formed sheet-metal parts, and fasteners [12]. The example was from the pioneering work of the Boeing Phantom Works Advanced Direct Digital Manufacturing group in retrofitting F-18 fighter jets with dozens of parts produced using PBF. Many of these parts replaced standard ducting components to deliver cooling air to electronics modules. Significant part reductions, elimination of fasteners, and optimization of shapes are illustrative of the advances made by Boeing. Through these methods, many part manufacturing tools and assembly operations were eliminated.

A second example, from Loughborough University, illustrates the advantages of reconceptualizing the design of a component based on the ability to avoid limitations of conventional manufacturing processes. Figure 17.8 shows a front plate design for a diesel engine [7]. The channels through which fuel or oil flow are gun drilled. As a result, they are straight; furthermore, plugs need to be added to plug up the holes through the housing that enabled the channels to be drilled. The redesign shown in Fig. 17.8b was developed by designing the flow channels to ensure efficient flows, then adding a minimal amount of additional material to provide structural integrity. As a result, the part is smaller, lighter, and has better performance than the original design.

17.5.2 Hierarchical Structures

The basic idea of hierarchical structures is that features at one size scale can have smaller features added to them, and each of those smaller features can have smaller features added, etc. Tailored nano/microstructures are one example. Textures added to surfaces of parts are another example. In addition, cellular materials (materials with voids), including foams, honeycombs, and lattice structures, are a third example of hierarchical feature. To illustrate the benefits of designing with hierarchical flexibility, we will focus on cellular materials in this section.

The concept of designed cellular materials is motivated by the desire to put material only where it is needed for a specific application. From a mechanical engineering viewpoint, a key advantage offered by cellular materials is high strength accompanied by a relatively low mass. These materials can provide good energy absorption characteristics and good thermal and acoustic insulation properties as well [13]. When the characteristic lengths of the cells are in the range of 0.1–10 mm, we refer to these materials as mesostructured materials. Mesostructured materials that are not produced using stochastic processes (e.g., foaming) are called designed cellular materials.

In the past 15 years, the area of lattice materials has received considerable attention due to their inherent advantages over foams in providing light, stiff, and strong materials [14]. Lattice structures tend to have geometry variations in three dimensions; as is illustrated in Fig. 17.9. As pointed out in [15], the strength of foams scales as $\rho^{1.5}$, whereas lattice structure strength scales as ρ, where ρ is the volumetric density of the material. As a result, lattices with a $\rho = 0.1$ are about three times stronger per unit weight than a typical foam. The strength differences lie in the nature of material deformation: the foam is governed by cell wall bending, while lattice elements stretch and compress. The examples shown in Fig. 17.9 utilize the octet-truss (shown on the *left*), but many other lattice structures have been developed and studied (e.g., kagome, Kelvin foam) [16, 17].

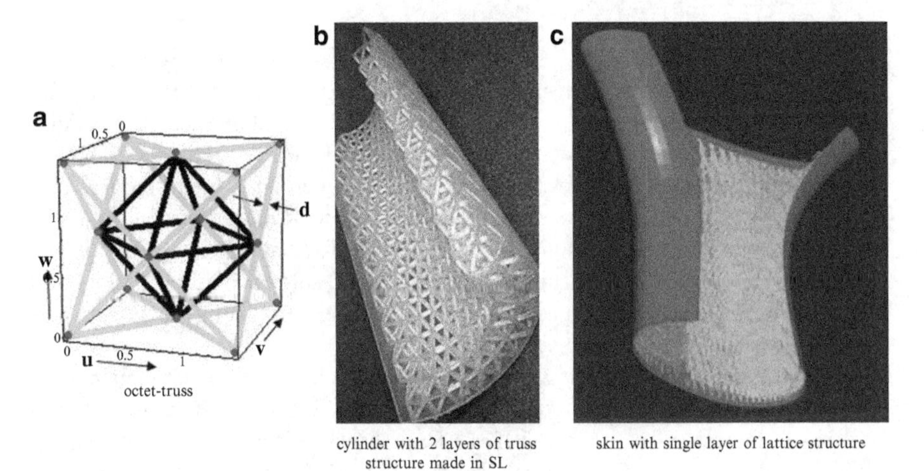

Fig. 17.9 Octet-truss unit cell and example parts with octet-truss mesostructures

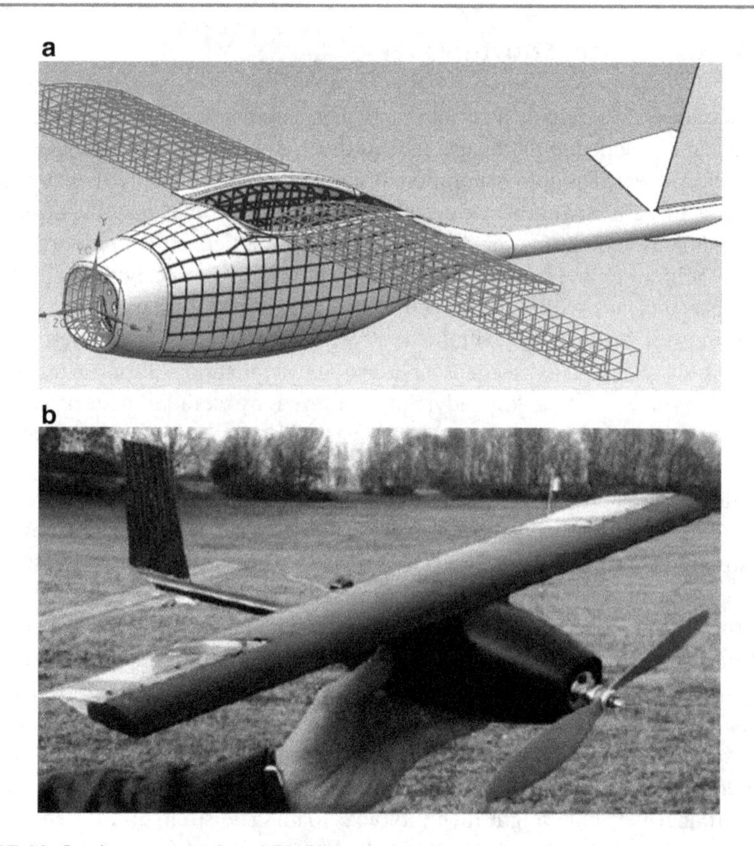

Fig. 17.10 Lattice structure-based UAV design. (**a**) lattice structure designs for fuselage and wings. (**b**) assembled UAV ready for test flight

The parts shown in Fig. 17.9b, c illustrate one method of developing stiff, lightweight structures, that of using a thin part wall, or skin, and stiffening it with cellular structure. Another method could involve filling a volume with the cellular structures. Using either approach often results in part designs with thousands of shape elements (beams, struts, walls, etc.). Most commercial CAD systems cannot perform geometric modeling operations on designs with more than 1,000–2,000 elements. As a result, the design in Fig. 17.9c, which has almost 18,000 shape elements, cannot be modeled using conventional CAD software. Instead, new CAD technologies must be developed that are capable of modeling such complex geometries [18]; this is the subject of Sect. 17.6.

Several groups designed unmanned aerial vehicles (UAV) components by applying various cellular structure design approaches. Figure 17.10 shows a hand-held UAV, the Streetflyer from AVID LLC, that was redesigned to utilize lattice structure reinforcement. The original design of the UAV utilized carbon fiber skins for the fuselage and wings, but required many assembly operations to add stiffeners, fastening features, and mounting features to the components. In contrast, by designing for AM, the lattice structure-based design had such features and stiffeners designed in. Experts at Paramount Industries, a 3D Systems company, fabricated the fuselage and wings in Duraform using PBF. Test flights demonstrated that the PBF-fabricated UAV performed well and, even though the UAV was not optimized, its performance approached that of the carbon fiber production version.

17.5.3 Industrial Design Applications

Some very intriguing approaches to product design have been demonstrated that take advantage of the shape complexity capabilities of AM, as well as some material characteristics. A leader in this field was a small company in The Netherlands called Freedom of Creation (FOC), founded by Janne Kytannen, which was purchased by 3D Systems in the early 2010s. See: http://www.freedomofcreation.com.

FOC began operations in the late 1990s. Their first commercial products were lamp shades fabricated in VP and PBF [20], an example of which is shown in Fig. 17.11a. They have since developed many families of lampshade designs. In 2003, they partnered with Materialise to market lampshades, which retail for 300 to 6,000 euros (as of 2009).

Many other classes of products have been developed, including chairs and stools, handbags, bowls, trays, and other specialty items. See Fig. 17.11b, c for examples of other products. Also, they have partnered with large and small organizations to develop special "give-aways" for major occasions, many of which were designed to be manufactured via AM.

In the early 2000s, they developed the concept of manufacturing textiles. Their early designs were of chain-mail construction, manufactured in PBF. Since then, they have developed several lines of products using similar concepts, including handbags, other types of bags, and even shower scrubs.

More recently, quite a few other companies have demonstrated very innovative designs of housewares, clothing, fashion accessories, and even shoes. Fashion shows have focused on AM-fabricated clothes, parts of clothing, and accessories. Some examples from 2014 include Anouk Wipprecht's electrified 3D printed dresses, hats/headpieces by Gabriela Ligenza, Ray Civello, and Stephen Ma, and dresses and accessories from Iris Van Herpen.

Another source of inspiration comes from browsing the virtual storefronts on shapeways.com and ponoko.com, where individual entrepreneurs and small companies can offer custom designs. Everything from jewelry to candle holders to bird houses can be found on these sites. Methods of manufacturing the designs offered on each storefront need to be provided and many times the only methods are through AM. Some sites provide design guidelines, suggestions, or even specially developed CAD tools.

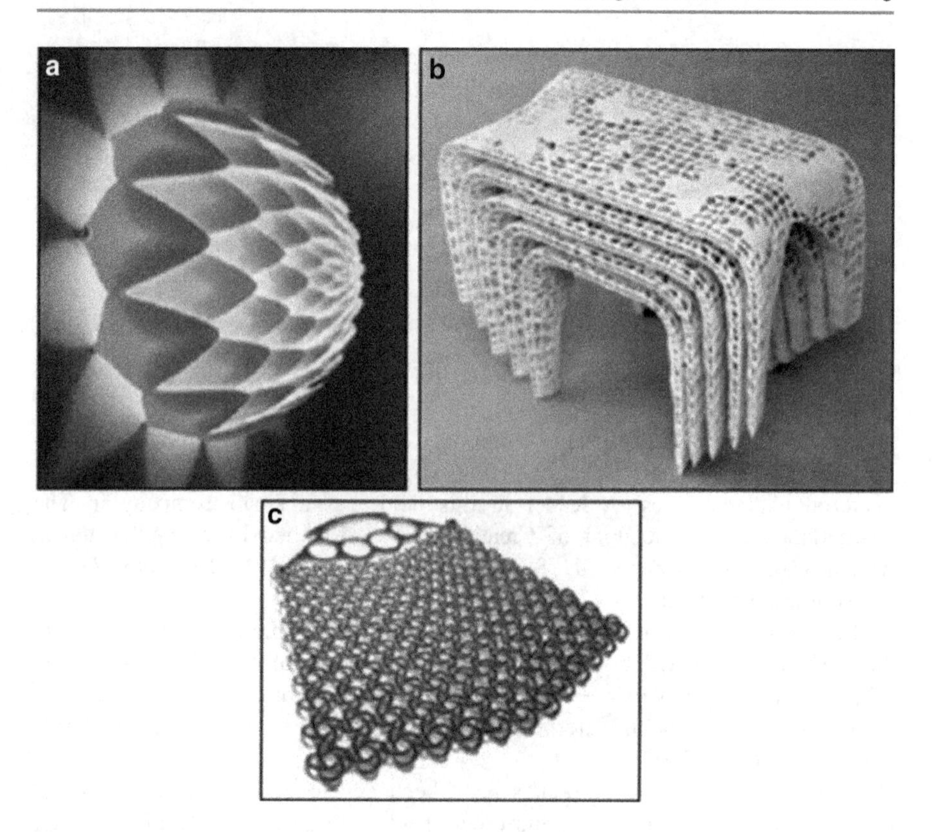

Fig. 17.11 Example products from Freedom of Creation: (**a**) a wall-mounted lampshade, Dahlia light, designed by Janne Kyttanen for Freedom Of Creation, (**b**) stacking footstools, Monarch Stools, designed by Janne Kyttanen for Freedom Of Creation, and (**c**) a handbag, Punch Bag, designed by Jiri Evenhuis and Janne Kyttanen for Freedom Of Creation

17.6 CAD Tools for AM

With tremendous design potential waiting for designers to explore, they need good tools to support their exploration. In this section, we present challenges and technologies associated with mechanical CAD systems.

17.6.1 Challenges for CAD

Current solid-modeling-based CAD systems have several limitations that make them less than ideal for taking advantage of the unique capabilities of AM machines. For some applications, CAD is a bottleneck in creating novel shapes and structures, in describing desired part properties, and in specifying material compositions. These representational problems imply difficulties in driving process

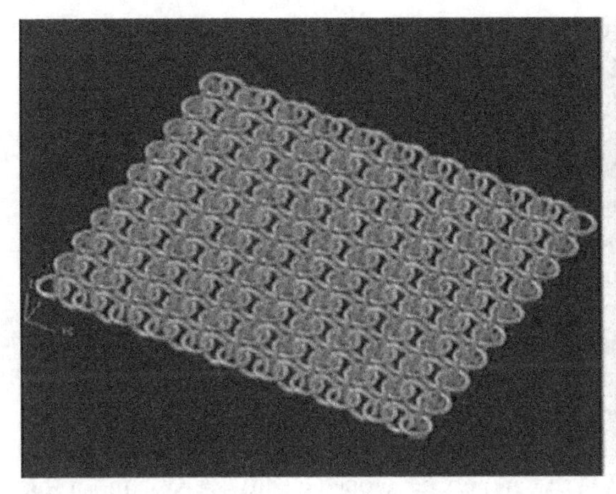
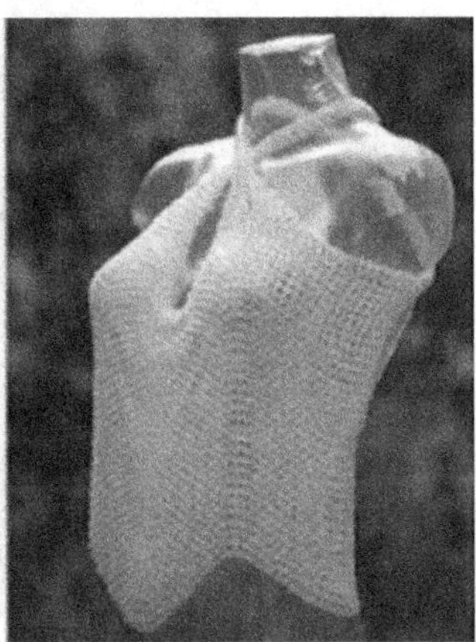

Fig. 17.12 Example of textiles produced using PBF

planning and other analysis activities. Potentially, this issue will slow the adoption of AM technologies for use in production manufacture. More specifically, the challenges for CAD can be stated as:

- Geometric complexity—need to support models with tens and hundreds of thousands of features.
- Physically based material representations—material compositions and distributions must be represented and must be physically meaningful.
- Physically based property representations—desired distributions of physical and mechanical properties must be represented and tested for their physical basis.

One example of the geometric complexity issue is illustrated by the prototype textile application, from Loughborough University and Freedom of Creation, shown in Fig. 17.12 [21]. On the left is a "chain mail"-like configuration of many small rings. On the right is an example garment fabricated on a PBF machine in a Duraform material. The researchers desired to fold up the CAD model of the garment so that it occupied a very small region in the machine's build platform, which would maximize the throughput of the PBF machine for production purposes. The Loughborough researchers had great difficulty modeling the collection of thousands of rings that comprise the garment in a commercial solid-modeling CAD system. Instead they developed their own CAD system for textile and similar structured surface applications over several years. However, having to develop custom CAD systems for specific applications will be a significant barrier to widespread adoption of AM.

Two CAD challenges can be illustrated by some simple examples. The Stratasys Connex 500 material jetting machine can deposit several different materials while building one part. To drive the machine, Stratasys needed to develop a new software tool that allows users to specify materials in different regions of STL files. It would be far better to be able to specify material composition in the original CAD system, so that vendor or machine-specific tools are not needed. The second example was a research project from the Fraunhofer Institute for Manufacturing and Advanced Materials (IFAM) in Bremen, Germany [22]. They developed a two-binder system for 3D Printing technology, where one binder is traditional and one is carbon laden. Their goal was to produce gradient strength steel parts by depositing the carbon according to a desired distribution of hardness. The model of hardness will be converted into a representation of carbon distribution, which will be converted into carbon-laden binder deposition commands for the 3DP system. After building, the part will be heat treated to diffuse the carbon into the steel. As a result, this application illustrates the need to represent distributions of mechanical properties (hardness) and material composition (carbon, steel), and relate these to processing conditions. The IFAM researchers developed a software system for this application.

17.6.2 Solid-Modeling CAD Systems

Parametric, solid-modeling CAD systems are used throughout much of the world for mechanical product development and are used in university education and research. Such systems, such as ProEngineer, Unigraphics, SolidEdge, CATIA, and SolidWorks, are very good for representing shapes of most engineered parts. Their feature-based modeling approaches enable fast design of parts with many types of typical shape elements. Assembly modeling capabilities provide means for automatically positioning parts within assemblies and for enforcing assembly relationships when part sizes are changed.

Commercial CAD systems typically have a hybrid CSG-BRep (constructive solid geometry—boundary representation) internal representation of part geometry and topology. With the CSG part of the representation, part construction history is maintained as a sequence of feature creation, operation, and modification processes. With the BRep part of the representation, part surfaces are represented directly and exactly. Adjacencies among all points, curves, surfaces, and solids are maintained. A tremendous amount of information is represented, all of which has its purposes for providing design interactions, fast graphics, mass properties, and interfaces to other CAD/CAM/CAE tools.

For parts with dozens or hundreds of surfaces, commercial CAD systems run with interactive speeds, for most types of design operations, on typical personal computers. When more than 1,000 surfaces or parts are modeled, the CAD systems tend to run very slowly and use hundreds of MB or several GB of memory. For the textile part, Fig. 17.12, thousands of rings comprise the garment. However, they have the same simple shape, that of a torus. A different type of application is that of hierarchical structures, where feature sizes span several orders of magnitude. An example is that of a multimaterial mold with conformal cooling channels, where the cooling channels have small fins or other protrusions to enhance heat transfer. The fins or protrusions may have sizes of 0.01 mm, while the channels may be 10 mm in diameter, and the mold may be 400 mm long. The central region of the mold may use a high conductivity, high toughness material composition, whereas the surface of the mold may have a high hardness material composition, where a conformal, gradient transition occurs within a region near the surface of the mold. As a result, the mold model may have many thousands of small features and must also represent a gradient material composition that is derived from knowledge of the geometric features. In addition, the range of size scales may cause problems in managing internal tolerances in the CAD system. Current CAD systems are incapable of representing the thousands of features or the graded material composition o f this mold example.

In summary, two main geometry-related capabilities are needed to support many emerging design applications, particularly when AM manufacturing processes will be utilized:

- Representation of tens or hundreds of thousands of features, surfaces, and parts.
- Managing features, materials, surfaces, and parts across size ranges of 4–6 orders of magnitude.

The ISO STEP standard provides a data exchange representation for solid geometry, material composition, and some other properties. However, it is intended for exchanging product information among CAD, CAM, and CAE systems, not for product development and manufacturing purposes. That is, the STEP representation was not developed for use within modeling and processing applications. A good assessment of its usefulness in representing parts with heterogeneous materials for AM manufacturing is given in reference [11], although at present the standards community is revisiting the potential usage of STEP for AM.

As mentioned above, the first challenge for CAD systems is geometric complexity. The second challenge for CAD systems is to directly represent materials, to specify a part's material composition. As a result, CAD models cannot be used to represent parts with multiple materials or composite materials. Material composition representations are needed for parts with graded interfaces, functionally graded materials, and even simpler cases of particle or fiber filler materials. Furthermore, CAD models can only provide geometric information for other applications, such as manufacturing or analysis, not complex multiple material information, which limits their usefulness. This type of limitation is clear when one considers the ink-jet printing examples mentioned so far (e.g., Stratasys multimaterial printer, IFAM carbon-laden inks). In the IFAM case, the addition of carbon to steel deals with the relatively well understood area of carbon steels. In other applications, novel material combinations that are less understood may be of interest. Two main issues arise, including the need to:

- Represent desired material compositions at appropriate size scales.
- Determine the extent to which desired material compositions are achievable.

Without a high fidelity representation of materials, it will not be possible to directly fabricate parts using emerging AM processes. Furthermore, DFM practices will be difficult to support. Together, these limitations may prevent the adoption of AM processes for applications where fast response to orders is needed.

The third challenge, that of representing physically based property distributions, is perhaps the most challenging. The IFAM example of relating desired hardness to carbon content is a relatively simple case. More generally, the geometry, materials, processing, and property information for a design must be represented and integrated. Without such integrated CAD models, it will be very difficult to design parts with desired properties. Analysis and manufacturing applications will not be enabled. The capability of utilizing AM processes to their fullest extent will not be realized. In summary, two main issues are evident:

- Process–structure–property relationships for materials must be integrated into geometric representations of CAD models.
- CAD system capabilities must be developed that enable designers to synthesize a part, its material composition, and its manufacturing methods to meet specifications.

17.6.3 Promising CAD Technologies

The challenges raised in the previous subsection are difficult and go against the directions of decades of CAD research and development. Some CAD technologies on the horizon, however, have promise in meeting these challenges. Two broad categories of technologies will be presented here, implicit modeling and multiscale modeling. Additional technologies can be combined to yield a CAD system that can be used to design components for a wide variety of purposes and with a wide variety of material compositions and geometric complexities.

17.6.3.1 Proposed DFAM System

Figure 17.13 shows one proposed DFAM system [19]. To the right in the figure, the designer can construct a DFAM synthesis problem, using an existing problem template if desired. For different problem types, different solution methods and algorithms will be available. Analysis codes, including FEA, boundary element, and specialty codes, will be integrated to determine design behavior. In the middle, the heterogeneous solid modeler (HSM) is illustrated that consists of implicit and multiscale modeling technologies. Heterogeneous solid modeling denotes that material and other property information will be modeled along with geometry. Libraries of materials and mesostructures enable rapid construction of design models. To the left, the manufacturing modules are shown. Both process planning and simulation modules are important in this system. After planning a manufacturing process, the idea is that the process will be simulated on the current design to determine the as-manufactured shapes, sizes, mesostructures, and

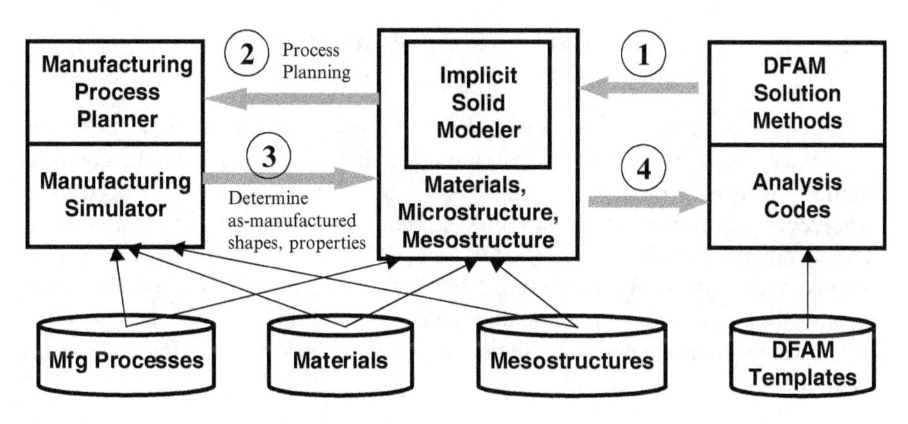

Fig. 17.13 DFAM system and overall structure

microstructures. The as-manufactured model will then be analyzed to determine whether or not it actually meets design objectives.

The proposed geometric representation is a combination of implicit, nonmanifold, and parametric modeling, with the capability of generating BRep when needed. Implicit modeling is used to represent overall part geometry, while nonmanifold modeling is used to represent shape skeletons. Parametric modeling is necessary when decomposing the overall part geometry into cellular structures; each cell type will be represented as a parametric model.

17.6.3.2 Implicit Modeling

Implicit modeling has many advantages over conventional BRep, CSG, cellular decomposition, and hybrid approaches, including its conciseness, ability to model with any analytic surface models, and its avoidance of complex geometric and topological representations [23]. The primary disadvantage is that an explicit boundary representation is not maintained, making visualization and other evaluations more difficult than with some representation types. For HSM, additional advantages are apparent. Implicit modeling offers a unified approach for representing geometry, materials, and distributions of any physical quantity. A common solution method can be used to solve for material compositions, analysis results (e.g., deflections, stresses, temperatures), and for spatial decompositions if they can be modeled as boundary value problems [24]. Furthermore, it provides a method for decomposing geometry and other properties to arbitrary resolutions which is useful for generating visualizations and manufacturing process plans.

In conventional CAD systems, parametric curves and surfaces are the primary geometric entities used in modeling typical engineered parts. For example, cubic curves are prevalent in geometric modeling; a typical 2D curve would be given by parametric equations such as

$$
\begin{aligned}
x(u) &= au^3 + bu^2 + cu + d \\
y(u) &= eu^3 + fu^2 + gu + h
\end{aligned}
\tag{17.1}
$$

These equations would have been simplified from their formulation as Bezier, b-spline, or NURBS (nonuniform, rational b-splines) curves [25]. In contrast, implicit functions are functions that are set equal to zero. Often, it is not possible to solve for one or more of the variables explicitly through algebraic manipulation. Rather, numerical methods must often be used to solve implicit equations. Frequently, sampling is used to visualize implicit functions or to solve them. More specifically, the general form of an implicit equation of three variables (assumed to be Cartesian coordinates) is presented along with the equation for a circle in implicit form:

$$
\begin{aligned}
z(x, y) &= 0 \\
z(x, y) &= \frac{1}{2r}\left[(x - x_c)^2 + (y - y_c)^2 - r^2\right]
\end{aligned}
\tag{17.2}
$$

where x_c, y_c are the x and y coordinates of the circle center and r is its radius.

Shapiro and coworkers have advanced the application of the theory of R-functions to show how engineering analyses [24] and material composition [26] can be performed using implicit modeling approaches. The advantage of their approach is the unifying nature of implicit modeling to model geometry, material composition, and distributions of any physically meaningful quantity throughout a part. Furthermore, from these models of property distributions, they can perform analyses using methods akin to the Boundary Element Method (BEM).

As an example, consider the 2D rectangular part shown in Fig. 17.14 with rectangular and circular holes. The implicit equations that model the boundaries of the part are presented in (17.3a–17.3d). Equations (17.3a and 17.3b) models the x-extents and y-extents of the part, while (17.3c) and (17.3d) models the rectangular hole and (17.3e) models the circular hole ($r = 0.6$, $x_c = y_c = 0.1$). Note that the equation for each boundary feature is 0-valued at the boundary, is positive in the part interior, and is negative in the part exterior. These equations were formulated using R-functions [26].

$$
w_1(x) = \frac{4 - x^2}{4} \tag{17.3a}
$$

$$
w_2(y) = \frac{8 + 2y - y^2}{9} \tag{17.3b}
$$

$$
w_3(x) = x^2 - 0.25 \tag{17.3c}
$$

$$
w_4(y) = 2(y - 3)^2 - 0.125 \tag{17.3d}
$$

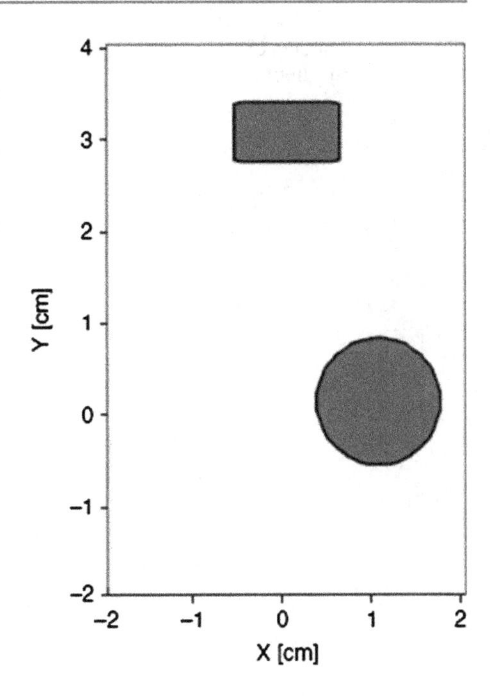

Fig. 17.14 Example part to illustrate implicit modeling

$$w_5(x,y) = \frac{1}{2r^2}\left[(x - x_c)^2 + (y - y_c)^2 - r^2\right] \tag{17.3e}$$

An equation for the entire part can be developed by combining the boundary functions using operators ∧ and ∨ which, in the simplest case, are functions "min" and "max," respectively; other more sophisticated expressions can be used. The part equation is

$$\mathbf{W} = w_1 \wedge w_2 \wedge (w_3 \vee w_4) \wedge w_5 \tag{17.4}$$

with the interpretation that the part is defined as Ω when **W** is greater than or equal to 0, $\boldsymbol{\Omega} = (\mathbf{W}(x,y) \geq 0)$

A plot of the part function is shown in Fig. 17.15, which shows contours of constant function value (17.4). Generalizing from the example, it is always the case that a single algebraic equation can be derived to represent a part using implicit geometry, regardless of the part complexity.

Additional, more sophisticated techniques can be applied to generate useful parameterizations of part models for modeling multiple materials or for applications in design, analysis, or manufacturing, but they will not be explored further here.

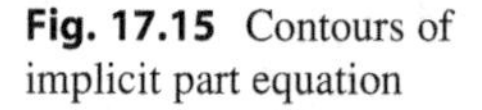

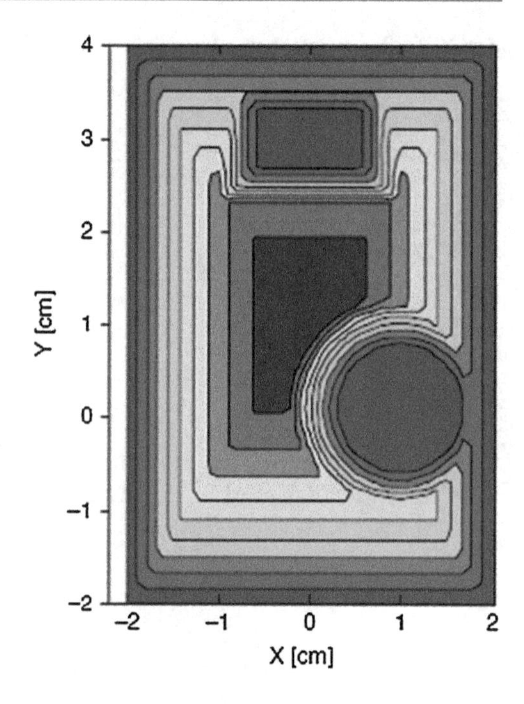

Fig. 17.15 Contours of implicit part equation

17.7 Synthesis Methods

The capabilities of AM processes have inspired many people to try to design structures so that they have minimum weight, without regard to geometric complexity. Quite a few researchers are investigating methods for synthesizing light weight structures, with the intention of fabricating the resulting structures using AM. The work has been extended in some cases to the design of compliant mechanisms, that is, one-piece structures that move. In this section, we provide a brief survey of some recent research in this area. A brief exploration of optimization methods will be covered, with an emphasis on the emerging area of topology optimization that promises to aid designers in efficiently exploring novel structures.

17.7.1 Theoretically Optimal Lightweight Structures

Several years ago, researchers rediscovered the pioneering work of AGM Michell in the early 1900s who developed the mathematical conditions under which structure weight becomes minimized [27]. He proved that structures can have minimum weight if their members are purely tension-compression members (i.e., are trusses) and derived the rules for truss layout. A typical Michell truss is shown in Fig. 17.16 for a common loaded plate structural problem. Note that the solution has a "wagon wheel" structure.

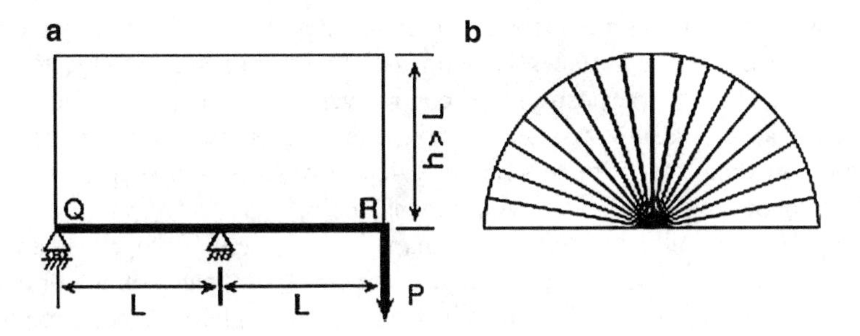

Fig. 17.16 Michell truss layout (**b**) for simple loaded plate example (**a**)

In general, it is difficult to compute optimal Michell truss layouts for any but the simplest 2D cases. Some researchers have developed numerical procedures for computing approximate solutions. At least one research group has proposed to fabricate Michell trusses using AM processes and has investigated multiple material solution cases [28]. For proposed synthesis algorithms for large complex problems, Michell trusses provide an excellent baseline against which solutions for more complicated problems can be compared.

17.7.2 Optimization Methods

In our context, optimization methods seek to improve the design of an artifact by adjusting values of design variables in order to achieve desired objectives, typically related to structural performance or weight, as well as possible without violating constraints. A variety of optimization problem formulations has been developed that vary based on type of objectives and scope of the problem. Good textbooks [29] and many research papers have been written on the subject. The three main types of optimization problems that have been explored for design for AM include, in order of increasing complexity and scope:

- Size optimization—where values of dimensions are determined
- Shape optimization—where shapes of part surfaces are changed
- Topology optimization—where distributions of material are explored

In size optimization, the values of selected dimensions are determined that best achieve the objectives while satisfying any constraints. For typical structural optimization applications, objectives could include the minimization of maximum stress, strain energy, deflection, or part volume or weight. One or more of these quantities may also be modeled as constraints. For many mechanical parts, a small number of size dimensions will be part of the optimization problem. However, for cellular structures, such as lattices, the number of design variables could number in the tens or hundreds of thousands.

Shape optimization is a generalization of size optimization. Typically the shape of bounding curves or surfaces is optimized to achieve similar objectives and constraints. As such, the positions of control vertices for curves or surfaces are often used as the design variables. Shape and size optimization are frequently combined in order to optimize structures that have free-form shapes, as well as standard shapes (e.g., cylinders) with dimensions.

In topology optimization, the overall shape, arrangement of shape elements, and connectivity of the design domain are determined. Again, part volume or compliance is minimized, subject to constraints on, for example, volume, compliance, stress, strain energy, and possibly additional considerations. The primary differences between topology optimization and shape or size optimization are in the starting geometric configuration and the choice of variables, which can lead to very significant improvements in structural performance. The recent interest in topology optimization as a design method for AM warrants a closer look into this technology.

17.7.3 Topology Optimization

Topology optimization (TO) methods determine the overall configuration of shape elements in a design problem. Often, TO results are used as inputs to subsequent size or shape optimization problems. As structural optimization methods, finite element analyses are performed typically during each iteration of the optimization method, which means that TO can be computationally demanding. Furthermore, TO solutions should result in structures that are nearly fully stressed, or have constant strain energy, throughout the structure geometry based on the specified loading conditions. Two main approaches have been developed for TO problems: truss-based and volume-based density methods.

17.7.3.1 Truss-Based Methods

In the truss-based approach, a mesh of struts among a set of nodes is defined in a volume of interest, where sometimes the mesh represents a complete graph (e.g., ground truss) and sometimes it is based on unit cells. Topology optimization proceeds to identify which struts are most important for the problem, determine their size (e.g., diameter), and remove struts with small sizes. Result quality is often a strong function of the starting mesh of struts. Results will resemble the lattice structures presented earlier, with variations in strut diameters evident.

In the first variations of truss-based methods, a ground truss was defined over a grid of nodes, with each node connected to every other node by a truss element. Each element's diameter was used as the design variable. As optimization proceeds, those elements whose diameters become small are deleted from the design. Although the methods worked well, they tended to be computationally expensive. Recently, more sophisticated methods have been developed that utilize a different problem formulation, involving background meshes and analytical derivatives for computation of sensitivities, for truss optimization methods [30]. Good results have

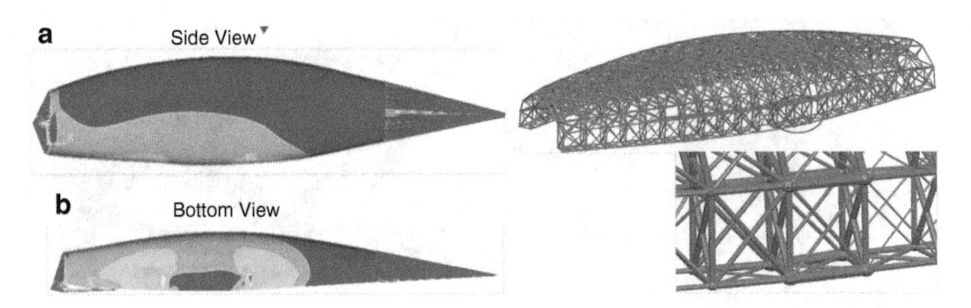

Fig. 17.17 SMS method results on UAV fuselage design problem

been achieved when both truss element size and position are used as design variables. Variations of these approaches have demonstrated the capability of achieving risk-based or reliability-related objectives [31].

Other synthesis methods utilize heuristic optimization methods in an attempt to greatly reduce the number of design variables in the optimization problem. For example, the Size Matching and Scaling (SMS) method starts with a conformal lattice structure (Sect. 17.5.2) but only requires two design variables, the minimum and maximum strut diameters, to optimize the structure [18]. The method works by performing a finite element analysis (FEA) on a solid body of the design. A conformal lattice structure is constructed that fits within the solid body. Local strain or stress values from the FEA results are used to scale struts in the lattice structure resulting in a set of relative strut size values. Size optimization is performed on the lattice structure to determine the values of the minimum and maximum strut diameters, using frame elements to model the lattice structure. Application of the SMS method to a simplified UAV fuselage design problem is illustrated in Fig. 17.17. Note that regions of high stress result in thick struts.

17.7.3.2 Volume-Based Density Methods

The second main approach is based on determining the appropriate material density in a set of voxels that comprise a spatial domain. The density-based TO method that is most common, and is used in the commercial software packages, is known as the SIMP (Solid Isotropic Material with Penalization) method. The starting geometry for the problem is a rectilinear block that is composed of a set of voxels. Each voxel has a density value which is used as its design variable. A density value of 1 indicates that the material is fully dense, while a value of 0 indicates that no material is present. Intermediate values indicate that the material need not be fully solid to support the local stress state in that voxel. Solutions are preferred that have voxels that are either fully dense or near 0 density, since typically partially dense materials are difficult to manufacture. Density values are used to scale voxel stiffness values in the FEA models that are used during the TO process.

The typical topology optimization problem is formulated as [32]:

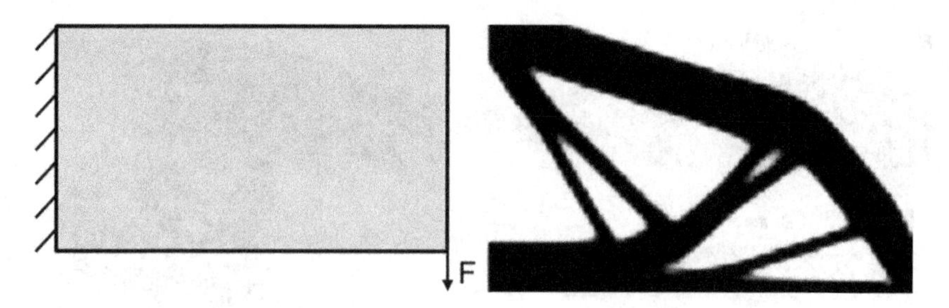

Fig. 17.18 Simple topology optimization example

$$\min_{x} L(u) = \int_{\Omega} f \cdot v dx + \int_{\Gamma} t \cdot v \, ds \tag{17.5}$$

such that

$$a(u, v) = L(v), \quad \text{all } v \in U \tag{17.6}$$

where

$$a(u, v) = \int_{\Omega} E_{ijkl} \varepsilon_{kl}(u) \varepsilon_{ij}(v) dx \tag{17.7}$$

x are points in the spatial domain of interest, U are admissible displacement fields, f are body forces, t are surface tractions. Equation (17.7) is known as the energy bilinear form. The design variables in this formulation are the elasticity tensors E_{ijkl}. In the SIMP method, the elasticity tensors are functions of density and sometimes orientation.

A typical example of topology optimization is shown in Fig. 17.18, which is a simple cantilever plate with a downward point load on its right side. Topology optimization algorithms can maintain the connectivity of material around the loading and boundary areas, and also ensure that these areas are connected. They can add an arbitrary number of holes or strut regions to the design domain. However, they often produce rough or undesirable part shapes. Although the design in Fig. 17.18 could be fabricated using AM, one would probably prefer smoother shapes and transitions between major shape elements. The example was computed using the popular 99-line TO Matlab code from Ole Sigmund [33], with inputs of 80×50 units in size, a volume fraction of 0.5, a penalization exponent of 3, and the rmin (filter size) term of 1.5.

Two popular commercial codes for TO are OptiStruct from Altair and Abaqus, which is marketed by Dassault Systemes. Both packages can solve a variety of TO problems. For example, OptiStruct provides a fairly general topology optimization capability for problems where the structural and system behavior can be simulated by finite element and/or multi-body dynamics analyses. As a result, both composite shells (layout of laminated construction by modifying ply thickness and angle) and mechanisms can be optimized.

Before describing each package in more detail, the general limitations of these commercial TO systems will be highlighted. First, topology optimization is based on approximate models of mechanics that can differ substantially from the actual part or material mechanics. Furthermore, the simple mechanics models are inadequate for cellular materials since their mechanics cannot be approximated by an isotropic solid. For example, assume that an element has a density of 30 % and the designer wants to use the octet truss construction in the region of that element. Placing an octet truss unit cell into the 30 % solid region will result in completely different mechanical behavior than the behavior assumed during topology optimization.

Second, the results of topology optimization are rarely manufacturable directly, even by AM. Typically, designs retain their meshed surfaces so that they are rougher (and tessellated) than would be desired. Part regions may become very thin between thick sections, introducing unwanted stress concentrations. Some topology optimization systems do a better job of producing designs with smoother surfaces (e.g., ABAQUS), but even so the user manuals typically recommend that topology optimization results be used to guide part design—they provide conceptual solutions—rather than be regarded as suitable for production usage. Furthermore, the models produced by topology optimization are typically not suitable for import back into a CAD system for refinement since they will be facetted (original CAD surfaces have been lost) and will not have any parameters associated with geometric shapes. As such, it is very difficult to modify or refine topology optimization models in CAD.

Abaqus is part of the Simulia brand of CAE software marketed by Dassault Systemes. Abaqus is generally considered an excellent FEA package with state-of-the-art nonlinear and plasticity analysis capabilities. Multi-physics simulation is provided with integration between structural, thermal, fluid flow, and other mechanics models. Additionally, Abaqus has an extensive library of material models that includes metals, polymers, rubbers, and even biological tissues. A wide array of physical properties is included, including standard mechanical, thermal, fluidic, acoustic, and diffusion, as well as user-defined materials.

The Abaqus Topology Optimization Module (ATOM) offers topology and shape optimization capabilities that utilize much of the simulation power of Abaqus. Specifically, topology and shape optimization is offered for single parts and assemblies, while leveraging advanced simulation capabilities such as contact, material nonlinearity, and large deformation.

The second commercial system to be discussed is from Altair, which offers their HyperWorks suite of CAE software that includes OptiStruct, their topology optimization package. More generally, OptiStruct is marketed as a structural analysis solver for linear and nonlinear structural problems under static or dynamic loadings. Structures can be optimized for their strength, durability, NVH (noise-vibration-harshness), thermal, and some acoustics characteristics. Altair claims that OptiStruct can solve optimization problems with thousands of design variables

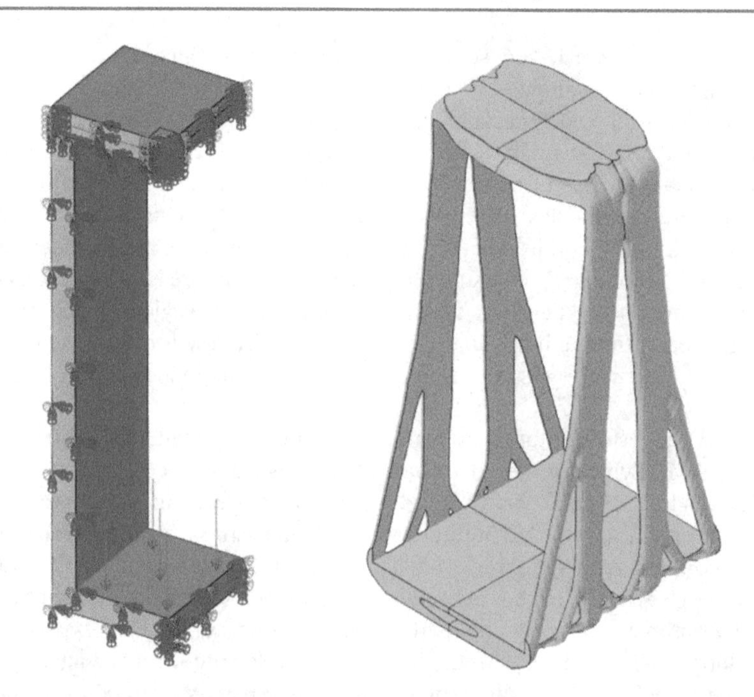

Fig. 17.19 3D cargo sling topology optimization example. Courtesy Mahmoud Alzahrani and Dr. Seung-Kyum Choi, Georgia Institute of Technology

and can combine topology, topography (e.g., vary thickness of a sheet), size, and shape optimization capabilities. Additionally, the composites models that can be generated in HyperMesh can be optimized in OptiStruct. Further, OptiStruct can optimize both flexible and rigid bodies during multi-body dynamic analysis. Fatigue-based concept design and optimization is also provided.

Also from Altair, solidThinking Inspire is a separate application that supports easy-to-use topology optimization capabilities. solidThinking is the name of the company that developed the software; they were acquired by Altair recently (2012–2013). From the company literature, Inspire does not seem to be integrated with HyperWorks or OptiStruct, but it should be only a matter of time before some integration is achieved.

As a second example, a more sophisticated 3D TO problem is shown in Fig. 17.19, which represents a cargo sling design problem. The design domain, shown in the left, is $3 \times 3 \times 6$ m in size with a material thickness of 0.3 m (a quarter model was used to take advantage of symmetry). A pressure load of 3 kPa was applied as shown by the arrows. Symmetry boundary conditions were used. The TO solution was computed in Abaqus for a volume constraint of 15 % of the initial volume in the design region, as shown on the right. The example demonstrates that reasonable solutions can be obtained using commercial TO systems in a reasonable amount of time (1 h on a standard PC).

TO remains a very active topic of research. Some of the research issues and directions under investigation include ensuring connectivity of regions in the resulting structures [34], improving the efficiency of TO methods by introducing the concept of a topological sensitivity [35], and exploring alternative solution approaches such as level sets and evolutionary structural optimization. The level set approach [36] models the distribution of material in a domain using an implicit function representation. The part boundaries are computed by finding the zero-level contour of this implicit function representation. Quite a lot of research is underway to develop efficient, robust level set solution algorithms, particularly for 3D problems. In contrast, evolutionary structural optimization methods [34] utilize stochastic, evolutionary optimization methods, such as genetic algorithms and particle swarm optimization, but with a problem formulation that is similar to SIMP. In these methods, typically elements are removed or added based on the sensitivity of an element or node as measured by the change in the structure's mean compliance of removing that element or node.

17.8 Summary

The unique capabilities of AM technologies enable new opportunities for designers to explore new methods for customizing products, improving product performance, cutting manufacturing and assembly costs, and in general developing new ways to conceptualize products. In this chapter, we compared traditional DFM approaches to DFAM. AM enables tremendous improvements in many of the considerations that are important to DFM due to the capabilities of shape, hierarchical, functional, and material complexity. Through a series of examples, new concepts enabled by AM were presented that illustrate various methods of exploring design freedoms. No doubt, many new concepts will be developed in future years. Challenges and potential methods for new CAD tools were presented to overcome the limitations of traditional parametric, solid-modeling CAD systems. A brief overview of optimization methods was given to illustrate some automated synthesis methods for designing complex structures. Several examples were given to illustrate the types of solutions that can be generated; the resulting geometries are complex enough to preclude fabrication using conventional manufacturing processes.

This chapter covered a snapshot of design concepts, examples, and research results in the broad area of DFAM. In future years, a much wider variety of concepts should emerge that lead to revolutionary ways of conceiving and developing products.

17.9 Exercises

1.–4. Describe in your own words the four AM unique design capabilities described in this chapter and give one example of a product that could be improved by the proper application of each design capability. The example products cannot be ones that were mentioned in this book.

5. What are three ways that current designers are trained that are at odds with the concept of DFAM?
6. Why is optimization a more challenging issue with DFAM than for DFM?
7. For one of the products identified in problems 1–4, draw in CAD the original design and your redesign based upon the application of DFAM principles.

References

1. Susman GI (1992) Integrating design and manufacturing for competitive advantage. Oxford University Press, New York
2. Bralla JG (ed) (1986) Handbook of product design for manufacturing. McGraw-Hill, New York
3. Boothroyd G, Dewhurst P, Knight W (1994) Product design for manufacture and assembly. Marcel Dekker, New York
4. Shah J, Wright PK (2000) Developing theoretical foundations of DFM. ASME design for manufacturing conference, Baltimore, MD
5. Rosen DW, Chen Y, Sambu S, Allen JK, Mistree F (2003) The rapid tooling testbed: a distributed design-for-manufacturing system. Rapid Prototyp J 9(3):122–132
6. 3D Systems, Inc. http://www.3dsystems.com
7. Hague RJM (2006) Unlocking the design potential of rapid manufacturing. In: Hopkinson N, Hague RJM, Dickens PM (eds) Rapid manufacturing: an industrial revolution for the digital age, Chap. 2. Wiley, Chichester
8. Mavroidis C, DeLaurentis KJ, Won J, Alam M (2002) Fabrication of non-assembly mechanisms and robotic systems using rapid prototyping. ASME J Mech Des 123:516–524
9. Kataria A, Rosen DW (2001) Building around inserts: methods for fabricating complex devices in stereolithography. Rapid Prototyp J 7(5):253–261
10. Binnard M (1999) Design by composition for rapid prototyping, 1st edn. Kluwer Academic Publishing, Norwell
11. Patil L, Dutta D, Bhatt AD, Lyons K, Jurrens K, Pratt MJ, Sriram RD (2000) Representation of heterogeneous objects in ISO 10303 (STEP). Proceedings of the ASME international mechanical engineering congress and exposition, Mechanical Engineering Division, Orlando, FL, November, pp 355–364
12. Boeing Corp. http://www.boeing.com
13. Gibson LJ, Ashby MF (1997) Cellular solids: structure and properties. Cambridge University Press, Cambridge
14. Ashby MF, Evans A, Fleck NA, Gibson LJ, Hutchinson JW, Wadley HNG (2000) Metal foams: a design guide. Butterworth-Heinemann, Woburn
15. Deshpande VS, Fleck NA, Ashby MF (2001) Effective properties of the octet-truss lattice material. J Mech Phys Solids 49(8):1747–1769
16. Wang J, Evans AG, Dharmasena K, Wadley HNG (2003) On the performance of truss panels with Kagome cores. Int J Solids Struct 40:6981–6988
17. Wang A-J, McDowell DL (2003) Optimization of a metal honeycomb sandwich beam-bar subjected to torsion and bending. Int J Solids Struct 40(9):2085–2099

18. Nguyen J, Park S-I, Rosen DW (2013) Heuristic optimization method for cellular structure design of light weight components. Int J Precis Eng Manuf 14(6):1071–1078
19. Kytannen J (2006) Rapid manufacture for the retail industry. In: Hopkinson N, Hague RJM, Dickens PM (eds) Rapid manufacturing: an industrial revolution for the digital age, Chap. 18. Wiley, Chichester
20. Rosen DW (2007) Computer-aided design for additive manufacturing of cellular structures. Comput Aided Des Appl 4(5):585–594
21. Additive Manufacturing and 3D Printing Research Group, Nottingham University, UK. http://www.nottingham.ac.uk/research/groups/3dprg/index.aspx
22. Beaman JJ, Atwood C, Bergman TL, Bourell D, Hollister S, Rosen D (2004) Assessment of European Research and Development in Additive/Subtractive Manufacturing, final report from WTEC panel. http://wtec.org/additive/report/welcome.htm
23. Ensz M, Storti D, Ganter M (1998) Implicit methods for geometry creation. Int J Comput Geom Appl 8(5, 6):509–536
24. Shapiro V, Tsukanov I (1999) Meshfree simulation of deforming domains. Comput Aided Des 31(7):459–471
25. Zeid I (2005) Mastering CAD/CAM. McGraw-Hill, New York
26. Rvachev VL, Sheiko TI, Shapiro V, Tsukanov I (2001) Transfinite interpolation over implicitly defined sets. Comput Aided Geom Des 18:195–220
27. Michell AGM (1904) The limits of economy of material in frame structures. Philos Mag 8:589–597
28. Dewhurst P, Srithongchai S (2005) An Investigation of minimum-weight dual-material symmetrically loaded wheels and torsion arms. ASME J Appl Mech 72:196–202
29. Baldick R (2006) Applied optimization. Cambridge University Press, Cambridge
30. Qi X, Wang MY, Shi T (2013) A method for shape and topology optimization of truss-like structures. Struct Multidisc Optim 47:687–697
31. Patel J, Choi S-K (2012) Classification approach for reliability-based topology optimization using probabilistic neural networks. Struct Multidisc Optim 45(4):529–543
32. Bendsoe MP (1989) Optimal shape design as a material distribution problem. Struct Optim 1 (193–202):1989
33. Sigmund O (2001) A 99 line topology optimization code written in Matlab. Struct Multidiscip Optim 21:120–127
34. Rozvany GIN (2009) A critical review of established methods of structural topology optimization. Struct Multidisc Optim 37:217–237
35. Suresh K (2013) Efficient generation of large-scale pareto-optimal topologies. Struct Multidisc Optim 47:49–61
36. Wei P, Wang MW (2009) Piecewise constant level set method for structural topology optimization. Int J Numer Meth Engng 78:379–402

Rapid Tooling 18

Abstract

This chapter discusses how additive manufacturing can be used to develop tooling solutions. Although AM is not well suited to high-volume production in a direct digital manufacturing sense, it does have some benefit when producing volume production tools. This can be from the perspective of using AM to create patterns for parts that are required using materials or properties not currently available using AM or for longer run tooling where AM may be able to simplify the process chain. Commonly referred to as rapid tooling, we discuss here how AM can contribute to the product manufacturing processes.

18.1 Introduction

The term "tooling" refers in this case to the use of AM to create production tools. The tool is therefore an impression, pattern, or mold from which a final part can be taken. There is a variety of different ways in which this can be achieved and these will be discussed in this chapter.

In recent years, as can be seen from other chapters in this book, there has been a tendency to attempt to use AM for production of parts directly from the machine. This is the so-called Direct Digital Manufacture (DDM) and there are numerous reasons why this can be a preferable approach to production. However, there are still a number of reasons for creating tooling rather than DDM:

- The larger the number of parts produced, the more cost-effective it may be to make a production tool, provided it is known how many parts can be made using such a tool.
- The material requirements for the final part may be very specific and not currently available as an AM material but may however be possible through the tooling route.

I. Gibson et al., *Additive Manufacturing Technologies*,
DOI 10.1007/978-1-4939-2113-3_18

- It may be that the product developer wants to understand the tooling process and thus use AM to create a prototype tool.
- This may actually be the quickest and most effective way to create the tooling according to the required specifications. This may be particularly relevant where short lead-times are important.

Tooling is often broken up into two types, referred to as "short-run" and "long-run" tooling. Although discussed in numerous articles like those by Pham and Dimov [1], there are no specific definitions for either of these. Therefore we will attempt to distinguish them here.

Short-run tooling may also be referred to as prototype tooling or soft tooling. The objective is to use techniques that achieve a tool quickly, at low cost and with few process stages. Quite often there are a number of manual steps in the process. It is understood that only a few parts are likely to result from use of the tool; possibly even just one or two parts up to around 100 or more. Every time the tool is used, it should be inspected for damage and viability. It may even be possible (or necessary) to repair the tool before it can be used again. It should be noted that if a tooling solution is required in a very short time (say in a few days), then AM-based short-run tooling may be the only way to arrive there.

Long-run tooling has greater emphasis on use of tooling for mass production purposes. Some injection molding tools can last for years and millions of parts. Although wear is always going to occur, the wear-rate is very low due to the relative hardness of the tool compared with the resulting parts that come from them. The processes required to create long-run tools from AM would still be chosen for their relative cost and lead-time, but in this case they are more likely compared with conventional (subtractive) manufacturing processes. Almost every AM-based long-run tooling solution is likely to involve a metal fabrication process.

The benefits of using a rapid tooling solution may be difficult to determine, but could be immense. Very rarely is a product created from a single tool and the more complex the product, the more difficult it is to plan. Consider the problem of bringing a new mass-produced car to the market. Some parts will already be available, some existing parts may require redesigning while others will require design from scratch. Some of these new parts will be relatively simple, while others will have significant performance specifications that could have very long lead-times. Now consider how you would create a plan to bring all these together so that the car is launched on schedule. Even the manufacture of a very simple part could delay the whole process. The use of AM-based short-run and long-run tooling can be extremely beneficial because of the short reaction times and simplified process chains. A car manufacturer may be able to plan more easily and react to disturbances in the process chain more efficiently. Even tooling that does not last very long (or, for that matter, DDM) can be used to bridge the gap to long-term tooling made using conventional methods. Delivery times can be met even though the entire mass production facility has yet to be completed.

The majority of rapid tooling solutions are focused on the creation of injection molding (IM) tooling. This is because there are a huge number of products made from polymers using this approach. We will go on in this chapter to discuss how we can directly fabricate IM parts using AM as a replacement for subtractive machining processes. Electron discharge machining (EDM) is an alternative to the more conventional abrasive metal cutting that is worth separate consideration in this chapter. Of course, not all products are made from polymer parts. There is a huge variety of metal, ceramic, and composite-based materials and related manufacturing methods. One method that fits very well into an AM process chain is investment casting, which we will discuss here, followed by some less mainstream AM-based approaches that have found niches for some manufacturers.

18.2 Direct AM Production of Injection Molding Inserts

Wikipedia describes injection molding as the most common modern method of manufacturing parts and that it is ideal to produce high volumes of the same part [2]. The general principle is quite straightforward in that molten polymer is forced into a metal mold. A simple IM machine diagram can be seen in Fig. 18.1. Once the polymer has cooled and solidified, the mold splits open to reveal the part which is then ejected and the process repeats. There are many texts that cover IM in varying levels of detail. An excellent online resource can be found from Bolur [4]. From these we can see that, similar to many processes, optimization and maximization of the output from IM can be very complex. As our demand for higher throughput, performance, quality, etc. increases so will the need for more cost-effective solutions.

Since the IM process requires a mold that can somehow separate for the part to be removed, there are a number of issues that require attention:

- A simple mold will have a cavity into which the polymer is injected. A core will form the other side of the mold, which is removed after the cooling process so that the part can be ejected. A mechanism (usually a set of ejector pins) is engaged to push the part out from the cavity. However, for this to be effective, the cavity walls usually have a slight slope (referred to as a "draft angle") that reduces shear forces between the polymer and the mold that would cause the part to stick.
- Not all molds can be easily split into a simple core/cavity to reveal the part. Complex geometry parts may require mold sets that separate into more than two segments. Parts may require very careful redesign so that the number of mold components is minimized. Even so, mold sets can be very complex.
- Filling the mold with molten polymer can also be problematic. The mold must be completely full before it starts to solidify, else there may be cavities. Parts that comprise many features, like thick or thin walls, ribs, bosses, etc. must be carefully analyzed to ensure the mold set is properly filled. Very complex parts may require multiple injection and venting points to ensure effective mold filling as well as fine-tuning of the temperatures, pressures and cycle

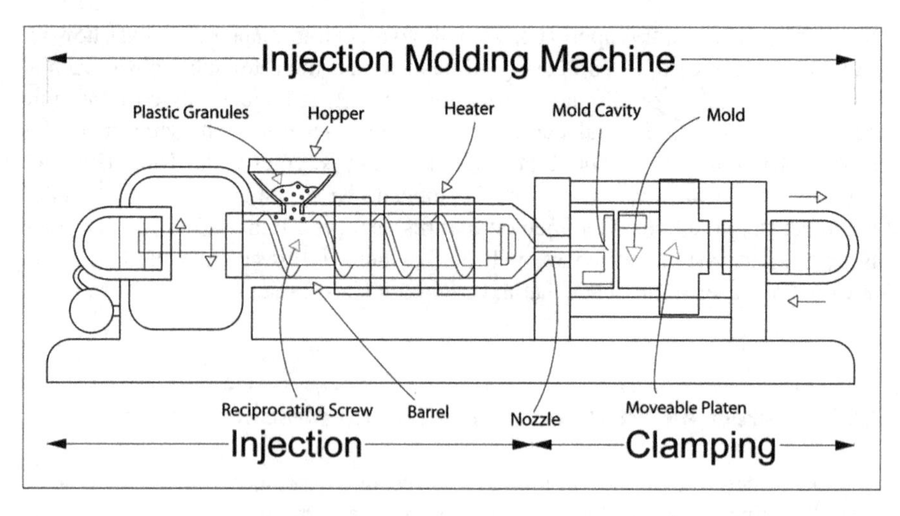

Fig. 18.1 A simple IM machine setup as drawn by Rockey [3]

operations within the IM machine. There are numerous softwares available for mold operation analysis, like Moldflow [5].

As can be seen in Figs. 18.1 and 18.2, an IM machine has a standard plate set into which mold sets are inserted. For these inserts, it is necessary to know where to locate the injection point, the ejector pins, risers, and other features that comprise a fully functioning mold solution. It is these inserts that effectively "customize" the process and where AM can therefore contribute towards a solution.

Inserts can be made using either metal or polymer AM technology. Polymer inserts are obviously less durable, but are much quicker and cheaper to make. In a white paper published by Stratasys, the Polyjet process was demonstrated to be effective for producing inserts for a variety of applications [6]. Figure 18.3 shows Polyjet inserts for a 2-cavity set, with a close-up of the ejector pin arrangement.

IM applications have been tested using the standard Polyjet materials. Best results were presented for the Digital ABS material. Parts were made in a conventional IM machines using a variety of materials, including polyamide, ABS, and polyethylene at temperatures up to 300 °C. Up to 100 cycles have been observed before the inserts broke. Similar results have been reported using SL and polymer laser sintered parts. It is important to note that the IM inserts made this way should be handled carefully so that they can achieve acceptable results. Even though the IM process operates above the heat deflection temperature for the AM materials, it is still possible to get acceptable molded parts. This is possible if the IM cycle is lengthened so that the parts can cool more inside the mold before separation and ejection. Note that this only really works for relatively simple core/cavity sets. For this type of application, the costs can be around half of similar aluminum molds, with significant reductions in lead-time. One can expect some hand-finishing of the resulting molded parts.

Fig. 18.2 A core/cavity mold set showing a central injection point and channeling to regions where 5 different parts are formed in one cycle

Fig. 18.3 Polyjet inserts for a two cavity mold set, showing a close-up of the ejector pins (courtesy Stratasys)

The primary concerns when making mold inserts using polymer AM are heat deflection, wear, and accuracy. Most AM processes can provide partial solutions to these problems, but generally the most accurate processes have low heat deflection temperatures and the highest temperature materials can be found in lower accuracy processes. A number of attempts have been made to develop materials for IM inserts with polymer AM processes. One material of note is the copper-polyamide material that was developed for the polymer powder bed fusion process. Adding a copper filler to the polyamide matrix material served to improve the heat transfer away from the surface when a mold is used in the IM machine. The copper also provided additional wear resistance, which increases the life of the mold. It is interesting to note however that this is not a widely used material as the copper-polyamide is not very useful for many other applications so only appropriate where a large number of these molds are needed.

A number of chapters in this book discuss AM of metal parts. One of the initial drivers for this technology was for IM mold inserts. AM can provide a near-net shape for the metal inserts. Several materials have been developed for metal AM that could be used for this, but the most widely used would be H13 tool steel. Almost every process that can achieve this is based on powder metal sintering. Near-net shape can be achieved up to an Ra surface roughness of 12–20 μm but this would generally not be acceptable for most applications and machining of the parting surfaces in particular would be necessary. If the mold surface also requires machine finishing, then very careful attention must be given to gaging so that all of the original part lies outside of the machining volume. Incorrect gaging could lead to some regions not having sufficient stock material to achieve an adequate surface. It is therefore common for designers to add material to the CAD model as a machining allowance. Figure 18.4 shows a tool set where the inserts were made using a powder metal system, with two parts that were molded from them.

Early metal powder AM machines were very expensive and suffered from problems with accuracy and consistent material properties. At that time there were a few alternative approaches to creating metal parts in the Rapid Steel [6] and KelTool [7] processes. While these approaches have become virtually obsolete, there was distinct advantage in that these processes could result in a fully metal part but using a conventional polymer AM machine. However, there was the need for additional furnace technology that added to the expense of the process.

Powder sintering could also be used to create parts that are a blend of polymer and metal powders. The polymeric material acts as a matrix that can hold the metal powder in place. The use of a high thermally conductive metal powder, like copper, would be the most ideal to use for the purpose of creating IM tooling inserts. The copper would cause heat energy to conduct away from the matrix polymer, thus allowing more rapid cooling during the IM process. The copper powder, being harder and more durable than the polymer, would also enable longer tool life.

One significant benefit to the use of AM for creation of injection mold tooling is the capability of creating conformal cooling channels. It is normal to run coolant through the IM inserts, facilitating the cooling of the plastic part following the injection of the molten polymer. This cooling process is very dependent on the geometry of the part being molded, with larger voluminous segments cooling slower than smaller, thinner sections. Greater flow of coolant close to the larger segments can enable faster and more regular cooling, which can also improve the part quality by preventing part warpage due to thermally induced stress. The geometric freedom that is a characteristic of AM can enable very complex cooling channels to be designed into the part. While the best way to achieve such conformal cooling is very much open to debate, benefits have been cataloged [8, 9]. An example of conformal cooling can be seen in Fig. 18.5. Note that this approach can be applied to both short- and long-run methods.

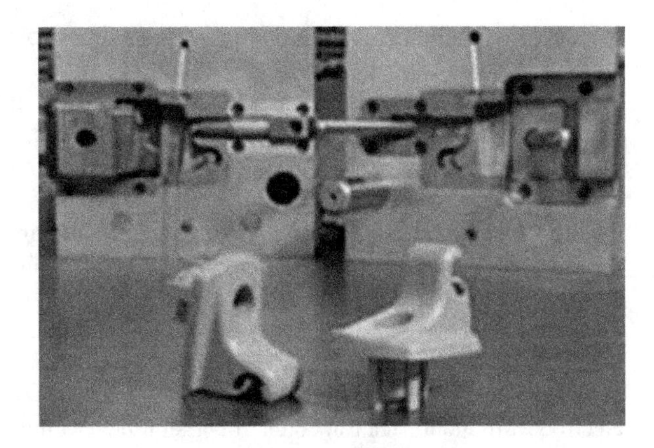

Fig. 18.4 A direct metal laser sintered tool set, with two parts that have been molded from them

Fig. 18.5 A tool insert design, showing the location of conformal cooling channels

18.3 EDM Electrodes

A number of attempts have been made to develop EDM electrodes by plating AM parts [10]. These electrodes could feasibly be used for die-sinking EDM for creating cavities for IM application. The most common method of plating the polymer AM parts would be by using electroless plating of copper. There are two major drawbacks to this plating approach. The first is that electroless plating is best suited to plating a thin layer of material on a surface. For EDM however, the electrodes are more effective with a thicker amount of conductive material deposited. It is difficult to deposit sufficiently thick material in a quick and easy manner and with controllable thickness. This leads to the second problem, which is that even if you can deposit sufficient material, the definition of the electrode will be compromised by this excessively thick layer of material. Although possible, it is not a very effective method of making electrodes.

While it may be possible to create an electrode using powder metallurgy methods from AM molds, possibly a more effective method would be to use direct metal fabrication. Stucker, et al. [11] used this approach to create electrodes using Zirconium diBoride (ZrB_2). This material was encapsulated in a copper matrix material, which was melted using a selective laser melting approach. The resulting metal matrix composite was observed to have good erosion characteristics, wearing approximately 1/16th the rate of a pure copper electrode.

Neither of the above approaches has achieved popularity and there appear to be much better ways of creating EDM electrodes. However, recent improvements in metal powder melting systems may revive this research and development since electrode production can account for a significant amount of the manufacturing costs.

18.4 Investment Casting

Investment casting is the process of generating metal parts from a nonmetal pattern. Figure 18.6 efficiently describes the investment casting process. The patterns are in some way assembled into a structure that can be coated with ceramic to produce a shell. The ceramic starts as a slurry into which the structure, referred to as a "tree" for obvious reasons, is dipped to produce a closely forming skin. Once this has dried, it is strengthened by applying more coats until it is strong enough to withstand the casting process. Prior to casting, the pattern is removed by burning out the material. Care must be taken at this stage to ensure all the material has been burned out of the shell, leaving no residue. The ceramic shell can withstand the high temperature of molten metal during the pouring process, which can then be left to cool before the shell is broken from the tree. The metal replicas of the original pattern are cut from the "trunk" of the structure prior to post-treatment.

The great advantage of this is that parts can be made in a wide range of materials, specific to the application. While powder metal AM systems can produce parts directly in metal, there is a much more limited range of metals available. Furthermore, this is an approach that can result in metal parts from a nonmetal AM technology. A number of AM processes are capable of directly making parts in wax, including material jetting and material extrusion. However, it is also possible to make investment casting patterns from other materials, including polycarbonate and ABS, which are available from a wide range of AM machines. The key is to ensure that the material does not expand rapidly during the burnout process, prior to the metal casting. One way to achieve this is to apply the honeycomb core approach, such as the SL QuickCast build style, rather than using a solid fill.

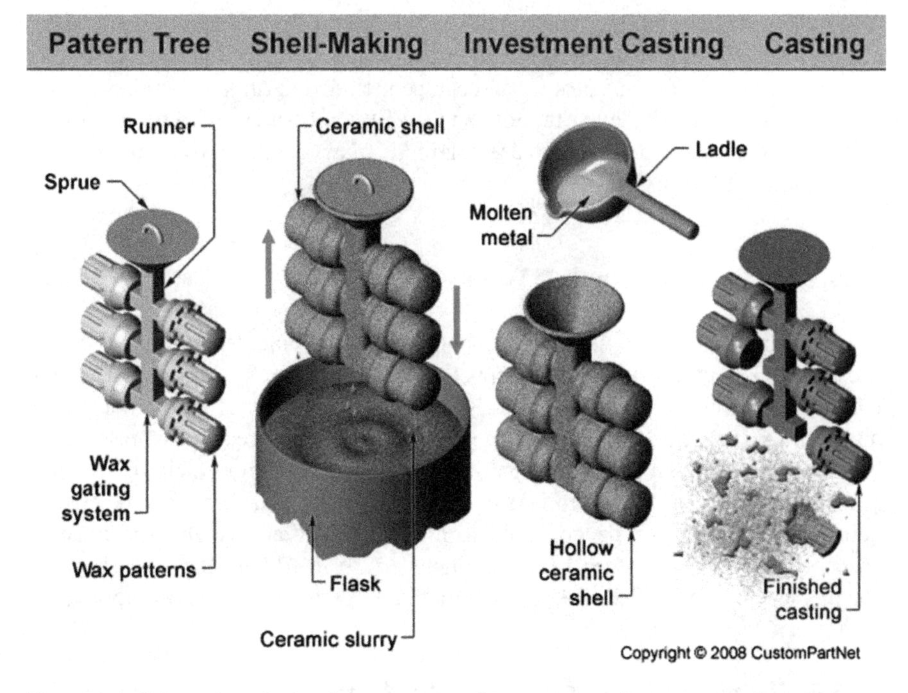

Fig. 18.6 Schematic of the investment casting process. Courtesy of CustomPart at custompartnet.com

18.5 Other Systems

The main purpose of this chapter is to describe the wide range of ways that AM can be used to enhance your manufacturing processes. While there is a push to using AM for direct digital manufacturing, there are still many products that require mass production and here we can see that AM can still contribute. Although injection molding and investment casting are probably the most widely used applications, there are numerous other approaches that have been considered. Below are a few other examples where AM can be used to help solve manufacturing problems.

18.5.1 Vacuum Forming Tools

Vacuum forming is commonly used in packaging, where plastic parts are formed from a flat sheet. A typical example is the clear blister packaging that is commonly used to display consumer products. Other examples include parts that form an outer shell for a product, like a plastic safety helmet for example. After the forming, it is common to cut away the material that surrounds the shaped plastic.

Heat and vacuum are applied when the sheet is placed over a tool, which has holes through which the vacuumed air is extracted. This allows the sheet to conform to the shape of the tool. If a small number of formed parts are required in a series plastic, then the tool could be fabricated using AM. Locating the vacuum holes would be a straightforward process and can be included during the build. Since the heat is not directly targeted at the tool and with the pressures and other forces not being very high, it is acceptable to use polymeric materials that are commonly used in AM, like ABS or nylon.

18.5.2 Paper Pulp Molding Tools

It is becoming quite popular to use paper pulp molding techniques to create packaging. The pulp is made from recycled paper and therefore very environmentally friendly. The forming process is also quite sustainable since it does not require much energy to create the shapes since they are primarily created by pressing out the excess water from the pulp. Again, if the packaging is for small volume part production, AM can be used to create the forming tools. The tools can be created quite quickly using a honeycomb fill to reduce build time, weight, and material costs. Furthermore, features can be included to facilitate the excess water channeling. Figure 18.7 shows a typical mold and part made using this approach.

18.5.3 Formwork for Composite Manufacture

Carbon and glass fiber composite is an increasingly popular material used to manufacture high performance items that require significant strength to weight ratios. This is particularly important for vehicles, where the reduction in weight can reduce the energy requirements to move it around. Use of AM can assist in this process, particularly where complex shapes are involved. Use of honeycomb core AM build methods can assist in the creation of lightweight patterns around which fiber reinforced composites can be wound. Alternatively, AM parts can act as molds into which carbon or glass fiber reinforced polymers (CFRP or GFRP) can be placed, either pre-impregnated (prepreg) or by applying the resin later. The AM part can be kept inside the composite part in some cases or the AM pattern can be separated from the composite after the resin curing (hardening) stage. The fact that some AM materials can be dissolved away could be useful at this stage. Figure 18.8 shows some parts that have been developed for constructing high performance UAVs (unmanned air vehicles) using CFRP. The white parts are all made using AM.

18.5.4 Assembly Tools and Metrology Registration Rigs

The majority of products made are assembled in some way from multiple components. Any technique that can simplify or accelerate the assembly process can be extremely beneficial to a mass market manufacturer. We discuss the benefits

Fig. 18.7 A paper pulp molding tool shown with a molded packaging component (Courtesy of RedEye Redeyeondemand.com)

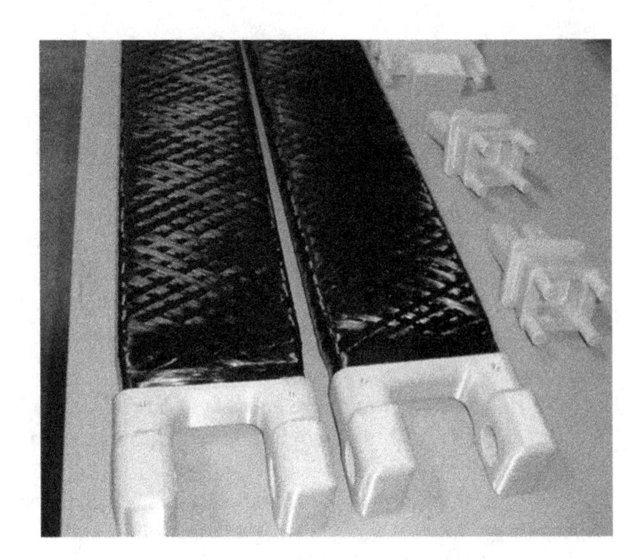

Fig. 18.8 Polymer melt extruded AM parts used as formwork for carbon composite manufacture

of DDM in terms of part simplification to reduce the assembly costs elsewhere in this book. However, even for assembly-based manufacture using conventionally made components AM can make a contribution. Some assembly processes benefit from the use of jigs that make it easier to perform the tasks by keeping some of the components in place as well as ensuring that all the components are present, like the example shown in Fig. 18.9. A variation of this approach can be seen with the metrology fixation system produced by Materialise to ensure automotive and

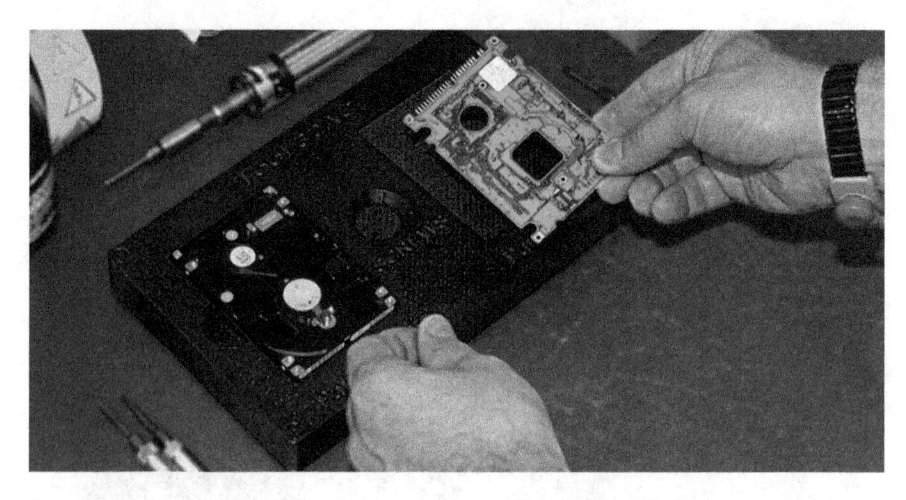

Fig. 18.9 A hard drive assembly jig (Courtesy of Javelin javelin-tech.com)

similar moldings are kept in place during the metrology process for quality assurance purposes [12].

18.6 Exercises

1. Are there any other reasons for using AM to create tooling other than the 4 mentioned at the beginning of this chapter?
2. What different IM flow analysis software can you find on the Internet?
3. Make a list of the different metal AM technologies that are available. What materials are available for creating IM tooling inserts? Can you find any examples of inserts that have been developed?
4. Find two examples of conformal cooling from the web. Can you identify which method is better? Why is it better?
5. Investigate the manufacture and use of EDM electrodes. What are the potential benefits and pitfalls surrounding the use of AM to directly fabricate these electrodes?
6. There are certainly other examples of mass-manufacturing processes that use AM technology. Build a portfolio of examples and use as a means to discuss how they can benefit in terms of time, cost, ease of use, etc.

References

1. Pham DT, Dimov SS (2003) Rapid prototyping and rapid tooling—the key enablers for rapid manufacturing. Proc IME C J Mech Eng Sci 217(1):1–23
2. commons.wikimedia.org/wiki/File:Injection_moulding.png, courtesy of Brockey
3. Rockey B. Created for Wikipedia in the article on Injection Molding

4. Bolur PC. A guide to injection moulding of plastics. pitfallsinmolding.com
5. Moldflow injection molding simulation. www.autodesk.com
6. 3dsystems.com. Rapid Tooling with the SLS® Process and LaserForm™ A6 Steel Material
7. 3dsystems.ru. 3D Keltool, How it works
8. Plasticstoday.com. Conformal cooling: why use it now?
9. Sachs E, Wylonis E et al (2000) Production of injection molding tooling with conformal cooling channels using the three dimensional printing process. Polym Eng Sci 40 (5):1232–1247
10. Arthur A, Dickens PM, Cobb RC (1996) Using rapid prototyping to produce electrical discharge machining electrodes. Rapid Prototyp J 2(1):4–12
11. Stucker BE, Bradley WL, Eubank PT, Bozkurt B, Norasetthekul S (1999) Manufacture and use of ZrB2/Cu composite electrodes. U.S. Patent 5,870,663
12. Materialise metrology system. manufacturing.materialise.com/3d-scanning-measuring-services

Applications for Additive Manufacture 19

Abstract

Additive manufacturing is coming into its third decade of commercial technological development. During that period, we have experienced a number of significant changes that has led to improvements in accuracy, better mechanical properties, a broader range of applications, and reductions in costs of machines and the parts made by them. In this chapter we explore the evolution of the field and how these developments have impacted a variety of applications over time. We note also that different applications benefit from different aspects of AM, highlighting the versatility of this technology.

19.1 Introduction

Additive manufacturing is coming into its third decade of commercial technological development. During that period, we have experienced a number of significant changes that has led to improvements in accuracy, better mechanical properties, a broader range of applications, and reductions in costs of machines and the parts made by them. Also in previous chapters, we have seen that AM technologies can vary according to the following nonexclusive list of parameters:

Cost: Since some machines employ more expensive technologies, like lasers, they will inevitably cost more than others.

Range of materials: Some machines can only process one or two materials, while others can process more, including composites.

Maintenance: With some machines being more complex than others, the maintenance requirements will differ. Some companies will add cost to their machines to ensure that they are better supported.

Speed: Due to the technologies applied, some machines will build parts faster than others.

I. Gibson et al., *Additive Manufacturing Technologies*,
DOI 10.1007/978-1-4939-2113-3_19

Versatility: Some machines have complex setup parameters where part quality can be balanced against other parameters, like build speed. Other machines have fewer setup variations that make them easier to use but perhaps less versatile.

Layer thickness: Some machines have a limitation on the layer thickness due to the material processing parameters. Making these layers thinner would inevitably slow the build speed.

Accuracy: Aside from layer thickness, in-plane resolution also has an impact on accuracy. This may particularly affect minimum feature size and wall thickness of a part. For example, laser-based systems have a minimum feature size that is based on the diameter of the laser beam.

Driven by the automotive, aerospace, and medical industries, AM has found applications in design and development within almost every consumer product sector imaginable. As AM becomes more popular and as technology costs inevitably decrease, this can only serve to generate more momentum and further broaden the range of applications. This momentum has been added to with the recent addition of commercial AM machines that can directly process metal powders.

This chapter discusses the use of AM for medical, aerospace and automotive applications which have consistently been the key industries driving innovation in AM. With aerospace and automotive industries, AM is valued mainly because of the complex geometric capabilities and the time that can be saved in development of products. With medicine, the benefit is primarily in the ability to include patient-specific data from medical sources so that customized solutions to medical problems can be found. We begin with a brief survey of historical developments in rapid prototyping (RP), rapid tooling, and other advances, with a focus mostly on aerospace and automotive industries.

19.2 Historical Developments

In the late 1980s, 3D Systems started selling their first stereolithography machines. The first five customers of the SLA-1 beta program were AMP Incorporated, General Motors, Baxter Health Care, Eastman Kodak, and Pratt & Whitney [1]. These companies represent the four largest industrial sectors, in terms of historical AM usage, including automotive (GM and AMP, their automotive and consumer business group was the customer), health care (Baxter), consumer products (Eastman Kodak), and aerospace (Pratt & Whitney). Texas Instruments, specifically their Defense Systems & Electronics Group, was also an early adopter who applied AM to the aerospace field. Similarly, one of the first customers of DTM was BF Goodrich, which is a supplier to the aerospace and automotive industries.

Focusing on the aerospace industry, many success stories were realized by design and manufacturing engineers who used AM for rapid prototyping purposes. In many cases, thousands of dollars and months of product development time were saved through the use of RP, since prototype parts did not have to be fabricated using conventional manufacturing processes. Additionally, many new applications for AM parts were discovered.

19.2.1 Value of Physical Models

Early adopters discovered that AM, through the rapid prototyping function, provided several benefits including enhanced visualization, the ability to detect design flaws, reduced prototyping time, and significant cost reductions associated with the ability to develop correct designs quickly. Of course, there were also significant costs associated with being an early adopter. AM machines were more expensive than conventional machine tools and people had to be hired and trained to run the AM machines. New post-processing equipment had to be installed and hazardous solvents used to clean SL parts. But, for those companies willing to take the risk, the significant investments in AM had a large return-on-investment when AM was integrated into their product development processes.

For example, in 1992, Texas Instruments reported several case studies demonstrating thousands of dollars and months of prototyping time saved through the use of stereolithography [1]. Furthermore, they were one of the first companies to explore the use of SL parts as patterns for investment casting.

Chrysler purchased two SLA-250 machines in early 1990 and reported that they fabricated over 1,500 parts in the first 2 years of usage, with the machines running virtually 24 h per day and 7 days per week [1]. They also reported significant time and cost savings particularly for form/fit and packaging assessments. They and other companies soon realized that they could greatly increase their chances of winning contracts to supply parts if they included RP parts with their quotes. By including physical prototypes, they can demonstrate that they understand the design requirements and both customer and supplier can identify potential problems early on.

In the medical industry, Depuy, Inc. was another early adopter of SL. They reported on a project that began in 1990 to develop a new line of shoulder implants with dozens of models for various component sizes [1]. They used SL models, fabricated on their in-house SLA-250 machines, of the implant components during several iterations of early project reviews, saved several months of development time, and avoided costly changes before production. Furthermore, they used SL masters for urethane tooling to make wax patterns for investment casting for the first 500 pieces of each size. As they noted, this allowed them to proceed with product launch as part of their development process.

19.2.2 Functional Testing

Engineers at aerospace, automotive, and medical device companies soon discovered that AM parts could be used for a variety of functional testing applications. Specifically flow testing was investigated by these companies, even with the early

SL resins that were brittle and absorbed water easily. As one example, Chrysler tested air flow through several cylinder head designs in early 1992. They built a model of the cylinder head geometry in SL, installed steel valves and springs, then ran the model on their flow bench. They achieved a 38 % improvement in air flow.

Other companies reported similar experiences. Engineers at Pratt & Whitney pioneered several new types of flow apparatus and experiments with SL in the early years with both air and water. A report from Porsche in 1994 described water flow testing in a series of engine models to study coolant flow characteristics [2]. By using SL and an early epoxy resin, they could successfully design, fabricate, and test engine models within about 1 week per iteration.

Also in 1994, Allied Signal reported on a study where SL models of turbine blades were used to determine their frequency spectra [2]. To study the use of SL models, they built SL models at full scale and at 3:1 scale, tested all three blades experimentally, and compared the results to finite element analysis. Theoretically, the full-scale SL models should have natural frequencies that are 35.7 % of those of the steel blades; experimentally, they determined that the SL blades exhibited frequencies 35 % less. Similarly, the 3:1 scale SL blades had natural frequencies that were 12 % of the steel blade frequencies, compared to a theoretical prediction of 11.9 %. In comparison, FEA predictions ranged from 3.6 % lower to 19.4 % higher than experimental results. As a consequence, Allied Signal had much more confidence in their use of SL models than FEA, since the SL models enabled much more accurate determinations of natural frequencies.

Concurrently, aerospace companies started using AM parts to perform wind tunnel testing. Wind tunnel models are typically instrumented with arrays of pressure sensors. Standard metal models required considerable machining in order to fabricate channels for all of the wiring to the sensors. With AM, the channels and sensor mounts could be designed into the model. Automotive companies also adopted this practice. For high-speed testing, or large aerospace models, rapid tooling methods were commonly used in order to fabricate stiffer metal wind tunnel models. With proper designs, engineers could design the channels and sensor mounts into the AM patterns that were subsequently used to produce the tooling.

19.2.3 Rapid Tooling

Prior to 1992, Chrysler experimented with a variety of rapid tooling processes with stereolithography master patterns. This included vacuum forming, resin transfer molding, sand casting, squeeze molding, and silicone molding. Many of these techniques were covered in the previous chapter. The point here is to put this activity in a historical context and realize how early in the AM field's development these applications were investigated.

An area of significant effort in both the aerospace and automotive industries was the use of SL parts as investment casting patterns. Early experiments used thin-walled SL patterns or hollow parts. Because SL resins expand more than investment casting wax, when used as patterns, the SL part tended to expand and crack the ceramic shell. This led to the development of the QuickCast™ pattern style in 1992, which is a type of lattice structure that was added automatically to hollow part STL files by SL machine pre-processing software. The QuickCast style was designed to support thin walls but not to be too strong. Upon heating and thermal expansion, the QuickCast lattice struts were designed to flex, collapse inward, break, but not transfer high loads to the part skins which could crack the shell.

The QuickCast 1.0 style worked, but not as well as desired. This led to the development of QuickCast 1.1 by 3D Systems in 1995 and then QuickCast 2.0 by Phill Dickens and Richard Hague at the University of Nottingham in the late 1990s. This was quickly adopted by many manufacturers and service bureaus and, arguably, revolutionized the investment casting industry.

Another interesting development in the early 2000s was the large-frame binder jetting technology by ExOne, where a sand material was developed that was suitable for use as sand casting dies. As mentioned in Chap. 8, ExOne marketed the S15 binder jetting machine for several years (the technology was purchased from a German company Generis GmbH in 2003). As one example, two of these sand machines were operating at the Ford Dunton Technical Center in England in the mid-2000s (they may still be operational) to support their design and development activities. Much of the Ford of Europe operations are housed here, including small car design, powertrain design and development, and some commercial vehicles. As of the end of 2005, ExOne had reportedly sold 19 S15 machines, each of which cost over $1M.

More recently, Boeing, Northrop-Grumman and other aerospace companies have used material extrusion technology to fabricate tooling. They developed tooling designs for composite part lay-up that were suitable for ME fabrication. Other reported tooling applications included drill guides and various assembly tools.

19.3 The Use of AM to Support Medical Applications

AM models have been used for medical applications almost from the very start, when this technology was first commercialized. AM could not have existed before 3D CAD since the technology is digitally driven. Computerized Tomography (CT) was also a technology that developed alongside 3D representation techniques. Figure 19.1 shows a CT machine, a model directly generated from this machine (shown as cross-sectional slices) and a model with all segments combined into a 3D image. CT is an X-ray-based technique that moves the sensors in 3D space relative to the X-ray source so that a correlation can be made between the position and the absorption profile. By combining multiple images in this way, a 3D image can be built up. The level of absorption of the X-rays is dependent on the density of the subject matter, with bone showing up very well because it is much denser than the surrounding soft tissue. What some people don't realize is that soft tissue images can also be created using CT technology. Clinicians use CT technology to create 3D

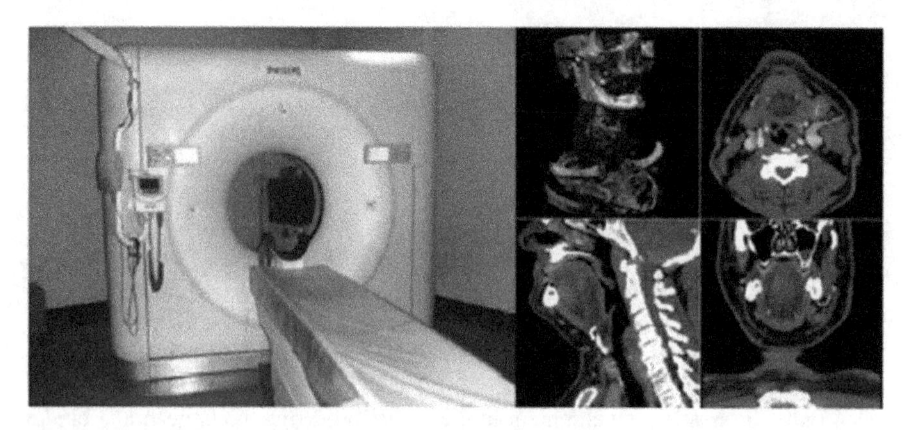

Fig. 19.1 A CT scanner with sliced images and a 3D image created using this technology

images for viewing the subject from any angle, so as to better understand any associated medical condition. Note that this is one of a number of developing technologies working in the 3D domain, including 3D MRI, 3D Ultrasound, and 3D laser scanning (for external imaging). With this increasing use of 3D medical imaging technology, the need to share and order this data across platforms has led to information exchange standards like DICOM [3], from the National Electrical Manufacturers Association in the USA, which allows users to view patient data with a variety of different software and sourced from a variety of different imaging platforms.

While originally used just for imaging and diagnostic purposes, 3D medical imaging data quickly found its way into CAD/CAM systems, with AM technology being the most effective means of realizing these models due to the complex, organic nature of the input forms. Medical data generated from patients is essentially unique to an individual. The automated and de-skilled form of production that AM provides makes it an obvious route for generating products from patient data.

AM-based fabrication contributes significantly to one or more of the following different categories of medical applications:

- Surgical and diagnostic aids
- Prosthetics development
- Manufacturing of medically related products
- Tissue Engineering

We will now go on to discuss how AM is useful to these application areas and some of the issues surrounding their implementation.

19.3.1 Surgical and Diagnostic Aids

The use of AM for diagnostic purpose was probably the first medical application of AM. Surgeons are often considered to be as much artists as they are technically proficient. Since many of their tasks involve working inside human bodies, much of their operating procedure is carried out using the sense of touch almost as much as by vision. As such, models that they can both see from any angle and feel with their hands are very useful to them.

Surgeons work in teams with support from doctors and nurses during operations and from medical technicians prior to those operations. They use models in order to understand the complex surgical procedures for themselves as well as to communicate with others in the team. Complex surgical procedures also require patient understanding and compliance and so the surgeon can use these models to assist in this process too. AM models have been known to help reduce time in surgery for complex cases, both by allowing the surgeons to better plan ahead of time and for them to understand the situation better during the procedure (by having the model on hand to refer to within the operating theater). Machine vendors have, therefore, developed a range of materials that can allow sterilization of parts so that models can be brought inside the operating theater without contamination.

Most applications relate to models made of bony tissue resulting from CT data rather than using soft tissue constructs. MRI data, which is more commonly used for soft tissue imaging, can also be used and cases with complex vascular models have been reported [4]. Bone, however, is more obvious because many of the materials used in AM machines actually resemble bone in some way and can even respond to cutting operations in a similar manner. AM models of soft tissue may be useful for some visualizations, but less can be learned from practicing surgery on them since they will not be compliant in the same way. Many models may benefit from having different colors to highlight important features. Such models can display tumors, cavities, vascular tracks, etc. Material extrusion and jetting technologies can both be used to represent this kind of part, but probably the most impressive visual models can be made using the colored binder jetting process from 3D Systems. Sometimes, these features may be buried inside bone or other tissue and so having an opaque material encased in a transparent material can also be helpful in these situations. For this, the Stereocol resin that was independently developed for SLA machines [5] or the Connex material from Stratasys [6] can be used to see inside the part. The Stereocol material no longer appears to be commercially available, however. Some examples of different parts that illustrate this capability can be seen in Fig. 19.2.

Some of the most noteworthy applications of AM as medical models were from well-publicized surgeries to separate conjoined twins. Surgeons reported that having multicolored, complex models of the head or abdomen areas were invaluable in planning the surgeries, which can take 12–24 h and involve teams of surgeons and support staff [7].

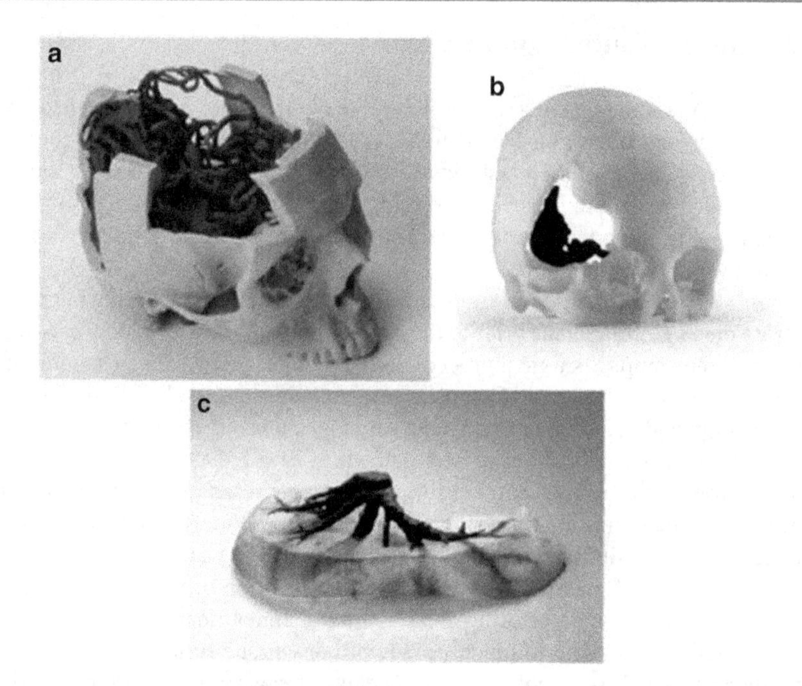

Fig. 19.2 Images of medical parts made using different colored AM systems. (**a**) 3DP used to make a skull with vascular tracks in a darker color. (**b**) A bone tumor highlighted using ABS. (**c**) Stratasys Connex process showing vascularity inside a human organ

19.3.2 Prosthetics Development

Initially, CT generated 3D data combined with the low resolution of earlier AM technology to create models that may have looked anatomically correct, but that were perhaps not very accurate when compared with the actual patient. As the technology improved in both areas, models have become more precise and it is now possible to use them in combination for fabrication of close-fitting prosthetic devices. Wang [8] states that CT-based measurement can be as close as 0.2 mm from the actual value. While this is subjective, it is clear that resulting models, when built properly, can be sufficiently precise to suit many applications.

Support from CAD software can add to the process of model development by including fixtures for orientation, tooling guidance, and for screwing into bones. For example, it is quite common for surgeons to use flexible titanium mesh as a bone replacement in cancer cases or as a method for joining pieces of broken bone together, prior to osteointegration. While described as flexible, this material still requires tools in order to bend the material. Models can be used as templates for these meshes, allowing the surgeon's technical staff to precisely bend the mesh to shape so that minimal rework is required during surgery. Figure 19.3 shows a maxillofacial model that has been used for this purpose [9].

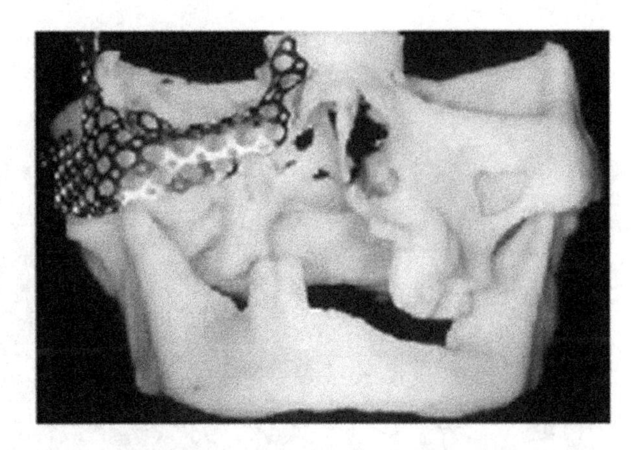

Fig. 19.3 Titanium mesh formed around a maxillofacial model

Alternatively, many AM processes can create parts that can be used as casting patterns or reference patterns for other manufacturing processes. Many prosthetics comprise components that have a range of sizes to fit a standard population distribution. However, this means that precise fitting is often not possible and so the patient may still experience some postoperative difficulties. These difficulties can further result in additional requirements for rehabilitation or even corrective surgery, thus adding to the cost of the entire treatment. Greater comfort and performance can be achieved where some of the components are customized, based on actual patient data. An example would be the socket fixation for a total hip joint replacement. While a standardized process will often return joint functionality to the patient, incorrect fixation of the socket commonly results in variable motion that may be a discomfort, painful and require extensive physiotherapy to overcome. Customized fixtures can be made directly in titanium or cobalt–chromium (both of which are widely used for implants) using powder bed fusion technology. Such custom devices would reduce the previously mentioned problems by making it possible to more precisely match the original or preferred geometry and kinematics. The use of metal systems provides considerable benefit here. While metal AM systems are not capable of producing the smooth surface finish required for effective joint articulation, the characteristic slight roughness can actually benefit osteointegration when placed inside the bone. Smooth joint articulation can be achieved through extensive polishing and use of coatings. Most metal systems may provide custom-shaped implants, but the use of highly focused energy beams will mean that the microstructure will be different and the parts may be more brittle than their equivalent cast or forged components; making brittle fracture from excessive impact loading a distinct possibility. An excellent example of this can be found in the case shown in Fig. 19.4, where Prof. J. Poukens led a multidisciplinary team to implant a complete titanium mandibular joint into an 83-year-old woman [10].

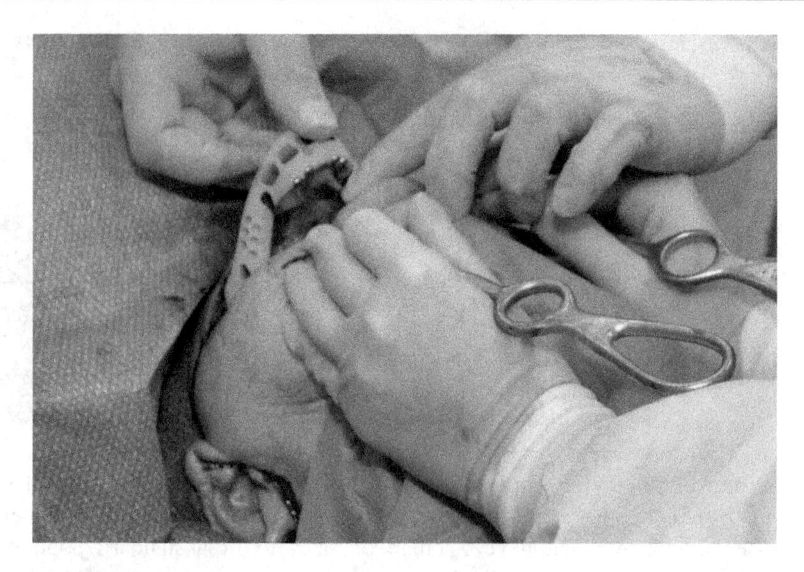

Fig. 19.4 Titanium jaw implant being located during surgery

19.3.3 Manufacturing

There are now examples where customized prosthetics have found their way into mainstream product manufacture. The two examples that are most well known in the industry are in-the-ear hearing aids from companies like Siemens and Phonak and the Invisalign range of orthodontic aligners as developed by Align Technology [11]. These examples are discussed in detail in Chap. 16. Both of these applications involve taking precise data from an individual and applying this to the basic generic design of a product. The patient data are generated by a medical specialist who is familiar with the procedure and who is able to determine whether the treatment will be beneficial. Specialized software is used that allows the patient data to be manipulated and incorporated into the medical device.

One key to success for customized prosthetics is the ability to perform the design process quickly and easily. The production process often involves AM plus numerous other conventional manufacturing tasks, and in some cases the parts may even be more expensive to produce; but the product will perform more effectively and can sell at a premium price because it has components which suit a specific user. This added value can make the prosthetic less intrusive and more comfortable for the user. Additionally, the use of a direct digital manufacturing makes it easier for manufacturers and practitioners.

19.3.4 Tissue Engineering and Organ Printing

The ultimate in fabrication of medical implants would be the direct fabrication of replacement body parts. This can feasibly be done using AM technology, where the materials being deposited are living cells, proteins, and other materials that assist in the generation of integrated tissue structures. However, although there is a great deal of active research in this area, practical applications are still in the main some way off. The most likely approach would be to use printing and extrusion-based technology to undertake this deposition process. This is because droplet-based printing technology has the ability to precisely locate very small amounts of liquid material and extrusion-based techniques are well suited to build soft tissue scaffolding. However, ensuring that these materials are deposited under environmental conditions conducive to cell growth, differentiation, and proliferation is not a trivial task. This methodology could eventually lead to the fabrication of complex, multicellular soft tissue structures like livers, kidneys, and even hearts. There are now even a number of 3D cell printers commercially available that can create simple layer-wise formations of cells, primarily for testing and experimental purposes.

A slightly more indirect approach that is more appropriate to the regeneration of bony tissue would be to create a scaffold from a biocompatible material that represents the shape of the final tissue construct and then add living cells at a later juncture. Scaffold geometry normally requires a porous structure with pores of a few hundred microns across. This size permits good introduction and ingrowth of cells. A microporosity is often also desirable to permit the cells to insert fibrils in order to attach firmly to the scaffold walls. Different materials and methods are currently under investigation, but normally such approaches use bioreactors to incubate the cells prior to implantation. Figure 19.5 shows a scaffold created for producing a mixture of bone and cartilage and then implanted into a rabbit [12]. The scaffold was a mixture of polycaprolactone (PCL) which acts as a matrix material, which is also biodegradable. Mixing tri-calcium phosphate (TCP) enhances the biocompatibility with bone to encourage bone regeneration and also enhances the compressive modulus of the scaffold. Even with this approach, it is still a challenge to maintain the integrity of the scaffold for sufficient lengths of time for healthy and strong bone to form. While using this approach to create soft tissue structures or load-bearing bone is also some way from reality, some non-load-bearing bone constructs have already been commercially proven [13].

19.4 Software Support for Medical Applications

There are a number of software tools available to assist users in preparing medical data for AM applications. Initially, such software concentrated on the translation from medical scanner systems and the creation of the standard STL files. Models made were generally replicas of the medical data. With the advent of the DICOM scanner standard, the translation tools became unnecessary and it became necessary for such systems to add value to the data in some way. The software systems therefore evolved to include features where models could be manipulated and measured and where surgical procedures like jawbone resections could be simulated in order to determine locations for surgical implants. These have further

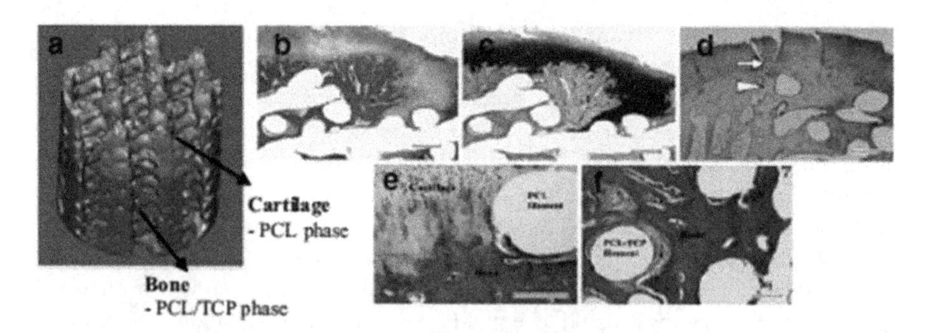

Fig. 19.5 Hybrid scaffolds composed of two phases: (**a**) Polycaprolactone (PCL) layer for cartilage tissue and bottom PCL/TCP (Tri-calcium phosphate) layer for bone. (**b**–**f**) Implantation in a rabbit for 6 months revealed formation of subchondral bone in the PCL/TCP phase and cartilage-like tissue in PCL phase. Bar is 500 μm in (**b**–**d**) and 200 μm in (**e**) and (**f**)

evolved to include software tools for inclusion of CAD data in order to design prosthetic devices or support for specific surgical procedures.

Consider the application illustrated in Fig. 19.6 [14]. In this application, a prosthetic denture set is fixed by drilling precisely into the jawbone so that posts can be placed for anchoring the dentures. A drill guide was developed using AM, positioning the drill holes precisely so that the orthodontist could drill in the correct location and at the correct angle. The software system allows the design of the drill guide to be created, based on the patient data taken from medical scans. AM models can also be used in the development of the prosthetic itself.

Most CAD/CAM/CAE tools are used by engineers and other professionals who generally have good computer skills and an understanding of the basic principles of how such tools are constructed. Clinicians have very different backgrounds and their basic understanding is of biological and chemical sciences with a deep knowledge of human anatomy and biological construction. Computer tools must therefore focus on being able to manipulate the anatomical data without requiring too much knowledge of CAD, graphics, or engineering construction. Software support tools for AM-related applications should therefore provide a systematic solution where different aspects of the solution can be dealt with at various stages so that the digital data are maintained and used most effectively, like the application in Fig. 19.6 where software and AM models were used at various stages to evaluate the case and to assist in the surgical procedure.

Tissue engineering is where AM is heading in the medical arena, leading to direct manufacture of medical replacement parts. Software tools that deal with these applications are likely to be very different from conventional CAD/CAM tools. This is because the data are constructed in a different form. Medical data are almost by definition freeform. If it is to be accurately reproduced, then these models require large data files. In addition, the scaffolds to be created will be highly porous, with the pores in specific locations. STL files are likely to be somewhat useless in these applications, plus if the STL files included the pore architecture they would be inordinately large. Figure 19.5a for example, would normally be made using an

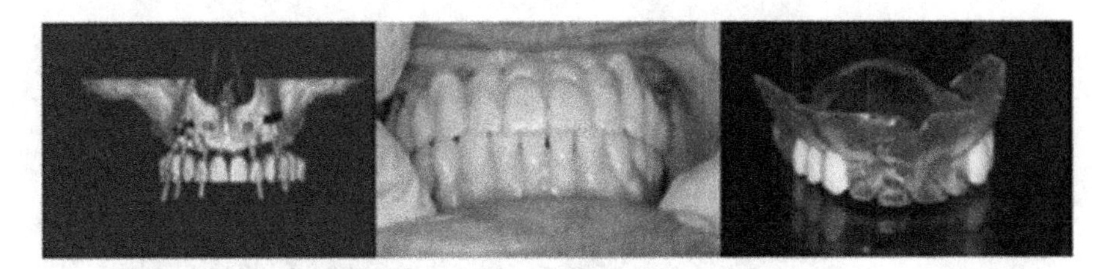

Fig. 19.6 Drill guides developed using AM-related software and machines

extrusion process similar to FDM, where each cross-member of the scaffold would normally correspond to an extrusion road. It would be somewhat pointless for every cross-member to be described using STL, since the slices correspond to the thickness and location of these cross-member features. Most scaffold fabrication systems, like the 3D-Bioplotter from EnvisionTEC [15], shown in Fig. 19.7, include an operating system that includes a library of scaffold fill geometries that include pore size and layer thickness rather than STL slicing systems.

19.5 Limitations of AM for Medical Applications

Although there is no doubt that medical models are useful aids to solving complex surgical problems, there are numerous deficiencies in existing AM technologies related to their use to generate medical models. Part of the reason for this is because AM equipment was originally designed to solve problems in the more widespread area of manufactured product development and not specifically to solve medical problems. Development of the technology has therefore focused on improvements to solve the problems of manufacturers rather than those of doctors and surgeons. However, recent and future improvements in AM technology may open the doors to a much wider range of applications in the medical industry. Key issues that may change these deficiencies in favor of using AM include:

- Speed
- Cost
- Accuracy
- Materials
- Ease of use

By analyzing these issues, we can determine which technologies may be most suitable for medical applications as well as how these technologies may develop in the future to better suit these applications.

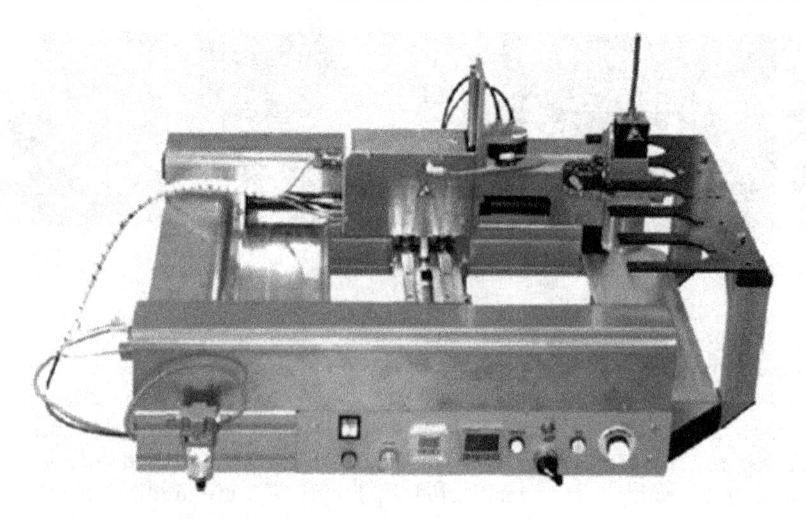

Fig. 19.7 The EnvisionTEC Bioplotter. Note the interchangeable extrusion head system and the extensive use of stainless steel in the fabrication

19.5.1 Speed

AM models can often take a day or even longer to fabricate. Since medical data needs to be segmented and processed according to anatomical features, the data preparation can in fact take much longer than the AM building time. Furthermore, this process of segmentation requires considerable skill and understanding of anatomy. This means that medical models can effectively only be included in surgical procedures that involve long-term planning and cannot be used, for example, as aids for rapid diagnosis and treatment in emergency operations.

Many AM machines now have excellent throughput rate, both in terms of build speed and post-processing requirements. A few more iterations towards increasing this throughput could lead to these machines being used in outpatient clinics, at least for more effective diagnosis. However, it must be understood that this use must be in conjunction with improvements in supporting software for 3D model generation that reduces the skill requirements and increases the level of data processing automation. For tissue engineering applications, the time frames are a lot longer since we must wait for cells to proliferate and combine in the bioreactors. However, the sooner we can get to the stage of seeding scaffolds with cells, the better.

19.5.2 Cost

Using AM models to solve manufacturing problems can help save millions of dollars for high-volume production, even if only a few cents are saved per unit. For the medical product (mass customization) manufacturing applications mentioned earlier, machine cost is not as important as perhaps some other factors. In comparison, the purpose of medical models for diagnosis, surgical planning, and prosthetic development is to optimize the surgeon's planning time and to improve quality, effectiveness, and efficiency. These issues are more difficult to quantify in terms of cost, but it is clear that only the more complex cases can easily justify the expense of the models. The lower the machine, materials, and operating costs, the more suitable it will be for more medical models. Some machines are very competitively priced due to the use of low-cost, high-volume technologies, like inkjet printing. Some other processes have lower-cost materials, but this relates to consumable costs, which can also be reduced with increase of volume output.

19.5.3 Accuracy

Many AM processes are being improved to create more accurate components. However, many medical applications currently do not require higher accuracy because the data from the 3D imaging systems are considerably less accurate than the AM machines they feed into. However, this does not mean that users in the medical field should be complacent. As CT and MRI technologies become more accurate and sophisticated, so the requirements for AM will become more challenging. Indeed some CT machines appear to have very good accuracy when used properly. Also, this generally relates to medical models for communication and planning, but where devices are being manufactured the requirements for accuracy will be more stringent. Applications which require precise fitting of implants are now becoming commonplace.

19.5.4 Materials

Only a few AM polymer materials are classified as safe for transport into the operating theater and fewer still are capable of being placed inside the body. Those machines that provide the most suitable material properties are generally the most expensive machines. Powder-based systems are also somewhat difficult to implement due to potential contamination issues. This limits the range of applications for medical models. Many AM machine manufacturers now have a range of materials that are clinically approved for use in the operating theater.

Metal systems, on the other hand, are being used regularly to produce implants using a range of technologies, as reported by Wohlers [16]. Of these, it appears that titanium is the preferred material, but Cobalt Chromium and Stainless Steel are both available candidates that have the necessary biocompatibility for certain applications.

19.5.5 Ease of Use

AM machines generally require a degree of technical expertise in order to achieve good quality models. This is particularly true of the larger, more complex, and more versatile machines. However, these larger machines are not particularly well suited to medical laboratory environments. Coupled with the software skills required for data preparation, this implies a significant training investment for any medical establishment wishing to use AM. While software is a problem that all AM technologies face, it doesn't help that the machines themselves often have complex setup options, materials handling, and general maintenance requirements.

19.6 Further Development of Medical AM Applications

It is difficult to say whether a particular AM technology is more or less suited to medical applications. This is because there are numerous ways in which these machines may be applied in this field. One can envisage that different technologies may find their way into different medical departments due the specific benefits they provide. However, the most common commercial machines certainly seem to be well suited to being used as communication aids between surgeons, technical staff, and patients. Models can also be suitable for diagnostic aids and can assist in planning, the development of surgical procedures, and for creating surgical tools and even the prosthetics themselves. Direct fabrication of implants and prosthetics is however limited to the direct metal AM technologies that can produce parts using FDA (The US Food and Drug Administration) certified materials plus the small number of technologies that are capable of non-load-bearing polymer scaffolds.

For more of these technologies to be properly accepted in the medical arena, a number of factors must be addressed by the industry:

- Approvals
- Insurance
- Engineering training
- Location of the technology

19.6.1 Approvals

While a number of materials are now accepted by the FDA for use in medical applications, there are still questions regarding the best procedures for generating models. Little is known about the materials and processes outside of the mainstream AM industry. Approval and certification of materials and processes through ASTM will certainly help to pave the way towards FDA approval, but this can be a very long and laborious process.

Those (relatively few) surgeons who are aware of the processes seem to achieve excellent results and are able to present numerous successful case studies. However, the medical industry is (understandably) very conservative about the introduction of these new technologies. Surgeons who wish to use AM generally have to resort to creative approaches based on trusting patients who sign waivers, the use of commercial AM service companies, and word of mouth promotion. Hospitals and health authorities still do not have procedures for purchase of AM technology in the same way they might purchase a CT machine.

19.6.2 Insurance

Many hospitals around the world treat patients according to their level of insurance coverage. Similar to the aforementioned issue of approvals, insurance companies do not generally have any protocols for coverage using AM as a stage in the treatment process. It may be possible for some schemes to justify AM parts based on the recommendations of a surgeon, but some companies may question the purpose of the models, requiring additional paperwork that may deter some surgeons from adopting that route.

Again, this issue may be solvable through a process of legitimizing the industry. In the past, AM was considered as a technology suitable mainly for prototypes in the early phases of product development. As we move more and more into mainstream manufacturing, the industry and consumers become more demanding. Part of the satisfying of this demand is the certification process. Insurance companies are also more likely to accept these technologies as part of the treatment process if there are effective quality control mechanisms in place. Also, the increasing number of successful applications using metal systems may lead to the polymer-based machines also becoming more acceptable.

19.6.3 Engineering Training

Creating AM models requires skills that many surgeons and technicians will not possess. While many of the newer, low-cost machines do not require significant skill to operate, preparation of the files and some post-processing requirements may require more ability. The most likely skills required for the software-based processing can be found in radiology departments since the operations for preparation of a software model are similar to manipulation and interpretation of CT and MRI models. However, technicians in this area are not used to building and manipulating physical models. These skills can however be found in prosthetics and orthotics departments. It is generally quite unusual to find radiology very closely linked with orthotics and prosthetics. The required skills are, therefore, distributed throughout a typical hospital.

19.6.4 Location of the Technology

AM machines could be located in numerous medical departments. The most likely would be to place them either in a laboratory where prosthetics are produced, or in a specialist medical imaging center. If placed in the laboratories, the manual skills will be present but the accessibility will be low. If placed in imaging centers, the accessibility will be high but the applications will probably be confined to visualization rather than fabrication of medical devices. Fortunately, most hospitals are now well equipped with high-speed intranets where patient data can be accessed quickly and easily. A separate facility that links closely to the patient data network and one that has skilled software and modeling technicians for image processing and for model post-processing (and associated downstream activities) may be a preference.

19.6.5 Service Bureaus

It can be seen that most of the hurdles for AM adoption are essentially procedural in nature rather than technical. A concerted effort to convince the medical industry of the value of AM models for general treatment purposes is, therefore, a key advancement that will provide a way forward.

There are small but increasing number of companies developing excellent reputations by specializing in producing models for the medical industry. Companies like Medical Modelling LLC [7] and Anatomics [17] have been in business for a number of years, not just creating models for surgeons but assisting in the development of new medical products. These service bureaus fill the skill gap between the medics and the manufacturers. At the moment, this technology is not well understood in the medical industry and it may be some time before it can be properly assimilated. Eventually, AM technology will become better suited to a wider range of medical applications and at this point, the hospitals and clinics may have their own machines with the inbuilt skills to use them properly. Furthermore, the large medical product manufacturers will also see the benefits of this technology in product development and DDM. As the technology becomes cheaper, easier to use and better suited to the application, such support companies may no longer be necessary to support the industry. This is something the AM industry has seen in other application sectors. In the meantime, these companies provide a vital role in supporting the industry from both sides.

19.7 Aerospace Applications

As mentioned, aerospace is another industry that has traditionally applied AM since it was introduced. The primary advantage for production applications in aerospace is the ability to generate complex engineered geometries with a limited number of processing steps. Aerospace companies have access to budgets significantly larger than most industries. This is, however, often necessary because of the high performance nature of the products being produced.

19.7.1 Characteristics Favoring AM

Significant advantages could be realized if aerospace components were improved with respect to one or more of these characteristics:

Lightweight: Anything that flies requires energy to get it off the ground. The lighter the component, the less energy is required. This can be achieved by use of lightweight materials, with high strength to weight ratio. Titanium and aluminum have traditionally been materials of choice because of this. More recently carbon fiber reinforced composites have gained popularity. However, it is also possible to address this issue by creating lightweight structures with hollow or honeycomb internal cores. This kind of topology optimization is quite easy to achieve using AM.

High temperature: Both aircraft and spacecraft are subject to high-temperature variations, with extremes in both high and low temperatures. Engine components are subject to very high temperatures where innovative cooling solutions are often employed. Even internal components are required to be made from flame retardant materials. This means that AM generally requires its materials to be specially tailored to suit aerospace applications.

Complex geometry: Aerospace applications can often require components to have more than one function. For example, a structural component may also act as a conduit or an engine turbine blade may also have an internal structure for passing coolant through it. Furthermore, geometric specifications for parts may be determined by complex mathematical formulae based on fluid flow, etc.

Economics: AM enables economical low production volumes, which are common in aerospace, since hard tooling is not needed. Designers and manufacturing engineers need not design and fabricate molds, dies, or fixtures, or spend time on complex process planning (e.g., for machining) that conventional manufacturing processes require.

Digital spare parts: Many aircraft have very long useful lives (20–50 years or longer) which places a burden on the manufacturer to provide spare parts. Instead of warehousing spares, or maintaining manufacturing tooling, over the aircraft's long life, the usage of AM enables companies to maintain digital models of parts. This can be much easier and less expensive than warehousing physical parts or tools.

19.7.2 Production Manufacture

All of the major aerospace companies in the USA and Europe have pursued production applications of AM for many years. Boeing, for example, has installed tens of thousands of AM parts on their military and commercial aircraft. Reportedly, over 200 different parts are flying on at least 16 models of aircraft [18]. Until recently, all of these were nonstructural polymer parts for military or space applications. For commercial aircraft, polymer parts need to satisfy flammability requirements, so their adoption needed to wait until flame retardant polymer PBF materials were developed. For metals, material qualification and part certification took many years to achieve. In addition to parts manufacturing, aerospace companies are also developing new higher-performance materials in both metals and polymers, as well as processing methods.

Some of the first large scale, metal part production manufacturing applications are emerging in the aerospace industry. GE purchased Morris Technologies in 2012 as part of a major investment in metal AM for the production of gas turbine engine components. The part that has received the most attention is a new fuel nozzle design for the CFM LEAP (Leading Edge Aviation Propulsion) turbofan engine, as shown in Fig. 19.8 [19]. The new fuel nozzle took the part consolidation concept to new levels by reportedly combining 18–20 parts into one integrated design and avoiding many brazed joints and assembly operations. This new design is projected to have a useful life five times that of the original design, a 25 % weight reduction, and additional cost savings realized through optimizing the design and production process. Additionally, the fuel nozzle was engineered to reduce carbon build up, making the nozzle more efficient.

Production manufacturing of the nozzles is scheduled to begin in 2015. Each engine contains 19 fuel nozzles and more than 4,500 engines have been sold to date, so production volume could exceed 100,000 total parts by 2020. This is claimed to save 1000 lb of weight out of each engine. The nozzles are fabricated using the cobalt-chrome material fabricated in EOS metal PBF machines. Parts are likely to be stress relieved while still in the powder bed, followed by hot isostatic pressing (HIP) to ensure that the parts are fully dense. An in-process inspection technology was developed jointly between GE Aviation and Sigma Labs for use in the EOS machines. Called PrintRite3D, the technology is used to inspect and verify metal parts while they are being fabricated. It consists of software for closed-loop control and data analysis to determine if parts are within specification, along with a set of sensors to that allow a controlled weld pool volume. The company claims that their technology enables the control of an alloy's microstructure by controlling the temperature history throughout the part.

Several metal PBF vendors offer a variety of titanium alloy materials for use on their machines. One recent development is a variant of titanium called Ti–6Al–4V ELI, which denotes a titanium alloy with about 6 % aluminum, 4 % vanadium, and Extra Low Interstitials (ELI), meaning the alloy has lower specified limits on iron and interstitial elements carbon and oxygen. The ELI variant has better corrosion resistance and mechanical properties, particularly at cryogenic temperatures, than standard Ti–6Al–4V. Due to these properties, the alloy has excellent biocompatibility and is of great interest in the medical industry. Its light weight, high strength, and high toughness properties mean that it is a good candidate for aerospace applications, as well. A recent ASTM standard addresses specifications for metal PBF parts fabricated from this alloy.

Fig. 19.8 GE Aviation fuel nozzle

As another example, Airbus has developed a second-generation aluminum-magnesium-scandium alloy, called ScalmalloyRP, for metal PBF processes. The material reportedly has mechanical properties that are twice as good as commercially available aluminum alloys, with high corrosion resistance and good fatigue properties [18].

Early efforts towards production manufacturing with polymer PBF systems were performed at Boeing. In 2002, a Boeing spin-off company, On Demand Manufacturing, was formed. Their first application was to manufacture environmental control system ducts to deliver cooling air to electronics instruments on F-18 military jets. They rebuilt several SLS Sinterstation machines in order to ensure that they could fabricate these parts reliably and repeatably. ODM was purchased by RMB Products, Inc., in 2005 but continues operations today.

Airbus investigated topology optimization applications in order to develop part designs that were significantly lighter than those suitable for conventional manufacturing processes. Shown in Fig. 19.9 is an A320 nacelle hinge bracket that was originally designed as a cast steel part, but was redesigned to be fabricated in a titanium alloy using PBF [20]. Reportedly, they trimmed 10 kg off the mass of the bracket, saving approximately 40 % in weight. This study was performed as part of a larger effort to compare life-cycle environmental impacts of part design.

Many more production applications of AM can be expected in the near future as materials improve and production methods become standardized, repeatable, and certified. New design concepts can be expected, such as the A320 hinge bracket, for

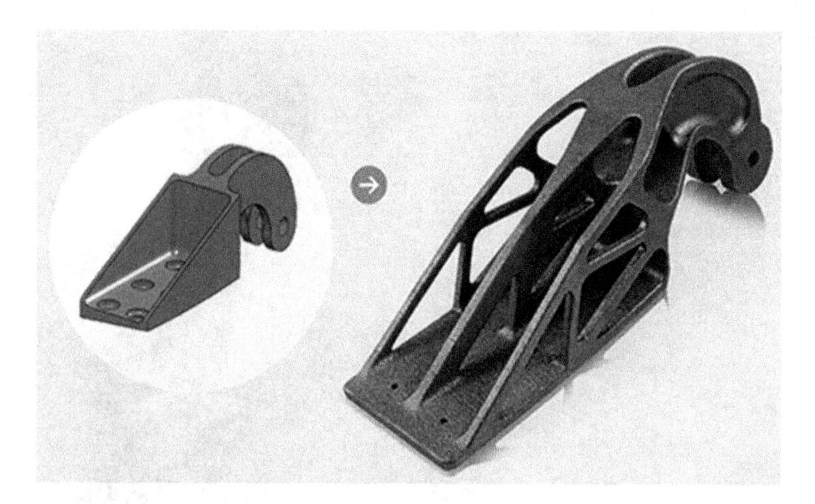

Fig. 19.9 A320 hinge bracket redesigned for AM. Courtesy EOS GmbH

not only piece parts, but entire modules. AM vendors are developing larger frame machines so that larger parts can be fabricated, opening up new opportunities for structural metal components and functional polymer parts.

19.8 Automotive Applications

As mentioned, the automotive industry was one of the early adopters of AM and personnel at these companies pioneered many types of AM applications in product development. Companies in this industry continue to be heavy users of AM, accounting for approximately 17 % of all expenditures on AM in 2013. This positions the automotive industry behind only industrial/business machines (18.5 %) and consumer products/electronics (18 %), which are very large and broad industries in terms of the largest users of AM.

Since production volumes in the automotive industry are often high (100,000s per year), AM has typically been evaluated as too expensive for production manufacturing, in contrast to the aerospace industry. To date, most manufacturers have not committed to AM parts on their mass-produced car models. However, there have been niche applications of AM that are worth exploring.

As mentioned in the Historical Developments section, a variety of rapid prototyping applications were developed by automotive companies and their suppliers. In addition to RP and rapid tooling, suppliers to this industry used AM parts to debug their assembly lines. That is, they used AM parts to test assembly operations and tooling to identify potential problems before production assembly commenced. Since model line change-over involves huge investments, being able to avoid problems in production yielded very large savings.

In the metal PBF area, Concept Laser, a German company, introduced their X line 1000R machine recently, which has a build chamber large enough to accommodate a V6 automotive engine block. This machine was developed in collaboration with Daimler AG. It is not clear if they intend their automotive customers to fabricate production engine blocks in this machine, but they claim the machine was developed with production manufacture in mind. According to Concept Laser, the 1000R is capable of building at a rate of 65 cm^3 per hour, which is fast compared to some other metal PBF machines. Additionally, the machine was designed with two build boxes (powder chambers) on a single turntable so that one build box could be used for part fabrication, while the other could be undergoing cool-down, part removal, pre-heating or other non-part-building activities.

For specialty cars or low-volume production, AM can be economical for some parts. Applications include custom parts on luxury cars or replacement parts on antique cars. The example of Bentley Motors was given in Chap. 17. Polymer PBF was used to fabricate some custom interior components, such as bezels, that were subsequently covered in leather and other materials. Typically, Bentley has production volumes of less than 10,000 cars for a given model, so this qualifies as low production volume.

Local Motors is a small company that is experimenting with crowdsourcing and other novel methods of new vehicle development. They participated in the DARPA FANG military vehicle development exercise, for example. They utilize AM when it makes sense for their applications. In a separate initiative, they conducted a crowd-sourced car design project, with the requirement that the majority of the car would be fabricated by AM. They plan to fabricate the body and structural components using a new, large-frame material extrusion machine from Oak Ridge National Laboratories at the International Machine Technology Show in September 2014.

Among the racing organizations, Formula 1 has been a leader in adopting AM. Originally using AM for rapid prototyping, some of the teams started putting AM parts on their race cars in the early to mid-2000s. These were typically nonstructural polymer PBF parts. Similarly to the aerospace industry, Formula 1 teams utilized AM models for wind tunnel testing of scale models, as well as parts for full size car models. Teams from other racing organizations, including Indy and NASCAR, have also made AM an integral aspect of their car development process.

19.9 Exercises

1. How does Computerized Tomography actually generate 3D images? Draw a sketch to illustrate how it works, based on conventional knowledge of X-ray imaging.
2. What are the benefits of using color in production of medical models? Give several examples where color can be beneficial.

3. Why might extrusion-based technology be particularly useful for bone tissue engineering?
4. What AM materials are already approved for medical applications and for what types of application are they suitable?
5. Consider the manufacture of metal implants using AM technology. Aside from the AM process, what other processing is likely to be needed in order to make a final part that can be implanted inside the body?
6. Why would AM be particularly useful for military applications?
7. How can the use of AM assist in the development of a new, mass-produced automobile?
8. Find some examples of parts made using AM in F1 and similar motorsports.

References

1. Jacobs PF (1992) Rapid prototyping & manufacturing: fundamentals of stereolithography. SME, Dearborn
2. Jacobs PF (1996) Stereolithography and other RP&M technologies: from rapid prototyping to rapid tooling. Society of Manufacturing Engineers and American Society of Mechanical Engineers, Dearborn
3. DICOM, Digital Imaging and Communications in Medicine, developed by the Medical Imaging and Technology Alliance Division of the National Electrical Manufacturers Association. www.medical.nema.org
4. Objet medical application case study on conjoined twin separation. www.objet.com/Docs/Med_Twins_A4_low.pdf
5. Cordis, discussion on the use of Stereocol resin for medical applications. www.cordis.europa.eu/itt/itt-en/97-4/case.htm
6. Connex, multiple material AM system. www.objet.com/3D-Printer/Connex500/
7. Medical Modeling Inc. http://www.medicalmodeling.com/
8. Wang J, Ye M, Liu Z, Wang C (2009) Precision of cortical bone reconstruction based on 3D CT scans. Comput Med Imaging Graph 33(3):235–241
9. Total jaw implant. www.xilloc.com/patients/stories/total-mandibular-implant
10. Gibson I, Cheung LK, Chow SP, Cheung WL, Beh SL, Savalani M, Lee SH (2006) The use of rapid prototyping to assist medical applications. Rapid Prototyp J 12(1):53–58
11. Align, clear orthodontic aligners using AM technology. www.invisalign.com
12. Shao XX, Hutmacher DW, Ho ST et al (2006) J Biomaterials 27(7):1071
13. Osteopore, tissue engineering technology. www.osteopore.com.sg
14. Materialise. www.materialise.com
15. EnvisionTEC, 3D-Bioplotter technology. www.envisiontec.com
16. Wohlers TT (2009) Wohlers report: rapid prototyping and tooling state of the industry; annual worldwide progress report. Wohlers, Fort Collins
17. Anatomics. www.anatomics.com
18. Wohlers T (2014) Wohlers report 2014: additive manufacturing and 3D printing state of the industry, annual worldwide progress report. Fort Collins, Wohlers
19. GE Aviation. http://www.ge.com/stories/additive-manufacturing
20. EOS. http://www.eos.info/eos_airbusgroupinnovationteam_aerospace_sustainability_study. Accessed 17 June 2014

Business Opportunities and Future Directions

20

Abstract

The current approach for many manufacturing enterprises is to centralize product development, product production, and product distribution in a relatively few physical locations. These locations can decrease even further when companies off-shore product development, production, and/or distribution to other countries/companies to take advantage of lower resource, labor or overhead costs. The resulting concentration of employment leads to regions of disproportionately high underemployment and/or unemployment. As a result, nations can have regions of underpopulation with consequent national problems such as infrastructure being underutilized, and long-term territorial integrity being compromised (Beale, Rural Cond Trends 11(2):27–31, 2000).

20.1 Introduction

The current approach for many manufacturing enterprises is to centralize product development, product production, and product distribution in a relatively few physical locations. These locations can decrease even further when companies offshore product development, production, and/or distribution to other countries/companies to take advantage of lower resource, labor or overhead costs. The resulting concentration of employment leads to regions of disproportionately high underemployment and/or unemployment. As a result, nations can have regions of underpopulation with consequent national problems such as infrastructure being underutilized, and long-term territorial integrity being compromised [1].

Because of recent developments in additive manufacturing, as described in this book, there is no fundamental reason for products to be brought to markets through centralized development, production, and distribution. Instead, products can be brought to markets through product conceptualization, product creation, and product propagation being carried out by individuals and communities in any geographical region.

I. Gibson et al., *Additive Manufacturing Technologies*,
DOI 10.1007/978-1-4939-2113-3_20

In this chapter, *conceptualization* means the forming and relating of ideas, including the formation of digital versions of these ideas (e.g., CAD); *creation* means bringing an idea into physical existence (e.g., by manufacturing a component); and *propagation* means multiplying by reproduction through digital means (e.g., through digital social networks) or through physical means (e.g., by distributed AM production).

Many companies already use the Internet to collect product ideas from ordinary people from diverse locations. However, these companies are feeding these ideas into the centralized physical locations of their existing business operations for detailed design and creation. Distributed conceptualization, creation, and propagation can supersede concentrated development, production, and distribution by combining AM with novel human/digital interfaces which, for instance, enable non-experts to create and modify shapes. Additionally, body/place/part scanning can be used to collect data about physical features for input into digitally enabled design software and onward to AM.

Web 2.0 is considered as the second generation of the Internet, where users can interact with and *transform* web content. The advent of the Internet allowed any organization, such as a newspaper publisher, to deliver information and content to anyone in the world. More recently, however, social networking sites such as Facebook, or auction web sites such as eBay, enable consumers of web content to also be content creators. These, and most new web sites today, fall within the scope of Web 2.0.

AM makes it possible for digital designs to be transformed into physical products at that same location or any other location in the world (i.e., "design anywhere, build anywhere"). Moreover, the web tools associated with Web 2.0 are perfect for the propagation of product ideas and component designs that can be created through AM. The combination of Web 2.0 with AM can lead to new models of entrepreneurship.

Distributed conceptualization and propagation of digital content is known as digital entrepreneurship. However, the exploitation of AM to enable distributed creation of physical products goes beyond just digital entrepreneurship. Accordingly, the term, *digiproneurship* was coined to distinguish distributed conceptualization, propagation, and creation of *physical* products from distributed conceptualization and propagation of just digital content. Thus digiproneurship is focused on transforming *digi*tal data into physical *pro*ducts using an entrepre*neurship* business model. Short definitions of the terms introduced in this section are summarized in Fig. 20.1.

Web 2.0 + AM has the potential to generate distributed, sustainable employment that is not vulnerable to off-shoring. This form of employment is not vulnerable to off-shoring because it is based on distributed networks in which resource costs are not a major proportion of total costs. Employment that is generated is environmentally friendly because, for example, it involves much lower energy consumption than the established concentration of product development, production, and distribution, which often involves shipping of products worldwide from centralized locations.

Definition of Terms
Conceptualization
formation and relating of ideas or concepts by individuals or communities
Creation
bringing something into existence through digitally-enabled design and production
Propagation
multiplying by distribution/reproduction through digitally-enabled networks
digiproneurship
digital to physical product entrepreneurship

Fig. 20.1 Definition of terms

As discussed throughout this book (particularly in Chaps. 17, 18 and 20), developments in AM offer possibilities for new types of products. Thus, there are many potential markets for the outputs of digiproneurship.

20.2 What Could Be New?

20.2.1 New Types of Products

Developments in AM, together with developments in advanced Information and Communication Technologies (aICT), such as more intuitive human interfaces for design, Web 2.0, and digital scanning, are making it possible for person-specific/location-specific and/or event-specific products to be created much more quickly and at much lower cost. These products can have superior characteristics compared to products created through conventional methods. In particular, AM can enable previously intractable trade-offs to be overcome. For example, design trade-offs such as manufacturing complexity versus assembly costs can be overcome (e.g., geometrically complex products can now be produced as one piece rather than having to be assembled from several pieces); material selection trade-offs such as performance requirements versus microstructures can be overcome (e.g., turbine blades can now have both high strength and high thermal performance in different locations); economic trade-offs such as person-specific fit and/or functionality versus production time and/or cost can be overcome (e.g., customized prosthetics, such as hearing aids with person-specific fit, can be produced rapidly).

When utilizing an additive approach to production, the consumption of non-value adding resources can be radically reduced during the creation of physical goods. Further, the amount of factory equipment needed and, therefore, factory space needed is reduced. As a result, opportunities for smaller, distributed (even mobile) production facilities increase. Some examples are provided in Table 20.1. Perhaps most importantly, the potential for radically reducing the size of production facilities enables production at point-of-demand.

Although digiproneurship is probably best enabled by AM, any digitally driven technology which directly transforms digital information into a physical good can

Table 20.1 Radical reductions in the consumption of non-value adding resources

Example	First order effect	Second order effect
No need for molds/dies	Less material consumption	Lower start-up costs
Fewer parts to join	Less joining equipment	Less capital tied up in infrastructure
Fewer parts to assemble	Less labor and less assembly equipment	No need to off-shore production to low labor cost markets
No spare parts are stocked	Less storage space	Reduced factory and warehousing size

fall within the scope of digiproneurship. This can include the fabrication of structures which enclose space (such as for housing) whereby each individual piece could be created using a digitally driven cutting operation and then assembled at the point of need into a usable dwelling.

It is very important to note that the limitations of manufacturing equipment and the need for expert knowledge of microstructures and material performance have previously restricted the value of direct consumer control over content. Thus, most examples of consumer-produced content are for nonphysical products [2]. For example, a person who reads a newspaper (consumer of the newspaper) walks down a street and sees something newsworthy. The person takes a photograph of it. The person sends the image to the newspaper. The photograph is included in the newspaper, and hence the person becomes a partial producer of what they consume. While such forms of consumer input are established, it is only recently that developments in aICT and AM make possible consumer input into a wide range of physical goods.

From an engineering and design standpoint, AM technologies are becoming more accurate, they can directly build small products (micron-sized) and very large products (building-sized). New materials have been developed for these processes, and new approaches to AM are being introduced into the marketplace. From a business-strategies standpoint, AM technologies are becoming faster, cheaper, safer, more reliable, and environmentally friendly. As each of these advancements becomes available within the marketplace, new categories of physical goods become competitive for production using AM versus conventional manufacturing. Combination of aICT with AM thus offers a wide range of opportunities for innovation in products and product services. Opportunities exist for individuals (e.g., at home), B2B (Business to Business), and B2C (Business to Consumer). Further, opportunities exist for creation of designs or creation of physical components. Thus a digiproneur could be someone who: (1) creates digital tools for use by consumers or other digiproneurs; (2) creates designs which are bought by consumers or businesses; (3) creates physical products from digital data; or (4) licenses or operates enabling software or machinery in support of digiproneurship.

By replacing concentrated product development, production, and distribution with distributed product conceptualization, creation, and propagation it is possible for individuals or communities to bring products to different types of consumers without needing to make large investment in market research, design facilities, production facilities, or distribution networks. The reasons for this are further explained in the following subsection.

20.2.2 New Types of Organizations

Web 2.0 technologies have spawned a convergence of traditional craft with technologies. One need only attend a local Maker Faire to see the gamut of entrepreneurs offering products made traditionally, with lots of electronics, and with AM content. To support the emerging communities of craftsmen and women, online portals, blogs, and repositories have proliferated. For example, some portals have been established to focus on 3D Printing (www.3ders.org) or more broadly on making (www.instructables.com). The number of blogs focused on 3D Printing, AM, and making is too numerous to do justice by listing only one or two. Since the creation of 3D digital content can be challenging several repositories of 3D content have been created, the most well-known being Thingiverse (www.thingiverse.com). Even traditional craft-based media have changed with, for example, Make Magazine adopting a synergistic combination of traditional paper distribution with online content and interaction. Each of these examples represents a business entity that was created by an entrepreneur who wanted to leverage Web 2.0 and add value to AM users, companies, or communities.

In traditional manufacturing industries, companies such as MFG.com have become very success as industry matchmakers, finding suppliers or customers for companies around the world. They provide services for establishing supply chains and handling logistics for companies. Their expansion into the AM field seems inevitable and may already have begun. This may cause existing service bureaus adapt their business models. They can join existing networks of parts suppliers or possibly try to build their own networks. They could choose to concentrate on their technology (within AM) or become more consumer focused, possibly becoming a supplier to a virtual storefront company.

From a different perspective, companies can utilize Web 2.0 technologies to engage with their customers to a much greater extent. Customer co-design and crowdsourcing are new terms that relate to this customer focus. Some consumer companies, such as Nike, Dell, and Home Depot, have been pioneers in providing web-based tools that enable customers to configure their own products. We can expect this trend to continue to grow exponentially. New opportunities will emerge for unprecedented levels of customer engagement. We are seeing new companies created to provide customer-designed products, for example sunglasses, that are fabricated locally using AM. One could image kiosks at local shopping malls that are equipped with 3D printers for near real-time fabrication.

The Local Motors crowdsourcing example has been mentioned in earlier chapters. They have been an early adopter of crowdsourcing for automotive vehicles. Many other organizations and companies are experimenting with crowdsourcing technologies and practices for the development of products. It will be interesting to see how a highly technical and integrated product (such as a car) can be developed by hundreds of geographically dispersed individuals who are contributing informally on irregular schedules. From marketing and decision-making perspectives, however, having all of these individuals critique and vote on design alternatives and become invested in the outcome of the group activity can have tremendous benefits in terms of sales. Products may become successful simply because they "went viral" due to high levels of involvement from vocal online communities.

Going beyond Web 2.0 technologies, the area of cloud computing has enabled the emergence of cloud-based design and manufacturing (CBDM) concepts. We are already seeing mechanical CAD companies, such as Dassault Systemes and AutoDesk, offer cloud-based CAD and engineering systems. Some companies are talking about cloud-based AM part fabrication services. CBDM represents a natural evolution of this trend. One challenge for cloud-based manufacturing is the need for hard tooling for part manufacture and product assembly. It is difficult to provide flexible, scalable, "produce anywhere" services if one has to first fabricate a lot of tooling. On the other hand, AM offers a flexible, scalable, "produce anywhere" solution for CBDM. It is likely that we will see CBD products and services incorporate some CBM aspects. Perhaps an obvious first step is a "3D Print" button in cloud-based CAD systems. A longer term possibility is a convergence between cloud-based CAD and supply chain service providers (e.g., MFG.com) so that engineering designers can include manufacturing, vendor, and supply chain consequences of design decisions quickly in a seamless online environment.

20.2.3 New Types of Employment

As summarized in Table 20.1, innovative combinations of AM and aICT help eliminate non-value adding consumption of resources and reduce energy consumption arising from transportation of finished goods. Creation of physical products at point-of-demand can overturn current comparative disadvantages in the creation of physical products for global markets. As an example, today Finland has the comparative *disadvantages* of limited natural resources, far distance from mass markets, and relatively high labor costs. However, aICT + AM has the potential to make Finland's comparative disadvantages become *unimportant* in global value networks. This is because centralized models of physical production can be replaced by distributed models of value creation. In distributed models, design can take place anywhere in the world, and production can take place anywhere else in the world. As a result, there are opportunities for many jobs to be created in Finland by meeting "derived demand" for the software, hardware, and consultancy needed by creation organizations in other parts of the world. This is in addition to the jobs that can be created in Finland by meeting "primary demand" for physical goods which are used in Finland, Russia, Nordic regions, and beyond; or unique designs which can be electronically delivered to consumers worldwide for their creation. Thus, the resources that become important in digiproneurship are creativity, technological savvy, and access to digiproneurship networks.

Innovative combinations of AM and aICT make it possible for creation of diverse product types by people without prior knowledge of design and/or production. Regions of persistent unemployment could be reduced by enabling a dynamic network of aICT + AM micro-businesses and Small and Medium-sized Enterprises (SMEs). These could be distributed among local individuals working from their homes, from their garages, from their small workshops, or from light industrial premises. They could be distributed among families and communities that have a generational investment and an abiding commitment to the regions in which they live. Accordingly, the jobs generated by digiproneurship are resistant to concentration and outsourcing.

The labor cost component of aICT + AM products is relatively low. Thus, these combinations of high technology and low labor input mean that there is little incentive to outsource to low labor cost economies. A summary is provided in Table 20.2 of the factors that can enable overturning of regional disadvantages which might occur in the creation of physical products for global markets.

A diverse range of people and businesses could offer products via digiproneurship. Some examples of these people and business are:

- Artistic individuals who want to create unique physical goods
- Hobby enthusiasts who understand niche market needs
- IT savvy people who are interested in developing novel aICT software tools
- Farmers wanting to diversify beyond offering B&B to the occasional tourist
- Under-employed persons looking to provide supplemental income for their families
- Unemployed people who are reluctant to uproot to major cities to look for work
- Machine shops wanting to diversify and/or better utilize their skilled workforce
- SMEs that want to introduce more customer-specific versions of their product offerings
- Multinational corporations seeking to streamline the design and supply of goods which will be integrated into their products

Thus, digiproneurship represents the intersection of conceptualization, creation, and propagation, as illustrated in Fig. 20.2.

20.3 Digiproneurship

Entrepreneurship involves individuals starting new enterprises or breathing new energy into mature enterprises through the introduction of new ideas. Entrepreneurship is associated with uncertainty because it involves introducing a new idea [3]. Well-known examples of digital entrepreneurship include Facebook, Google, and YouTube. By taking digital entrepreneurship one step further, into the creation of physical goods, digiproneurship represents the next logical step.

Table 20.2 Overturning regional disadvantages in the creation of physical products

Typical location-based disadvantages	aICT + AM potential
Lack of natural resources	Products make use of relatively small quantities of high-quality engineered materials procurable worldwide
High labor costs	Labor content is smaller, but networking and technology integration content is higher
Distance from markets	aICT + AM products can be designed anywhere, propagated digitally and produced at the point-of-need. Shipping costs are minimized

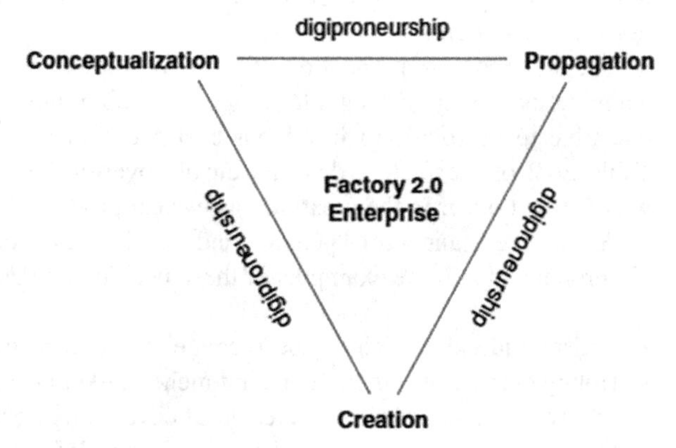

Fig. 20.2 Digiproneurship involves the creation of a business enterprise by connecting conceptualization, propagation, and/or creation

Distributed conceptualization and propagation can reduce the risks traditionally associated with entrepreneurship. In particular, digitally enabled conceptualization and propagation of new concepts and designs for physical products can eliminate the need for costly conventional market research. Further, digitally enabled propagation of product designs to point-of-demand AM facilities can eliminate the need for physical distribution facilities such as large warehouses, costly tooling such as injection molds, and difficult to manage distribution networks. Together, digitally enabled conceptualization, propagation, and creation can eliminate many of the uncertainties and up-front expenses that have traditionally caused many entrepreneurial ventures to fail.

Digiproneurship transcends traditional design paradigms by facilitating the emergence of enterprise through the self-expression of personal feelings and opinions. New digital interfaces which enable non-experts to capture their design intent as physically producible designs could radically transform the way products are conceived and produced.

One of the earliest enterprises that could be considered a digiproneurship enterprise was Freedom of Creation (www.freedomofcreation.com), as discussed in Chap. 20. Subsequently to Freedom of Creation, numerous other digiproneurship activities have been started, including FigurePrints (www.figureprints.com) for creation of World of Warcraft figures and virtual storefronts such as Shapeways (www.shapeways.com) and Ponoko (www.ponoko.com), which are online communities where digiproneurs can sell designs, services, and products.

Digiproneurship opportunities are now being considered early in the conceptualization stage for new products. The Spore game and Spore Creature Creator (www.spore.com) were designed such that Spore creatures, created by the game players, are represented by 3D digital data that can be transferred to an AM machine for direct printing using a color binder jetting process. This is unlike the original World of Warcraft figures, which appear 3D on-screen but are not 3D solid models; and thus require data manipulation in order to prepare the figure for AM. Spore Sculptor (sporesculptor.com) has been set up as a portal for Spore game users to purchase physical representations of their Spore creatures.

In the future, inexpensive, intuitive solid modeling tools, such as Google SketchUp (sketchup.google.com), may be used widely by consumers to design their own products. For many products, safety or intellectual property concerns will likely lead to software which will enable consumers to modify products within expert-defined constraints so that consumers can directly make meaningful changes to products while maintaining safety or other features that are necessary in the end-product.

The success of digiproneurship enterprises is due to their recognition of market needs which can be fulfilled by imaginative product offerings enabled through innovative combinations of aICT and AM. Although pioneers have demonstrated that successful enterprises can be established, the potential for digiproneurship extends significantly beyond the scope of today's technological capabilities and business networks. In particular, as aICT and AM progress, and new business networks are established, the opportunities for successful digiproneurship will expand.

Several research and development priorities for aICT and AM are crucial for further realization of digiproneurship:

Digiproneurship-related research and development priorities for aICT

- Development of geometric manipulation tools with intuitively understandable interfaces which can be used readily by non-experts.
- Application of expert-defined constraints (such as through shape grammars and computational semantics) to enable experts to create versatile parameters for digiproneurship products. These parameters conform to criteria for, e.g., safety and brand, but facilitate the creation of person-specific, location-specific, and/or event-specific versions by non-experts.
- Web-based digiproneurship tools which can enable non-experts to set up and operate their own digitally driven enterprise. These web-based tools encompass market opportunities and business issues; as well as technology characteristics and material properties.
- Additional web-based digiproneurship tools to assist in setting up virtual enterprises that integrate supplier identification, supply chain configuration, marketing, and product delivery.

Digiproneurship research and development priorities for AM

- Continuing the current trend to lower-cost equipment and materials
- Automating and minimizing post-processing of products after production, so that parts can go directly from a machine to the end customer with little or no human interaction
- Continuing the current trend to increasing diversification of machine sizes, speeds and accuracies
- Interfaces to automatically convert multimaterial and multicolor user-specified requirements directly into digital manufacturing instructions without human intervention

As digiproneurship matures, there will be a need for an increasing number of creation facilities that enable digiproneurs to reach customers irrespective of their location. Some of these creation facilities will be the 3D corollary to today's local copy centers. As such, they may even offer AM alongside 2D printers. Further, companies in all sectors may lease AM equipment in the same way that they lease document printers today. AM creation facilities could be located within department stores (e.g., for customer-specific exclusive goods such as jewelry); large hospitals (e.g., for patient-specific prosthetics); home improvement stores (e.g., for family-specific furnishings); and/or industrial wholesalers (e.g., for plant-specific upgrade fittings). Competition and cooperation among creation facilities that provide services to digiproneurs will be enabled by aICT. Those who establish these creation facilities will themselves be digiproneurs and aid other digiproneurs in creating physical products. Development of digiproneurship infrastructure will lead to an increasing ability by digiproneurs to conceptualize, create, and propagate competitive new products, resulting in a sustainable model for distributed employment wherever digiproneurship is embraced. This, then, will be "Factory 2.0." As Web 2.0 has seen the move from static web pages to dynamic and shareable content; Factory 2.0 will see the move from static factories to dynamic and shareable creation. To make this possible, Factory 2.0 will draw upon Web 2.0 and the distributed conceptualization and propagation which it and AM enables.

Since the advent of the industrial revolution, the creation of physical goods has become an ever more specialized domain requiring extensive knowledge and investment. This type of highly concentrated and meticulously planned factory production will continue. However, Factory 2.0 will likely flourish alongside it. This will enable production by consumers, as envisioned 40 years ago [4]. Thus, the innate potential of people to create physical goods will be realized by fulfilling the latent potential of Web 2.0 combined with AM in ever more imaginative ways. Additionally, for the first time since the industrial revolution began, the trends towards increasing urbanization to support increasingly centralized production may begin to reverse when the opportunities afforded by Factory 2.0 are fully realized.

Conclusions

There is no longer any fundamental reason for products to be brought to markets through centralized product development, production, and distribution. Instead, products can be brought to markets through product conceptualization, creation, and propagation in any geographical region. This form of digiproneurship is built around combinations of aICT and advanced manufacturing technologies.

Digiproneurship offers many opportunities for a reduction in the consumption of non-value adding resources during the creation of physical goods. Further, the amount of factory equipment needed and, therefore, factory space is reduced. As a result, opportunities for smaller, distributed, and mobile production facilities will increase. Digiproneurship can eliminate the need for costly conventional market research, large warehouses, distribution centers, and large capital investments in infrastructure and tooling.

Creation of physical products at point-of-demand can make regional disadvantages unimportant. A wide range of people and businesses could offer digiproneurship products, including artists, hobby enthusiasts, IT savvy programmers, underemployed and unemployed people who are reluctant to uproot to major cities to look for work, and others.

Novel combinations of aICT and AM have already made it possible for enterprises to be established based on digitally driven conceptualization, creation, and/or propagation. The success of these existing enterprises is due to their recognition of market needs which can be fulfilled by imaginative, digitally enabled product offerings. As aICT and AM progress, and new creation networks are established, CBDM will become a reality, the opportunities for successful digiproneurship will expand and Factory 2.0 will come into being.

As digiproneurship expands and Factory 2.0 becomes a reality, AM could come to have a substantial impact on the way society is structured and interacts. In much the same way that the proliferation of digital content since the advent of the Internet has affected the way that people work, recreate, and communicate around the world, AM could 1 day affect the distribution of employment, resources, and opportunities worldwide.

20.4 Exercises

1. Do you think AM has the potential to change the world significantly? If so, how? If not, why not?
2. In what ways could AM's future development mirror the development of the Internet?
3. Find and describe three examples of digiproneurship enterprises which are not mentioned in this book.
4. How would you define Factory 2.0?
5. Based upon your interests, hobbies, or background, describe one type of digiproneurship opportunity that is not discussed in this chapter.

References

1. Beale CL (2000) Nonmetro population growth rate recedes in a time of unprecedented national prosperity. Rural Cond Trends 11(2):27–31
2. Fox S (2003) Recognizing materials power: how manufacturing materials constrain marketing strategies. Manuf Eng 81(3):36–39
3. Drucker PF (1993) Innovation and entrepreneurship: practice and principles. HarperCollins, New York
4. Toffler A (1970) Future shock. Bantam Books, New York

Index

I. Gibson et al., *Additive Manufacturing Technologies*,
DOI 10.1007/978-1-4939-2113-3

CPSIA information can be obtained
at www.ICGtesting.com
Printed in the USA
LVHW081656300619
622778LV00001B/2/P